Melvyn Conroy

Nazi Eugenics

Precursors, Policy, Aftermath

Melvyn Conroy

NAZI EUGENICS

Precursors, Policy, Aftermath

ibidem-Verlag
Stuttgart

Bibliografische Information der Deutschen Nationalbibliothek
Die Deutsche Nationalbibliothek verzeichnet diese Publikation in der Deutschen Nationalbibliografie; detaillierte bibliografische Daten sind im Internet über http://dnb.d-nb.de abrufbar.

Bibliographic information published by the Deutsche Nationalbibliothek
Die Deutsche Nationalbibliothek lists this publication in the Deutsche Nationalbibliografie; detailed bibliographic data are available in the Internet at http://dnb.d-nb.de.

∞

Gedruckt auf alterungsbeständigem, säurefreien Papier
Printed on acid-free paper

ISBN-13: 978-3-8382-1055-1

Printed in Germany

To Bernice—*In Memoriam*

To Doreen

We are ordinarily unaware of the degree to which our being treated as civilized, decent, autonomous, moral agents depends on our ability to look and act like such agents. To the degree that we make it impossible for other people to look and act that way, we make it easy to treat them as less than human.

– John Sabini[1]

Only one judgement is passed on the executioner—he ceases to be a human being. Though looking on his victim as less than human, he becomes his own executioner; he executes the human being inside himself. But the victim—no matter what the executioner does to kill him—remains a human being forever.

– Vasily Grossman[2]

1 James Waller, *Becoming Evil. How Ordinary People Commit Genocide and Mass Killing* (Oxford: Oxford University Press, 2002), p. 248.

2 Vasily Grossman, *Everything Flows,* (London: Harvill Secker, 2010), p. 128.

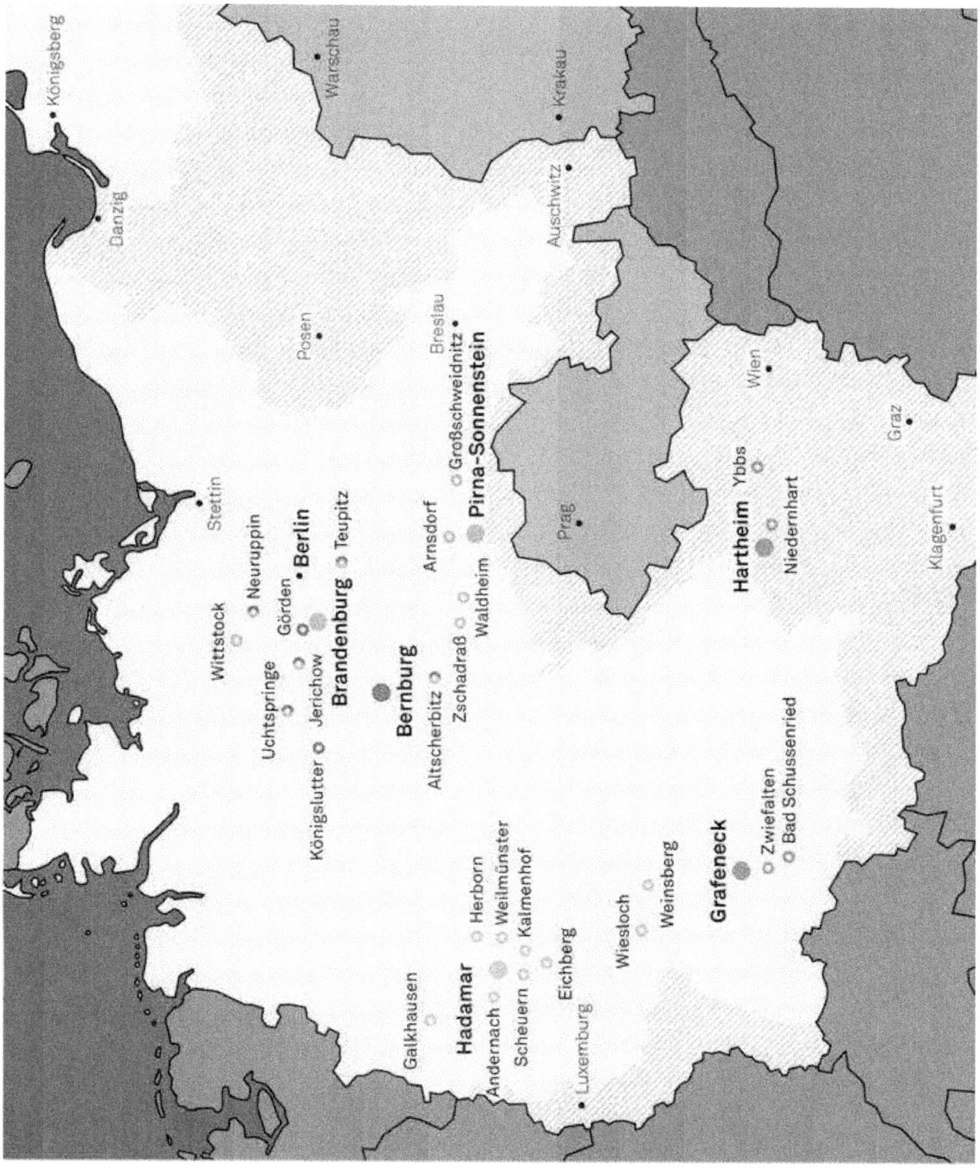
Königsberg
Danzig
Warschau
Krakau
Auschwitz
Posen
Breslau
Stettin
Neuruppin
Berlin
Teupitz
Wittstock
Görden
Brandenburg
Uchtspringe
Jericho
Königslutter
Bernburg
Altscherbitz
Zschadraß
Waldheim
Arnsdorf
Großschweidnitz
Pirna-Sonnenstein
Prag
Wien
Graz
Ybbs
Hartheim
Niedernhart
Klagenfurt
Zwiefalten
Bad Schussenried
Grafeneck
Weinsberg
Wiesloch
Galkhausen
Hadamar
Andernach
Scheuern
Herborn
Weilmünster
Kalmenhof
Eichberg
Luxemburg

CONTENTS

ACKNOWLEDGEMENTS

More than anything else, creating this book became a much needed diversion at what was a very stressful time in my life. Reading, researching, and writing about the subject matter under consideration may not immediately strike most people as being the usual response to a traumatic event. But it worked for me, providing the stimulus needed during many difficult days.

I realise now that writing it was in effect a kind of therapy, furnishing the impetus required to concentrate my mind on something other than my immediate concerns. When I began it I certainly never anticipated the result being published. That it has is due to the advice, influence, and support of many people, not least the numerous authors whose works I have extensively drawn upon. If space prohibits my thanking each one individually, I take this opportunity of expressing my appreciation to all of them for their scholarship and expertise. Whilst acknowledging the inspiration they have provided, unless otherwise attributed all opinions expressed in this book are my own.

Although thanks are due to many, there are some special colleagues without whose contribution publication of this book would not have been possible.

Valerie Lange and Florian Bölter at ***ibidem*** were helpful and considerate editors. They have guided me through the intricacies of developing my manuscript into the finished article, and did so with great courtesy and thoughtfulness, showing apparently limitless reserves of patience in dealing with a lengthy and complex text.

Joyce Field and Lance Ackerfeld at Jewish Gen are long-established friends whose support and advice has been invaluable over many years, but never more so than in connection with this book. It was Dr Robin O'Neil who introduced me to Jewish Gen, and it was Robin's tireless and invaluable investigation into the gestation of *Aktion Reinhard* and the tragic fate of Galician Jewry that encouraged me to explore more fully the historical background to a subject that has been of abiding interest to me for most of my life.

Dr Dick de Mildt, whose published works are an indispensable source to any student of this subject, was unfailingly courteous and helpful, particularly in guiding me through the intricacies of the German and Austrian legal systems, whilst Dr Tudor Georgescu was kind enough to read the draft manuscript and provide vital guidance. I am fortunate that so distinguished a scholar agreed to assume responsibility for such an important contribution. Natasha Neary took on the task of compiling the index, the preparation of which was a terrifying prospect for me.

Professor Matthew Feldman is a prized friend of many years standing, a truly talented academic with whom I have worked on numerous occasions. His faith in this book has often exceeded my own, and it is due to his determination and steadfast belief that it has finally seen publication. I hope this acknowledgement of the immense contribution he has made goes some way towards expressing the depth of my appreciation.

Finally, today and every day, my thoughts are of Bernice and Doreen, without whose love and devotion nothing was, or is, possible.

Melvyn Conroy
March 2017

INTRODUCTION

Eugenics is the religion of the future and it awaits its prophets.
– Nikolay Konstantinovich Koltsov[3]

The subject of intense global debate in the 100 years between the mid-nineteenth and mid-twentieth centuries in both Europe and the United States, it is doubtful if any discipline first acquired and then lost adherents with greater rapidity than the newly named pseudo-science of 'eugenics'. The concept proved neither easy to define with any degree of unanimity nor to implement, that is until the empowerment of Adolf Hitler, a political eugenicist more radical and extreme than even the most enthusiastic pioneer of social engineering could have imagined.

Throughout the ages disabled individuals had been universally viewed as flawed beings. Rather than being treated with the charity and compassion we today regard as their entitlement, they were regarded as second-class citizens. In the United States, for example, until the middle years of the twentieth century the disabled were interned against their will, sterilised involuntarily, and denied education, transportation, employment—even the right to vote.[4] Anti-Semitism based upon religious doctrine had an even longer and considerably more violent history. But as appalling as this tapestry of indifference and cruelty undoubtedly had been, it paled into insignificance when compared to the Nazi's attitude towards and treatment of those deemed by that regime to be "worthless" or "racially inferior".

The kind of supposedly scientific racism that in time produced a more-or-less comprehensible eugenic theory had a long history. If the Age of Enlightenment suggested a new dawning of reason, with freedom and equality for all, there were many who, whilst proclaiming the fundamental rights of man, were eager to qualify those very same rights. They believed that if all human beings were equal, some were certainly more equal than others. A number of the early proponents of what became known as social Darwinism and its offspring, eugenics, defended slavery; others believed that women were intellectually inferior to men and thus did not merit the same privileges; still others held what today would be seen as straightforwardly racist views, in one case going so far as to suggest that the intelligence of the Negro was on a par with that of a parrot.[5] These kinds of opinions do not by any means appear outrageous when compared to a variety of the pronouncements made by certain of the "scientists" of the nineteenth and twentieth centuries, as we shall see.

---

3 Michael Burleigh, *Eugenic Utopias and the Genetic Present* (Totalitarian Movements and Political Religions: Vol. 1, No. 1, 2000), p. 62. Koltsov was a Russian biologist and geneticist. Following denunciation of his theories as "fascistic nonsense" by disciples of Stalin's favourite, the agronomist, supporter of the theory of the inheritance of acquired characters, and charlatan, Trofim Lysenko, Koltsov was allegedly poisoned by the NKVD in 1940. Under Stalin a complete ban on the practice and teaching of genetics, condemned as a "bourgeois perversion", was imposed,.

4 Hugh Gallagher, *What the Nazi "Euthanasia Program" Can Tell Us About Disability Oppression* (Journal of Disability Policy Studies, Vol.12, No.2, 2001) 96–99, p. 96.

5 Robert N Proctor, *Racial Hygiene: Medicine Under the Nazis* (Cambridge Massachusetts: Harvard University Press, 1989), pp. 10–13.

It would thus be wrong to think that there was any degree of homogeneity among the acolytes of this secular religion. As with its traditional counterpart, orthodox religion, there were many schisms—Lamarckism (an organism can pass on characteristics that it acquired during its lifetime to its offspring), Darwinism (all species of organisms arise and develop through the natural selection of small, inherited variations that increase the individual's ability to compete, survive, and reproduce), Weismannism (all heritable characteristics are transmitted by the reproductive cells and acquired characteristics that cannot be inherited), Mendelism (a set of primary tenets relating to the transmission of hereditary characteristics from parent organisms to their offspring), among others. But all did at least share a common premise—the inviolability of heredity. This perspective might or might not be influenced by the component of environment, depending on the convictions of the believer, but of the fundamental tenet there could be no doubt.[6]

Eugenics was seen as a "scientific" debating chamber, a new kind of creed where any theory could be proposed, opposed, argued, or endorsed. Believers could find whatever they sought, for there were few absolutes in this new faith. And like any religion, its advocates could never be disproved, for just as to empirically demonstrate the existence of hell, or the paradise that awaited a steadfast believer, it was first necessary for he or she to die, at which point a lifetime's sacrifice and devotion might prove either hugely beneficial or completely pointless, so eugenicists could only offer a vision of the future in which poverty, disease, and ignorance had been banished. Eugenics did not, because it could not, offer a solution to the perceived problems of today, only to those of tomorrow. The benefits of its application would only be appreciated by future generations. But what a world those generations would inherit from their prescient forefathers, for in effect eugenics would become a secular religion, promising to deliver the utopian chimera of a "scientifically" created perfect society, in which sickness, crime, and anti-social behaviour would be eradicated forever. It was a tantalizing, if entirely unfeasible prospect.

Few of the proponents of the new social hygiene were great humanitarians, especially where they perceived a conflict with the immutable bedrock of the faith, natural selection. The Hungarian physician, József Madzsar, wrote in 1910: "The present form of social charity is even more dangerous because in most cases it obstructs the perishing of elements which are most burdensome and dangerous for society and it encourages their proliferation."[7] In words echoing those of Adolf Jost fifteen years earlier, he wrote: "If the state has the right to deprive citizens of their freedom, of their life even, it undoubtedly has the right to sterilise as well, especially when this can be executed without any other unpleasant consequences for the individual."[8] Like many others, Madzar went further. The aim should be not only to restrict the number of those deemed imperfect, but simultaneously to stimulate the growth of those considered the best. Quantity was to go hand in hand with quality. He concluded that eugenics was

6 Richard Weikart, *From Darwin to Hitler. Evolutionary Ethics, Eugenics, and Racism in Germany* (New York: Palgrave MacMillan, 2004), pp. 36–37.

7 Marius Turda, *'A New Religion'? Eugenics and Racial Scientism in Pre-First World War Hungary* (Totalitarian Movements and Political Religions: Vol.7, No. 3, September 2006), p. 308.

8 Ibid., p. 308.

therefore not simply a matter of social hygiene and heredity, but more importantly of political will. "A eugenic religion will take shape in the realm of ideologies," he stated. "This religion will forbid all forms of sentimental charity which are damaging to the species; it will enhance kinship and increase love for the family and for the race. In brief: eugenics is a manly, promising religion which calls upon the noblest feelings of our nature."[9]

This modern religion acquired disciples from every scholarly field—physicians, biologists, geneticists, sociologists, anthropologists, psychologists, economists—the list was both impressive and comprehensive. Nor was it limited to the scientific and academic community. Journalists and politicians also added to the mix. Together, they subscribed to the view that, like Pangloss in Voltaire's Candide, all would be for the best in the best of all possible worlds. Such an idyllic prognosis was irresistible to these individuals. However, the passing of what the eugenicists considered the most valuable genes from one life cycle to the next in a never ending process of racial betterment, could not, on its own, provide the complete answer. As Madzar had recognized, attaining a eugenic nirvana also required prevention of the transmission of defective genetic material by those deemed socially and racially inferior. The application of the means and ends of these beliefs to Nazi Germany, the positive and the negative, are the subject of this study.

For Alfred Ploetz , physician, biologist, eugenicist, founder of the German Society for Race Hygiene in 1905 (the first eugenics organization in the world),[10] and a pivotal influence on Nazi racial doctrine, there was no doubting that the future of the race was more important than the health of the individual.[11] This concept was at the core of National Socialism. As Telford Taylor commented in his opening remarks at the Nuremberg Medical Trial, it represented "that sinister undercurrent of German philosophy that preaches the supreme importance of the state and the complete subordination of the individual."[12] Without necessarily endorsing Nazism, some zealous enthusiasts still find that kind of eugenic theory attractive today. But for most rational beings, those fundamentalist principles were dead in the water after National Socialism. They had been directed by the Nazis along the road of discriminatory and racist bio-politics that had been their destiny even before the term "eugenics" was first coined by Sir Francis Galton.[13] In his *Inquiries into Human Faculty and its Development* of 1883, Galton wrote that he had been searching for

> a brief word to express the science of improving the stock, which is by no means confined to questions of judicious mating but which, especially in the case of man, takes cognizance of all the influences that tend

9 Ibid., p. 309.

10 Weikart, *From Darwin to Hitler*, p. 15.

11 Turda, *'A New Religion'?*, p. 324, note 46.

12 Ulf Schmidt, *Karl Brandt: The Nazi Doctor—Medicine and Power in the Third Reich* (London: Hambledon Continuum, 2007), p. 362.

13 Most sources attribute first use of the word to Galton in his 1883 work "*Human Faculty*", although the Shorter Oxford English Dictionary dates its first appearance to 1833.

in however remote a degree to give the most suitable races or strains of blood a better chance of prevailing speedily over the less suitable than they otherwise would have had.[14]

Despite these cautionary introductory words, this is not a polemic against eugenics, social hygiene, bioethics, or any other manifestation of biomedicine as envisioned in a post-Nazi world, certain aspects of which are now part of everyday life. Putting theological debate to one side, many would accept that some modern applications of eugenics, of which the Human Genome Project is perhaps the best example, represent a valuable scientific resource. Stem-cell research into the causes of and potential cure for certain genetic diseases and the contribution this would make towards the alleviation of human suffering, has the potential to be hugely beneficial. Particular contemporary applications of eugenics, for example oral contraception, have revolutionized latter-day lifestyles. Others may feel less easy about specific examples of genetic engineering, such as cloning. The difference, of course, is that these exemplars, unlike eugenics as originally conceived, represent the fruit of practical scientific studies, as opposed to abstract and often pernicious theories. But as Michael Burleigh has eloquently argued: "It is virtually impossible to discuss abortion, genetic engineering, in vitro fertilisation, negative or positive eugenics, euthanasia, organ transplants, psychiatry, sterilisation of the mentally incompetent or treatment of antisocial or violent individuals, without someone invoking the history of Nazi Germany to break, rather than make, a contemporary case."[15] The same concerns have been expressed by other commentators:

> The development of new genetic technologies has resulted in comparisons being drawn between the many horrendous atrocities once perpetrated in the name of eugenics and what might happen in the future. In the minds of most people, eugenics is usually associated with enforced sterilisation, racism, restrictive immigration policies and Nazi concentration camps. There is a danger that the public... will look at the new genetics and simply claim that it is unacceptable to them because of the past. The history of eugenics in the twentieth century suggests that this is a legitimate fear that needs to be addressed.[16]

Yet who can deny the mother of three children suffering from sickle-cell anaemia, when at a 1983 conference on gene therapy, she protested: "I am angry that anyone presumes to deny my children the essential genetic treatment of a genetic disease. I see such persons as simplistic moralists."[17] The past cannot always be a template for the future.

The fundamental problem accompanying these scientific advances may be expressed in another way. The history of the eugenic movement suggests that it cannot be taken for granted that these potential improvements in healthcare and living standards will be to the benefit of all. The Elysian ideals of a disease free, poverty free, crime free society remain the same eugenic objectives they have always been. But by its very nature, eugenics is elitist, aiming to produce a genetically flawless population. Not all can, nor in the eugenicists frequently expressed view, should, be the beneficiaries of these

---

14 Rachel Iredale, *Eugenics and its Relevance to Contemporary Health Care*, (Nursing Ethics, 7 (3), 2000), p. 206.

15 Michael Burleigh, *Ethics and extermination: Reflections on Nazi genocide*, (Cambridge: Cambridge University Press, 1997), p. 144.

16 Iredale, *Eugenics and its Relevance to Contemporary Health Care*, p. 206.

17 Daniel J Kevles, *In the Name of Eugenics: Genetics and the Uses of Human Heredity* (Berkeley: University of California Press, 1986), p. 291.

advances. What of those who slip through the net, those inevitable and unavoidable genetic accidents? If perfection is the ideal, what is to be done with those who, through no fault of their own, are somehow less than perfect? The true strength of any society is evidenced by the extent to which it protects its weakest members. Historically, what has eugenics offered them? And what will it offer them in the future? It is the answers to these questions that give rise to disquiet.

Much of this work is devoted to what "euthanasia" came to mean under National Socialism as the natural culmination of a near century of thinking that indissolubly linked racism and eugenics. The concern here is not with the ongoing debate about euthanasia as such, a subject concerning which individuals will doubtless have opinions of their own, but rather with what "euthanasia" meant in practice during the lifetime of the Third *Reich*. The quotation marks are important, for they are intended to indicate that what is being dealt with here is not the "mercy death" offered to the suffering out of compassion and empathy which was, and is, the meaning normally ascribed to the expression, but rather the killing of the helpless and vulnerable as part of a racially driven ideology coupled with a callously pragmatic economic imperative, which is how the Nazis chose to interpret the term. Indeed, it would be perfectly accurate to substitute the word "murder" for "euthanasia" throughout this text.

Nor is this a dissertation on the history of eugenics. It is an attempt to present the background to and application of Nazi ideology to the subject of eugenics as well as social and racial hygiene, where that was to lead, and the mind-set and subsequent fate of some of its advocates and exponents. This subject has been exhaustively, one might even say microscopically examined by German scholars over the last thirty years, but relatively little of this immense body of work has found its way into the English language, and it is towards the English speaking reader that this work is primarily directed. Some distinguished authors have provided a growing and important corpus of written material on the subject in English, and frequent reference is made to their efforts herein. But no two individuals will approach any issue in exactly the same way, and it is hoped that the reader will gain a somewhat different perspective on the topic from the manner in which this text is constructed. It begins with a brief history of the gestation of that Nazi eugenic ideology in order to better understand its application and consequences. Thus the first three chapters attempt to broadly illustrate the nature of thinking on matters of eugenics and social hygiene during the last years of the nineteenth and the first forty years of the twentieth centuries, particularly so far as Germany was concerned, for the National Socialists practised, albeit *in extremis*, what many others had long preached.

The fourth chapter examines the application of that ideology to Germany during the years from 1933 (the *Machtergreifung*, the Nazi seizure of power) to 1939 (the commencement of the "euthanasia" programme and the outbreak of the Second World War), and in particular focuses on the issue of sterilisation. Fifthly, the practical application of "euthanasia" through the organisation dedicated to it—T4—is described. In the sixth chapter, some of the killing centres in which "euthanasia" was carried out are considered. The seventh chapter is concerned with activities following the official cessation of the killing programme, the time of so-called "wild euthanasia," also known more accurately as "decentralised euthanasia". Eighthly, the application of "euthanasia" to concentration camp inmates, known by the coded term "14f13", is reviewed, while in the ninth chapter the importance of the "euthanasia" experience to the planned extermination of Jews in *Aktion Reinhard* is examined. Heinrich Himmler's many involvements in eugenic affairs make up the tenth chapter, with post-war trials and perceptions appraised in chapter eleven and some conclusions proposed in chapter twelve.

Finally, a lengthy appendix contains brief biographical details of some of the individuals involved in the theory and practice of the eugenic programme. Since a precise chronological delineation of the events described is neither possible nor indeed desirable, there is some inevitable overlapping of subject matter. As will hopefully be demonstrated, although given the then prevailing *Zeitgeist*, the *Shoah* was ultimately a logical consequence of Nazi eugenic thinking and practice (although by no means the sole cause of that catastrophe), this manuscript is not primarily Holocaust-centric. Nonetheless, the importance of the contribution eugenics made towards that tragedy will be a recurring theme, even if a comprehensive examination of the plan to eradicate the Jews of Europe and beyond represented by the policy named in memory of the newly deceased chief of the *Reichssicherheitshauptamt* (RSHA), *Aktion Reinhard*, is beyond the scope of this work.[18]

For many years there has been a lively debate among historians concerning the virtues, on the one hand, and pitfalls, on the other, of making moral judgements with regard to historical events. In 1895, Lord Acton instructed his listeners that they should "suffer no man and no cause to escape the undying penalty which history has the power to inflict on wrong."[19] Others are of the opinion that it is simply the historian's duty to epistemologically present the unvarnished facts and allow the reader to decide on the moral issues involved. As Richard Evans expressed the point; "The element of moral judgement, insofar as it is exercised at all, is in the end extraneous to the research rather than being embedded in the theory or methodology of it."[20] Whilst it is hoped that the error of presenting matters in simplistic absolutist terms of good and evil has been avoided, there is no doubt that so far as the subject of this text is concerned, moral judgements are not only unavoidable, but essential. Above all else, the practical application of Nazi eugenics represented the abandonment of a universally accepted set of moral principles that simply cannot be ignored. It is therefore the dictum of Lord Acton that rules here.

A few cautionary words. As Henry Friedlander pointed out, the terms "handicapped" or "disabled" were unknown both before and during the Nazi period. Instead, distasteful and inaccurate terms such as "idiots", "crazies", "cripples", and "feeble-minded" were in common use.[21] In the same manner, Sinti and Roma were described as "Gypsies", a wholly erroneous word which has acquired derogatory overtones. In order to

---

18 Reinhard Heydrich, who died on 4 June 1942 following an assassination attempt by Czechoslovakian patriots nine days earlier, held this position.

19 Philip Boobbyer, *Moral Judgements and Moral Realism in History* (Totalitarian Movements and Political Religions, Vol. 3, No. 2, 2002), p. 85.

20 Ibid., p. 86.

21 Henry Friedlander, *The Origins of Nazi Genocide: From 'Euthanasia' to the Final Solution* (Chapel Hill: University of North Carolina Press,1995), p. xxii

maintain some kind of consistency the original terminology has been quoted where appropriate throughout this work. No offense is intended thereby.

Experience teaches that dates and numbers are frequently problematical in a work of this nature. Sources do vary, often widely. Wherever precise evidence on such matters is not available, or in cases where the exact evidence cannot be subjected to corroboration from a second source (and sometimes even when it can), the reader will find use of words such as "about", "approximately", "early", or "late". Personal names too can present difficulties—Arthur/Artur, Walter/Walther, and so on. In such cases, the most commonly accepted spelling has been used. As for place names, anglicized versions (where they exist) are employed—so Nuremberg rather than Nürnberg, Munich rather than München, Warsaw rather than Warszawa. Where place names have changed, on their first appearance both are shown—Kovno/Kaunus, Posen/Poznan, etc. Thereafter, the contemporaneous name is utilised.

Diacritics in general present particular difficulties in a work of this nature. The English language contains few words with diacritic markings; Polish, by contrast, often seems to consist of little else. Such diacritics have normally been used when quoting the names of people and places, as well as where otherwise appropriate (except as already noted), in German, French and Hungarian, but not in Polish. Apologies are offered for the inconsistency, but it is hoped that the reader will have no difficulty, for example, in recognizing Belzec as being an anglicized version of Bełżec, or Lodz for Łódź.

Advantage has been taken of the extraordinary resource now provided by the internet to any student or researcher, whereby an overwhelming amount of detailed information on almost any given subject is accessible at the click of a mouse. The World Wide Web had exponentially expanded the availability of knowledge; the more cynical might suggest that it is has also performed the same service for ignorance, for there is much that is contentious, misleading, or simply wrong and/or ill-intentioned in cyberspace. Thus the same criteria need to be applied when utilising the internet as would be employed when consulting any other source. But it is self-evident that it is impossible for each and every individual to consult all of the primary documentation on the issues of interest. To that extent, the internet is no different than every other written authority. In researching any subject, most of us are dependant, to a greater or lesser degree, upon the translation and interpretation, literal and metaphorical, of talented individuals whose scholarship we trust. No apologies are therefore offered for the use of material from the internet, which in many cases would otherwise only be accessible with great difficulty, if available at all. The same care has been taken in evaluating and including such data as has been applied in utilising more traditional sources.

Finally, it is only proper to mention that in a number of instances there has been some minor editing of direct quotations in the interest of improving either punctuation or grammar, but hopefully in no case has the intended sense of any quotation been lost. If it has, of course the blame for that or any other errors in the text lies exclusively at the author's door.

CHAPTER 1: TOWARDS UTOPIA

> If a twentieth part of the cost and pains were spent in measures for the improvement of the human race that is spent on the improvement of the breed of horses and cattle, what a galaxy of genius might we not create! We might introduce prophets and high priests of civilization into the world, as surely as we can propagate idiots by mating crétins. - Francis Galton (1865)[22]

> In today's society applied eugenics is not a utopia anymore, and it will be even less so in the society of the future.
> – Jenő Vámos (1911)[23]

Charles Darwin published his *On the Origins of Species by Means of Natural Selection, or the Preservation of Favoured Races in the Struggle for Life* in 1859, creating a controversy from which Judeo-Christian theology in particular has since never recovered.[24] For all of its unintended subsequently malign influence, the book contained no racial theories. The suggestion that he would have endorsed such ideas would have horrified Darwin, who was a notably tolerant person. In fact, except by implication, the work contained little reference to *Homo sapiens*, essentially being concerned with the evolution of other animal life and that of plants.[25]

Long before Darwin shook the foundations of religious belief, there had been much supposedly scientific debate concerning the development of human characteristics. In the 1820's what became known as recapitulationist theory emerged. This posited that childhood in the white race was the equivalent of savagery or primitivism in evolutionary terms. Thus adults of "inferior" groups were deemed to be at the mental and emotional level of white male children or adolescents. Nor was this pseudo-science to be applied solely to non-white races. The "inferior" groups could (and did) consist of the anti-social, criminals, women, and any other disliked nationality the defining group chose to include. Although overtaken by Darwinism, the influence of such thinking on later eugenic theorists will become apparent.[26]

Two years before Darwin's *Origins of Species* appeared, in 1857 the French psychiatrist, Benedict Augustin Morel, had published his hugely influential work *Traite des degenerescence physique, et intellectuelles et morales de l'espece humaine* (*Treatise on the*

---

22 Michael W Perry, (ed), *G K Chesterton—Eugenics and Other Evils: An Argument Against the Scientifically Organised State* (Seattle: Inkling Books, 2001), p. 124. A convert to Catholicism, Chesterton managed to be simultaneously both anti-eugenic and anti-Semitic, thus proving that one is not necessarily the prerequisite of the other. [Richard S Levy (ed), *Antisemitism: A Historical Encyclopedia of Prejudice and Persecution* (Santa Barbara: ABC-CLIO Ltd, 2005), pp. 115–116].

23 Marius Turda and Paul J Weindling (eds.), "*Blood and Homeland*". *Eugenics and Racial Nationalism in Central and Southeast Europe, 1900–1940*, (Budapest: Central European University Press, 2007), p. 196.

24 There has been an enormous volume of literature published on this subject during the last 150 years posing arguments, some vehemently for, others equally vehemently against, Darwin's evolutionary principle, as well as the innumerable variations on specific issues raised by that topic. The concern of this text is primarily the impact and consequences of this debate on the development and implementation of specific Nazi policies derived from and subsequently endorsed by the then prevalent general eugenic theory.

25 Michael Burleigh, and Wolfgang Wipperman, *The Racial State: Germany 1933-1945* (Cambridge: Cambridge University Press, 1991), p. 28.

26 John W Dower, *War Without Mercy: Race and Power in the Pacific War* (New York: Pantheon Books, 1986), pp. 153–154.

Physical, Intellectual, and Moral Degeneracy of Mankind), in which he proposed a theory of "degeneration." Morel suggested that many illnesses, whether physical, intellectual, or moral, were all caused by a single process: degeneration. He concluded that most illnesses are the result of an incurable hereditary disorder. Allowing those suffering from such disabilities to reproduce presented a genetic risk to the nation. The causes of such a condition were, in Morel's view, the over-consumption of alcohol, tobacco, and opium. Morel believed that those damaged by overindulgence in these and other appetites developed illnesses which weakened their heredity value. This weakened state was passed on to future stock, for the effects were cumulative, so that after approximately four generations the degenerate line would end because the children of the ultimate generation would be born sterile imbeciles.

Degeneracy was a 'one size fits all' theory that neatly explained a wide variety of diseases. If the symptoms of those diseases were ostensibly different, to Morel they were simply alternative expressions of a single underlying disorder: degenerate heredity. The importance of Morel's theory on the development of eugenic thinking cannot be over emphasized, for as Morel wrote: "We are not dealing with the individual, the single human being, but with society as a whole and the means to such important an end have to be measured accordingly".[27] In this sentence can be seen the rationale for all that was to follow.

The war of words Darwin's thesis prompted only increased in intensity when in 1871 he published *The Descent of Man and Selection in Relation to Sex*, applying his evolutionary paradigm to human physical, intellectual, and moral development.[28] Deriving its impetus from Darwin's theory, the hypothesis that came to be known as social Darwinism proposed that competition in society, whether individual, national, or international, is the fundamental component of biological and social evolution.[29] Superior individuals or groups survived and flourished even as the weaker disappeared. This became a justification for imperialism and racism while discouraging liberalism and reform. Although bearing his name, social Darwinism was not limited to a specific interpretation of Darwin's writings, drawing as it did on the works of many others, including the philosopher and political and sociological theorist, Herbert Spencer, the demographer and political economist, Thomas Malthus, and perhaps most significantly, polymath Francis Galton.

Spencer postulated that species evolved via a process of natural selection, and that it was only by eradicating the weakest elements and promoting the strongest that survival of an organism was assured. In Spencer's view, what the eugenicists later defined as the "inferior" specimens were to be excluded from society in order to enable the "valuable" components to fulfil that society's ultimate potential for growth and happiness. Moreover, it was perfectly acceptable to utilise the "inferior" as a kind of slave labour in order to promote the well-being of the "valuable." From there, it was but a small step to suggesting that the destruction of the weakest in society was not only natural, but emi-

27 http://www.irren-offensive.de/speech_engl_treusch_dieter.htm (Accessed 21 August 2008).

28 Richard Weikart, *From Darwin to Hitler. Evolutionary Ethics, Eugenics, and Racism in Germany* (New York: Palgrave MacMillan, 2004), p. 1.

29 The term first appeared in an 1879 article by Oscar Schmidt.

nently desirable. Only the fittest would, and should, survive.[30] Or as Spencer brutally put it: "If they are sufficiently complete they should live. If they are not sufficiently complete to live, they die, and it is best they should die."[31]

Social Darwinism, with its application of Darwin's "struggle for survival" to human affairs, in turn gave birth to the "science" of eugenics. To all intents and purposes the two were to become virtually synonymous in the public's mind, although while there was much that they shared in common, the two concepts were not interchangeable.[32] The term "eugenics", a thesis which has no scientific basis, was coined by Galton, a cousin of Darwin, and was derived from the Greek "eugenes", meaning well-born or of good stock. Galton described it as "the science of the improvement of the human race by better breeding."[33] More precisely, this was to be "a science which deals with all influences that improve and develop the inborn qualities of a race."[34]

This took Spencer's concept of "survival of the fittest", a belief that was to become a basic element of Nazi ideology, to its logical conclusion.[35] Its antonym was dysgenics, defined as the progressive evolutionary weakening or genetic deterioration of a population of organisms relative to their environment, often due to the relaxation of natural selection or the occurrence of negative selection. The Hungarian zoologist, István Apáthy, believed that:

> The goal of eugenics is to call attention upon the fact that each social order, each habit, fashion or morality which acts against the selection of the best individuals in fact undermines the future of the entire society and it trades evolution, the salvation of future generations for the pleasures of the moment, for the individual comfort of the present generation.[36]

Or as the American sociologist and social Darwinist, William Graham Sumner, commented at around the turn of the twentieth century:

> The sentimentalists are always greatly outraged by the notion of the survival of the fittest which is produced by liberty. If we do not like the survival of the fittest, we have only one alternative and that is the survival of the unfittest. If A, the unfittest to survive, is about to perish and somebody interferes to make B, the fittest, carry and preserve A, it is plain that the unfittest is made to survive and that he is main-

---

30 Sylvia Anne Hoskins, *Nurses and National Socialism—a Moral Dilemma: One Historical Example of a Route to Euthanasia* (Nursing Ethics, 12 (1), 2005), p. 80.

31 Dennis Sewell, *The Political Gene—How Darwin's Ideas Changed Politics* (London: Picador, 2009), p. 35.

32 Mike Hawkins, *Social Darwinism in European and American Thought 1860–1945: Nature as model and nature as threat* (Cambridge: Cambridge University Press, 1998), p. 7.

33 Henry Friedlander, *The Origins of Nazi Genocide: From 'Euthanasia' to the Final Solution* (Chapel Hill: University of North Carolina Press, 1995), p. 4.

34 Rachel Iredale, *Eugenics and its Relevance to Contemporary Health Care*, (Nursing Ethics, 7 (3), 2000), p. 206.

35 "Survival of the fittest", perhaps the term most associated with social Darwinism, is often incorrectly attributed to Darwin himself. In fact, Darwin wrote that "It is not the strongest of the species that survives, nor the most intelligent that survives. It is the one that is the most adaptable to change," a doctrine which might be termed the "survival of the most adaptable", a concept much more difficult to politicise than Spencer's dangerous interpretation of natural selection.

36 Marius Turda, *'A New Religion'? Eugenics and Racial Scientism in Pre-First World War Hungary* (Totalitarian Movements and Political Religions: Vol.7, No. 3, September 2006), p. 316.

tained at the expense of B, who is curtailed and restrained by just so much. This process, therefore, is a lowering of social development and is working backwards, not forwards.[37]

Thus the fundamental tenet of the eugenics movement was that restricting the ability of "inferior" people to procreate (so-called "negative eugenics") whilst maximizing that of "superior" individuals (so-called "positive eugenics"), would benefit society. To Galton, the creation of better citizens through better breeding was an obvious and inevitable outcome if the principle of positive eugenics he espoused was followed:

> We can clearly observe the extreme diversity of character in children. Some are naturally generous and open, others mean and tricky; some are warm and loving, others cold and heartless; some are meek and patient, others obstinate and self-asserting; some few have the tempers of angels, and at least as many have the tempers of devils. In the same way... [that] by selecting men and women of rare and similar talent, and mating them together, generation after generation, an extraordinarily gifted race might be developed, so a yet more rigid selection, having regard to their moral nature, would, I believe, result in a no less marked improvement of their natural disposition.[38]

As for negative eugenics, Galton skirted around the issue. The inferior, "refuse" as he termed them, were to be discouraged and "retarded" from marriage and reproduction. Given time, in this way the superior would multiply and flourish while the "refuse" would wither and eventually be eliminated. Galton suggested a programme of financial incentives intended to encourage the desirable to propagate,[39] but the question of just how the undesirable were to be motivated to volunteer for the obverse remained unanswered. It may be supposed that, so far as both positive and negative eugenics were concerned, the notion of compulsion was never a consideration for Galton. Society, he felt, would resolve these issues as a more refined and vigorous praxis was established following recognition of the validity of his theories.

The obvious and immediate problem was clearly one of definition. What did "superior" and "inferior" signify? And who was to do the defining? These were, of course, questions which were (and are) incapable of satisfactory answers, but to the committed eugenicists the reasoning was obvious. "Valuable" life was represented by the racial or social class to which the eugenicist belonged, or with which he felt an affinity; "worthless" life was everybody else. In 1880, Robby Kossman, a German zoologist who later became a medical professor, succinctly expressed the eugenicist viewpoint:

> We see that the Darwinian world view must look upon the present sentimental conception of the value of life of the human individual as an overestimate completely hindering the progress of humanity. The human state also, like every animal community of individuals, must reach an even higher state of perfection, if the possibility exists in it, through the destruction of the less well-endowed individual, for the more excellently endowed to win space for the expansion of its progeny...The state only has an interest in preserving the more excellent life at the expense of the less excellent.[40]

This passage combines several elements of what was to become recognizably Nazi ideology; the individual is nothing, the nation (*Volk*) is everything; a person's worth can only measured by his or her contribution to the community; the Judeo-Christian ethic of

37 William Graham Sumner, *The Challenge of Facts and Other Essays*, (New Haven: Yale University Press, 1914). p. 423.

38 David Masson, *Macmillan's Magazine, Vol XII* (London: Macmillan and Co, 1865). pp. 324–325.

39 Richard Lynn, *Eugenics: A Reassessment* (Westport: Praeger Publishers, 2001), pp. 215–216.

40 Weikart, *From Darwin to Hitler*, p. 78.

the sanctity of life was moribund; the Malthusian concept of limiting population growth was paramount.[41] There was, however, a vital difference between Malthus' theory and the eugenicists' interpretation of it. Malthus advocated unrestricted birth control in order to manage the planet's resources. To the racial hygienists this was a dangerous over-simplification; a high birth rate among the genetically "valuable" with a matching lowered birth-rate among the genetically "worthless" was what was really important. Thus the Nazis proposed and practiced selective breeding as part of a policy of geopolitical aggrandisement and expansion, rather than from any altruistic or humanitarian motive. It would not be the meek who would inherit the earth, but the hereditarily valuable. It was for this reason that Nazi medical theorists disowned Malthus' ideas, often forcibly.[42] In National Socialist Germany the "valuable" were to be encouraged to reproduce; the "worthless" were to be forcibly prevented from doing so. Nowhere was this dichotomy better illustrated than in the Honour Cross of the German Mother (*Ehrenkreuz der deutschen Mutter*), an award instituted in 1938 to acknowledge the fecundity of 'Aryan' mothers possessing appropriate hereditary credentials. Whilst recognizing the fertility of women considered genetically valuable, the regime was simultaneously sterilising tens of thousands of others deemed biologically valueless. It would be unfair to lay the blame for this exclusively at Galton's door, but in essence it was a logical, if unforeseeable, consequence of his proposals.

The true lethality of this kind of reasoning only began to emerge with the practical linking of the theories of eugenics with the *Völkisch* movements plaguing the United States and many countries in Europe during the latter part of the nineteenth and early years of the twentieth centuries. Basing his conclusions on the questionable results of intelligence tests given to United States military recruits, in 1923 the psychologist Carl Brigham wrote in his book, *A Study of American Intelligence,* that the decline in the level of intelligence in American males was largely due to the immigration of "inferior" racial groups.[43] Eugenics and the purported issues of race became inseparable. The naturalist Zsigmond Fülöp, editor of Darwin's works in Hungary, expressed contemporary reasoning well: "Eugenics represents the application of Darwinism to society with the scope of improving the qualities of the race, especially social qualities".[44]

It was easy to bandy such expressions about, but what exactly did the eugenicists mean by terms such as "race" or "social qualities"? Having first acknowledged the in-

---

41 Darwin's view of natural selection through the struggle for existence was derived from Thomas Malthus' concept of an organism's tendency towards overpopulation, one consequence of which was the propensity of humans to reproduce faster than their food supply. Ergo, the racial hygienists concluded, the need of the stronger nation to expand its living space (*Lebensraum*) at the expense of the weaker, even as it bred allegedly superior stock to populate the living space gained and at the same time restricted the ability of its supposedly inferior stock to procreate. (Weikart, *From Darwin to Hitler,* p. 185). Malthus set out his argument in his 1798 treatise, *Essay on the Principle of Population.* Some consider Malthus a proto-eugenicist, going so far as to label him "the founding father of scientific racism." [Patrick Brantlinger, *Dark Vanishings: Discourse on the Extinction of Primitive Races, 1800–1930* (Ithaca: Cornell University Press, 2003), p. 17)].

42 Robert N Proctor, *Racial Hygiene: Medicine Under the Nazis* (Cambridge Massachusetts: Harvard University Press, 1989), p. 342, note 30.

43 James W Trent, *"Who shall say who is a useful person?" Abraham Myerson's opposition to the eugenic movement* (History of Psychiatry, No. 12, 2001), p. 42.

44 Turda, *'A New Religion'?,* p. 309.

herent difficulties in arriving at any wholly acceptable conclusions, in 2002 the American Psychological Association attempted to arrive at a definition not just of race, but also of ethnicity and culture. They determined that "the definition of race is considered to be socially constructed, rather than biologically determined. Race, then, is the category to which others assign individuals on the basis of physical characteristics, such as skin colour or hair type, and the generalizations and stereotypes made as a result." Ethnicity, it was suggested, was "the acceptance of the group mores and practices of one's culture of origin and the concomitant sense of belonging." And culture was "the embodiment of a worldview through learned and transmitted beliefs, values, and practices, including religious and spiritual traditions. It also encompasses a way of living informed by the historical, economic, ecological, and political forces on a group."[45] If anything these attempted definitions confirm the difficulties inherent in endeavouring to establish parameters for these essentially indefinable terms. Be that as it may, it is doubtful if many eugenic theorists were anything like as specific in their interpretation of these matters. "Race" in particular could have whatever meaning the speaker or author sought to ascribe to it, so that in 1956, Winston Churchill, as a younger man an enthusiastic eugenicist, could produce a book about the British entitled *The Island Race*, as if that nation possessed some kind of specific racial characteristics. Nearly thirty years later, Margaret Thatcher felt able to repeat the words of her hero when suggesting that the inhabitants of the Falklands were, like the British, "*an island race*", thus misusing a word that by then had been recognized as being not only worthless but, in the wake of the Nazi experience, had sinister connotations.[46]

This is not simply a matter of semantics. As the eminent geneticist, Richard Lewontin has observed:

> We should stop talking about major races because to talk about major races gave the impression that there were big differences between these groups in things that mattered—I mean, skin colour, after all, doesn't matter except in some vague aesthetic sense—rather than the things that really mattered: people's characters, their intelligence, their behaviour, whether they're going to compete with other people or not and so on. The evidence then became that there weren't any interesting differences in such things, and so we should stop talking about race.[47]

But of course we have not. If we are still reluctant to abandon the concept of 'race' or any of its linguistic variations, often using it as a kind of shorthand in place of more accurate terminology, for the eugenicists this wholly fallacious notion represented the very cornerstone of their beliefs.

If, as has been suggested, "every German had his own idea of race", it is certainly true that racial identity was an issue that the Nazis were never able to resolve to their satisfaction. For some supposed "experts" it was a matter of physical characteristics, whilst others dismissed this notion as "unscientific", since there were few individuals who could claim to be free of at least a degree of racial mixing. The regime tied itself into knots in trying to find an answer to this self imposed problem, consistent only in its

---

45 http://www.apa.org/pi/oema/resources/policy/multicultural-guidelines.aspx (Accessed 16 November 2009).

46 Kathleen Wilson, *The Island Race: Englishness, empire and gender in the eighteenth century* (London: Routledge, 2003), p. 54.

47 http://www.pbs.org/race/000_About/002_04-background-01-04.htm (Accessed 5 December 2009).

inconsistency for, with the exception of the Jews, for whom a strictly applied legal definition was in place, the determination of questions of national and racial identity varied greatly within the various regions under Nazi hegemony, being extremely selective in some places yet quite the opposite in others. As with so many other eugenic matters, much of course depended upon who was doing the defining, for such guidelines as existed were quite imprecise, and few of these racial "experts" were in agreement on the criteria to be applied.[48] In short, Nazi racial policy remains the best example we have yet had of the absurdity of attempting to identify either nations or individuals by such means, and of promoting the values purportedly inherent in one race over those of another.

The perception of eugenics as a new religion was recognized by Galton in a1904 address, in the course of which he stated: "It [eugenics] has, indeed, strong claims to become an orthodox religious tenet of the future, for eugenics co-operates with the workings of nature by securing that humanity shall be represented by the fittest races. What nature does blindly, slowly, and ruthlessly, man may do providently, quickly and kindly."[49] Galton believed, as Lincoln Steffens was to say in a quite different context, "I have seen the future and it works." However, Galton continued, society was not yet ready to accept eugenics. It would only be so "when a sufficiency of evidence shall have been collected to make the truths on which it rests plain to all. That moment has not yet arrived"[50] In other words, it was first necessary to educate the public concerning the immense benefits this modern faith would bring before its precepts could be implemented. Moreover, Galton's new theology would turn traditional religious ethics, together with the concepts of *Liberté, égalité, fraternité* and The Declaration of the Rights of Man, on their collective heads. For Galton was neither a convinced Christian, nor a democrat, writing as he did:

> It is the obvious course of intelligent men—and I venture to say it should be their religious duty—to advance in the direction whither Nature is determined they shall go, that is towards the improvement of the race....But it [Democracy] goes farther than this, for it asserts than men are of equal value as social units, equally capable of voting, and the rest. This feeling is undeniably wrong and cannot last.[51]

Problems associated with the so-called "feebleminded" (a wholly inaccurate term utilised to include everything from mental disability to alcoholism, sexual promiscuity, and other deemed anti-social behaviour) became the bedrock of eugenic theorizing.[52]

---

48 Mark Mazower, *Hitler's Empire—Nazi Rule in Occupied Europe* (London: Allen Lane, 2008), pp. 182–183.

49 Turda and Weindling, "*Blood and Homeland*", p. 438.

50 Karl Pearson (ed), *The Life, Letters and Labours of Francis Galton, Volume Three*, (Cambridge: Cambridge University Press, 1930), p. 321

51 Karl Pearson (ed), *The Life, Letters and Labours of Francis Galton, Volume Two*, (Cambridge: Cambridge University Press, 1924). p. 119.

52 Herbert Henry Goddard approvingly defined feeble-mindedness as a "state of mental defect existing from birth or from an early age and due to incomplete or abnormal development in consequence of which, the person affected is incapable of performing his duties as a member of society in the position of life to which he is born." [Henry Herbert Goddard, *Feeble-mindedness: Its Causes and Consequences*

Labelled as idiots, imbeciles, morons, or degenerates, the existence of such individuals was used to promote the concept that there existed an explicit relationship between low intelligence and both immorality and crime. The cause of the social problems of the time was deemed to be inherited feeblemindedness, and the concomitant poverty a product of hereditary degeneracy. It was concluded "Not all criminals are feebleminded, but all feeble-minded persons are at least potential criminals. That every feeble-minded woman is a potential prostitute would hardly be disputed by anyone."[53]

These were global perceptions. U.S. President Theodore Roosevelt was simply reflecting a majority of contemporary opinion when he commented: "Someday we will realize that the prime duty, the inescapable duty, of the good citizen of the right type is to leave his or her blood behind him in the world; and that we have no business to permit the perpetuation of citizens of the wrong type"[54] A few years later Calvin Coolidge, at that time vice-President of the United States, declared: "America must be kept American. Biological laws show...that Nordics deteriorate when mixed with other races."[55] The American psychologist, Henry Goddard, believed it was obvious that no mentally impaired person should be allowed to marry or become a parent. Moreover, it was the duty of the "intelligent" section of society to enforce this rule. Oliver Wendell Holmes, Goddard's eminent fellow American jurist concluded:

> It is better for all the world, if instead of waiting to execute degenerate offspring for crime, or to let them starve for their imbecility, society can prevent those who are manifestly unfit from continuing their kind. The principle that sustains compulsory vaccination is broad enough to cover cutting the Fallopian tubes. Three generations of imbeciles are enough.[56]

Darwin had written in *The Descent of Man*: "Both sexes ought to refrain from marriage if they are in any marked degree inferior in body or mind...Everyone does a good service who aids towards this end.[57] In Victorian Britain the implication of marriage was, of course, procreation. If not proposing voluntary sterilisation, Darwin was suggesting at least voluntary abstinence. Others, equally eminent, were prepared to go further. As Home Secretary, Winston Churchill called for a "simple surgical operation (sterilisation) so the inferior could be permitted freely in the world without causing much inconvenience to others," and wrote to Herbert Asquith, then Prime Minister, "I am convinced that the multiplication of the Feeble-Minded, which is proceeding now at

---

(New York: Arno Press, 1973), p. 4)]. This could (and was) interpreted to mean almost anybody who did not fit a particular set of established parameters.

53 Friedlander, *The Origins of Nazi Genocide*, p. 6.

54 Samuel Totten, William S Parsons, Israel W Charny (eds.), *Century of Genocide: Eyewitness accounts and Critical Views* (New York: Routledge, 2004), p. 213.

55 Daniel J Kevles, *In the Name of Eugenics: Genetics and the Uses of Human Heredity* (Berkeley: University of California Press, 1986), p. 97.

56 Friedlander, *The Origins of Nazi Genocide*, p. 8–9. More than 65,000 Americans were subject to involuntary sterilisation during the inter-war years. [Philip Zimbardo, *The Lucifer Effect: How Good People Turn Evil* (London: Rider Books, 2007), p. 313].

57 Sheila Faith Weiss, *Race Hygiene and National Efficiency: The Eugenics of Wilhelm Schallmayer* (Berkeley: University of California Press, 1987), p. 29.

an artificial rate, unchecked by any of the old restraints of nature, and actually fostered by civilised conditions, is a terrible danger to the race."[58] In 1903, H. G. Wells wrote:

> The conclusion that if we could prevent or discourage the inferior sorts of people from having children, and if we could stimulate and encourage the superior sorts to increase and multiply, we should raise the general standard of the race is so simple, so obvious, that in every age I suppose there have been voices asking in amazement why the thing is not done.[59]

One year earlier, Wells had been even more outspoken in his promotion of eugenics:

> The men of the New Republic ... will rout out and illuminate urban rookeries and all places where the base can drift to multiply; they will contrive a land legislation that will keep the black, or yellow, or mean-white squatter on the move; ... so that childbearing shall cease to be a hopeful speculation for the unemployed poor; ... This thing, this euthanasia of the weak and sensual, is possible. On the principles that will probably animate the predominant classes of the new time, it will be permissible, and I have little or no doubt that in the future it will be planned and achieved.[60]

Wells and Churchill were not alone in their concern for the future of Anglo-Saxons, nor was Wells a voice crying out in the wilderness with his dire prophecies. In a lecture to the Eugenics Education Society[61] in 1910, George Bernard Shaw commented:

> We should find ourselves committed to killing a great many people whom we now leave living, and to leave living a great many people whom we at present kill. We should have to get rid of all ideas about capital punishment ... A part of eugenic politics would finally land us in an extensive use of the lethal chamber. A great many people would have to be put out of existence simply because it wastes other people's time to look after them.[62]

Shaw's defenders suggest that his remarks were tongue-in-cheek, a satirical reference to the policies of the more extreme eugenicists, but by 1934 Shaw was surely no longer joking when he wrote:

> The moment we face it frankly we are driven to the conclusion that the community has a right to put a price on the right to live in it ... If people are fit to live, let them live under decent human conditions. If they are not fit to live, kill them in a decent human way. Is it any wonder that some of us are driven to prescribe the lethal chamber as the solution for the hard cases which are at present made the excuse for dragging all the other cases down to their level, and the only solution that will create a sense of full social responsibility in modern populations?[63]

The "lethal chamber", a phrase in common usage around the turn of the century, and whose meaning was clear to all, was just five years away at the time of these reflections.

The nature of the rabid bigotry that often drove this kind of thinking is well illustrated by the comments of the Briton, Robert Reid Rentoul, a member of the Royal College of Surgeons, the General Medical Council of Education, the Medico-Legal Society and the Society for the Study of Inebriety—a man clearly considered a physician of dis-

---

58 Desmond King, *In the Name of Liberalism: Illiberal Social Policy in the United States and Britain*, (Oxford: Oxford University Press, 2002), p. 69, note 24.

59 H G Wells, *Mankind in the Making*, (Charleston: BiblioBazaar, 2006), p. 35.

60 H G Wells, *Anticipations*, (Charleston: BiblioBazaar, 2007), pp. 187–188.

61 In 1926 'Education' was dropped from the Society's title. (Sewell, *The Political Gene*, p. 247, note 3).

62 Dan Stone, *Breeding Superman: Nietzsche, Race and Eugenics in Edwardian and Interwar Britain* (Liverpool: Liverpool University Press, 2002), p. 127.

63 Ibid., p. 130.

tinction. In his 1906 book *Race Culture; or Race Suicide? (A Plea for the Unborn)*, Rentoul wrote: "The intermarriage of British with foreigners should not be encouraged. A few of us know the terrible monstrosities produced by the intermarriage of the white man and the black . . . From the standpoint of race culture it is difficult to understand the action of those who advocate the naturalization of foreigners".[64]

In a book entitled *Heredity and Eugenics in Relation to Insanity* published by the British Eugenics Society in 1912, Frederick Walker Mott, the chief pathologist of the London County Council made clear the racism inherent in contemporary eugenic reasoning: "The alien Jew and Irish Roman Catholic have large families, as their religion prohibits restrictions...25% of our population, made up mainly of the above mankind poor types, is producing 50% of our children and if this goes on it must lead to degeneracy."[65] In his fellow-countryman Charles Wicksteed Armstrong's *The Survival of the Unfittest,* published in 1927, there was more than a whiff of *The Protocols of the Learned Elders of Zion.* After claiming that "England, possessing the finest human stock in the world, is at the present time doing all in her power to destroy it", Armstrong managed to contrive an alleged confluence of Bolshevism and anti-Semitism worthy of Hitler himself: "The deliberately devilish policy of these Russian Jews is to use eventually the whole of Asia's immense resources in population and wealth for the furtherance of their aim — world revolution, or the suppression of civilization".[66] When the long-forgotten British eugenicist Anthony Mario Ludovici felt able, in several books published between the two world wars, to promote not only compulsory sterilisation, but also infanticide, incest, and what amounted to "euthanasia"—"The time has come to recognize the inevitability of violence and sacrifice, and consciously to select the section or elements in the world or the nation that should be sacrificed"—it becomes evident that racial hygiene was very far from being a solely German preoccupation.[67]

It is evident that all of these notions—sterilisation of the "inferior" and, by either implication or direct proposal, their removal from society, the prohibition of marriage between those considered unsuitable or unworthy, a direct association between mental deficiency and crime, racism as a guiding political principle—were symptomatic of much contemporary eugenic ratiocination, and all found their way into Nazi ideology. Zsigmond Fülöp had asked rhetorically, "whether we need to initiate a eugenic social policy, or in other words: is there a need for the creation of a human stock bodily and mentally stronger and more valuable than the present one? The answer to this question is nothing but yes."[68] By 1935, the French Nobel Prize winner, Alexis Carrel, felt able to write that the criminal and the insane should be "humanely and economically disposed of in small euthanasia institutions supplied with proper gases."[69] This was the "lethal chamber" writ large. The 1936 German introduction to Carrel's book *Man, The Unknown* included an enthusiastic endorsement of Nazi policies: "The German govern-

---

64 Dan Stone, *Race in British Eugenics* (European History Quarterly, Vol.31, no.3, 2001), p. 399.

65 David Pilgrim, *The Eugenic Legacy in Psychology and Psychiatry* (International Journal of Social Psychiatry, Vol.54, No.3, 2008), p. 274.

66 Stone, *Race in British Eugenics*, p. 400.

67 Ibid., p. 401.

68 Turda, *'A New Religion'?*, p. 312.

69 Proctor, *Racial Hygiene*, p. 180.

ment has taken energetic measures against the propagation of the defective, the mentally diseased, and the criminal. The ideal solution would be the suppression of each of these individuals as soon as he has proven himself to be dangerous."[70] It was hardly surprising that Carrel became the favoured eugenicist of Vichy France, and in recent years has been quoted with approval by Jean-Marie le Pen, former leader of the French far-right 'Front National' party.

It has been suggested that the worldwide popularity of eugenics in the first half of the twentieth century can be explained by the combination of a number of factors; science and scientists were held in high esteem, established class and racial inequalities and prejudices were vindicated by the new discipline, and perceived failings in social welfare policies addressed. Moreover, refuge could be found from the Victorian concept of environment and conditioning as the determining factor in human behaviour. If every society was plagued with poverty, crime, prostitution, alcoholism and disease, all believed to be genetic in nature, the chimera of eugenics appeared to provide an answer to these otherwise unfathomable problems.[71] Even so, the illogicality inherent in Galton's comparison of the breeding of animals or plants with that of human beings was self evident. Capable of an immensely greater variety of intellectual achievements than any other of earth's life forms, mankind was more than just another type of mammal. And as should have been painfully obvious to Galton when he looked around him, the mating of attractive, healthy or academically distinguished parents did not guarantee genetically improved or even equally high standards in their progeny. It did not in itself necessarily increase the probability of such a result. Healthy parents had sick children too. As for maintaining the supposed purity of a racial bloodline, this fallacy had everything to do with prejudice and nothing at all to do with science. The only certainty was that any society implementing the theories of Galton and his disciples would eventually result in that community giving its blessing to the perpetration of criminal acts against those deemed "unfit" to breed—or indeed to live.

70 Andrés Horacio Reggiani, *God's Eugenicist: Alexis Carrel and the Sociobiology of Decline*, (New York: Berghahn Books, 2007), p. 71.

71 Iredale, *Eugenics and its Relevance to Contemporary Health Care*, p. 207

CHAPTER 2: RACIAL HYGIENE

> What profit does humanity derive from the thousands of cripples who are born each year, from the deaf and dumb, from cretins, from those with incurable hereditary defects etc who are kept alive artificially and then raised to adulthood?...What an immense aggregate of suffering and pain these depressing figures represent for the unfortunate sick people themselves, what a fathomless sum of worry and grief for their families, what a loss in terms of private resources and costs to the state for the healthy! How much of this loss and suffering could be obviated, if one finally decided to liberate the totally incurable from their indescribable suffering with a dose of morphia. - Ernst Haeckel (1904)[72]

Amongst the most fervent of German disciples of Darwin and Galton were Ernst Haeckel (another polymath), the aforementioned Alfred Ploetz, and the physician Wilhelm Schallmayer. Whatever their other differences of interpretation may have been, these and many other influential Darwinians shared a common credo; all observable events, whether religious, ethical or otherwise were subject to scientific diktat. Everything was explicable in terms of simple cause and effect, in the inexorable application of the laws of nature.[73] Accordingly, there was no place for God or divine intervention in their doctrine.

Haeckel named his philosophy monism, which in essence proposed that *Homo sapiens* were not separate and different from the rest of nature. Everything—all inorganic matter and all life, including that of humans—had material, energetic, and metaphysical aspects. Life was different from inorganic matter only in its degree of organization. If Darwin suggested that all life might have originated from a single organism, Haeckel proposed that life emerged from non-living matter.[74] Haeckel's views on most subjects would be considered somewhat extreme today. He favoured eliminating criminals (killing them was "a good deed for the better part of society"),[75] as well as removing the mentally and physically ill from the hereditary chain.[76] Like many of the eugenicists, he regarded advances in medicine with disdain:

> The progress of modern medical science, although still little able really to cure diseases, yet possesses and practises more than it used to do the art of prolonging life during lingering, chronic diseases for many years. Such ravaging evils as consumption, scrofula, syphilis, and also many forms of mental disorders, are transmitted by sickly parents to some of their children, or even to the whole of their descendants. Now, the longer the diseased parents, with medical assistance, can drag on their sickly existence, the more numerous are the descendants who will inherit incurable evils, and the greater will be the number of individuals, again, in the succeeding generation, thanks to that artificial medical selection, who will be infected by their parents with lingering, hereditary disease.[77]

---

72 Michael Burleigh and Wolfgang Wipperman, *The Racial State: Germany 1933–1945* (Cambridge: Cambridge University Press, 1991), p. 31.

73 Richard Weikart, *From Darwin to Hitler. Evolutionary Ethics, Eugenics, and Racism in Germany* (New York: Palgrave MacMillan, 2004), pp. 12–13.

74 Haeckel sets out his theory in detail in *The Riddle of the Universe: At the Close of the Nineteenth Century* (Charleston: BiblioBazaar, 2009). For example: "Only thus can we attain a clear knowledge of the long scale of psychic development which runs unbroken from the lowest, unicellular forms of life up to the mammals, and to man at their head." (p. 103).

75 Weikart, *From Darwin to Hitler*, p. 160.

76 Ibid., p. 148.

77 Sheila Faith Weiss, *Race Hygiene and National Efficiency: The Eugenics of Wilhelm Schallmayer* (Berkeley: University of California Press, 1987), pp. 34–35

Haeckel was a racist of the most extreme variety, believing as he did that racial extermination was not only inevitable, but beneficial for the evolutionary process. He asserted that no woolly-haired race had ever achieved anything of historical importance.[78] In his 1868 book, *The Natural History of Creation*, Haeckel wrote that Europeans were taking over the entire world, driving other races to extinction. Whilst that may have been true in contemporary geopolitical terms, his conclusion that "even if these races were to propagate more abundantly than the white Europeans, yet they would sooner or later succumb to the latter in the struggle for life,"[79] has mercifully proved less accurate. As to whether Haeckel was or was not anti-Semitic, there are powerful views for and against.[80] Given the times, it might be considered that this was really a question of degree. Haeckel was opposed to all orthodox religions, and harboured a not uncommon German dislike of *Ostjuden* (literally Eastern Jews, i.e. Jews originating from eastern European states such as Russia and Poland), a view which he considered (with some justification) was shared by assimilated German Jews. However, comments such as: "When we make a careful anthropological study of the personality of Christ [the] characteristics which distinguish his high and noble personality...are certainly not Semitical; they are rather features of the higher Aryan race," suggest a more than passing familiarity with anti-Semitic reasoning.[81] Himself an archetypal "Aryan" in appearance, Haeckel's 1899 book *The Riddle of the Universe* was to have enormous influence on German eugenic thinking.

Until Haeckel's death in 1919, the Monist League, founded by him in 1906, remained a significant influence within the eugenic movement. Its relevance to Nazism, however, is disputed. Thus, on the one hand it is suggested that the Monist League showed hostility to the National Socialists by opposing Nazism's suppression of individual freedom: "Haeckel and the Monist League promoted many social reforms that were anathema to Hitler, such as homosexual rights, feminism, and pacifism"[82] Others present a differing view:

> If one surveys the origins of the *Völkisch* movement in Germany during the three or four decades prior to the First World War it is apparent that Haeckel played an influential, significant, indeed a decisive role in its genesis and subsequent development. An impressive number of the most influential *Völkisch* writers, propagandists, and spokesmen were influenced by or involved in some way with either Haeckel or his Monist followers. In the development of racism, racial eugenics, Germanic Christianity, nature worship, and anti-Semitism, Haeckel and the Monists were an important source and a major inspiration for many of the diverse streams of thought which came together later on under the banner of National Socialism.[83]

What is indisputable is that on attaining power, the Nazis banned the Monist League. It seems probable the contrasting opinions quoted above are examples of Hitler's propensity to draw on whatever aspects of political or ethical notions suited his

---

78 Richard J Evans, *The Coming of the Third Reich* (London: Allen Lane, 2003), p. 36.

79 Mike Hawkins, *Social Darwinism in European and American Thought 1860–1945: Nature as model and nature as threat* (Cambridge: Cambridge University Press, 1998), p. 140.

80 http://home.uchicago.edu/~rjr6/articles/Haeckel-antiSemitism.pdf (Accessed 21 March 2008).

81 Jonathan Peter Spiro, *Defending the Master Race: Conservation, Eugenics, and the Legacy of Madison Grant*, (Burlington: University of Vermont Press, 2008) , p. 123.

82 Weikart, *From Darwin to Hitler*, p. 70.

83 Daniel Gasman, *The Scientific Origins of National Socialism* (New Brunswick: Transaction Publishers, 2004), p. 147.

worldview. Nazism was what might be termed a cafeteria ideology—serve yourself from the notions on offer. For there can be no doubt that some features of Monism—eugenics, euthanasia, and racism—did feature prominently in National Socialism. Given the prevailing societal climate, this was hardly coincidental.[84] However, pronouncement's such as:

> What in the world has the doctrine of descent to do with socialism ? It has already been abundantly proved on many sides, and long since, that these two theories are about as compatible as fire and water...Socialism demands equal rights, equal duties, equal possessions, equal enjoyments for every citizen alike ; the theory of descent proves, in exact opposition to this, that the realisation of this demand is a pure impossibility, and that in the constitutionally organised communities of men, as of the lower animals, neither rights nor duties, neither possessions nor enjoyments have ever been equal for all the members alike nor ever can be,[85]

were hardly likely to endear Haeckel politically to the Nazis, or for that matter to the communists.

Among those influenced by Haeckel's theories was Friedrich Ratzel, who as a young journalist had spent a good deal of time in the United States, Cuba, and Mexico during the 1870s. Originally a student of zoology, on his return to Germany he became a lecturer in geography, eventually holding a chair in the subject, first at the Technical High School in Munich, then at the University of Leipzig. In the last years of the nineteenth century, Ratzel developed the concept of *Lebensraum* ("living space"), which held that the conquest of regions outside of Europe by the white race and the banishment (or worse) of the indigenous population which inevitably accompanied it, was a natural consequence of Darwin's "struggle for existence", coupled to mankind's eternal quest for territorial expansion. It was this constant migratory impulse, Ratzel believed, that drove human history onward. Not to follow it held out the possibility of being overwhelmed by other races better equipped for survival. In other words, imperialism was good, even essential. The disappearance of subjugated "inferior" races was of no account; their extermination could be justified in social Darwinist terms as a clear example of the "survival of the fittest."[86] It was a proposition to which first Kaiser Wilhelm II and subsequently Adolf Hitler were only too eager to subscribe.

Alfred Ploetz was another heavily influenced by Haeckel,[87] and although opposed to the Nazis prior to their attainment of power, was happy enough to subsequently support them. The Swiss psychiatrist August Forel, who has been nominated as the grandfather of the German eugenic movement, also played a significant role in Ploetz's development.[88] An early advocate of eugenics, Forel proposed that mental illness was not only hereditary but was increasing with the advance of civilization, and he became a major influence on many other eugenicists, including Ernst Rüdin, a classmate and some-

---

84 Weikart, *From Darwin to Hitler*, p. 70.

85 Ernst Heinrich Philipp August Haeckel, *Freedom in Science and Teaching* (Charleston: BiblioBazaar, 2009) , pp. 90–91.

86 David Olusoga and Casper Erichsen, *The Kaiser's Holocaust—Germany's Forgotten Genocide and the Colonial Roots of Nazism* (London: Faber and Faber, 2010), pp. 108–111.

87 Weikart, *From Darwin to Hitler*, p. 15.

88 Forel was a founder member of the Monist League. Haeckel tried unsuccessfully to persuade him to become president of the organisation. (Bernhard Kuechenhoff, *The psychiatrist August Forel and his attitude to eugenics* (History of Psychiatry, No. 19(2), 2008), p. 218).

time brother-in-law of Ploetz.[89] Writing an 1885 paper on mental asylums, Forel concluded: "I will leave open the question of whether it is the best and most humane way to eradicate such disgusting specimens of [the] human brain by painless death."[90] He proposed that sterilisation of the insane was a national sacrifice comparable to that made by a soldier in time of war,[91] and was also enough of a racist to be quoted with approval by Houston Stewart Chamberlain:

> Professor August Forel, the well-known psychiatrist, has made interesting studies in the United States and the West Indian Islands, on the victory of intellectually inferior races over higher ones because of their greater virility. 'Though the brain of the Negro is weaker than that of the white, yet his generative power and the predominance of his qualities in the descendants are all greater than those of the whites. The white race isolates itself [therefore] from them more and more strictly, not only in sexual but in all relations, because it has at last recognised that crossing means its own destruction'. Forel shows by numerous examples how impossible it is for the Negro to assimilate our civilisation more than skin-deep, and how so soon as he is left to himself he everywhere degenerates into the 'most absolute primitive African savagery... And Forel, who as a scientist is educated in the dogma of the one, everywhere equal, humanity, comes to the conclusion: 'Even for their own good the blacks must be treated as what they are, an absolutely subordinate, inferior, lower type of men, incapable themselves of culture. That must once for all be clearly and openly stated'.[92]

It is illuminating to record that Forel's promotion of racism, enforced sterilisation, and "euthanasia" was conveniently overlooked when in 1986 and 1988 his undoubted other contributions to real science were acknowledged by exhibitions in Zürich and Berne, accompanied by effusive paeans to his standing as "a man who even today can act as a role model and whose work still challenges us."[93]

Ploetz was elitist, hoping that the Society for Race Hygiene he had jointly founded in 1905 would draw its membership from the upper quartile of the population, those who were supposedly biologically "superior".[94] However, since there was no objective manner of determining this, he had to settle for a pledge from members to undergo a medical examination before marriage, in order to establish their suitability to reproduce.[95] In his own words, Ploetz believed that "whereas social hygiene aims for the substantial development of social institutions, the goal of race hygiene is to maintain and even strengthen egoism, as it is advantageous for the individuals in their struggle for existence."[96] To this end, conception was "not to be left to accident, or to an over-excited

---

89 Weikart, *From Darwin to Hitler*, p. 86.

90 Kuechenhoff, *The psychiatrist August Forel*, p. 220.

91 Robert N Proctor, *Racial Hygiene: Medicine Under the Nazis* (Cambridge Massachusetts: Harvard University Press, 1989), p. 96. Forel later wrote: "By 1885, without knowing the work of Galton, I had suggested negative eugenics, sometimes under the pretence of medical reasons, but in reality to prevent those disgusting species of mankind from reproducing." (Kuechenhoff, *The psychiatrist August Forel*, p. 220).

92 Houston Stewart Chamberlain, *Foundations of the Nineteenth Century, Volume I* (Port Chester: Adegi Graphics, 1977), p. 290.

93 Kuechenhoff, *The psychiatrist August Forel*, pp. 215–216.

94 Membership of the Society grew rapidly. In 1905 there were 32 members; by 1930 there were more than 1,300. The growth in the number of branches was equally impressive. By 1935 there were as many branches in Germany and Austria as there had initially been individual members. (Proctor, *Racial Hygiene*, p. 18).

95 Weikart, *From Darwin to Hitler*, p. 93.

96 Marius Turda and Paul J Weindling (eds.), "*Blood and Homeland*". *Eugenics and Racial Nationalism in Central and Southeast Europe, 1900–1940*, (Budapest: Central European University Press, 2007), p. 290.

moment, but rather regulated according to the principles which science has determined for the circumstances and time." If, despite these strictures, pregnancy should result in an incapacitated child, a "college of physicians...should prepare a gentle death...through a small dose of morphia."[97]

Ploetz's personal racism was apparent from his decision to name the German eugenics movement "race hygiene" (*Rassenhygiene*)[98], because, as he put it: "The hygiene of the entire human species coincides with that of the Aryan race, which, except for a few smaller races, like the Jewish—which in any case is mostly Aryan—represents the civilized race par excellence; to further it [the "Aryan" race] is the same as furthering all of humanity."[99] Yet in an apparently contradictory opinion, Ploetz did not subscribe to the view that there were 'pure' races anywhere on the globe. All races had been interbreeding for aeons, and the results had not necessarily been deleterious, providing the races in question were genetically similar.[100] Ploetz did not appear to have any difficulty in reconciling these diametrically opposed opinions.

Disciples of Ploetz's interpretation of social Darwinism were not restricted to Germany. Dr. Alois Scholz, who later became chairman of the *Wiener Gesellschaft für Rassenpflege* (founded 1924), the Austrian (Viennese) branch of Ploetz's German Society for Race Hygiene, explained racial hygiene as follows: "Just as the term implies, it deals with...the care for the gene pool of the people. ...Only if we promote the strong and that which is able to live and wipe out that which is unable to live, as demanded by nature, are we promoting that hygiene, which is useful to the whole." Some went even further. Professor Dr. Otto Reche, director of the Department of Anthropology at the University of Vienna (of whom more later), commented in 1925: "Racial hygiene must be the basis of all domestic policy and at least a part of foreign policy as well."[101]

As early as 1798, the German philosopher Christoph Meiners divided humanity into two categories—the fair-skinned "beautiful", who were attractive both physically and intellectually, and the dark-skinned "ugly", who were not merely unattractive, but also semi-civilized.[102] The eminent British philosopher, Bertrand Russell, wrote in 1927: "It seems on the whole fair to regard negroes as on the average inferior to white men, although for work in the tropics they are indispensable, so that their extermination (apart from questions of humanity) would be highly undesirable."[103] Although Russell's comments were not made in a eugenic context, they illustrate contemporaneous thinking; Caucasians stood at the summit of civilization. This view was common to a multitude of

97 Burleigh and Wipperman, *The Racial State*, p. 32.

98 István Apáthy defined racial hygiene in the following terms: "Public hygiene is concerned with the improvement of public life conditions and public health; racial hygiene fights against certain maladies which endanger not only the survival of isolated individuals but the survival of the entire species". (Marius Turda, *'A New Religion'? Eugenics and Racial Scientism in Pre-First World War Hungary* (Totalitarian Movements and Political Religions: Vol.7, No. 3, September 2006, p. 315).

99 Weikart, *From Darwin to Hitler*, p. 118.

100 Proctor, *Racial Hygiene*, p. 21.

101 http://www.doew.at/information/mitarbeiter/beitraege/rachyg.html (Accessed 23 November 2008).

102 Burleigh and Wipperman, *The Racial State*, p. 24

103 Proctor, *Racial Hygiene*, p. 179.

eugenicists—and not just in Germany. In the last years of the nineteenth century, Theodore Roosevelt produced a four volume history of the American frontier entitled *The Winning of the West* which included the following: "...but it is of incalculable importance that America, Australia, and Siberia should pass out of the hands of their red, black, and yellow aboriginal owners, and become the heritage of the dominant world races."[104]

In evaluating individuals on the basis of worth, racism was never far from the minds of the eugenicists, and was an inherent component of the supposed science.[105] The "darker peoples of southern Europe and the Slavs of eastern Europe are less intelligent than the fair peoples of western and northern Europe" wrote the Harvard psychologist Robert M. Yerkes, adding that the "Negro lies at the bottom of the scale" of intelligence. Harry Hamilton Laughlin, director of the Eugenics Record Office in the United States, compared "human racial crossing with mongrelisation in the animal world" and argued that "immigrants from southern and eastern Europe, especially Jews, were racially so different from, and genetically so inferior to, the current American population that any racial mixture would be deleterious."[106] The eminent English radical socialist, Sydney Webb, wrote in 1907:

> In Great Britain at this moment, when half, or perhaps two-thirds of all the married people are regulating their families, children are being freely born to the Irish Roman Catholics and the Polish, Russian and German Jews on the one hand, and the thriftless and irresponsible... on the other. This can hardly result in anything but national deterioration; or, as an alternative, in this country gradually falling to the Irish and the Jews.[107]

But as was so often to be the case, in Germany this racism went just a little further. If European culture was superior, "Aryan" mores reigned supreme. Ploetz was so convinced of this that he formed a secret organisation, The Nordic Ring, to disseminate a more racist form of eugenics, the object of which, he stated, was "a Nordic-Germanic race hygiene." Ploetz's considered Africans "inferior", as were Poles, and black Americans unintelligent and immoral.[108] Since for most eugenicists of the early twentieth century, anti-Semitism was a *sine qua non,* it is hardly surprising that it featured in Ploetzs' rationale, if initially in a somewhat less virulent form than in certain of his contemporaries.[109] Although thereafter somewhat modifying his views, he was hardly reticent about euthanasia either, writing:

> In relation to the persistently weak, i.e. the elderly, the incurable, and those who are otherwise defective, a society will be preserved all the better, the more these are disposed of. For their preservation requires sacrifice on the part of the strong and thereby reduces the ability of the whole to preserve itself...Any

104 Olusoga and Erichsen, *The Kaiser's Holocaust,* pp. 106–107.

105 Turda and Weindling, "*Blood and Homeland*", p. 283.

106 Henry Friedlander, *The Origins of Nazi Genocide: From Euthanasia to the Final Solution* (Chapel Hill: University of North Carolina Press, 1995), p. 6.

107 Daniel J Kevles, *In the Name of Eugenics: Genetics and the Uses of Human Heredity* (Berkeley: University of California Press, 1986), p. 74.

108 Weikart, *From Darwin to Hitler,* p. 119.

109 The term 'anti-Semitism was coined by the Hamburg journalist, Wilhelm Marr, in 1879. The meaning ascribed to it was racial rather than religious, although baptism could, in some circumstances, make acceptance in academic and professional circles somewhat easier. Gustav Mahler, for example, famously converted from Judaism to Catholicism in 1897 in order to secure the position as director of the Vienna Court Opera (later renamed the Vienna State Opera).

faulty and defective individuals still produced [later] can only be disposed of by annihilation or expulsion.[110]

At the very least, Ploetz was callous about the value of individual life. "The State is not there to see that the individual gets his rights, but to serve the race," he stated.[111], and he was equally unfeeling in proposing a policy of killing weak or deformed children.[112] Medical care should not be provided for those of poor genetic quality, for in so doing their survival and, worse, their ability to reproduce was facilitated. This was a common view among racial hygienists; traditional medicine might cure the "worthless" individual, but only at the expense of the betterment of the race. Thus certain diseases were the eugenicists' friends, since they only attacked the weakest in society.[113]

Ploetz's views on the desirability of war were no less harsh. All young men should serve in the army, so that during the campaign the "specially assembled [biologically] bad elements" could be brought "to the place where one needs primarily cannon fodder."[114] This was far from an outrageous view from the eugenicists' perspective; most of them were militarists, believing that war was a major contributor to human progress, which, if might was indeed right, was logical, if nothing else. Conventional wisdom dictated that war encouraged the development of mankind by benefitting the fit at the expense of the weak. Friedrich Hellwald, for example, commented: "Science knows no 'natural right.' In nature only one right reigns, which is no right, the right of the stronger, violence. But violence is also the highest source of law [or right], since without it [i.e.violence] legislation is unthinkable....Properly speaking the right of the stronger has also been valid at all times in human history."[115] In Britain, Karl Pearson, later the first holder of the Galton Chair of Eugenics at the University of London, claimed in 1900, at the height of the Boer War: "This dependence of progress on the survival of the fittest race, terribly black as it may seem to some of you, gives the struggle for existence its redeeming features; it is the fiery crucible out of which comes the finer metal." If there was an end to war, "mankind would no longer progress." Only through war could the advance of the "inferior" be checked and the law of natural selection prevail.[116]

The American eugenicist, David Starr Jordan, believed that war could be seen as the killing off of only the physically fit male members of the population, since the disabled stayed safely at home. It followed that war was only reprehensible when those dying were of "superior" race. Then it became dysgenic. But if warfare succeeded in destroying "inferior" specimens, it became an important contributor to the evolutionary process. It was for this reason most eugenicists regarded the Great War with horror, for it was overwhelmingly the highest race, the Caucasians, who were killing each other, leaving the "inferior" races to thrive.[117] In Germany this had become a commonly accepted philosophy by the turn of the century. If the anthropologist Ludwig Woltmann could

---

110 Weikart, *From Darwin to Hitler*, p. 152.

111 Robert Jay Lifton, *The Nazi Doctors: Medical Killing and the Psychology of Genocide* (London: Papermac, 1990), p. 24.

112 Weikart, *From Darwin to Hitler*, p. 152.

113 Proctor, *Racial Hygiene*, p. 15

114 Weikart, *From Darwin to Hitler*, p. 177.

115 Ibid., p. 169.

116 Margit Szöllösi-Janze (ed), *Science in the Third Reich* (Oxford: Berg, 2001), p. 188.

117 Ibid., p. 165

propose that the "Germanic race has been selected to dominate the earth", others viewed that coming domination as a consequence of a genetic imperative which would find its natural expression in warfare. "Without war, inferior or decaying races would easily choke the growth of healthy budding elements, and a universal decadence would follow," declared General Friedrich von Bernhardi.[118]

Ploetz's overall reasoning posed a quandary, one that Darwin himself had acknowledged. How was it possible to reconcile the two disparate faces of a society which on the one hand sought to improve the lot of the weak, the poor, and the unhealthy, whilst on the other was dedicated to bettering the quality of the race? Was it not impossible to benefit one except to the detriment of the other?[119] To Ploetz the answer was simple; it lay in racial hygiene, with all that that implied. This was simply a matter of ends and means. If the 'valuable' were to thrive, the 'worthless' would perforce fall by the wayside.

Wilhelm Schallmayer, originally a student of law and philosophy who turned instead to natural science and medicine,[120] first came to prominence with his 1903 book *Heredity and Selection in the Life of Nations: A Study in Political Science on the Basis of the New Biology*, winner of the prestigious Krupp Prize competition.[121] In the introduction to his book he wrote: "This view [Darwinism] had an especially powerful influence on ethics. It not only produced new views about the origin and evolution of ethical commands and thus new foundations for them, but it also led to the call for a partial alteration of presently valid ethical views."[122] Schallmayer believed that Christian morality was not simply outmoded, it was a hindrance to evolutionary progress. What was needed was an abandonment of out-of-date ethics based on religious belief, for like life itself, moral values too must evolve. Whosoever was first to adopt these evolutionary ethics would have an advantage and thus a better chance of prevailing in the never-ending struggle for existence. "The views of Christianity, insofar as they are at all influential, do not have the tendency to improve selection, either consciously or unconsciously, but rather—naturally unconsciously—has the opposite tendency."[123] So those who after appropriate medical examination were deemed suitable, were not merely to be permitted to marry and reproduce, but positively encouraged to do so by means of financial and other incentives. Polygamy was to be allowed for especially racially valuable male specimens By way of contrast, those who failed the proposed examination would be prevented from reproducing through compulsory sterilisation.[124] Given such a rejection of established moral codes, anything was possible in terms of "negative eugenics", that is to say the elimination, by one means or another, of those hindering evolutionary progress. Thus, in Schallmayer's view, advances in medicine were not improving the quality of life—they were destroying it:

---

118 Evans, *The Coming of the Third Reich*, p. 35.

119 Mark B Adams (ed), *The Wellborn Science: Eugenics in Germany, France, Brazil, and Russia* (New York: Oxford University Press, 1990), p. 16.

120 Weiss, *Race Hygiene and National Efficiency*, pp. 39–41

121 Weikart, *From Darwin to Hitler*, p. 15.

122 Ibid., p. 16.

123 Ibid., p. 51.

124 Burleigh and Wipperman, *The Racial State*, p. 31.

Of all our cultural achievements which impede natural selection with respect to the human race, medicine was the first to arouse my suspicion. This question was bound to trouble me since I was a medical student at the time. In fact, the joy I derived from my profession subsided as I became more and more convinced that, on the whole, the therapeutic application of medical science not only did not contribute to the perfection of the human species, but often even damaged it.[125]

For Schallmayer, the concept of human rights no longer existed (indeed if it ever had). As a contemporary, Alexander Tille, commented, "Against the rights of the stronger, every historical right is completely invalid."[126] The individual was valueless; only the continuation and evolution of the species was important. Nor was all life uniformly worthy. "Making the unequal equal can only be an ideal of the weak," Schallmayer wrote.[127] He and his disciples were opposed to providing medical assistance to women in childbirth, for such aid enabled those who would otherwise not reproduce to do so—and they, inevitably, were "inferior" specimens. Tille approved of killing the disabled, and like Ploetz and Schallmayer, believed that sick children should receive no medical care. Instead they should be left to die, for to do so would effectively strengthen the hereditary link.[128] There was an inherent contradiction in an argument which bemoaned a declining birth rate, yet at the same time wished to encourage that decline. Of course, it all depended upon who was giving birth, a polarity that was to become evident as state policy under Hitler.[129]

If less racist than many of his fellow eugenicists, and judged by the standards of the day, not notably anti-Semitic, Schallmayer was no believer in racial equality either. In rejecting Ploetz's Nordic Aryanism, he still believed that black Africans were mentally and culturally inferior to Europeans.[130] Wars between unequal races were of positive value, because the "inferior" would inevitably be annihilated. "On the whole the influence of war on human evolution should still be considered overwhelmingly favourable."[131]

Nowhere was the racist imperialism of Wilhelmine Germany better illustrated than in the Herero campaign in South West Africa (today Namibia), commonly described as the first act of genocide of the twentieth century.[132] As the influence of social Darwinism spread, so with it the concept of *Lebensraum* developed. The new colonies, chaotic and dangerous, were to become "civilized." South West Africa was deemed particularly suitable for major colonization, and so became the crucible in which German social hygienic theory was first tested.[133]

As early as 1837, the German philosopher, Georg Wilhelm Friedrich Hegel, had written in *The Philosophy of History*:

125 Weiss, *Race Hygiene and National Efficiency*, p. 44
126 Weikart, *From Darwin to Hitler*, p. 35.
127 Burleigh and Wipperman, *The Racial State*, p. 93.
128 Evans, *The Coming of the Third Reich*, p. 36.
129 Proctor, *Racial Hygiene*, pp. 18–19.
130 Burleigh and Wipperman, *The Racial State*, p. 119.
131 Weikart, *From Darwin to Hitler*, p. 177.
132 Ibid., p. 205.
133 Dan Stone (ed), *The Historiography of Genocide* (Basingstoke: Palgrave Macmillan, 2008), p. 325.

The Negro, as already observed, exhibits the natural man in his completely wild and untamed state...From these various traits it is manifest that want of self-control distinguishes the character of the Negroes. This condition is capable of no development or culture, and as we see them at this day, such have they always been...it [Africa] is no historical part of the world; it has no movement or development to exhibit...What we properly understand by Africa, is the Unhistorical, Undeveloped Spirit, still involved in the conditions of mere nature, and...on the threshold of the World's History.[134]

One hundred years later opinions had not markedly changed, and were applied by the Nazis not to Africa and Africans, but to the inhabitants of the vast "uncivilized" territories of eastern Europe and beyond. Accordingly, Hitler could state in the latter geographical context that "the Germans...will have to constitute among themselves a closed society...the least of our stable-lads must be superior to any native."[135]

In South West Africa in 1904–1908, native forced labour became an essential component of German colonization, thereby setting the precedent for the Nazi conquest and occupation of eastern Europe in the1940s.[136] In search of Kaiser Wilhelm II's "place in the sun", and as an intrinsic part of this exercise in empire building, Germany waged a succession of exterminatory wars of colonization in Africa during the late nineteenth and early twentieth centuries. Theodor Leutwein, Governor of South West Africa 1894–1905, was explicit concerning his policies, even claiming that native exploitation accorded with eugenic principles, as well as being financially beneficial:

Stripped of all idealistic and humanitarian impediments, the final objective of all colonization is to make money. The colonizing race has no intention of bringing happiness to the aboriginal people, the kind of happiness that the latter perhaps expects. In the first instance, the conquerors seek their own advantage. Such objectives correspond to human egotism, and therefore accord with nature. Colonial policy must therefore be determined by the expected profits.[137]

The man appointed Commissioner for Settlement in South West Africa, Paul Rohrbach, was even more unequivocal in defining the ethos of German colonialism:

It is not right either among nations or among individuals that people who can create nothing should have a claim to preservation. No false philanthropy or race-theory can prove to reasonable people that the preservation of any tribe of nomadic South African Kaffirs...is more important for the future of mankind than the expansion of the great European nations, or the white race as a whole...[138]

The parallels between German military behaviour in the African wars of the late nineteenth century and in the former Soviet Union after *Barbarossa* in 1941 are demonstrably acute. In both cases there was a refusal to grant the enemy the status of an equal opponent entitled to any fundamental rights. Prisoners were allowed to die through malnutrition, disease, and neglect, or were often simply executed. In South West Africa, Herero and Nama prisoners, including women and children, were imprisoned in concentration camps where the mortality rate was 30–50 percent. The annihilation of civilian populations and a "scorched earth" policy were common to both the

134 Jürgen Zimmerer, *The Birth of the Ostland out of the Spirit of Colonialism: a Postcolonial Perspective on the Nazi policy of Conquest and Extermination*, Patterns of Prejudice, Vol.39, No 2, 2005), p. 205.

135 Ibid., p. 206.

136 Ibid., p. 207.

137 L H Gann and Peter Duignan, *The Rulers of German Africa, 1884–1914* (Stanford: Stanford University Press, 1977), p. 44.

138 Olusoga and Erichsen, *The Kaiser's Holocaust*, p. 112.

campaigns in Africa and those in the Soviet Union forty years later, as was ruthless economic exploitation.[139]

Following a German policy which had the effect of reducing the Herero to the status of second-class citizens by denial of any fundamental human rights, an uprising broke out on 12 January 1904.[140] The authorities in Berlin considered attempts by Leutwein to suppress the insurrection disappointing. A firmer hand was needed to crush the natives and bring total victory. Having already waged a brutal war of suppression in German East Africa, and been a member of the German forces sent to aid in the crushing of the so-called 'Boxer Rebellion' in China on 1900–01, General Lothar von Trotha was appointed Commander in Chief of the German army in South West Africa in May 1904. The Kaiser's message to his troops on their departure for China had been:

> When you meet the enemy, you will beat him; you will give no pardon and take no prisoners. Those whom you capture are at your mercy. As the Huns a thousand years ago under King Etzel made a name for themselves that has lasted mightily in memory, so may the name "Germany" be known in China, such that no Chinese will ever again even dare to look askance at a German.[141]

Von Trotha and his comrades did not disappoint their Emperor, any more than the forces of the United States, Great Britain, Russia, France, Japan, Italy and Austria sent to crush the rebellion displeased their respective governments. This was the heyday of imperialism—"inferior" indigenous life was valueless.

Arriving in South West Africa on 11 June 1904, von Trotha assumed the dictatorial powers granted to him by the Kaiser, overriding the civilian authorities, including Leutwein. Since he believed that a war in Africa could not be fought in accordance with the principles of the Geneva Convention (again, a parallel with Hitler and *Barbarossa*), von Trotha concluded that only the extermination of the Herero and later of the Nama would resolve the racial war he was fighting. He was therefore quite prepared to destroy "the rebellious tribes in streams of blood."[142] Following the battle of Waterberg, his genocidal intentions were made clear in his proclamation of 2 October 1904:

> ...The Herero people must however leave the land. If the populace does not do this, I will force them out with (cannon). Within the German borders, every Herero, with or without a gun, with or without cattle, will be shot. I will no longer accept women and children, I will drive them back to their people or I will let them be shot at...[143]

A convinced racist, Von Trotha justified his actions in eugenic terms: "The philanthropic disposition will not rid the world of the...law of Darwin's, the 'struggle of the fittest'".[144] A newspaper reported him as uttering words which were to become familiar forty years later: "Against 'nonhumans' [*Unmenschen*] one cannot conduct war 'humanely'".[145] The official war history of the campaign made it clear that "the waterless

139 Zimmerer, *The Birth of the Ostland out of the Spirit of Colonialism*, p. 209.

140 Stone, *The Historiography of Genocide*, p. 326.

141 Isabel V Hull, *Absolute Destruction—Military Culture and the Practices of War in Imperial Germany* (Ithaca: Cornell University Press, 2006), p. 135.

142 Stone, *The Historiography of Genocide*, p. 326.

143 Zimmerer, *The Birth of the Ostland out of the Spirit of Colonialism*, p. 209.

144 Weikart, *From Darwin to Hitler*, p. 206

145 Hull, *Absolute Destruction*, p. 33.

Omaheke [desert] would complete what German weapons had begun: the destruction of the Herero people."[146] It would be difficult to find a clearer example of unashamed mass murder, a practical application of the racist eugenic theories so prevalent at the time, as a result of which an estimated 65,000, or 81 percent, of Hereros died.[147] A year later von Trotha began another campaign against the Nama (some of whom had been his allies in the war against the Herero), which eventually resulted in a further 10,000 African dead. He called on the Nama to surrender whilst boasting:

> The Nama who chooses not to surrender and lets himself be seen in the German area will be shot, until all are exterminated. Those who, at the start of the rebellion, committed murder against whites or have commanded that whites be murdered have, by law, forfeited their lives. As for the few not defeated, it will fare with them as it fared with the Herero, who in their blindness also believed that they could make successful war against the powerful German Emperor and the great German people. I ask you, where are the Herero today?

Yet in the end, this *Vernichtungsstrategie* (strategy of annihilation) became self-defeating. For as Leutwein, himself considered a relative moderate,[148] remarked:

> We have expended several hundred million marks and the lives of several thousand German soldiers. As a result we have.... totally destroyed the pastoral industry of our colony. We have destroyed two-thirds of our native labour. Worse still, we have as yet [by March 1906] been unable to restore peace.[149]

Although von Trotha was eventually ordered to change his policies, and was recalled to Germany in November 1905, this was due less to the application of humanitarian principles than to the unfavourable publicity engendered by German atrocities in Africa. Nor did the killing cease with von Trotha's departure. On the contrary, through internment in concentration camps and an intentionally exterminatory policy of forced labour, what is now acknowledged as a clear act of genocide, in the most literal sense, continued unabated.[150] In words that might also have been uttered by Hitler decades later, Alfred Graf von Schlieffen, Chief of the General Staff stated: "The race war, once commenced, can only be ended by annihilation or the complete enslavement of one party." Von Schlieffen found it a cause of regret that von Trotha had not been permitted to personally complete his policy of extermination.[151]

In an eerie precursor of the activities of Julius Hallevorden and Carl Schneider nearly forty years later, the camp physician of the Shark Island concentration camp, Dr Bofinger, removed the brains of deceased Nama prisoners and forwarded them to the Institute of Pathology in Berlin, where Christian Fetzer attempted to prove the similarities between the Nama and various species of ape. Bofinger was allegedly greatly feared by the prisoners. According to a missionary's report it "was never the case that even a single person recovered in the Lazarett [Field Hospital]." This was hardly surprising, as

---

146 Zimmerer, *The Birth of the Ostland out of the Spirit of Colonialism*, p. 209.

147 Weikart, *From Darwin to Hitler*, p. 206

148 It has been suggested that Leutwein's policy of cultural genocide would ultimately have been as destructive as von Trotha's physical version. (Stone, *The Historiography of Genocide*, pp. 323–343, particularly note 23).

149 Gann and Duignan, *The Rulers of German Africa*, p. 75.

150 Olusoga and Erichsen, *The Kaiser's Holocaust*, p. 360.

151 Stone, *The Historiography of Genocide*, p. 327.

Bofinger was also conducting medical experiments on living prisoners in a worthless attempt to find a cure for scurvy.[152]

Lest it be thought that imperialist practices of this murderous nature were an exclusively German phenomenon in Africa, mention should be made of Cecil Rhodes and his British South African Company, responsible for the deaths of thousands of men, women, and children in the course of stealing vast areas of land and huge numbers of cattle from the Mashona and Ndebele peoples of Southern Rhodesia (now Zimbabwe) during the last decade of the nineteenth century. In 1896, Lord Jarvis, a supporter of Rhodes, wrote: "I hope the natives will be pretty well exterminated . . . our plan of campaign will probably be to . . . wipe them out ". The following year Lord Grey wrote that even the missionary Father Biehler felt: "The only chance for the future of the [Mashona] race is to exterminate the whole people, both male and female, over the age of 14!" Other European nations behaved no better, and in some cases immeasurably worse. In the Belgian Congo, between 1885 and 1920 ten million Africans were killed or worked to death in the ivory and rubber trades. Joseph Conrad called it "the vilest scramble for loot that ever disfigured the history of human conscience and geographical exploration".[153] However, although these and other examples provided Hitler with no shortage of precedents for colonial style exploitation on his invasion of eastern Europe, it was the genocidal Wilheminian behaviour in South West Africa that provided the most apposite analogy. As the German anti-Nazi political economist Moritz Julian Bonn commented, National Socialist racial policies were merely amplifications of von Trotha's own murderous racism: "The Nazi creed is based on the same cheap conception of Darwinism, and like their colonial predecessors, they do not believe in the unaided working of this supposed law of nature."[154]

Within this context, the contribution of Hans Grimm, a once popular author who had spent his formative years in South Africa, is worthy of note. Grimm created a political slogan with his 1926 novel *Volk ohne Raum* (*Nation Without Space*), accentuating the need for German colonial expansion.[155] Regarded as a classic by the Nazis, the book was a bestseller in the 1930s and 1940s, becoming compulsory reading in German schools.[156] No less than 500,000 copies of the 1,344 page single volume edition had been sold by 1943. Although never a party member, Grimm was predictably right-wing in his political views and supported the Nazis. His first-hand experience of colonial life during the years of the Herero and Nama genocide, as reflected in his writing, was an influential factor in formulating National Socialist expansionist foreign policy propaganda.

---

152 Olusoga and Erichsen, *The Kaiser's Holocaust*, p. 226.

153 Benjamin Madley, *From Africa to Auschwitz: How German South West Africa Incubated Ideas and Methods Adopted and Developed by the Nazis in Eastern Europe* (European History Quarterly, Vol. 35 No.3, 2005), p. 431

154 Olusoga and Erichsen, *The Kaiser's Holocaust*, p. 330.

155 Gann and Duignan, *The Rulers of German Africa*, p. 41.

156 Zimmerer, *The Birth of the Ostland out of the Spirit of Colonialism*, p. 217.

Haeckel, Ploetz, Schallmayer and their fellow eugenicists may have shared a common belief in the inviolability of heredity, but they enjoyed few other opinions in common. Amongst their number were some for and some against abortion and birth control, a number who were pacifists and others who were militarists, some who supported involuntary euthanasia and others who opposed it, differing degrees of racism and anti-Semitism—in fact it seems difficult to find a subject on which there was complete agreement between any two or more parties. Among the most important of these disagreements were debates on two issues; were alleged social "afflictions" such as alcoholism or sexually transmitted diseases hereditary?; and secondly, could the transmission of hereditary traits be affected by external influences, such as environment, education and so on?[157]

Today, the views of some of these eugenicists appear very strange indeed. The Italian psychiatrist Cesare Lombroso believed that criminals were throwbacks to an earlier stage of human evolution.[158] In his view, crime was a disease, and criminality an inherited trait; a 'born criminal' could be identified by the presence of certain physical defects, such as a flattened or upturned nose, or fleshy lips. Like Schallmayer, Christian von Ehrenfels suggested polygamy as the answer to biological improvement, with the more valuable males (of whom he was doubtless one) having several wives.[159] August Weisman considered reliance on technology to be harmful to evolution; therefore the myopic should be discouraged from wearing spectacles, and visits to the dentist prohibited, since those with poor eyesight and weak teeth would produce biologically sounder offspring if their conditions were left untreated.[160] Karl Vogt claimed that certain mentally sick individuals were closer to animals than humans (a not uncommon view amongst eugenicists), regarding those so afflicted as a virtual missing link between ape and man.[161] Bartholomäus von Carneri wrote that "entire human tribes stand lower than animals...The mental activity of the elephant, horse, and dog [are] significantly better than the lowest human species."[162] Franz Carl Müller-Lyer was of the opinion that producing a child with a sick constitution was a crime almost as bad as murder.[163] Eugen Wolfsdorf viewed caring for sick family members to be a waste of money and energy.[164] Alfred Hegar made no distinction between the mentally disabled and criminals. One was mentally "inferior", the other morally "inferior"—why not kill both?[165] It is difficult to refute the pithy conclusion expressed by the *Illustrated London News* in 1910:

---

157 Marius Turda, *'A New Religion'? Eugenics and Racial Scientism in Pre-First World War Hungary* (Totalitarian Movements and Political Religions: Vol.7, No. 3, September 2006), p. 314.

158 Weikart, *From Darwin to Hitler*, p. 37. Since he was Jewish, Lombroso was infrequently quoted by Nazi eugenicists, despite the fact that his views largely coincided with their own. (Proctor, *Racial Hygiene*, p. 202).

159 Weikart, *From Darwin to Hitler*, p. 141.

160 Ibid., p. 84.

161 Ibid., p. 95

162 Ibid, pp. 110–111.

163 Ibid., p. 139

164 Ibid., p. 151.

165 Ibid., p. 155.

"The only daring suggestion for the improvement of the human race that Eugenics suggests to us is that the world would be a jollier place if there were fewer cranks in it."[166]

This was a proposition that could certainly have been applied to a man who has become, in effect, the patron saint of post-Second World War right-wing extremism, the Italian philosopher Julius Evola, whose eccentric views were such as to make him too *outré* even for Mussolini's Fascist Party. Although no "scientific" eugenicist like others mentioned here, he surely is deserving of a place in their company. Not only anti-Semitic, but anti-Christian and anti-modernist as well, Evola peddled a preposterous concept of "Aryanism" that made him a natural bedmate of Heinrich Himmler, whose undercover agent in Italy he was. No Darwinian either, Evola (who seemed to have been opposed to almost everything) wrote: "We do not believe that man is derived from the ape by evolution. We believe that the ape is derived from man by involution." Darwin's theories "promulgate a distorted and mutilated concept of man."[167] In February 1940 Evola had published an article on "Jews and Mathematics" in which he concluded that Judaism existed in complete opposition to "Aryan civilization".[168] It therefore comes as no surprise that Evola moved to Germany after Italy's surrender in September 1943, there finding a natural home in Himmler's *Ahnenerbe* (see chapter 10) along with a host of other crackpots.

The list of eccentric viewpoints is lengthy. However, the cornerstone of heredity was common not only to those who might be considered to have held somewhat unconventional opinions, but to global perceptions of the new religion. The eugenics movement was, after all, international,[169] (the world's first professorial chair in eugenics was established in 1911 at University College London), but as has been illustrated, it became particularly influential in Germany. Specific individual sources for Hitler's eugenic thinking are generally difficult to identify with certainty, and have been the subject of much scholarly debate,[170] but there are some influences which are indisputable. Prime among these is Fritz Lenz.

Lenz was born in 1887 in Pflugrade, Pomerania, now Redlo, Poland. As a medical student at the University of Freiburg he was a pupil of the racial anthropologist Eugen Fischer, author in 1913 of a highly successful book entitled *Rehobother Bastard und das Bastardisierungsproblem beim Menschen* (*The Rehoboth Bastards and the Bastardization problem Among Humans*), a study of the descendants of mixed marriages between Eu-

---

166 Dan Stone, *Breeding Superman: Nietzsche, Race and Eugenics in Edwardian and Interwar Britain* (Liverpool: Liverpool University Press, 2002), p. 128.

167 Julius Evola, *Eros and the Mysteries of Love: The Metaphysics of Sex* (Rochester: Inner Traditions International Ltd, 1991), p. 9.

168 Joshua D Zimmerman, *Jews in Italy Under Fascist and Nazi Rule, 1922–1945* (Cambridge: Cambridge University Press, 2005), p. 139.

169 National eugenic societies, frequently involving physicians and psychiatrists, were established in Germany (1905), Great Britain (1907–08), the USA (1910) and France in 1912. In that year an International Congress for Eugenics was also inaugurated. [Michael Burleigh, *Eugenic Utopias and the Genetic Present* (Totalitarian Movements and Political Religions: Vol. 1, No. 1, 2000), p. 62)]. Other European countries formed eugenic societies in the years that followed [Turda and Weindling,"*Blood and Homeland*", p. 2–3]. The American Eugenics Society, founded in the United States in 1922, was one of the later arrivals on the scene.

170 See Weikart, *From Darwin to Hitler*, pp. 209–227 for a brief analysis of the origins of Hitler's ethic, if such it can be called.

ropean men and African women in a German South West African community. Needless to say, Fischer disapproved of such racial mixing, and supported segregation in German colonies.[171] A complete lack of evidence was no impediment to his claim that the children of mixed marriages were of "lesser racial quality".[172] In this he merely reiterated the fallacies expounded by "scientists" like Felix von Luschan, Director of the Berlin Museum of Ethnology, who had proposed that in mixed race populations the blood of the "inferior" race would increasingly dominate in successive generations, until a point would eventually be reached when the progeny of such inter-breeding would regress to a point where only pure-blooded members of the "inferior" race were produced.[173] This kind of inverted Spencerian nonsense, a theory which might be termed "the survival of the unfittest", was fully endorsed by Fischer, who continued that, so far as "inferior" races were concerned:

> One should grant them the amount of protection that an inferior race confronting us requires to survive, no more and no less and only for so long as they are of use to us—otherwise free competition, that is, in my opinion, destruction.[174]

Ignoring his prognosis and continuing racial mixing in the manner of the Rehoboth Bastards was a recipe for disaster so far as the "superior" race was concerned. After all, it was in Fischer's view irrefutable that

> ...every European people that has adopted the blood of inferior races—and that Negroes...and many others are inferior only mad people would deny—has, without exception, atoned for the adoption of these inferior elements with their mental and cultural downfall.[175]

In the prevailing climate Fischer was pushing against an open door, his "research" merely reinforcing existing prejudices and stereotypes. As a reward for his book, and in a triumph of style over substance, Fischer was made a full professor, and was later appointed director of the *Kaiser-Wilhelm Institut für Anthropologie, menschliche Erblehre und Eugenik* (Kaiser Wilhelm Institute of Anthropology, the Teaching of Human Heredity and Eugenics).[176] From 1933–35 he served as chancellor of Berlin University,[177] where in his inaugural address he proudly boasted: "What Darwin was not able to do, genetics has achieved. It has destroyed the theory of the equality of man."[178]

Lenz, who also studied under Ploetz, was a convinced racist eugenicist, publishing the magazine *Archiv für Rassen- und Gesellschaftsbiologie* (*Archive for Racial and Social Biology*) between 1913 and 1933. In 1917 he also published an essay, republished in 1933 with a preface stating that this work contained the essence of Nazism: "Everything comes from the ideal of the race: culture, evolution, personality, happiness, re-

---

171 Weikart, *From Darwin to Hitler*, p. 121.

172 Burleigh and Wipperman, *The Racial State*, p. 38.

173 Olusoga and Erichsen, *The Kaiser's Holocaust*, p. 245.

174 Friedlander, *The Origins of Nazi Genocide*, pp. 11–12

175 Olusoga and Erichsen, *The Kaiser's Holocaust*, p. 250.

176 The institution was founded in 1927, an important step in the scientific recognition of racial hygiene in Weimar Germany. (Proctor, *Racial Hygiene*, p. 39).

177 Zimmerer, *The Birth of the Ostland out of the Spirit of Colonialism*, p. 213.

178 Proctor, *Racial Hygiene*, p. 345, note 74.

demption...With every activity and with every inactivity we have to ask ourselves: Does it benefit our race? And to make our decision accordingly."[179]

Together with Erwin Baur and Eugen Fischer, in 1921 Lenz authored *Grundriss der menschlichen Erblichkeitslehre und Rassenhygiene* (*Outline of Human Genetics and Racial Hygiene*)[180] This work and his theory of "race as a value principle" promoted Lenz and his two co-authors to the position of Germany's leading racial theorists, since their ideas provided "scientific" justification for Nazi ideology, particularly the emphasis on "Aryan" superiority, and the desirability of eliminating "inferior" humans. The Baur-Fischer-Lenz opus went through five editions by 1940, and was considered serious scientific research at the time—and not only in Germany.[181] It was received favourably in Britain, the Netherlands, Italy, and many other countries. In the United States it was hailed as "the standard textbook of human genetics", which given some of the odd conclusions it contained now seems difficult to believe.[182]

Lenz was appointed as the first associate professor of race hygiene in Germany at the University of Munich in 1923.[183] At that time he was critical of what he perceived to be German sloth in dealing matters of sterilisation, complaining that the medical provisions of the Weimar Constitution prevented the undertaking of an adequate number of vasectomies. Moreover, Germany had no legislation equivalent to the American laws preventing marriage between people suffering from conditions such as epilepsy or mental illness, or between individuals of different races. Lenz's only criticism of America was a perceived insufficient focus on "Aryan", as opposed to white, supremacy,[184] since he believed the Nordic race creatively and intellectually superior to any other.[185] Lenz claimed that the revolutionary turmoil in Germany after 1918 was caused by inferior racial elements. He warned that the nation's racial superiority was threatened, stating: "The German nation is the last refuge of the Nordic race...before us lies the greatest task of world history".

For Lenz, compulsory sterilisation was a sacred mission.[186] "As things are now," he pronounced in 1934, "it is only a minority of our fellow citizens who are so endowed that their unrestricted procreation is good for the race."[187] Like Schallmayer and Ehren-

179 Weikart, *From Darwin to Hitler*, p. 121

180 Turda and Weindling, "*Blood and Homeland*", p. 227. A copy of Lenz's jointly-authored book was presented to Hitler while he was imprisoned in Landsberg Prison in 1924 (Ibid., p. 267), and is now housed in the US Library of congress. It bears a dedication to the *Führer* by the Nazi publisher, Julius Friedrich Lehmann. [Gretchen E.Schafft, *From Racism to Genocide: Anthropology in the Third Reich* (Urbana: University of Illinois Press, 2004). pp. 61–62]. In 1931 Lenz claimed that "many passages in it [the book] are mirrored in Hitler's expressions". (Weikart, *From Darwin to Hitler*, p. 223). Whether Hitler actually read the book is a matter of conjecture, although Lenz's son, Widukind stated that his father was told that Hitler had read it whilst in Landsberg. (Proctor, *Racial Hygiene*, p. 352, note 51). The question of the literary influences on Hitler's *Weltanschauung* (World View) is the subject of ongoing debate. It is reasonable to assume that he would certainly have read the crackpot racist writings of authors like Jörg Lanz, Artur Dinter, Willibald Hentschel and many others, all freely available in post-First World War Germany and Austria. (Burleigh and Wipperman, *The Racial State*, p. 37).

181 Weikart, *From Darwin to Hitler*, p. 223.

182 Proctor, *Racial Hygiene*, pp. 50–59.

183 Turda and Weindling, "*Blood and Homeland*", p. 24.

184 Lifton, *The Nazi Doctors*, p. 23.

185 Weikart, *From Darwin to Hitler*, p. 223.

186 Lifton, *The Nazi Doctors* p. 24.

187 Lisa Pine, *Nazi Family Policy 1933–1945* (Oxford: Berg, 1999), p. 105.

fels, he favoured polygamy (that is for those suitable), but realized the impossibility of its adoption in the foreseeable future.[188] And like so many other eugenicists he opposed the protection of "inferior" individuals, and supported the killing of disabled children.[189] But there were times when Lenz came across characters with opinions even more radical than his own. At a committee meeting attended by Himmler, Lenz opposed equality for illegitimate children because he believed it would have a negative hereditary effect. Himmler disagreed, arguing that illegitimacy was no disgrace. On the contrary, equality was necessary for the illegitimate in order to ensure a high birth rate and ward off the proliferation of homosexuality and abortion.[190] Himmler at least practiced what he preached, fathering two illegitimate children.

Both Himmler and Lenz would probably have agreed with the Hungarian, Jenő Vámos, who theorized that

> If we want to improve the Caucasian race, those individuals are a priori suitable who bear the typical characteristics of the race, provided they serve a social purpose. On the other hand, for instance such individuals, who show characteristics contrary to their sexual nature—like the homosexuals—are a priori unsuitable, even if, in other aspects, they do befit a social purpose.[191]

Himmler indicated his obsession with these matters in 1936 with the creation of a Central Office to Combat Homosexuality and Abortion, an organisation which, in true Orwellian fashion, maintained records of both women who had had abortions as well as those members of the medical profession who had procured them. The introduction to his secret directive of 10 October 1936 on the subject began:

> The serious danger to population policy and public health represented by the still relatively high number of abortions which are a major violation of the fundamental National Socialist worldview, as well as the homosexual activity of a not inconsiderable layer of the population which poses one of the greatest dangers to youth, requires more than before the effective combating of these public scourges.[192]

Despite some initial consequential prosecutions, ultimately Himmler was no more successful in curtailing perceived undesirable traits in human sexuality than those who attempted to do so both before and after him.[193]

Lenz's views on motherhood were, if anything, even more extreme than his opinions on sterilisation and illegitimacy. When reviewing an article by Eugen Fischer for a 1913 dictionary, in which Fischer proposed that a healthy woman ought to give birth to eight or nine children during the course of her lifetime, Lenz declared that Fischer's submission was quite inadequate:

> It is a fact that a woman is capable of giving birth for a period of nearly thirty years. Even when we consider a woman giving birth only once every two years, this means a minimum of fifteen births per mother. Anything less than this must be considered the result of unnatural or pathological causes.[194]

---

188 Weikart, *From Darwin to Hitler*, p. 143.

189 Ibid., p. 223.

190 Friedlander, *The Origins of Nazi Genocide*, p. 19.

191 Turda, *'A New Religion'?*, p. 320.

192 Michael Berenbaum, *Witness to the Holocaust* (New York: HarperCollins, 1997) p. 110.

193 Jill Stephenson, *Women in Nazi Germany* (Harlow: Pearson Education Limited, 2001), p. 39.

194 Proctor, *Racial Hygiene*, p. 124.

Such was the judgement of Nazism's leading eugenicist. As a reward for his promotion of racist eugenic views, in 1933 the Nazis appointed Lenz director of the eugenics section at the aforementioned Kaiser Wilhelm Institute and a professor at the University of Berlin.[195] Initially, Lenz's anti-Semitism appeared to be of the less extreme variety. In 1931 he wrote: "We must of course deplore the one-sided anti-Semitism of National Socialism. Unfortunately, it seems that the masses need such 'anti' feelings." Predictably, however, he then attempted to rationalise the racism he had just deplored:

> ...we cannot doubt that National Socialism is honestly striving for a healthier race. The question of the quality of our hereditary endowment is a hundred times more important than the dispute over capitalism or socialism, and a thousand times more important than that over the (Weimar Republic.)[196]

Notwithstanding his earlier moderation, he subsequently found no difficulty in justifying the Nuremberg Laws of 1935:

> As important as the external features for their evaluation is the lineage of individuals. A blond Jew is also a Jew. Yes, there are Jews who have most of the external features of the Nordic race, but who nevertheless display Jewish mental tendencies. The legislation of the National Socialist state therefore properly defines a Jew not by external race characteristics, but by descent.

By now firmly toeing the Party line, in 1940 Lenz described the replacement of Jews and Slavs in occupied eastern Europe to be "the most weighty task of racial politics." Settlers there should be chosen "according to the criteria used by the SS to select racial settlers."[197]

After the war, Lenz continued to work as a professor of genetics at the University of Göttingen. When questioned, he stated that the Holocaust would undermine the study of human genetics and racial theory, and continued in his belief that the eugenic theories of racial differences had been scientifically proven. He died in 1976 in Göttingen. Ironically, his son Widukind was a rather more distinguished geneticist, being among the first to recognize the dangers to unborn children of the drug thalidomide. Unlike his father, Widukind Lenz, who died in 1995, was recognized and respected during his lifetime as an eminent physician and humanitarian.

In truth, there was little, if anything, original in Nazi ideology, if indeed National Socialism as a political credo can be considered ideological at all in any meaningful sense. As has been observed: "[National Socialism] had no doctrine in the proper sense of the word...The ideas—few in number and crude as well as shallow—that formed the hard core of the national socialist creed had hardly sufficient logical coherence to deserve the name of ideology..."[198] *Lebensraum*, eugenics, racism, anti-Semitism, fascism, nationalism, militarism, anti-capitalism and the like, all had significant antecedents in

195 Turda and Weindling, "*Blood and Homeland*", p. 24.

196 Benno Müller-Hill, *Murderous Science: Elimination by Scientific Selection of Jews, Gypsies, and Others in Germany 1933–1945* (Woodbury: Cold Spring Harbor Laboratory Press, 1998), p. 9.

197 Proctor, *Racial Hygiene*, p. 350, note 9.

198 Arthur L Caplan, (ed), *When Medicine Went Mad: Bioethics and the Holocaust* (Totowa: Humana Press, 1992,) p. 331, note 2.

Germany. As has been observed: "Hitler did not invent much. Most of the time he was content to take up ideas that were already in the air and to carry them to their ultimate conclusion. Euthanasia and profound meditations on 'lives that did not deserve to be lived' were commonplaces at that time."[199] The uniqueness of Nazism was to pull the miasma of these disparate elements together into a noxious totalitarian entirety.

Despite the inflammatory rhetoric, many of Nazism's extreme policies appear to have had limited appeal to Germans as a whole. Having peaked at 37.4 percent in the election of 31 July 1932, the Nazi share of the national vote had declined to 33.1 percent in the final "free" election of 6 November 1932,[200] which would seem to indicate that, to take but a single element, the NSDAP's[201] vehemently antagonistic Jewish policy did not, as some have suggested, signify a singularly Germanic eliminationist anti-Semitism.[202] In fact, so far as this particular issue is concerned, there is scant evidence that German (as opposed to Nazi) anti-Semitism was eliminationist at all, except in its consequences. Negative racial stereotyping is hardly uncommon in most societies, and anti-Semitism was so commonplace in pre-World War II Europe and the United States that its appearance in publications of the day, both popular and academic, and in public pronouncements, was considered no more than the norm.[203] However, the extent to which Jews were represented in many professions provided fertile soil for the Nazis to plough; in Germany between 1918 and 1933 Jews made up 0.78 percent of the population, but numbered 16 percent of the doctors, 15 percent of the dentists, 25 percent of the lawyers, 50 percent of the theatre directors, and occupied 80 percent of the leading positions on the Berlin stock exchange. In Vienna in 1936 it was estimated that 62 percent of all lawyers were Jewish, as were 47 percent of physicians, and nearly 29 percent

---

199 Enzo Traverso, *The Origins of Nazi Violence* (New York: The New Press, 2003), p. 127.

200 Alan Bullock, *Hitler and Stalin: Parallel Lives* (London: Harper Collins, 1991), p. 1083.

201 NSDAP = National Socialist German Workers Party (*Nationalsozialistische Deutsche Arbeiterpartei*), acronym 'Nazi'.

202 See in particular Daniel Jonah Goldhagen, *Hitler's Willing Executioners—Ordinary Germans and the Holocaust* (London: Little, Brown and Company, 1996), where this thesis is explored at considerable length. For example: "Whatever else Germans thought about Hitler and the Nazi movement, however much they might have detested aspects of Nazism, the vast majority of them subscribed to the underlying model of Jews and in this sense (as the Nazis themselves understood) were 'Nazified' in their views of Jews. It is, to risk understatement, no surprise that under Nazi dispensation the vast majority of Germans continued to remain anti-Semitic, that their anti-Semitism continued to be virulent and racially grounded, and that their socially shared 'solution' to the 'Jewish Problem' continued to be eliminationist." (pp. 87–88). Such conclusions have been refuted by many distinguished scholars. One of these, Eberhard Jäckel, described Goldhagen's book as "a failure of dissertation, faultily researched through and through...It is not state of the research, it does not live up even to mediocre standards, it is simply bad." [Robert R Shandley (ed), *Unwilling Germans—The Goldhagen Debate* (Minneapolis: University of Minnesota Press, 2001), p. 87].
This is not to say that there are not others who support Goldhagen's position. These ultra-intentionalists are typified by observations such as: "...indifference was more than acquiescence. It was a willed desire throughout the German population, for the elimination and extermination of persons of the Jewish race." [James M Glass, *"Life Unworthy of Life"—Racial Phobia and Mass Murder in Hitler's Germany*, (New York: Basic Books, 1997], p. 3. As the eminent historian, Julius H. Schoeps observed: "It is absurd to blame 'the Germans' in their totality for Nazi crimes. The accusation of collective guilt, which was levelled immediately after the war...is no more insightful just for being taken up again in 1996." (Shandley, *Unwilling Germans*, p. 79).

203 James Waller, *Becoming Evil. How Ordinary People Commit Genocide and Mass Killing* (Oxford: Oxford University Press, 2002), pp. 36–49.

of university tutors.[204] Yet at the same time, little more than one-quarter of one percent of Austrian government employees were Jewish—an example of a centuries-old endemic anti-Semitism, if religious rather than racial in character.[205] On 27 April 1933, the *Deutsche Allgemeine Zeitung* newspaper concluded:

> A self-respecting nation cannot, on a scale accepted up to now, leave its higher activities in the hands of people of racially foreign origin...Allowing the presence of too high a percentage of people of foreign origin in relation to their percentage in the general population could be interpreted as an acceptance of the superiority of other races, something decidedly to be rejected.[206]

The 'what if' school of history is normally to be avoided at all costs, but with regard to the manner in which racial hygiene came to be applied by the Nazis, a modest degree of speculation may be permissible. Given the National Socialist's declining share of the electoral vote, it is interesting, but of course ultimately futile, to speculate about the direction German politics might have taken if Hitler had not been invited to form a government on 30 January 1933. Still, most aspects of the eugenic policies the Nazis were to pursue were inherited from the Weimar Republic, and merely required their implementation by this most radical of regimes.[207] In view of the negative eugenic strategy being advocated and often followed in the United States, Sweden, and elsewhere at this time, it is not unreasonable to conclude that the adoption of at least some form of compulsory sterilisation might have been embraced in Germany whatever the political persuasion of the newly inaugurated government.[208] Moreover, it would be mistaken to assume that in the first decades of the twentieth century, racial hygiene or indeed racism of the crudest kind was exclusively a fascist, or even a notably right-wing obsession. It was embraced with equal enthusiasm by many on the left of the political spectrum, by social democrats as well as conservatives. In 1931 the German Communist party supported sterilisation of psychiatric patients under certain conditions, and between 1931 and 1938 Germany and the Soviet Union shared a joint Institute for Racial Biology in Moscow.[209] But of course, in this as in so many other matters, the Nazis adopted the most extreme position.

Nor should it be assumed that the National Socialist eugenic *Weltanschauung* (world view) was derived solely from Darwinism and the supposedly "scientific" reasoning to which it gave birth. There was no shortage of contributors to the melange of *fin-de-siècle* German political philosophy. The Nietzschean rejection of Judeo-Christian morali-

---

204 Richard Lynn, and Satoshi Kanazawa, *How to explain high Jewish achievement: The role of intelligence and values* (Science Direct, Personality and Individual Differences, Vol. 44, No.4, 2007, p. 802). There were, of course, sound historical reasons for this ostensibly disproportionate number of Jews in certain professions, but then as now, the statistics provided useful propaganda for anti-Semites, something that could be used as evidence of an alleged 'world-wide Jewish conspiracy' to dominate and control all of society, however preposterous that idea may have been (and is) in reality.

205 David S Wyman (ed), *The World Reacts to the Holocaust* (Baltimore: The John Hopkins University Press, 1996), pp. 476–477.

206 Saul Friedländer, *The Years of Persecution: Nazi Germany and the Jews 1933–39* (London: Phoenix, 2007), pp. 30–31.

207 Michael Burleigh (ed), *Confronting the Past: New Debates on Modern German History* (London: Collins & Brown Limited, 1996), p. 100.

208 Even in the Weimar years advocates of positive eugenics were gradually losing the argument with the proponents of negative eugenics. (Friedländer, *The Years of Persecution*, p. 39).

209 Proctor, *Racial Hygiene*, p. 22–23.

ty, indeed of religion itself ("God is Dead") found many adherents. Nietzsche proclaimed himself an anti-Darwinist, although there was sufficient ambiguity in his writings to suggest otherwise for Darwinian enthusiasts. In *The Will to Power*, Nietzsche wrote:

> The Biblical prohibition 'thou shalt not kill!' is a piece of naiveté compared with the seriousness of the prohibition of life to decadents: 'thou shalt not procreate'—Life recognizes no solidarity, no 'equal rights' between the healthy and the degenerate parts of an organism: one must exercise the latter—or the whole will perish.—Sympathy for the decadents, equal rights for the ill-constituted—that would be the profoundest immorality, that would be antinature itself as morality.[210]

No eugenicist could have expressed the concept better. Moreover there are many other passages in Nietzsche which indicate, at the very least, sympathy for the negative eugenic viewpoint.[211] However, the hypothesis for which he is perhaps best remembered today, his vision of the 'Superman' as a creative genius who would dominate others by intellectual rather than biological superiority, was subject to wilful misinterpretation, particularly by the Nazis, who adapted Nietzsche's notion to serve their own rationale of the *Übermensch* as a racial prototype.[212] In this they also conveniently overlooked Nietzche's approval of the inclusion of the Jews as part of what he termed "the strongest possible European mixed race," a notion that hardly sat well with Hitler's maniacal Judeophobia.[213] Through careful selection of his writings, the Nazis felt confident enough of their interpretation of his philosophy to assert "...that only a conscious National Socialist can fully comprehend Nietzsche."[214] He was, they claimed, one of their own, an uncompromising anti-Semite.[215] This adoption conveniently overlooked the irony that his insanity would actually have made Nietzsche singularly suitable for inclusion in the Nazi's own "euthanasia" programme.[216]

A major influence on Anglo-German relations in the years preceding the First World War, the philologist and philosopher, Alexander Tille, was the first to translate and edit Nietzche's works in English. A conservative social Darwinist, Tille was a prominent member of the burgeoning movement, with observations such as: "When marriage to a sick wife produces sick cripples, then that deed deserves condemnation, it is an immoral act—even if conventional ethics praises it as a deed of heroic, altruistic self-sacrifice." Tille found natural selection functioning in some unusual places and ways:

> Nature, acting ineluctably, eliminates human beings who have degenerated into animals from amongst the ranks of the rest, thus East London operates to an extent as a national sanatorium; any attempt to help the "unfortunate" only reduces Nature's enormous significance in this regard.[217]

---

210 Steven E Aschheim, *The Nietzsche Legacy in Germany, 1890–1990* (Berkeley: University of California Press, 1992), p. 243.

211 Weikart, *From Darwin to Hitler*, pp. 48–49

212 Ibid., p. 46.

213 Michael Burleigh, *The Third Reich: A New History* (London: Pan Books, 2001), p. 98.

214 Aschheim, *The Nietzsche Legacy in Germany*, p. 237.

215 Ibid., pp. 244–245.

216 Ernst Klee, "*Euthanasie*" *im NS-Staat: Die "Vernichtung lebensunwerten Lebens"* (Frankfurt am Main: Fischer Taschenbuch Verlag, 1991), p. 17.

217 Fred Bridgham, (ed), *The First World War as a Clash of Cultures*, (Rochester: Camden House, 2006), pp. 215–218.

The ruling Social Democrat Party in Prussia unsuccessfully attempted to force through a National premarital form of eugenic certification in 1919–20.[218] However, the regional governments of Saxony and Thuringia were able to convince the Reich government into revising the clause in the Reich Criminal Code pertaining to offences against the person so as to permit doctors to carry out voluntary eugenic sterilisations. In Saxony, the district physician of Zwickau, Gustav Boeters, had drafted a law entitled "The prevention of unworthy life through operative measures". He then lobbied the regional government unceasingly for the introduction of such legislation—in short, a compulsory Sterilisation Law for those deemed "inferior". In a letter of 3 December 1923 he boasted that he and other physicians were already sterilising disabled individuals against their will:

> To my knowledge, I am the first German medical official who has dared to translate the aims of practical racial hygiene into action in his area of professional responsibility. We in Zwickau have undertaken sterilisation operations on mental defectives and others, under the aegis of our highest public authorities...since in many cases the consent of parents and others is not to be had at any price, even though the necessity of an operation is clearly evident for anyone not himself a mental defective, I urge the introduction of legislative coercion.[219]

In 1928, the Criminal Law Committee of the Reichstag debated the voluntary sterilisation of 'hereditarily ill criminals' as the price to be paid for their early release from custody. Again, this debate produced no legislation. Eventually, the Prussian Health Council formed a commission, which in October 1932 presented a draft Sterilisation Law to the Prussian government. Both Prussia and Saxony then lobbied the Reich government to introduce this at federal level. By the time this bill had passed through the legislative process, the Nazis were in power, with their own distinctive brand of eugenics.[220]

---

218 Although such genetic screening has generally fallen out of favour, it is practised in modern-day China. By 1914 it was also being enforced in 50 percent of U.S. states. (Burleigh, *Eugenic Utopias and the Genetic Present*, p. 69.)

219 Horst Biesold, *Crying Hands: Eugenics and Deaf People in Nazi Germany* (Washington: Gallaudet University Press, 1999), pp. 16–17.

220 Burleigh, *Eugenic Utopias and the Genetic Present*, p. 69

CHAPTER 3: "EUTHANASIA"

> The fundamental principles of criminal procedure were seriously affected inasmuch as expert physicians declare that in all good conscience they could no longer give a diagnosis in dubious cases of the increased insanity of accused persons in order to establish a basis for their confinement in a sanatorium or asylum because such commitment, in its result, was equivalent to the execution of a death sentence without a previous trial in court. – Franz Schlegelberger, acting Minister of Justice (March 1941)[221]

> What is interesting *and* important about the killing programme is not the mad-dog killers, but rather the careful, orderly, and quite methodical manner by which the full German medical and scientific establishment proceeded to kill its patients over a period of years. – Hugh Gallagher[222]

The Shorter Oxford English Dictionary defines euthanasia as "the action of inducing a quiet and easy death." Derived from the Greek prefix "*eu*" (goodly, or well) and suffix "*thanatos*" (death), this grant of a "mercy death" may occur with the consent of the individual concerned, is then termed "voluntary" euthanasia, and was the meaning originally attributed to the expression in ancient Greece and Rome, where it was frequently practised. In his book "Utopia", in 1516 Sir Thomas More described the ending of the life of the incurably ill as an act of mercy. Essentially, More suggested that such action should never be undertaken without the authority of the sufferer. It is undoubtedly true that for many people, the option of ending one's life in the face of unbearable pain and anguish is the preferable of two equally dreadful alternatives. In an admittedly small survey of adults suffering from a variety of potentially incurable medical conditions, 96 percent felt it was worse to be kept alive under hopeless circumstances with the certainty of impending death than it would be to die, 82 percent felt that total loss of independence would be worse than death, and 73 percent were so averse to a life of unremitting pain and suffering that they would rather be dead.[223] It may be reasonable to assume that these responses typify public attitude toward the question of voluntary euthanasia.

Centuries before More, Plato had rather more callously suggested in *The Republic* that "defective offspring... will be quietly and secretly disposed of", and that the state "will provide treatment for those...citizens whose physical and psychological constitution is good; as for the others, it will leave the unhealthy to die."[224] Eugenicists were (and are) fond of quoting both men. They are less eager to quote Adolf Hitler, who in 1928 wrote in an unpublished manuscript:

> While nature only allows the few most healthy and resistant out of a large number of living organisms to survive in the struggle for life, people restrict the number of births and then try to keep alive what has

221 http://nuremberg.law.harvard.edu/php/pflip.php?caseid=HLSL_NMT01&docnum=588&numpages=5&startpage=1&title (Accessed 12 December 2009)

222 Hugh Gregory Gallagher, *By Trust Betrayed: Patients, Physicians, and the License to Kill in the Third Reich* (St. Petersburg: Vandamere Press, 1995).

223 http://members.tripod.com/~Amis_Lee/fallingtree/eu.html (Accessed 20 January 2007) For a broader summary of medical research and sources regarding end-of-life issues, see http://search.apa.org/search?query=end (Accessed 20 January 2007), amongst many others.

224 Rachel Iredale, *Eugenics and its Relevance to Contemporary Health Care*, (Nursing Ethics, 7 (3), 2000), p. 206.

been born, without consideration of its real value and its inner merit. Humaneness is therefore only the slave of weakness and thereby in truth the most cruel destroyer of human existence.[225]

In words that presaged Hitler, the French philosopher, Henri Lichtenberger had written in 1898:

> There are unfortunates whom it is inhuman to relieve. There are degenerates whose death should not be delayed...The earth must not be a lazar-house inhabited by the sick and discouraged, or else the healthy man will perish from disgust and pity. To spare future degenerations the depressing sight of misery and ugliness, let us kill all those who are ripe for death, let us have the courage not to retain those among us who are falling, but let us push them so that they may fall even more quickly.[226]

Such thinking led to the concept of "involuntary" euthanasia, as for example when a patient is suffering from an incurable and painful disease or terminal illness, or is in a coma and is considered unlikely to regain consciousness. In such circumstances, a third party or parties may determine to put an end to the patient's suffering. The circumstances are, in general, that the person involved is no longer capable of making up his or her mind and/or of expressing his or her ultimate wish.[227]

Lest it be thought that such notions are outdated, in another experimental survey conducted among 570 students at the University of Hawaii in the early 1970s, 90 percent of those participating agreed that there would always be some people fitter for survival than others; 91 percent agreed that under extreme circumstances elimination of those most dangerous to the general welfare was justified, with 29 percent supporting this option even if it applied to their own families. Moreover, if required by law to do so, no fewer than 89 percent were prepared to participate in the life or death decision-making process, and 9 percent were prepared to assist with the killing, or with both the decision and the elimination of the condemned.[228] Given the ease with which Nazi Germany was able to recruit individuals ranging from university professors to labourers in order to implement its policies, there is no reason to suppose that these are unrepresentative figures.

Nazi "euthanasia" was, in fact, devoid of any humanitarian or compassionate reasoning, but rather was symptomatic of a mercilessly applied pseudo-scientific theory coupled to a ruthless economic policy. The Nazis destroyed "life unworthy of life" (*lebensunwertes Leben*) as they termed it, not as an act of mercy, but as part of a strategy to murder that part of the population least able to defend itself. That policy came to be directed not only at German citizens, but at those of other eastern European countries which fell under Nazi hegemony. The "euthanasia" programme had nothing to do with concern for the sick and suffering. Rather it formed an essential part of the evolving Na-

225 Ian Dowbiggin, *A Concise History of Euthanasia: Life, Death, God, and Medicine* (Lanham: Rowman & Littlefield, 2007), p. 93

226 Dan Stone, *Breeding Superman: Nietzsche, Race and Eugenics in Edwardian and Interwar Britain*, (Liverpool: Liverpool University Press, 2002), p. 71.

227 Another to broach the subject of involuntary euthanasia for those suffering from incurable diseases was the American author W. Duncan McKim, who in his 1899 book *Heredity and Human Progress*, wrote that its application would additionally perform the eugenic function of "artificial selection", thereby "improving the human race." [Enzo Traverso, *The Origins of Nazi Violence* (New York: The New Press, 2003), p. 122.]

228 Philip Zimbardo, *The Lucifer Effect: How Good People Turn Evil* (London: Rider Books, 2007), pp. 284–285.

zi policy of extermination on a massive scale. That policy reached its apogee with the industrialised murder of the Jews, but had the programme arrived at its intended conclusion, the eventual death toll would have been immeasurably greater. *Generalplan Ost* (General Plan for the East), the Nazi scheme for the reorganization of the demographic and economic structure of eastern Europe, mainly through depopulation, envisaged the deportation or murder of between 31 and 51 million people, depending upon the ambition of the planner.[229] As it was, deaths attributable in one form or another to the Nazi regime did eventually total countless millions. At the root of many of those deaths lay an obsession with the bogus science of eugenics, and its eventual godchild, state-sponsored murder.

Among those contributing to this concept of a eugenically defined Nazi paradise was Otto Reche, an anthropologist and associate of Eugen Fischer and Fritz Lenz. Reche's speciality was the study of blood types, a matter of supreme importance to the regime. He was an adviser to the SS Race and Settlement Main Office (*SS Rasse-und Siedlungshauptamt*), to whom he suggested that in order to maintain racial purity, Jews and Poles should be deported from regions where it was intended that Germans were to settle. Shortly after the invasion of Poland in 1939 he wrote to a colleague:

> ... we need space (*Raum*) but no Polish lice in our fur. I am absolutely sure that the racial-scientific matter determines the solution of all of these questions, since in the future we do not want to build a Germany in the East that would only be linguistically German but a racial mishmash, with strong Asiatic elements and Polish in character. That would be no German nation, nor a corner stone for a German future!...[230]

Amongst other things, Reche maintained that *Volksdeutsche* returning to the *Reich* should first be screened in order to establish that they were acceptable on racial hygienic grounds.[231] He was imprisoned for a brief period after the war, but thereafter, despite his Nazi associations (he had been a member of the Party since 1937), he went on to enjoy a distinguished post-war career, being awarded the Austrian Honorary Cross for Science and Art First Class in 1965.

Adolf Jost had argued in his 1895 book, *The Right to Death* (*Das Recht auf den Tod*), that if the state demanded the sacrifice of thousands of individuals in wartime, it had the same "right" in times of peace to demand the sacrifice of the impaired and non-productive, who were draining the state of its resources.[232] During the Great War that theory of peacetime civilian immolation became a wartime reality as mortality rates in German asylums soared to unprecedented levels. It has been estimated that approximately 30 percent of the pre-1914 asylum population, or more than 71,000 people,

---

229 Götz Aly and Susanne Heim, *Architects of Annihilation: Auschwitz and the Logic of Destruction* (London: Phoenix, 2003), p. 255. There is evidence that similar plans for massive resettlements in western Europe were also contemplated [Jürgen Zimmerer, *The Birth of the Ostland out of the Spirit of Colonialism: a Postcolonial Perspective on the Nazi policy of Conquest and Extermination*, Patterns of Prejudice, Vol.39, No 2, 2005, p. 200, note 9].

230 http://www.drustvo-antropologov.si/AN/PDF/2009_2/Anthropological_Notebooks_XV_2_Geisenhainer (Accessed 12 February 2008).

231 Ingo Haar and Michael Fahlbusch, *German Scholars and Ethnic Cleansing, 1919–1945*, (New York: Berghahn Books, 2005), p. 15.

232 Robert Jay Lifton, *The Nazi Doctors: Medical Killing and the Psychology of Genocide* (London: Papermac, 1990), p. 46.

died as a result of malnutrition, sickness or neglect.[233] The British blockade and consequent rationing, coupled with a lack of medication and clothing, poor sanitary conditions and overcrowding all contributed towards the death toll, but attitudes towards the institutionalised had also changed. In 1920, Karl Bonhoeffer, chairman of the German Psychiatric Association observed how

> ...we were forced by the terrible exigencies of war to ascribe a different value to the life of the individual than was the case before...we had to get used to watching our patients die of malnutrition in vast numbers, almost approving of this, in the knowledge that perhaps the healthy could be kept alive through these sacrifices.[234]

Dr Georg Ilberg, director of the Sonnenstein asylum, commented in 1922:

> If one surveys the higher death rate among the mentally ill in psychiatric institutions during the war, the great sacrifice of death that our innocent patients had to offer is naturally painfully regrettable...At any rate, it was not possible and not justified to give the mentally ill more food than the healthy.[235]

In his 1961 evidence to a Frankfurt court, Hans Hefelmann, responsible for handling petitions at the KdF (*Kanzlei des Führers der NSDAP*—Chancellery of the *Führer*), testified to the impact this kind of reasoning allegedly had on the decision to introduce "euthanasia" in 1939:

> Asylum directors pointed out that there had been an alarming mortality rate during the First World War. It had been much higher than the number of extremely severe, incurable cases that would have been considered for euthanasia...During that time, many doctors and nurses had been called up for military service. Food and pharmaceuticals had become scarce, leading to a great increase in the mortality rate, because what was available had to be distributed equally among the curable and the incurable...If euthanasia were [now] administered to the most serious cases, it would ensure a peacetime level of therapeutic care for less severe cases by making available a relatively large number of nurses within the institution...The doctors stressed that the numerous deaths, confirmed by all parties, that had occurred after even greater suffering during the First World War would provide justification for the recommended euthanasia measures.[236]

Whether this evaluation, that in times of national crisis some lives could be assessed as being worthier of preservation than others, provided at least a part of the necessary

---

233 In Prussia alone an estimated 45,000 mental patients died, while in Bavaria the mortality rate in asylums reached 70 percent. [Dick de Mildt, *In the Name of the People: Perpetrators of Genocide in the Reflection of Their Post-War Prosecution in West Germany. The 'Euthanasia' and 'Aktion Reinhard' Trial Cases* (The Hague: Martinus Nijhoff Publishers, 1996), p. 335, note 19]. In some institutions in Baden it was not until 1925 that the mortality rate returned to its pre-war level. [Bronwyn Rebekah McFarland-Icke, *Nurses in Nazi Germany* (Chichester: Princeton University Press), p. 23].

234 Michael Burleigh, *Death and Deliverance : 'Euthanasia' in Germany c. 1900–1945* (London: Pan Books, 2002), pp. 11–12. Bonhoeffer also realised where the risk lay in rationalising such thinking, for he went on to say:
"But in emphasising the right of the healthy to stay alive, which is an inevitable result of periods of necessity, there is also a danger of going too far: a danger that the self-sacrificing subordination of the strong to the needs of the helpless and ill, which lies at the heart of any true concern for the sick, will give ground to the demand of the healthy to live." (Ibid.)
Karl Bonhoeffer is today regarded by many with something less than approbation. He was the father of the founder of the Confessing Church, Dietrich Bonhoeffer.

235 McFarland-Icke, *Nurses in Nazi Germany*, p. 35.

236 Götz Aly, Peter Chroust, Christian Pross, *Cleansing the Fatherland: Nazi Medicine and Racial Hygiene* (Baltimore: The Johns Hopkins University Press, Baltimore, 1994) , p. 157.

stimulus for doctors to adopt and practice the principle of "euthanasia", seems plausible, if ethically deplorable. Certainly in the years immediately following the First World War, there were German psychiatrists who believed that whilst patients might be entitled to enjoy the *same* rights as everybody else, they were not entitled to *better* rights than others, a curious inversion of the concept of healthcare. Instead of defending and preserving the lives of the incapacitated and helpless, these physicians reduced everything to a simple question of economics. If times were hard, they were to be equally hard for all.[237] The most notorious example of this logic was the prize winning essay of 1931 by Emil Bratz, entitled *Can the Care of the Mentally Ill Be Arranged More Cheaply, and How?*, wherein it was proposed that patients with favourable prognoses receive the best treatment, whilst those with less positive diagnoses be removed to inferior institutions, where their care would only cost half as much.[238] On purely economic grounds a distinction was to be made between those mental patients who could be considered "curable" and those who could not. The quality of care each category of patient so assessed was to receive would be decided solely by the application of a financial benchmark.[239]

Yet at the same time, nurses were adjured to respect the dignity of patients, to "treat them in the same friendly, courteous way as other people, and as we ourselves wish to be treated,"[240] and psychiatrists were themselves encouraged "to support with particular love and enthusiasm those to whom nature has not been so kind, so that the way to a life worthy of living is also opened to them."[241] This dichotomy, contrasting the ruthless application of financial constraints with the traditional caring nature of physicians and nurses, was to become ever more apparent with the advent of National Socialism and the associated application of the tenets of racial hygiene.

Whatever the rationale, there seems little doubt that the events of 1914–1918 did have a profound impact on German thinking so far as the legalization of so-called "mercy killing" was concerned. Nowhere was this better illustrated than in a short (62 page) but extraordinarily influential booklet published in 1920. Twenty-five years after Jost, in *Der Freigabe der Vernichtung lebensunwerten Leben* (*The Permission to Destroy Life Unworthy of Life*), Karl Binding, a criminal lawyer, and Alfred Hoche, a psychiatrist, proposed that "unworthy life" included not only the incurably sick, but many of the mentally ill and feebleminded, as well as retarded and deformed children. Killing such people was "*an allowable, useful act.*" History, they believed proved their thesis, and in time society would adopt it:

> There was a time, now considered barbaric, in which eliminating those who were born unfit for life, or later became so, was taken for granted. Then came the phase, continuing into the present, in which, finally, preserving every existence, no matter how worthless, stood as the highest moral value. A new age will

237 McFarland-Icke, *Nurses in Nazi Germany*, pp. 36–38.

238 Ibid., p. 64.

239 Mark P Mostert, *Useless Eaters: Disability as Genocidal Marker in Nazi Germany* (Journal of Special Education. Vol. 36, No. 3, 2002), p. 156

240 McFarland-Icke, *Nurses in Nazi Germany*, p. 60.

241 Ibid., p. 62.

> arrive—operating with a higher morality and with great sacrifice—which will actually give up the requirements of an exaggerated humanism and overvaluation of mere existence.[242]

To the eugenicists, this supposed human detritus appeared to have less intelligence, higher levels of antisocial behaviour, and, accordingly, less value than worthier individuals, such as, naturally, the eugenicists themselves. They were "burdensome existences" (*Ballastexistenzen*).[243] There had been many others who wrote in a similar vein, but for Binding and Hoche in particular, the right to life was not simply an entitlement—it demanded justification; it was a privilege earned only by being a fully functioning member of society, specifically in economic terms.[244] Writing of those with mental disabilities, and urging the introduction of "involuntary euthanasia", Binding continued:

> Their life is absolutely pointless, but they do not regard it as being unbearable. They are a terrible, heavy burden upon their relatives and society as a whole. Their death would not create even the smallest gap—except perhaps in the feelings of their mothers or loyal nurses.[245]

Although both Binding and Hoche were right-wing nationalists, somewhat ironically, Hoche was privately critical of Nazi eugenic laws, pointing out that their implementation would have precluded the birth of Goethe, Schopenhauer, and Beethoven, among others.[246] However, the unimaginable had occurred; physicians were being encouraged, not to save life, but to take it. And Binding and Hoche had unwittingly introduced both the arguments for and vocabulary of "euthanasia."[247]

By no means all of the medical profession of that era concurred with the view expressed by Binding and Hoche. Many eugenicists pointed out that their concern was with well-birth, not with well-dying, that is positive rather than negative eugenics. In essence, this was the conceptual difference between sterilisation and "euthanasia." As reprehensible as compulsory sterilisation may have been, it was conceived of as being anti-cacogenic, not outright murder, except in the limited sense of the murder of unborn future generations, the possibility of whose birth could in any event be adjudged undesirable for one perceived reason or another. Dr E Baege commented in 1931 that "for eugenics, the occasionally occurring thought about extermination of life unworthy of living is of course out of the question. This would undoubtedly be the surest method of hindering offspring; it is just irreconcilable with ethics."[248] Another doctor, M. Beer, had already written in 1914: "Once respect for the sanctity of human life has been diminished by introducing voluntary mercy killing for the mentally-healthy incurably ill, and involuntary killing for the mentally ill, who is going to ensure that matters stop

---

242 Roderick Stackelberg, *The Routledge Companion to Nazi Germany* (London: Routledge, 2007), p. 73.

243 Henry Friedlander, *The Origins of Nazi Genocide: From Euthanasia to the Final Solution* (Chapel Hill: University of North Carolina Press), p. 15.

244 Mostert, *Useless Eaters*, p. 157.

245 Burleigh, *Death and Deliverance*, p. 17. Binding also proposed euthanasia as a means of accelerating the death of the terminally ill or mortally wounded, as well as its application to the mentally healthy who, unconscious or in a coma, would have approved of their own euthanasia had they been able. Of necessity, the latter employment would have required a considerable degree of mind reading on the part of those administering euthanasia.

246 Ibid., p. 16. Hoche was opposed to the Nazi "euthanasia" programme, although his own writings had been a major inspiration for the policy.

247 de Mildt, *In the Name of the People*, p. 53.

248 McFarland-Icke, *Nurses in Nazi Germany*, p. 65.

there?"[249] This is the "slippery slope" argument. In essence the concern is that once such a process has been started, where will it end? Such disquiet could lead to statements such as that of the Catholic Bishop, Joseph V Sullivan, who in 1989 had obviously been inspired by the crimes of the Nazi regime:

> If voluntary euthanasia were legalised, there is good reason to believe that at a later date another bill for compulsory euthanasia would be legalised. Once the respect for human life is so low that an innocent person may be killed directly even at his own request, compulsory euthanasia will necessarily be very near. This could lead easily to killing all incurable cancer patients, the aged who are a public care, wounded soldiers, all deformed children, the mentally afflicted, and so on. Before long the danger would be at the door of every citizen...That is why euthanasia under any circumstances must be condemned.[250]

Dr Beer's was a pertinent question, for others in Germany were increasingly less troubled by ethical concerns. In 1922, the popular writer Ernst Mann had demanded the extermination of the mentally and terminally ill, as well as that of crippled and incurably ill children and the killing of the 'exhausted.' At the time this was considered an abnormal and extreme view; in little more than a decade it had become official government policy.[251] A year after Baege's comments, another doctor, Berthold Kihn, in a subsequently published lecture entitled "The eradication of the less valuable from society", suggested that not only were the state's resources being squandered in supporting irresponsible individuals who reproduced without regard to the economic consequences, but that advances in medical care were "keeping beings alive whose value to society is at least regarded as very debatable." He was able to calculate that killing the mentally ill would save 150 million *Reichsmarks* and release a substantial pool of labour dedicated to their care. "In the battle against inferiority," he continued, "every measure that seems inexpensive and effective is permitted."[252] However, he concluded, such options were unrealistic. Such a change in either the legal or in "somewhat over-developed ethical sensibilities" was unfeasible in the near future.[253] He was a poor judge of the times, for Kihn was himself to become a major player in both sterilisation and "euthanasia" affairs shortly after passing these remarks. Given his published views, this was hardly surprising.

In 1922, an unemployed down-and-out named Florian Huber, who had been severely wounded in the First World War and was a holder of the Iron Cross, was convicted of armed robbery and murder in Bavaria. A psychological examination determined that Huber "demonstrated some physical evidence of degeneracy." Rather than concluding that Huber might therefore be unfit to stand trial, his "degeneracy" was taken as evidence that he was probably hereditarily damaged and thus beyond redemption. Huber was executed. A year later, Theodor Viernstein established a "Criminal-Biological Information Centre" in Bavaria to collect details of all known criminal offenders, their families and circumstances. In this way it was believed that hereditary abnormalities could be scientifically pinpointed. Similar organisations soon appeared in other German states. A significant number of those in a position of authority believed that the "aso-

249 Burleigh, *Death and Deliverance*, p. 15.
250 David Lamb, *Down the Slippery Slope: Arguing in applied ethics* (London: Routledge, 2003), p. 2.
251 Burleigh, *Death and Deliverance*, p. 21.
252 McFarland-Icke, *Nurses in Nazi Germany*, p. 66.
253 Burleigh, *Death and Deliverance*, p. 41.

cials"[254] thereby identified, as such troublesome individuals came to be termed, should, at the very least, be compulsorily sterilised in order to prevent their progeny contaminating future society.[255] It was symptomatic of much contemporary thinking. To many psychiatrists logic dictated that since it had been determined to their satisfaction that so much mental disease was hereditary in nature, and the bearers could not be persuaded to voluntarily prevent the passing on of their illness by avoiding reproduction, compulsory sterilisation provided a convenient and permanent answer to the perceived problem. Some went further. A meeting of Bavarian psychiatrists held in 1931 debated the sterilisation and "euthanasia" of persons with chronic mental illnesses.[256] Well before the *Machtergreifung* there were psychiatrists advocating, and in some cases even initiating the killing of their patients, often on eugenically justified cost-cutting grounds. It can thus be seen that whilst the advent of National Socialist government undoubtedly intensified and accelerated the process, notions of compulsory sterilisation and "euthanasia" as a curative were already well-known and perfectly acceptable to certain elements within the medical profession—and elsewhere.[257]

It is natural to pose the question of how a nation so apparently civilized and singularly cultured came to accept such callous, inhuman, and immoral ideas as the norm, for without the approval of the populace, tacit or otherwise, none of what was to follow would have been possible. To provide an answer, analysts habitually consider the political, social, and economic background of the times. As has been illustrated, many of the medical notions circulating in the early years of the twentieth century concerning the issues under discussion were international. But it is suggested that the circumstances pertaining to their implementation in Germany were unique. The nation-state had itself only been in existence since 1871.[258] Prior to the abdication of Kaiser Wilhelm II in 1918 and the creation of the Weimar Republic in the following year, there had been no centuries-long tradition of democracy. The Weimar government was born out of the disaster of defeat in the Great War, and in the course of the next fifteen years suffered a series of crushing blows. The terms of the humiliatingly punitive treaty of Versailles[259]—the crippling reparations, the loss of 13 percent of Germany's territory, a

254 The Ministry of the Interior defined "asocials" in the following terms: Persons who through minor, but repeated, infractions of the law demonstrate that they will not adapt themselves to the natural discipline of the National Socialist state, e.g, beggars; tramps (Gypsies); alcoholics; whores with contagious diseases, particularly sexually transmitted diseases, who evade the measure taken by the health authorities. A further category, the "work shy" were persons against whom it can be proven that on two occasions they have, without reasonable grounds, turned down jobs offered to them, or who, having taken a job, have given it up after a short while without a valid reason. [Saul Friedländer, *The Years of Persecution: Nazi Germany and the Jews 1933–39* (London: Phoenix, 2007), pp. 203–204.]

255 Richard J Evans, *The Coming of the Third Reich* (London: Allen Lane, 2003), pp. 144–145.

256 Leo Alexander, *Medical Science Under Dictatorship* (The New England Journal of Medicine 241, 1949), p. 39.

257 Michael Burleigh, *Ethics and extermination: Reflections on Nazi genocide*, (Cambridge: Cambridge University Press, 1997), pp. 114–117.

258 Evans, *The Coming of the Third Reich*, p. 2

259 There is an argument that the Versailles treaty can be viewed as being either too severe on Germany, or alternatively as not being severe enough—too harsh for Germany not to seek its revocation, yet not

tenth of its population, and all of its overseas colonies[260]—was followed by hyper-inflation in 1923, and world-wide economic depression in 1929. There had been a brief period of relative stability between the two latter events, but the new democracy was a fragile creature, wracked by bitter street fighting between the political parties of the extreme left and right, with numerous politically motivated murders. By 1932, on the eve of Hitler's attainment of power, unemployment had reached a staggering 6 million, representing almost 25 percent of the workforce. If to this number was added the dependants of the unemployed, about 20 per cent of the entire population, some 13 million people, were effectively on the breadline.[261] Scapegoats had to be found for this seemingly endless chaos. And as the American psychologist, Ervin Staub, writes: "Blaming others, scapegoating, diminishes our own responsibility. By pointing to a cause of the problems, it offers understanding that, although false, has great psychological usefulness. It promises a solution to problems by action against the scapegoat."[262]

Although all of these factors undoubtedly played some part in the rise of Nazism and the implementation of the National Socialist manifesto, this simplistic explanation contains a fundamental flaw, for it presupposes a set of ethics and morality common to all. It suggests that what occurred during the Nazi era was an example of aberrant human behaviour. Today, we naturally view the actions of the Nazis from the perspective of a liberal democracy. However, history teaches that such high minded principles are not the rule. The concept of 'natural law' will be examined later, but ethics and morality have always meant differing things to different people in different places at different times. What if, as has been suggested:

> During the National Socialist era a large proportion of the Germans were committed to a moral code, which instead of condemning, actually demanded the degradation and persecution of other people, and which among other things prescribed that it was necessary and good to kill?[263]

To some extent this echoes Goldhagen's thesis quoted earlier, but without Goldhagen's specificity relating it solely to Jews. In other words, what today seems to us immoral and repugnant was, at the time, perceived by the perpetrators to be perfectly moral, essential and most of all, desirable. This is not as outlandish as it sounds. When Atzec priests ripped the hearts out of living prisoners in ritual sacrifice, or when the Inquisition ordered heretics burned at the stake, they doubtless felt their actions were morally justified. If indeed, eugenics was the new creed, and Nazism its natural offspring, it was easy to rationalize any act of inhumanity on similar quasi-religious grounds.

The American sociologist, William Isaac Thomas proposed that "if people define situations as real, they are real in their consequences." People's behaviour is dependent on the manner in which they perceive and define situations; this in turn means that it is

---

harsh enough to prevent its eventual emendation. [Donald Bloxham, *The Final Solution: A Genocide* (Oxford: Oxford University Press, 2009), p. 134].

260 Evans, *The Coming of the Third Reich*, p. 62.

261 Ibid., p. 236.

262 James Waller, *Becoming Evil. How Ordinary People Commit Genocide and Mass Killing* ((Oxford: Oxford University Press, 2002), p. 256.

263 Harald Welzer, *Mass murder and moral code: some thoughts on an easily misunderstood subject* (History of the Human Sciences, Vol.17 Nos 2/3, 2004, p. 17.)

conditions that characterize what is considered to be absolutely right in one set of circumstances appearing in another set of circumstances to be absolutely wrong—and vice-versa. This theory also suggests that while the individual may perceive themselves to be perfectly moral and their actions completely justified, viewed from another time and place their behaviour may seem murderously incomprehensible. If this proposition does accurately reflect individual behaviour, it eliminates any suggestion that it was first necessary to overcome moral or ethical scruples before embarking on a policy of mass murder. To the perpetrator, their actions *were* perfectly moral and ethical in the then prevailing circumstances, so that in later quite different circumstances they were able to see themselves as victims rather than criminals, and self-pityingly proclaim their distress at the duties they had had no option but to perform.[264] A variation on this concept has been suggested by the ethicist Dr. Arthur Caplan. In his belief, physicians did not set aside their ethics during the Nazi era but saw their actions as synonymous with their ethical commitment to heal the people through the elimination of undesirable elements.[265]

Yet all of these arguments raise doubts. If the creators and executors of Nazi eugenic policies were so convinced at the time of committing these acts of being able to justify their actions, why did so many either commit suicide, obtain new identities, emigrate, or simply vanish on the collapse of the regime, if not out of the knowledge of certain guilt? It seems rather that individuals are able to compartmentalise their actions, segregating and rationalizing the most extreme conduct, then revising their perception of that conduct as circumstances change. Or as has been suggested: "Actions must be capable of being endowed with meaning for the person who carries them out and in some way of being incorporated into a self-concept, which does not seriously undermine the subject's personal feeling of moral integrity."[266] This subject is considered in greater depth in later chapters of this work.

Whatever individual motivation may have been (and there were as many reasons for becoming a perpetrator as there were perpetrators), fertile ground had been prepared for the Nazis by the nineteenth century writings of racists such as the Frenchman, Joseph Arthur, Comte de Gobineau, and the Englishman, Houston Stewart Chamberlain, among many others, with their dissemination of the idea of "Aryan" supremacy and the worthlessness of the *Untermenschen*, ideas enthusiastically embraced by many of the influential eugenicists of the day. Gobineau was a diplomat and writer who, in his seminal 1851 book *Essai sur l'Inégalité des Races Humaines* (*The Inequality of Human Races*), had neatly categorized humanity into three races—the white, the yellow, and the black. Of these the lowest was the black; "His mental faculties...are dull or even non-existent...He kills willingly, for the sake of killing." The yellow was somewhat higher up the evolutionary ladder, but not by much; "He tends to mediocrity in everything...The brutish hordes of the yellow race seem to be dominated by the needs of the body." By far the most superior was the white race, which possessed "a monopoly of beauty, intelligence, and strength." And the best of the best? Why of course the "Aryan", defined as a

---

264 Ibid., pp. 22–23.

265 Susan Benedict, and Jochen Kuhla, *Nurses' Participation in the Euthanasia Programs of Nazi Germany* (Western Journal of Nursing Research, Vol. 21(2), 1999),p. 258.

266 Welzer, *Mass murder and moral code*, p. 29.

nebulous race of Indo-Iranian origin whose descendants were responsible for everything "great, noble, and fruitful in the works of man on this earth." But even the "Aryan" was in a state of inexorable degeneration because of miscegenation with inferior races. Gobineau had acted first as secretary to Alexis de Tocqueville, then as chief of cabinet when the latter was appointed foreign minister, and it is perhaps best left to de Tocqueville to pass definitive judgement on his protégé's maliciously irrational racist stereotyping. In 1853 he wrote to Gobineau: "I believe that [your doctrines] are probably quite false; I know that they are certainly very pernicious."[267]

If the dyspeptic Gobineau was not notably anti-Semitic, regarding all of humanity other than the "Aryan" with an equal degree of revulsion, the same could not be said of Chamberlain, to whom all things German were holy. His book *Die Grundlagen des Neunzehnten Jahrhunderts* (*The Foundations of the Nineteenth Century*), published in 1899–1900, had sold 100,000 copies by 1914, and had run to no less than twenty-eight editions by 1938. The beneficiary of an inheritance that obviated the necessity of his working for a living, in 1885 Chamberlain moved to Dresden, where he became acquainted with Richard Wagner's widow, Cosima. So enamoured with the recently deceased arch anti-Semite Wagner and his family did Chamberlain become, that in 1906 he divorced his wife, married Wagner's daughter Eva, and settled in Bayreuth, a town forever to be associated with the composer. In 1901 a meeting had been arranged between Chamberlain and the Kaiser that was to last two days, at the end of which Chamberlain had convinced Wilhelm that it was the Kaiser's special mission to combat the decadence of contemporary civilization and raise the banner of a revitalizing Aryan racism.[268] Central to Chamberlain's argument was an all-embracing anti-Semitism.

Chamberlain's perception of history was essentially simple; it was no more than an eternal conflict between the forces of good (the "Aryan-Christian worldview") and evil ("Jewish materialism"). Building upon the foundations laid by Gobineau, Chamberlain went even further than his predecessor by claiming that it was the Teutonic branch of the "Aryan" race that reigned supreme. Germans and Jews were locked in an unending struggle to decide the future of humankind.[269] It is not difficult to see the appeal of such a hypothesis to the nascent fascist, Adolf Hitler, or for that matter the allure of the future *Führer* to the ideologue Chamberlain, who in October 1923 concluded what amounted to a love letter to Hitler in the following terms:

> My faith in Germandom has not wavered for a moment, though my hopes were—I confess—at a low ebb. With one stroke you have transformed the state of my soul. That Germany, in the hour of her greatest need, brings forth a Hitler—that is proof of her vitality...May God protect you![270]

For some of the populace it was but a short step from the pseudo-scientific theorizing of Chamberlain and his successors such as Alfred Rosenberg to wholesale approval of National Socialist policies. If the traditional purity of the Nordic/ Teutonic/"Aryan"

---

267 Jonathan Peter Spiro, *Defending the Master Race: Conservation, Eugenics, and the Legacy of Madison Grant*, (Burlington: University of Vermont Press, 2008), pp. 103–106

268 Paul Weindling, *Health, Race, and German Politics between National Unification and Nazism, 1870–1945* (Cambridge: Cambridge University Press, 1993), p. 108.

269 Richard S Levy (ed), *Antisemitism: A Historical Encyclopedia of Prejudice and Persecution* (Santa Barbara: ABC-CLIO Ltd, 2005), pp. 113–115.

270 Stackelberg, *The Routledge Companion to Nazi Germany*, p. 85.

race had been maintained, if in the past the resources of the state had not been squandered on worthless individuals, could much anguish have been avoided? And could the despair felt by so many of the population be alleviated now and in the future? Who better to answer these questions than Adolf Hitler, the demagogue who had consistently preached fanatical support for this kind of reasoning, and had been an early and enthusiastic supporter of the prevalent biological gobbledegook, in the process providing deceptively simple answers to immeasurably complex problems? As the eminent postwar German geneticist, Benno Müller-Hill commented, Nazi ideology regarding eugenic matters was easily explained: human differences were based upon an individual's blood, that is to say their genes, and could never be changed. A Jew would always remain a Jew, an anti-social an anti-social, and so on. Those of inferior blood could not receive rights equal to those of superior blood. Moreover, as the possibility existed that the inferior might breed more rapidly than the superior, the inferior must be identified, isolated, and eventually eliminated by one means or another in order to protect 'civilization'[271] For, in the words of Victor Klemperer, "...a man can change his coat, his customs, his education and his belief, but not his blood."[272] Or as a 1937 *'Handbook for the Hitler Youth'* expressed it: "Environmental influences have never been known to bring about the formation of a new race. That is one more reason for our belief that a Jew remains a Jew, in Germany or any other country. He can never change his race, even by centuries of residence among another people."[273]

The extent to which the Nazis were prepared to take this ludicrous reasoning is well illustrated by the case of an SA man in southern Germany who was struck by a car close by a Jewish hospital, to where he was taken. There he was given a transfusion of 'Jewish blood'. Subsequently the man appeared before an SA tribunal to determine whether his blood had been so tainted as a consequence that he should be ejected from the SA. It was ruled that, although a strict interpretation decreed that the man was now racially impure and should therefore no longer be permitted to remain a member of the organisation, the Jewish donor had fought at the front in the Great War. In the circumstances his contaminated blood was acceptable. The donee was allowed to remain in the SA.[274]

In 1942, when the regime's genocidal activity was at its most deadly, a biology textbook for girls was published which made no secret of where Nazi eugenics led. In a discussion of insect life, the following was observed:

> The instinctual state of...ants corresponds to the leadership state among mankind; however, the principles of a perfect insect state give people cause to think. They have preserved bees and ants in the struggle for survival and thereby proved their validity. We earlier noted the following truths...:

---

271 Benno Müller-Hill, *Murderous Science: Elimination by Scientific Selection of Jews, Gypsies, and Others in Germany 1933–1945* (Woodbury: Cold Spring Harbor Laboratory Press, 1998), p. 23.

272 Victor Klemperer, *The Language of the Third Reich: LTI—Lingua Tertii Imperii—A Philologist's Notebook* (London, Continuum, 2006), p. 163.

273 Robert N Proctor, *Racial Hygiene: Medicine Under the Nazis* (Cambridge Massachusetts: Harvard University Press, 1989), pp. 37– 38.

274 Ibid., p. 149.

1. The work of the individual has only one purpose: to serve the whole group.
2. Major accomplishments are possible only by the division of labour.
3. Each bee risks its life without hesitation for the whole.
4. Individuals who are not useful or are harmful to the whole are eliminated.
5. The species is maintained by producing a large number of offspring.

It is not difficult for us to see the application of these principles to mankind.[275]

[275] http://www.calvin.edu/academic/cas/gpa/textbk01.htm (Accessed 1 September 2008). This website is an invaluable resource, consisting as it does of many examples of Nazi propaganda.

CHAPTER 4: A MARCHING COLUMN

Whatever proportions these crimes finally assumed, it became evident to all who investigated them that they had started from small beginnings. The beginnings at first were merely a subtle shift in emphasis in the basic attitude of the physicians. It started with the acceptance of the attitude, basic in the euthanasia movement, that there is such a thing as life not worthy to be lived. This attitude in its early stages concerned itself merely with the severely and chronically sick. Gradually the sphere of those to be included in this category was enlarged to encompass the socially unproductive, the ideologically unwanted, the racially unwanted and finally all non-Germans. But it is important to realize that the infinitely small wedged-in lever from which this entire trend of mind received its impetus was the attitude toward the non-rehabilitable sick.

– Leo Alexander[276]

Jews do not transform themselves into Germans by writing books on Goethe

– Fritz Lenz[277]

In Germany the term "Race Hygiene" was in use long before the label of "eugenics" became common, and the German Society for Race Hygiene (*Deutsche Gesellschaft für Rassenhygiene*) was to eventually represent all eugenicists. In 1931, two years before Hitler's assumption of power, Fritz Lenz, in his capacity of professor of race hygiene at the University of Munich, declared: "Hitler is the first politician with truly wide influence who has recognized that the central mission of all politics is race hygiene and who will actively support this mission."[278] And support it he did, for example in a speech to the National Socialist Physicians League in which he announced that he could manage without lawyers, engineers, and builders, but "you, you National Socialist doctors, I cannot do without you for a single day, not a single hour. If not for you, if you fail me, then all is lost. For what good are our struggles, if the health of our people is in danger?"[279] It was a typically melodramatic pronouncement, and Hitler was preaching to the converted, but it doubtless had the intended impact on his audience, filling them with an enhanced sense of their self-importance. Hitler was not simply firing up a group of fanatics. He spoke from conviction. After a meeting with him in June 1937, Gerhard Wagner, the *Reich* Doctors' Leader reported that "the Führer considered the cleansing of the medical profession more important than for example that of the civil service, as the task of the physician was in his opinion one of leadership, or should be such."[280] By "cleansing", Hitler of course meant the removal of Jews.

In 1931 Ernst Rüdin, a Swiss psychiatrist, who together with Ploetz had been one of the founding members of the Society for Race Hygiene, and was a long time associate of Lenz, had become director of the Kaiser Wilhelm Institute for Psychiatry in Munich. In a speech delivered one year later, Rüdin stated: "People who are themselves mentally ill

---

276 Leo Alexander, *Medical Science Under Dictatorship* (The New England Journal of Medicine 241, 1949), p. 44.

277 Robert N Proctor, *Racial Hygiene: Medicine Under the Nazis* (Cambridge Massachusetts: Harvard University Press, 1989), p. 55.

278 Henry Friedlander, *The Origins of Nazi Genocide: From Euthanasia to the Final Solution* (Chapel Hill: University of North Carolina Press, 1995), p. 12.

279 Proctor, *Racial Hygiene*, p. 64.

280 Saul Friedländer, *The Years of Persecution: Nazi Germany and the Jews 1933–39* (London: Phoenix, 2007), p. 226.

or hereditarily feebleminded should not have children."[281] Rüdin had been a long term advocate of a radical eugenic policy, proposing the compulsory sterilisation of alcoholics, among others,[282] as well as emphasising that "[an] avalanche-like explosion of financial burdens" were inevitable because of "the increase of necessary care for defectives and sick persons of all types."[283] In 1933 he was appointed by the Nazis to head the Society for Race Hygiene. Author of the 1933 Nazi Sterilisation Law, in 1934, Rüdin stated: "The psychiatrist and the healthy person are allies against the genetically defective. The psychiatrist must render his service to the ultimate aim of a hereditary pure, able and superior race."[284] Rüdin joined the Nazi Party in 1937, somewhat later than might have been anticipated. In 1939 Hitler personally awarded him the Goethe Medal for Art and Science, and in 1942 bestowed on him the title of "Pioneer of Racial Hygiene". Unstinting in his praise of Hitler and Nazism, in 1943 Rüdin commended both for their "decisive...path-breaking step toward making racial hygiene a fact among the German people...and inhibiting the propagation of the congenitally ill and inferior." Moreover both the laws for "preventing the further penetration of the German gene pool with Jewish blood," as well as the SS for "its ultimate goal, the creation of a special group of medically superior and healthy people of the German Nordic type," were *sans pareil*.[285]

There were exceptions to the prevalent *zeitgeist*. Samuel Beckett pointed out the irony in this Nazi obsession with the creation of a race of "supermen." An "Aryan", he wrote, must be blonde like Hitler, thin like Göring, handsome like Goebbels, virile like Röhm—and be named Rosenberg.[286] Nor did every physician advocate unbridled sterilisation and euthanasia as being entirely salutary. At a 1931 conference of Bavarian psychiatrists in Munich, Professor Oswald Bumke made some remarkably prescient and morally impeccable observations:

> I should like to make two additional remarks. One of them is, please for God's sake leave our present financial needs out of all these considerations. This is a problem which concerns the entire future of our people, indeed, one may say without being over-emotional about it, the entire future of humanity. One should approach this problem neither from the point of view of our present scientific opinion nor from the point of view of the still more ephemeral economic crises. If by sterilisation we can prevent the occurrence of mental disease then we should certainly do it, not in order to save money for the government but because every case of mental disease means infinite suffering to the patient and to his relatives. But to introduce economic points of view is not only inappropriate but outright dangerous because the logical consequence of the thought that for financial reasons all these human beings, who could be dispensed with for the moment, should be exterminated, is a quite monstrous logical conclusion; we would then have to put to death not only the mentally sick and the psychopathic personalities but all the crippled in-

281 Michael Burleigh, *Death and Deliverance : 'Euthanasia' in Germany c. 1900–1945* (London: Pan Books, 2002), p. 45.

282 Horst Biesold, *Crying Hands: Eugenics and Deaf People in Nazi Germany* (Washington: Gallaudet University Press, 1999), p. 15.

283 Wolfgang Uwe Eckart (ed), *Man, Medicine, and the State: The Human Body as an Object of Government Sponsored Medical Research in the 20th Century* (Stuttgart: Franz Steiner Verlag, 2006), p. 76.

284 http://www.irren-offensive.de/kleespeech.htm (Accessed 10 January 2007).

285 Robert Jay Lifton, *The Nazi Doctors: Medical Killing and the Psychology of Genocide* (London: Papermac, 1990), p. 28.

286 Michael Burleigh, *The Third Reich: A New History* (London: Pan Books, 2001), p. 356. In terms of physique, colouring and facial features, Heinrich Himmler was about as far removed from the Nordic ideal (a long skull, narrow forehead, blonde hair, blue eyes) as it was possible to be. [Peter Padfield, *Himmler: Reichsführer-SS* (London: Papermac, 1990), pp. 104–105].

cluding the disabled veterans, all old maids who do not work, all widows whose children have completed their education, and all those who live on their income or draw pensions. That would certainly save a lot of money but the probability is that we will not do it.

The second point of advice is to use utmost restraint, at least until the political atmosphere here in this country shall have improved, and scientific theories concerning heredity and race can no longer be abused for political purposes. Because, if the discussion about sterilisation today is carried into the arena of political contest, then pretty soon we will no longer hear about the mentally sick but, instead, about Aryans and non-Aryans, about the blonde Germanic race and about inferior people with round skulls. That anything useful could come from that is certainly improbable; but science in general and genealogy and eugenics in particular would suffer an injury which could not easily be repaired again.[287]

Yet three years later at another convention, in reviewing the Weimar Republic years, Bumke stated:

Gentlemen, today we have gathered in quite another Germany. Today once again each German heart is filled with hope. But this time around it is not the last onset of a people slowly tiring, but an uprising that has definitively lifted the whole of Germany out of the timidity of the post-war years, an uprising that will fortify us once more, after earnest application, internally as well as externally.[288]

Admittedly not a wholehearted endorsement of the regime's eugenic policies, but hardly a staunchly anti-Nazi line either. And even in his 1931 remarks, Bumke had effectively prescribed involuntary sterilisation in certain circumstances. The war over, he was quick to disassociate himself from National Socialist excesses, stating in June 1945 that other than the Nazi "hotheads" among them, he and other psychiatrists "never really cooperated with the sterilisation law...we have sabotaged the law...sterilisation fell gradually into disuse, especially since 1943."[289]This was at best disingenuous. With widespread "euthanasia" available, there was no longer any need for sterilisation, as Bumke well knew. In fact, only about 5 percent of all sterilisations occurred after 1939.[290] The fact that adult "euthanasia" began in the same year is hardly coincidental.

At the 1929 party rally, Hitler had suggested that countless lives could be eliminated by racial measures:

If Germany was to get a million children a year and was to remove 700–800,000 of the weakest people, then the final result might even be an increase in strength. The most dangerous thing is for us to cut off the natural process of selection and thereby rob ourselves of the possibility of acquiring able people. The first born are not always the most talented or strongest people... As a result of our modern humanitarianism we are trying to maintain the weak at the expense of the healthy.[291]

At another 1931 conference, this time of the Inner Mission's Standing conference on Eugenics,[292] the chairman, Hans Harmsen, called for the eugenic sterilisation of the

---

287 http://www.ess.uwe.ac.uk/genocide/Medical%20Case%20Trial%20Prosecution4.htm (Accessed 25 January 2008). Bumke ignored his own prophetic remarks. Within a few years he had become an enthusiastic advocate of Nazi eugenic legislation. (Burleigh, *Death and Deliverance*, p. 44.)

288 Michael H Kater, *Doctors Under Hitler* (Chapel Hill: University of North Carolina Press, 2000), p. 135.

289 http://www.ess.uwe.ac.uk/genocide/alexrep1.htm#Sterilisation (Accessed 25 January 2008).

290 Proctor, *Racial Hygiene*, p. 114.

291 Burleigh, *The Third Reich*, p. 382.

292 'The Inner Mission' was the principal Protestant health and welfare organisation. [Michael Burleigh, *Ethics and extermination: Reflections on Nazi genocide*, (Cambridge: Cambridge University Press, 1997), p. 130].

"less valuable."[293] Only one courageous voice dissented from the general acceptance of the proposal. "Sterilisation will become a fashion lacking any foundations", he declared. "We know nothing about this, drawing our conclusions only from experiments with animals and plants...Operations such as sterilisation may well be admissible in individual cases, [but] on the whole they are to be rejected." He continued:

> Should a doctor be obliged in these instances to execute such a death sentence, or should it be the state's responsibility to compel the doctor to do so? Or is the state prepared to form a special class of professional executioners which carries out these matters? These considerations should be energetically brought to the attention of those in public office. It would imply that a certain group of doctors would have to be entrusted with it. The medical profession would then not only be a profession of helpers, but also of hangmen.

From whom did this outspoken opposition emanate? None other than Carl Schneider, a man who over the next decade became one of the leading lights in the "euthanasia" programme, adjudicating over the right to life of countless individuals, as well as devoting himself to research at his Heidelberg clinic on the freshly removed brains of murdered children.[294] How men like Schneider could attempt to justify their conversion to barbarism will be examined later.

Schneider's objections notwithstanding, with some exceptions the Protestant Church was happy to embrace compulsory sterilisation, seeking only to conduct operations within their own institutions rather than incurring the cost of sending patients to state-run clinics or hospitals. Even some prominent Catholics found the application of the new law quite acceptable. Whilst Christian organisations of all denominations, not surprisingly, welcomed the advent of National Socialist government as a "bulwark against Bolshevism," both Protestant and Catholic leadership was much more circumspect about so-called "euthanasia", although there were a number of eugenic enthusiasts in positions of authority in both faiths. But there were also those who were not prepared to become accomplices to murder, and fought to preserve the lives of helpless individuals condemned out of hand by a merciless regime.[295]

It required a quite specific political mindset to produce the Nazi state: "Racism and sexism are necessary components of a repressive eugenics policy, but they are not sufficient. In Germany, it took not only a racist ideology, but a totalitarian state with the will and ability to impose this ideology ruthlessly and murderously on its people."[296]

And so, very soon after attaining power in 1933, the Nazis commenced an extensive propaganda campaign with the object of acquainting the German nation with the benefits of first, mass sterilisation, and subsequently, "euthanasia". Goebbels, a master in the manipulation of public opinion, used newspapers, magazines, radio, and film to suggest

---

293 Ibid.

294 Dick de Mildt, *In the Name of the People: Perpetrators of Genocide in the Reflection of Their Post-War Prosecution in West Germany. The 'Euthanasia' and 'Aktion Reinhard' Trial Cases* (The Hague: Martinus Nijhoff Publishers, 1996), p. 302. Emphasis in original.

295 Burleigh, *Ethics and extermination*, pp. 130–141.

296 Arthur L Caplan, (ed), *When Medicine Went Mad: Bioethics and the Holocaust* (Totowa: Humana Press, 1992), p. 302

that life would be so much better for the worthwhile many if the worthless few, who were such an encumbrance to the *Volksgemeinschaft* (racial community), were prevented from reproducing their genetic fault line. The logical extension of that reasoning was that the valueless (as suitably defined) should not just be simply prevented from propagation; they should be eradicated. Would it not be preferable if an end was put to the misery and suffering of their hopeless lives?

Film in particular played an important role in preparing the public for the acceptance of compulsory sterilisation of the mentally and physically handicapped. Five 16mm silent films were made by the Racial and Political Office in the mid 1930s.[297] These films consistently equated sickness with criminality, whilst stressing how the resources spent on asylums and prisons could be put to so much better use. They contrasted the allegedly luxurious conditions in which mental patients were kept with the miserable existence of slum dwellers in the cities. Why waste money on such hopeless cases? Subtle, the message was not. Far more sophisticated was the sound film *Opfer der Vergangenheit* (*Victims of the Past*), a 1936 joint production of the Racial and Political Office and the *Reich* Propaganda Ministry. On Hitler's instructions, the film was shown in all 5,300 German cinemas. It went much further than its predecessors, speaking ominously of Jewish mental patients and the consequences of the abandonment of the law of natural selection. It contrasted "healthy German citizens" (that is, girls performing gymnastic exercises) with asylum inmates, then provided the solution to the surmised problem—compulsory sterilisation.[298]

Even more insidious were the films planned to endorse "euthanasia", by then already in progress. Hermann Schweninger, a recruit to the killing programme with some experience of filmmaking, was commissioned to visit a number of asylums in autumn 1940.[299] He shot some10,000 metres of material for editing, much of it with sound. In September 1940, Jewish patients at the Eglfing-Haar psychiatric hospital were filmed as part of the film *Abschaum der Menschheit* (*Scum of Humanity*), a self-explanatory title. With filming completed, the patients involved were transferred to the Brandenburg killing centre and gassed later the same month.[300] The principal films resulting from Schweninger's odyssey were *Dasein ohne Leben* (*Existence without Life*) and a documentary, *Geistig Kranke* (*Mentally Ill*). Both films were unequivocal in their depiction of medically sanctioned murder. Believed lost for ever, eight of the twenty-three rolls of

---

297 Burleigh, *Death and Deliverance*, p. 182. The films were *Sünden der Väter* (*Sins of the Fathers*, 1935); *Abseits vom Wege* (*Off the Path*,1935); *Erbkrank* (*Hereditarily Ill*, 1936); *Alles Leben ist Kampf* (*All Life is a Struggle*, 1937); *Was du ererbt* (*What You Have Inherited*, date unknown). Far from being concerned by these manifestations of National Socialist propaganda, American eugenicists organised screenings of an English-language version of *Erbkrank* at the Carnegie Institute, and the film was then distributed to churches and high schools. [Michael Burleigh, *Eugenic Utopias and the Genetic Present* (Totalitarian Movements and Political Religions: Vol. 1, No. 1, 2000), p. 63.]

298 Lifton, *The Nazi Doctors*, p. 48; Mark P Mostert, *Useless Eaters: Disability as Genocidal Marker in Nazi Germany* (Journal of Special Education. Vol. 36, No. 3, 2002), p. 160.

299 Schweninger, a school friend of Viktor Brack (of whom more later), was employed by *Gekrat* (see below) as a transport leader, and had been involved in the deportation of thousands of patients from Southern Germany to killing centres. [Götz Aly, Peter Chroust, Christian Pross, *Cleansing the Fatherland: Nazi Medicine and Racial Hygiene* (Baltimore: The Johns Hopkins University Press, Baltimore, 1994), p. 27].

300 Proctor, *Racial Hygiene*, p. 207.

film shot by Schweninger were discovered in Potsdam by Michael Burleigh in 1989–1990.

But the most effective film treatment of "euthanasia" was *Ich klage an* (*I Accuse*) scripted by Schweninger in 1941, an adaption of the 1936 novel *Sendung und Gewissen* (*Mission and Conscience*) by the ophthalmologist, Hellmuth Unger, a man destined to become a leading figure in "euthanasia" affairs. The message was more understated than the blatant propaganda of the earlier films, and consequently all the more potent. "Would you, if you were a cripple, want to vegetate forever?" rhetorically asks a doctor in *Ich Klage an*.[301] In picturing euthanasia as a voluntary action, actually going so far as to have a character hypocritically suggest that the most important precondition was that the patient personally request euthanasia, the film was a blatant distortion of real Nazi policy.[302]

In January 1942 the SD[303] collated public reaction to the film (by then seen by some 18 million people),[304] which, with a relatively small number of exceptions, was acceptance of the film's central theme—people suffering from incurable illnesses or diseases should be granted a quick, legally sanctioned death.[305] Some, however, were more sceptical, as the SD report made clear, quoting the comments of one viewer:

> You can think what you like about this, but who is going to guarantee that there won't be any abuses? As soon as laws like this are introduced it will be easy for the government to have anyone they consider undesirable declared incurable by a commission for any reason at all and eliminate them. And moreover, people with enough influence or money to criticise others will soon have somebody declared insane.[306]

By this time the "euthanasia" programme, officially halted in August 1941, had entered an even more sinister phase, but *Ich klage an* can be seen as part of the attempt to pave the way for general acceptance of an eventual "euthanasia" law. After the war, Schweninger continued his motion picture career, this time as a maker of corporate and advertising films. Others involved in the making of *Ich Klage an* were equally successful in continuing with their cinematic endeavours in the new Federal Republic.[307]

Leonardo Conti expressed this mystical obsession with the purity of the *Volk* in yet another film made in about 1942, long after both the sterilisation and "euthanasia" policies had been determined:

> We have identified a decreasing birth rate, negative selection and race-mixing as the major dangers which can lead people into the abyss; but this does not describe in full the dangerous influences affecting a civilised people...The destiny of a people is never at rest, it must either rise or fall; a people must be either hammer or anvil—the value of their blood, the value of their race is decisive. The German people have proven their value. They have to become the core of a Great German Reich; then we will succeed in

301 Mostert *Useless Eaters*, p. 160.

302 Lifton, *The Nazi Doctors*, p. 48

303 SD= *Sicherheitsdienst*, Security Service, the intelligence arm of the SS.

304 Michael Burleigh, and Wolfgang Wipperman, *The Racial State: Germany 1933–1945* (Cambridge: Cambridge University Press, 1991), p. 157.

305 Burleigh, *Death and Deliverance*, p. 189 ff.

306 Burleigh and Wipperman, *The Racial State*, p. 159.

307 Ibid., p. 167

proving that although the individual human being is mortal and his or her life ends with death, a people can renew itself and become more beautiful and stronger from generation to generation, an eternal life.[308]

Or as Reinhard Heydrich ominously defined Nazi dogma in 1936, "[National Socialism] no longer begins with the state but rather with the *Volk*...Accordingly, we National Socialists recognise only enemies of the *Volk*. They are always the same, and they will always remain the same. They are the opponents of the racial, *völkisch* and spiritual substance of our *Volk*."[309]. Central to Nazi ideology was this concept of the *Volksgemeinschaft,* the "racial community", a fantastic creation imagined as a homogeneous nation consisting entirely of racially, politically, and physically desirable subjects. Although there is no doubt that Heydrich regarded Jewry as the principal opponent of the *Volksgemeinschaft,* given his definition the list of potential enemies of the Nazi state was clearly terrifyingly long, and their fate equally clearly preordained.

In the same year that Heydrich's *völkisch* comments were published, Dr Wilhelm Hinsen had attended a conference at which Fritz Bernotat, the district administrator responsible for asylums in the Wiesbaden area, had stated: "If I was a doctor, I would do away with these patients." To his credit, Hinsen replied if that was the case "German medicine can congratulate itself that you are not a doctor."[310] Although there was no doubting the eventual aim, like the persecution of the Jews with which to some extent it ran parallel, "euthanasia" was only gradually and carefully introduced, Bernotat's comments notwithstanding. Asylums were deliberately neglected during the 1930s. Specialist wards were closed, and conditions deliberately exacerbated to the extent that basic hygiene and feeding, let alone treatment, often became impossible.[311] As the level of funding for mental hospitals decreased, the quality of psychiatric help declined with it. A significant number of psychiatrists were SS members, already imbued with contempt rather than compassion for their patients. It was often the case that those responsible for the administration of hospitals were no more than SA thugs or simply card-carrying members of the Party, men who owed their escape from unemployment and the dole not to any degree of expertise, but rather to political influence. They were totally unsuited for the tasks at hand.[312] Many asylums developed into little more than freak shows as the general populace were admitted to witness the often disturbing

---

308 Ulf Schmidt, *Karl Brandt: The Nazi Doctor—Medicine and Power in the Third Reich* (London: Hambledon Continuum, 2007), pp. 189–190. This kind of 'either-or 'reasoning was typical of Hitler, who only saw matters in terms of absolutes. It is not surprising to see it repeated in the pronouncements of his minions.

309 Michael Wildt, *An Uncompromising Generation—The Nazi Leadership of the Reich Security Main Office* (Madison: University of Wisconsin Press, 2009), p. 9.

310 Burleigh, *Death and Deliverance,* p. 51. Bernotat's sole interest was in making economies. When Dr Friedrich Mennecke, himself responsible for countless "euthanasia" deaths, protested about overcrowding in mental hospitals, Bernotat responded, "strike them all dead and then you'll have space." (Ibid, p. 53.)

311 Burleigh, *Ethics and extermination,* pp. 117–118.

312 Michael Burleigh (ed), *Confronting the Past: New Debates on Modern German History* (London: Collins & Brown Limited, 1996), p. 101.

conditions of the patients.[313] This all formed part of the deliberate manipulation of public opinion towards acceptance of "euthanasia" as a beneficial answer to many of society's problems. Beginning with handicapped or sick young children, "euthanasia" was systematically extended to include "impaired" older children, then "defective" adults, and finally individuals selected on racial rather than medical grounds.[314] Hitler had set out his thoughts on the subject quite clearly in *Mein Kampf*:

> The right of personal freedom recedes before the duty to preserve the race. There must be no half measures. It is a half measure to let incurably sick people steadily contaminate the remaining healthy ones. This is in keeping with the humanitarianism which, to avoid hurting one individual, lets a hundred others perish. If necessary, the incurably sick will be pitilessly segregated — a barbaric measure for the unfortunate who is struck by it, but a blessing for his fellow men and posterity.[315]

With the early introduction of legislation (the "Law for the Restoration of the Professional Civil Service" of 7 April 1933), the Nazis intended to thoroughly purge long-established ethical and administrative public supervisory bodies.[316] The decree not only affected the lives of Germany's 1.5 million civil servants, but also that of tens of thousands of others employed in public service. The purge, however, was not entirely effective; by late 1935, some of those who had been dismissed were being considered for re-employment, and the exigencies of war made the acceleration of that process inevitable.[317] On 1 June 1933, the "Law for the Reduction of Unemployment" came into effect. This popular measure was intended not only to achieve the purpose stated in its title, but also to force women out of the workplace and back into the home in their "natural" capacity of wives and mothers. It did so by providing loans to newly married couples, which could be repaid through fecundity. The debt was reduced with each birth, and considered repaid on the arrival of the fourth child.[318] The loans were also conditional upon the female partner ceasing paid employment. Loans could be refused "if one of the prospective marriage partners is suffering from a hereditary or mental physical illness which renders their marriage undesirable to the whole national community." By a subsequent amendment, all applicants for such loans were to be subjected to a medical examination prior to approval. Further tax benefits and other measures introduced in 1934 and 1935 to stimulate procreation were denied to those considered of "lesser ra-

---

313 Burleigh, *Death and Deliverance*, p. 47 ff.

314 Padfield, *Himmler*, p. 261.

315 Robert E Conot, *Justice at Nuremberg* (New York: Harper & Row, 1983), p. 204.

316 The legislation of 7 April 1933 was of fundamental importance, since it represented the first attempt by the Nazis to legally define a Jew. A "non-Aryan" was "anyone descended from non-Aryan, particularly Jewish, parents or grandparents. It suffices if one parent or grandparent is non-Aryan." (Friedländer, *The Years of Persecution,* p. 27). The requirement to produce evidence of 'Aryan' ancestry had a long history in Germany. By the end of the nineteenth century, some organizations had introduced bylaws excluding Jews on the basis of "race". The use of this so-called "Aryan paragraph" expanded exponentially after the First World War. For example, in 1920 the nationalist student fraternities (*Burschenschaften*) adopted a rule that required prospective members to provide evidence of "Aryan" descent as far back as their grandparent's generation, and prohibited marriage to Jewish or "coloured" women. [Richard S Levy (ed), *Antisemitism: A Historical Encyclopedia of Prejudice and Persecution* (Santa Barbara: ABC-CLIO Ltd, 2005), pp. 39–40]. In this respect , as in so much else, the Nazis were unoriginal, simply more radical.

317 Bronwyn Rebekah McFarland-Icke, *Nurses in Nazi Germany* (Chichester: Princeton University Press, 1999), p. 96.

318 Burleigh, *Ethics and extermination,* p. 163.

cial value."[319] Depending upon one's point of view, this public welfare legislation can be viewed as an exemplar of either positive or negative eugenics, since it contained elements of both.

In 1932, the Weimar government had drafted a law for the voluntary sterilisation of the handicapped. On attaining power, the Nazis wasted little time in implementing a statute which accorded with their interpretation of birth control. "We want to prevent the poisoning of the entire bloodstream of the race" commented Göring's legal assistant.[320] On 14 July 1933, the "Law for the Prevention of Genetically Diseased Offspring" was approved by Hitler's cabinet. Amended on 7 December 1933, it came into force on 1 January 1934.[321] The new law decreed not voluntary, but compulsory sterilisation for persons characterized by a wide variety of disabilities.[322] As part of the new system Hereditary Health Courts were established to decide whether compulsory sterilisation was appropriate for such citizens.[323] These courts, normally forming part of local civil courts, consisted of a lawyer and two doctors, one of whom was a supposed expert on genetic pathology.[324] The verdict of the lower court could only be challenged in a Higher Hereditary Health Court. If an appeal failed, sterilisation took place irrespective of the wishes of the victim or their family.[325] Importantly, this legislation was intended to be explicitly eugenic in nature, that is to say no criminal act had to have been committed for which punishment was prescribed.[326] Since eugenicists had devoted so much time and effort attempting to prove a link between perceived hereditary flaws and criminality, it is to be supposed that this principle was not always applied, and that conviction of a criminal offence was of itself sometimes sufficient to result in a ruling for sterilisation. In the first year that the new law was in operation, the Hereditary Health Courts confirmed sterilisation in more than 90 percent of cases heard. If an appeal to the Higher Hereditary Health Court failed (during the entire Nazi era only about 3 percent of appealed decisions were reversed), those still refusing to report for sterilisation were usually sent to concentration camps.[327]

The size and scope of the sterilisation programme had changed the very nature of German medicine. Every physician was perforce a genetic doctor, supposedly able to provide an expert opinion on matters of hereditary racial characteristics. These opin-

---

319 Burleigh and Wipperman, *The Racial State*, pp. 46–48.

320 Clarence Lusane, *Hitler's Black Victims: The Historical Experiences of Afro-Germans, European Blacks, Africans, and African Americans in the Nazi Era* (London: Routledge, 2003) p. 134.

321 The law was further amended in 1935. According to Hans Heinrich Lammers, even at this early stage Hitler was contemplating the killing of mental patients (Burleigh, *Death and Deliverance*, p. 99).

322 Alleged hereditary diseases were defined as including congenital mental deficiency, schizophrenia, manic-depressive insanity, hereditary epilepsy, Huntington's Chorea, hereditary blindness, hereditary deafness, and severe hereditary physical malformation. (Ibid., p. 61)

323 In 1935 there were 205 lower courts and 31 higher courts. Although a much higher number of courts had been anticipated, it is unlikely that these figures were exceeded. (Proctor, *Racial Hygiene*, p. 361, note 33).

324 Proctor, *Racial Hygiene*, pp. 102–103. In June 1938 membership of the Heredity Health Court was expanded to include two laypersons "of German or related blood" who "have an understanding of family life." (Ibid, p. 116).

325 Burleigh and Wipperman, *The Racial State*, p. 48. The new law even provided for the use of force in cases where this was considered necessary (Ibid, p. 137).

326 Proctor, *Racial Hygiene*, p. 102.

327 Ibid., pp. 106–107.

ions were sometimes based on peculiar techniques. Matters such as the shape of teeth, gums and tongue, the form and structure of the ear, the manner of walking and the posture when resting, as well as patients' answers to dubious IQ questions, could often form part of the evidence produced by these "experts" before the health courts, frequently determining sterilisation issues.[328] Psychiatric patients considered eligible for release were offered a stark choice—either submit to sterilisation or have further institutionalisation paid for by themselves or their family.[329] This option, which was in effect no option at all, helped convince those most susceptible to accepting their ideas from those in positions of authority (in this case nurses from doctors), that patients could be divided into the "good", that is those who voluntary accepted sterilisation, and the "bad", those who had to be coerced into it. The extent to which this mindset may have influenced individual nurses' subsequent actions when "euthanasia" was introduced is worthy of consideration.[330]

One practitioner enthused about how the new law freed the psychiatrist "in the realm of eugenics from the emotional difficulties that we have found more and more embarrassing over the last ten years."[331] Typical of the fervour with which the sterilisation law was embraced was the action of Werner Villinger, chief physician of the nursing home at Bethel, who registered 1,700 of his 3,000 patients as being suitable for the operation.[332] Having first ordered the sterilisation of a young female inmate of the Kaufbeuren mental home, in his capacity as a Hereditary Health Court judge the psychiatrist Hermann Pfannmüller (of whom more later) investigated a further twenty-one allegedly "degenerate" members of the woman's family, recommending the highly urgent sterilisation of ten of them on the grounds that "the danger of reproduction appeared imminent."

School teachers were instructed to encourage children to construct their own family trees in order to discover any eugenically suspect family members.[333] The biologist and fervent Nazi, Paul Brohmer, devoted his 1933 book *Biologieunterricht und völkische Erziehung* (*Biology Classes and Ethnic Education*) to outlining the manner in which the young were to be educated in matters of racial hygiene. Since they were already familiar with the symptoms of degeneration in animals and plants from their own everyday observations, it did not require a great deal of initiative for pupils to also find similar indications of decline and degeneracy in man. Thus racial eugenics could be introduced at an appropriate point, not just in the study of zoology and biology, but in geography and history too. He continued:

> Not one elementary school pupil should leave school without having internalized the iron command that he is to bear part of the responsibility for the fate of his fatherland, without the awareness that he is only a link in the chain of his ancestors and descendants and the carrier of future generations...It can be suggested that the student draw up a genealogical chart of his family as far back as he can go. In addition, he can be asked questions about the physical characteristics of his parents and other forbears as far as they

328 Ibid., pp. 110–112. Hitler reputedly believed that Jews could be identified on the basis of whether the earlobes were or were not attached to the neck. (Ibid., p. 150).

329 McFarland-Icke, *Nurses in Nazi Germany*, p. 133.

330 Ibid., p. 133–151.

331 Ibid., p. 131.

332 de Mildt, *In the Name of the People*, p. 50.

333 Burleigh, *Ethics and extermination*, p. 119.

can be determined (size, figure, shape of head and face, colour of hair and eyes, form of nose, etc.), about their intellectual and characteroligical qualities, their special achievements(for example, rescues during the war, scientific or literary publications, compositions), their life span and cause of death. In given cases, deformities and hereditary diseases should also be reported. The number of children produced by the student's ancestors should be determined.[334]

The promotion of racial hygiene was remorseless. A women's magazine article confronted the accusation that the sterilisation policy was "going against the laws of nature" in the following terms:

> Until National Socialist rule, the German people neglected the laws of nature...It not only disregarded the laws of heredity, of selection and of eradication, but directly opposed them by not only keeping the unfit alive at the cost of the healthy, but even guaranteeing their procreation...Every hereditarily sick German woman will, once she realises this, take this operation upon herself in order to keep her whole race healthy. "But doesn't that mean she's sinning against life?"...What does life mean then? Just go to a lunatic asylum...[335]

Within the medical profession it was not only physicians who were encouraged to embrace the tenets of National Socialism. In 1933, Dr. Friedrich Bartels, deputy leader of the health department of the *Reich* Ministry of the Interior, called on nurses to concentrate less on individual patient care and more on collective prevention. It was their duty to "secure and promote a genetically sound, valuable race and...not to expend an exaggerated effort on the care for genetically or racially inferior people...at the cost of the more valuable people." Nurses should improve the health of the sick and weak "only so far as their inherited biological predisposition allows."[336] On completion of their training under the new regime, nurses were required to swear an oath of allegiance:

> I solemnly swear that I will be steadfastly faithful and obedient to Adolf Hitler, my *Führer*. I promise to fulfil my duties, wherever I may be designated to work, faithfully and conscientiously as a National Socialist nurse in service to the national community, so help me God.[337]

The *Frankfurter Zeitung* reported in October 1936 that National Socialist nurses had sworn this oath in the Cologne-Aachen district for the first time. Ominously, the article continued: "The future would present tasks which could properly be performed only by men and women fully imbued with the philosophical attitude of National Socialism."[338] The nature of that 'philosophical attitude' soon became apparent.

It is impossible to be certain of the extent to which this politicalization of nursing helped to influence later individual behaviour, but it is not unreasonable to think that it would have had an affect on the actions of at least some of those taking this oath, as it did in similar circumstances with members of the *Wehrmacht*, who found themselves bound by the oath of allegiance to Hitler that they had been made to swear. Kate Gumbmann, a nurse at Hadamar testified: "I thought it was assumed by a nurse that she followed all orders that were given her."[339] Whilst refuted as a defence in early Allied

---

334 George Lachmann Mosse, *Nazi Culture: Intellectual, Cultural and Social Life in the Third Reich* (Madison: University of Wisconsin Press, 2003), pp. 89–90.

335 David F Crew, (ed), *Nazism and German Society, 1933–1945* (London: Routledge, 1994), p. 120.

336 Mary D Lagerway, *Nursing Ethics at Hadamar* (Qualitative Health Research, Vol.9, No.6, 1999), p. 764.

337 Ibid., p. 765.

338 Mosse, *Nazi Culture*, p. 261.

339 Lagerwey, *Nursing Ethics at Hadamar*, p. 767.

post-war legal proceedings, there is little doubt that this blind acceptance of superior authority was a trait deeply imbedded in the German psyche.

In at least one respect, the National Socialist regime was merely heir to an existing racist policy of sterilisation. In the wake of the First World War, French colonial troops, many of them black, were among those occupying the Rhineland. Weimar government propaganda was quick to portray these troops in the worst possible (and wholly untrue) light, accusing them of spreading every kind of disease. Friedrich Ebert, the first president of the Weimar Republic, spoke of "the black scourge", and a committee was formed in Munich with the intent of "educating the public concerning the dangers facing the white race."[340] The children of the inevitable coupling of the troops and local women were referred to as 'Rhineland Bastards', their number grossly inflated; in truth they never represented the slightest threat to the "purity of the Germanic race." By 1935 there were estimated to be 500–800 such individuals. Fifteen years earlier their insignificant number had not prevented a journal protesting:

> Shall we silently accept that in future instead of the beautiful songs of white, pretty, well-formed, intellectually developed, lively, healthy Germans, we will hear the raucous noise of horrific, broad skulled, flat nosed, ungainly, half-human, syphilitic half-castes on the banks of the Rhine?

Needless to say, Hitler discovered a Jewish conspiracy behind all this; in *Mein Kampf* he wrote that the Jews had "brought the negroes into the Rhineland" with the intention "of ruining the hated white race by the necessarily resulting bastardisation." The matter troubled the Nazis for years. In 1927 the Bavarian Ministry of the Interior had unsuccessfully suggested sterilising the children, an idea that the Nazis took up again once in office.[341] But it was not until 1937 that the *Gestapo* finally arranged for the compulsory sterilisation of many of these children, on no grounds other than their parentage. Others were subjected to medical experiments, or simply disappeared.[342]

The adoption of enforced eugenic sterilisation on a national scale was by no means limited to Nazi theorists, or for that matter solely to Germany, although it was only the Nazis who linked the subject to a racist ideology. Nowhere was the concept of compulsory sterilisation more enthusiastically endorsed than in the United States, where 30,000 people had been sterilised before 1939, half of them in California alone.[343] By 1 January 1950 the total of sterilised American citizens had risen to 50,707.[344] Eventually more than 65,000 individuals were sterilised in 33 states under compulsory sterilisation programmes. After 1945 such legislation was revoked or if it remained on the statute books, largely ignored. Some American eugenicists, like Harry Hamilton Laughlin, had been extraordinarily naïve when discussing the subject. In 1929 he declared: "The possibility that sterilisation might become a political tool that might be used by one race against another, or by one religion against another, or by one social class against another, is extremely remote."[345] Yet it is self-evident that enforced sterilisation, by its

---

340 Proctor, *Racial Hygiene*, p. 112.

341 Burleigh and Wipperman, *The Racial State*, pp. 128–130.

342 http://www.ushmm.org/wlc/article.php?lang=en&ModuleId=10005479 (Accessed 24 May 2008).

343 Burleigh, *Eugenic Utopias*, p. 63.

344 Proctor, *Racial Hygiene*, p. 360, note 9.

345 Ibid., p. 295.

very nature, would inevitably be aimed at the poor and under privileged, as well as the racially and genetically "inferior". It would always be those most in need of protection that would provide the victims of such a policy.

A further example of the prevailing American sentiment was the opinion of Charles M Goethe, successful entrepreneur, founder of California State University, Sacramento, and a proponent of compulsory sterilisation and racially motivated discrimination, who wrote in 1935: "However one abhors dictatorship, one is also impressed that Germany, by sterilisation, and by stimulating birth rates among the eugenically high-powered, is gaining an advantage over us as to future leadership." His enthusiasm for the Nazi eugenic model was still evident in October 1937, when he wrote: "

> The Germans are forging far ahead of us in this matter of accumulated data. They say they have already 4 Nobel prizes to 1 of ours, population considered, and that if we do not accept their methods, they will run away from us with world leaders.[346]

Such views were far from uncommon. The eugenic movement attracted many prominent and influential followers in the United States between 1908 and 1925, their number including scientists, educators, industrialists, and politicians.[347] Although the movement's influence had begun to wane somewhat thereafter, by 1928 racial hygiene was being taught in almost three-quarters of all American colleges and universities.[348] "The American Eugenics Society" was what nowadays might be termed a group of Nazi fellow-travellers. Clarence G Campbell, the Society's Honorary President, was even more vehement than Charles Goethe in his support of events then occurring in Germany, writing as he did in 1936:

> The German nation has adopted a policy of biological improvement in its racial quality as its major national objective, to which all other objectives are regarded as subsidiary...This policy is not the creation of political opportunists designed to flatter national vanity or to engender racial antagonisms, but it is the integration of the well-considered conclusions of its anthropologists, its biologists, and its sociologists...No earnest eugenicist can fail to give approbation to such a national policy.

And what if the German template was not adopted by other nations? Campbell concluded his article with a dire warning:

> If ...they permit a progressive deterioration in their own racial quality and survival value, and if in subsequent generations they fall behind and fail in the inescapable competition for racial survival, the present generation could only be deemed to bear a heavy responsibility for such a failure.[349]

The preceding year Campbell had represented the American eugenic movement at an international conference on population science held in Berlin, where he was effusive in his praise of Nazi policies. Having first commended the efforts of a number of non-German eugenicists, he continued:

> It is from a synthesis of the work of all such men that the leader of the German nation, Adolf Hitler, ably supported by the Minister of the Interior, Dr Frick, and guided by the nation's anthropologists, its eugeni-

---

346 http://www.csus.edu/cshpe/eugenics/docs/goethe_report.pdf (Accessed 31 August 2008).

347 James W Trent, *"Who shall say who is a useful person?" Abraham Myerson's opposition to the eugenic movement* (History of Psychiatry, No. 12, 2001), p. 33.

348 Proctor, *Racial Hygiene*, p. 400, note 11.

349 http://www.eugenicsarchive.org/eugenics/ (Accessed 31 August 2008).

cists, and its social philosophers, has been able to construct a comprehensive race policy of population development and improvement that promises to be epochal in racial history.[350]

To the undoubted satisfaction of his hosts Campbell declared that the "difference between the Aryan and the Jew was as insurmountable as that between black and white."[351] It is hardly surprising to learn that as the most frequently quoted non-German "scientist", Campbell was the darling of the Nazi press.[352]

In an act of mutual admiration, in 1932 the German historian of science, Reinhold Müller, effusively praised American pioneering in matters of racial hygiene. Whilst until 1926, he wrote, discussion of the subject had been purely academic and scientific in nature, it had been the Americans who, through massive research, had "proved" Galton's thesis that qualities of the mind were as heritable as qualities of the body. Others quoted specific examples of American eugenically based legislation. In the same year as Müller, Walter Schultze, the Bavarian Health Inspector, admired racial laws placing restrictions on the immigration of Jews, Poles, and southern Europeans to the United States. Based upon its adoption of sterilisation laws and its restrictive immigration policies, the United States was a country where racial policy and thinking were much more the norm than elsewhere, declared Schultze.[353] This was something to be commended—and emulated.

As late as 1940, the American eugenicist and racist, Lothrop Stoddard, spent four months in Germany, during the course of which he was invited by the Nazis to serve for a time as a judge in a Hereditary Health Court. Stoddard concluded that the sterilisation law was "being administered with strict regard for its provisions and that, if anything, judgements were almost too conservative." The "worst strains in the Germanic stock" were being weeded out "in a scientific and truly humanitarian way." The "Jews problem" was "already settled in principle and soon to be settled in fact by the physical elimination of the Jews themselves from the Third Reich." Stoddard was well connected, counting Himmler, Darré, Ribbentrop, and even Hitler among his acquaintances, as well as leading German racial hygienists like Günther, Fischer, and Lenz.[354] Given what is now known, it is natural to speculate exactly what Stoddard meant by "the physical elimination of the Jews". In 1940, the enforced emigration of the Jews from Germany was still on the agenda, and it almost certainly this to which Stoddard referred, but it was nonetheless a revealing observation.

In Great Britain in 1931, A G Church, a Labour Member of Parliament and General Secretary of the National Union of Scientific Workers, introduced a bill in the House of Commons proposing the voluntary sterilisation of "mental defectives." The consent of the patient and their family would be required, and the process would be subject to government and judicial approval. The example of California was enthusiastically cited. Unless such legislation was adopted, it was suggested, the mentally disabled would re-

---

350 Stefan Kühl, *The Nazi Connection: Eugenics, American Racism, and German National Socialism* (New York: Oxford University Press, 2002), p. 34.

351 Dennis Sewell, *The Political Gene—How Darwin's Ideas Changed Politics* (London: Picador, 2009), p. 145.

352 Kühl, *The Nazi Connection*, p. 35.

353 Proctor, *Racial Hygiene*, pp. 97–101.

354 Kühl, *The Nazi Connection*, pp. 61–63.

main free to breed, thereby posing a threat to the future of the nation. The bill was defeated.[355] But sterilisation laws were introduced in Denmark (1929), Norway (1934), Sweden (1934), Finland (1935), Iceland (1938), and a number of eastern European countries. More than 100,000 people were sterilised in Denmark, Norway and Sweden alone, the majority *after* 1945. The 1982 Swedish Nobel Peace Prize winner, Alva Myrdal, commented in 1941:

> In our day of highly accelerated social reforms the need for sterilisation on social grounds gains new momentum. Generous social reforms may facilitate home-making and child-bearing more than before among the groups of less desirable as well as more desirable parents. This may not be regretted in itself as the personal happiness of these individuals and the profitable rearing of their children already born are not to be neglected. But the fact that community aid is accompanied by increased fertility in some groups hereditarily defective or in other respects deficient and also the fact that infant mortality among the deficient is decreasing demands some corresponding corrective.[356]

Some were prepared to go further. Hildur Nygren, a Social Democrat Member of Parliament, and in 1951 Minister of Ecclesiastic Affairs, favoured compulsory sterilisation, stating: "We must not be so taken up with the idea of freedom and civil rights for each and every person in this generation, that we forget the just demands of the next"[357] Another Social Democrat, Karl Johan Olsson, commented: "I think it is better to go a little too far than to risk bringing unfit and inferior offspring into the world."[358]

In a complete *volte-face*, in August 1997 the Swedish government denounced the policy of its predecessors. Margot Wallstrom, the Swedish Minister for Social Policy, issued a statement, saying: "What went on is barbaric and a national disgrace." She pledged to create a law ensuring that involuntary sterilisation would never again be used in Sweden, and promised compensation to the victims of sterilisation legislation.[359]

In another example of what was then contemporary thought, Traian Herseni, racial "expert" in the Iron Guard, the Romanian version of the Nazi party, enthusiastically endorsed what had been German practice for many years when he wrote in 1941:

> We need eugenic laws and eugenic practices. Reproduction cannot be left to chance...Certain people must not be allowed to reproduce, and inferior races must be completely isolated from the main [ethnic] group. The sterilisation of certain categories of people must not be conceived stupidly as a violation of human dignity, but as a tribute to beauty, morality, and perfection.[360]

Not everybody bought into this interpretation of inherited hereditary characteristics. As early as 1918, the American psychiatrist Abraham Myerson had pointed out that the results of his studies suggested that whilst insanity occasionally reappeared over

---

355 Matthew Thomson, *The Problem of Mental Deficiency: Eugenics, Democracy, and Social Policy in Britain, c.1870–1959* (Oxford: Oxford University Press, 1998), pp. 60–61.

356 Gunnar Broberg and Nils Roll-Hansen, *Eugenics and the Welfare State: Sterilization Policy in Denmark, Sweden, Norway, and Finland* (East Lansing: Michigan State University Press, 2005), p. 105

357 Burleigh, *Eugenic Utopias*, pp. 66–67.

358 Broberg and Roll-Hansen, *Eugenics and the Welfare State*, p. 107.

359 http://www.paulbogdanor.com/left/eugenics.html (Accessed 18 July 2010). The minister went on to say that the government was "anxious to show that similar policies were pursued in the US, Germany, Austria and Finland, among others."—a classic example of *tu quoque*.

360 David S Wyman (ed), *The World Reacts to the Holocaust* (Baltimore: The John Hopkins University Press, 1996), p. 230.

generations, mentally ill people often had perfectly normal children and siblings, and were themselves the children of normal parents.[361] "There is no such thing as insanity, no such thing as feeble-mindedness, no such thing as delinquency," Myerson wrote in 1922.[362] If mental deficiency was related to anything, it was to itself.[363] Environmental factors (meaning nurture) were more important than nature in attempting to arrive at an understanding of human development.[364] In Myerson's view, individuals suffering from mental illness should avoid marriage, not because they might produce disabled progeny, but simply because they could not afford to support their children, disabled or not. Instead of punishment, the concern should be with social improvement: "Better diet, more sunshine, more leisure, more freedom from stress and strain, better education, and sound mental hygiene."[365] At the time he was swimming against the tide, but in his persistent opposition to the more outlandish claims of the eugenicists he proved completely justified.

Initially, the methods of sterilisation prescribed in Germany were standard surgical procedures, but the second amendment to the law, issued on 4 February 1936, provided that under certain conditions the Reich Ministries of Interior and Justice could prescribe a non-surgical method, and the regulation of 25 February 1936 permitted the use of X-rays for the sterilisation of women over the age of thirty-eight, or in cases where a surgical procedure might prove hazardous.[366] Whatever the method, it has been estimated that by 1939, at least 300,000 men and women had been sterilised, many of whom subsequently became victims of the "euthanasia" programme.[367] A probable 75,000 further persons were sterilised after 1939, largely in areas newly incorporated within the *Reich*.[368] Between 1934 and 1937, an estimated 480 individuals died as a result of the operation.[369] How many women fell victim to cancers induced by exposure to radiation as a method of sterilisation is unknown.[370]

Franziska Schwarz was a 16 year-old deaf girl, whose younger sister suffered from a similar disability, as did some of her father's brothers. Her parents too, were hard of hearing. One day in 1935 the family received a letter instructing Franziska and her mother to report to a health office to arrange for their sterilisation. Her uncle appealed

361 Trent, "*Who shall say who is a useful person?*", p. 37.
362 Ibid., p. 40.
363 Ibid., p. 45.
364 Ibid., p. 41.
365 Ibid., p. 43.
366 Robert N Proctor, *The Nazi War on Cancer* (Princeton: Princeton University Press, 2000), p. 91.
367 The number was equally divided between males and females. The same source suggests a total of 320,000 sterilisations between 1934 and the outbreak of war.(Burleigh, *Racial State*, p. 253).
368 Friedlander, *The Origins of Nazi Genocide*, p. 25 ff. The total of 375,000 sterilised represented about 0.5 percent of the *Reich's* population—an extraordinarily high percentage. A total of 400,000 sterilisations is quoted in other sources, (for example Burleigh, *Eugenic Utopias*, p. 70.) Even this was nowhere near the figure Fritz Lenz had in mind of 10–15 percent of the population. [George J Annas and Michael A Grodin, *The Nazi Doctors and the Nuremberg Code: Human Rights in Human Experimentation* (New York: Oxford University Press, 1992), p. 21].
369 Burleigh and Wipperman, *The Racial State*, p. 253.
370 Proctor, *The Nazi War on Cancer*, pp. 90–91.

on their behalf to the Hereditary Health Court, pointing out that deafness was not necessarily hereditary. He himself had perfect hearing. Moreover, Franziska's mother was menopausal, and promised to have no further children. The court therefore ruled that her mother did not have to be sterilised; however, the order for Franziska's sterilisation was confirmed. She was ordered to report to the Women's Hospital in Munich. In great pain after the operation she applied for sick pay—which was refused. As matters transpired the operation had been unsuccessful, and in 1938 Franziska found herself pregnant. When she reported to a gynaecologist for consultation, he ordered the forcible termination of her pregnancy—which was carried out. Unable to marry the father of the aborted child without a certificate confirming that she had been sterilised, in 1941 she consented to a further operation resulting in her permanent sterilisation.[371] It was a typical case. Those who were in less than perfect physical condition, the promiscuous, the inefficient, the careless and others who were simply deemed "socially feeble minded", all fell victim to the sterilisation measures.[372]

The example of Else K. is another that illustrates the working of both the compulsory sterilisation policy and the marriage health law. Before being granted permission to marry, Else was required to undergo a medical examination, part of which included an intelligence test, which she failed. Furthermore it was discovered that her deceased brother had suffered from schizophrenia, and had been confined to an asylum. As a consequence of these two factors she was not only sterilised, but also prohibited from marrying the man of her choice, since unlike Else he was "hereditarily fit". Because she had been sterilised their union could produce no offspring; if the marriage were permitted to proceed, the potential husband's valuable genetic material would therefore be lost to the nation.[373]

In 1999, Dorothea Buck described the lifelong traumatic effect of enforced sterilisation:

> While I was a patient at the evangelical church institution in Bethel for three-quarters of a year when I was just 18 years old, I had not one single consultation with a doctor, neither before nor after the forced sterilisation, by which I was in fact completely surprised and which was carried out without giving me any reasons beforehand why it was about to take place or afterwards why it had already taken place. At that time a person aged 19 was not yet responsible for her or himself. During her very first visit my mother was given the choice: either she agreed to my sterilisation or I would be put into an institution up to the age of 45. "This would be very much worse!" In a state of shock, she agreed. She told me this later. The clergyman at the Bethel institution cited words from the Bible at our beds without ever speaking to us personally. The most anguishing experience of human indignity, which remains indelibly in me to this day, is that the doctors and ministers in the Bethel institution didn't deem us worthy or capable of a discussion and in addition kept us in custody without any useful activity under a barrage of biblical words. The Bethel doctors and clergymen left us completely alone with our concerns about being labelled as "inferior hereditarily ill", the life-long consequences of the marriage ban, and the rigorous restrictions in our educational and occupational possibilities.[374]

---

371 Ina R Friedman,., *The Other Victims: First-Person Stories of Non-Jews Persecuted by the Nazis* (New York: Houghton Mifflin Harcourt, 1995), pp. 67–75.

372 Burleigh and Wipperman, *The Racial State*, pp. 252–258.

373 Panikos Panayi (ed), *Weimar and Nazi Germany: continuities and discontinuities* (White Plains: Longman, 2000), p. 206.

374 http://www.irren-offensive.de/buckspeech.htm (Accessed 19 August 2008).

As early as August 1933, Professor Walter Schultze, the State Health Commissioner for Bavaria, had made the Nazi government's intentions clear. "Coddling of [the] physically and mentally hopelessly worthless" was a waste of resources. Sterilisation alone was not the answer. Positive selection was part of the solution, but what was really needed was a policy of extermination; he added: "This policy has already partially been set in motion in our present-day concentration camps."[375]

In November 1933 the Nazis enacted the "Law Against Dangerous Habitual Criminals", thereby further obscuring the distinction between actual criminal activities and the unacceptable social behaviour that typified many people with disabilities. The law stipulated that in addition to the penalties already provided by the penal code, these criminal "asocials" (*Asozialen*) could be committed to state asylums, held in protective custody for an indeterminate period, and, in the case of sex offenders, officially castrated. In addition the law provided for the prohibition of defendants practicing their professions or occupations.[376] Among those held to be "asocial" were many Gypsies, a large number of whom were taken into protective custody in 1938.[377] In September 1934, Gerhard Wagner, the *Reich* Doctors' Leader, advised physicians that Hitler would sanction abortions performed where the union of either parent was likely to produce a child with a hereditary complaint. Moreover, it was now permissible for abortion and sterilisation to be performed concurrently, thereby ensuring that a potential problem pregnancy would not occur again.[378] But when it came to "the natural fertility of the German people", that is those racially and genetically acceptable to the regime, the Nazis were totally anti-abortion. In 1937, doctors performing unauthorized abortions on women of "valuable" stock could be sentenced to lengthy terms of imprisonment. When war began, terminating such a pregnancy could in some circumstances result in a death sentence for the physician involved.[379]

A succession of such laws and decrees prepared the ground for the Nuremberg Laws of 1935,[380] which, whilst aimed primarily at Jews, were also a precursor of legislation concerning marriage among people with disabilities. On 18 October 1935, the "Law for the Protection of the Hereditary Health of the German Nation" was enacted, the so-called "Marriage Health Law" (*Ehegesundheitsgesetz*), prohibiting any marriage if either of the parties suffered from some form of mental disability, had a "hereditary disease"

375 de Mildt, *In the Name of the People*, p. 52

376 Friedlander, *The Origins of Nazi Genocide*, pp. 23–24

377 Burleigh and Wipperman, *The Racial State*, p. 122.

378 Friedlander, *The Origins of Nazi Genocide*, p. 30.

379 Proctor, *Racial Hygiene*, pp. 121–123.

380 The 'Law for the Protection of German Blood and German Honour' of 15 September 1935 began: "Imbued with the realisation that the purity of German blood is a prerequisite for the continued existence of the German people..." This was Nazi racist ideology in a nutshell. The law went on to ban marriage or sexual relations between Jews and "Aryans", prohibit the employment in Jewish households of "Aryan" females under 45 years of age, and forbid Jews to fly the *Reich* flag. [Reinhard Rürup (ed), *Topography of Terror: Gestapo, SS and Reichssicherheitshauptamt on the "Prinz-Albrecht-Terrain." A Documentation* (Berlin: Verlag Willmuth Arenhövel, 2000), pp. 114–116.] " In Germany," declared a 1936 publication, " the Jewish question is simply the race question." The first supplementary decree to the Law, issued on 14 November, broadened the prohibition on sexual relations and marriage to include not only those between Germans and Jews, but also those between Germans and persons of "alien blood". Shortly thereafter "alien blood" was defined as that flowing through the veins of "Gypsies, Negroes, and their bastards". [Friedländer, *The Years of Persecution*, pp. 151–153].

as legally defined, or suffered from a communicable infection, especially tuberculosis or venereal disease.[381] Possession of a "certificate of fitness to marry", issued by a local health authority, was made compulsory for any couple wishing to wed.[382] Although the law stated that every individual intending to marry should first possess such a certificate, resources for so doing were inadequate. In practice, only those suspected of significant "genetic defects" were examined thoroughly. For most applying for the certificate, the medical examination was brief and superficial.[383]

These laws did not spring fully formed from the minds of inventive Nazi jurists. They had a long pre-history, having been the subject of debate in Germany since the late nineteenth century. In 1905 the geographer Paul Langhan had proposed the prohibition of mixed marriages, that is marriages between "healthy" and "unhealthy" partners, and the adoption of involuntary sterilisation. Fifteen years later the last German governor in Africa, Heinrich Schnee, was an early proponent of apartheid in advocating strict racial segregation and the avoidance of sexual relations with the autochthonous population so as to prevent "the regression of Europeans to the cultural level of a native." Nor were these exclusively German concerns, shared as they were by all of the western imperial powers, who treated the "inferior races" of their recently formed empires with a disdain that soon evolved into a policy of extermination. As has been illustrated above, the roots of Nazi racism can be found in the attitudes common to many nations as part of the exploitation of their newly acquired territories. Combining this racist precedent with the dogma of social hygiene was to produce a lethal brew under National Socialism.[384]

Much of the eventual Nazi legislation had been pre-empted by Heinrich Himmler in 1931, when he had set his stall out regarding marriage so far as it involved members of the SS and their partners, and the anticipated subsequent parturitions. On 31 December of that year, a Race Office of the SS (*Rasseamt der SS*) was established to ensure that all SS marriages (and subsequently all sexual relations involving SS-men) were in accordance with Nazi eugenic and racial principles.[385] Inspired by the *Blud und Boden* (Blood and Soil) "expert", Walther Darré, Himmler's "Engagement and Marriage Decree" thus read:

1. The SS is an association of German men of Nordic determination selected on special criteria.
2. In conformity with the National Socialist *Weltanschauung* and recognizing that the future of our people depends on the selection and retention of racially and hereditarily sound good blood, I establish with effect from 1 January 1932 the "marriage consent" for all unmarried members of the SS.
3. The goal striven for is the hereditarily sound, valuable extended family of German, Nordically determined type.
4. Marriage consent will be granted and denied solely and exclusively on the criteria of race and hereditary health.

381 Friedlander, *The Origins of Nazi Genocide*, p. 31.

382 Burleigh and Wipperman, *The Racial State*, p. 49.

383 Proctor, *Racial Hygiene*, p. 141.

384 Enzo Traverso, *The Origins of Nazi Violence* (New York: The New Press, 2003), pp. 52–54.

385 Ibid., p. 137.

5. Every SS-man who intends to marry has to apply for this purpose to the *Reichsführer*-SS for the marriage consent.
6. SS members who marry despite having been denied the marriage consent will be dismissed from the SS; they will be given the option of resigning.
7. The appropriate processing of marriage requests is the task of the Race Office of the SS.
8. The Race Office of the SS-manages the "Clanbook of the SS", in which the families of SS members will be entered after the granting of the marriage consent or approval of the request for registration.
9. The *Reichsführer*-SS, the head of the Race Office, and the specialists of this office are bound to secrecy on their word of honour.
10. The SS is clear that with this order it has taken a step of great significance. Derision, scorn, and misunderstanding do not affect us; the future belongs to us.[386]

Both SS-men and their prospective brides were required to genealogically prove the purity of their bloodline as far back as five generations. Girls had to provide medical evidence of their own and their family's good physical and mental health, undergo a comprehensive medical examination, and be measured and assessed to ensure that they conformed to the required "Nordic" criteria.[387]

Side by side with the negative eugenics exemplified by the expanding application of sterilisation policy, the Nazis practised positive eugenics by attempting to encourage a reversal of the declining "Aryan" birth rate. In the wake of the terrible death toll of the Great War, The National League of Large Families (*Reichsbund der Kinderreichen*—RdK), had been established long before the Nazi attainment of power as part of a policy intended to stimulate child bearing within German society as a whole. Under Nazi government it became a very different organisation. No longer were those deemed "valueless" by the regime to be encouraged to reproduce—quite the reverse.[388] The League's objectives under Nazism were made clear by its leader, Wilhelm Stüwe: "The more hereditarily healthy families a nation possesses, the more certain its future is." The organisation was "the storm troop in the field of National Socialist population policy."[389] Since it was believed that, on the basis of probability alone, the larger the family the more likely it was to produce an outstanding individual, perhaps even a genius, much time and effort was spent in propagandising the ideals of the RdK. Stüwe pronounced: "The phrase 'father of a large family' or 'mother of a large family' must again become a title of honour, as it always was in the best periods of German history [before] the cult of Liberalism...with Marxism and Jewry as its apostles" had eroded those traditional values.[390]

There were some benefits arising from membership of the RdK, although these usually took the form of added status rather than anything more tangible. The costs of utilities or transportation were sometimes reduced for large families, and legislation provided for some favourable treatment regarding taxation and child allowances. Most of

386 Padfield, *Himmler*, p. 103.
387 Ibid., p. 104.
388 Pine, Lisa, *Nazi Family Policy 1933–1945* (Oxford: Berg, 1999), p. 88.
389 Ibid., p. 91.
390 Ibid., p. 101.

all, motherhood was glorified; three children entitled a woman to preferential shopping rights, whilst from 1938, fecundity entitled receipt of the Honour Cross of the German Mother[391]—four children meant bronze, six, silver, and eight, gold, an idea probably stolen from the Berlin Olympiad of two years earlier. Nor were mothers the only beneficiaries. Karl Frenzel, a *Brenner* (burner) at three "euthanasia" killing centres and a particularly brutal guard at Sobibor extermination camp, fathered five children, which not only entitled him to the honorary title of *'kinderreicher Familienvater'* (literally 'a child-rich family father'), but ensured his exemption from military service.[392]

The British psychiatrist Eliot Slater, a student of Ernst Rüdin, published an article in the *Eugenic Review* in 1936 entitled *German Eugenics in Practice*, which neatly summarized the prevailing climate in Germany:

> With successive hammer-blows the German citizen is driven into a swastika-shaped hole. The atmosphere of compulsion pervades the whole of his life. The fact that he and his fellow men are now to be selected and bred like a herd of cattle seems to him hardly more distasteful than a hundred other interferences in his daily life. There is little doubt that these measures will have at least a partial success. If commanded with authority the German docilely obeys. The command now is to breed.[393]

On 1 September 1939 the health courts were ordered to only authorize sterilisation where there would otherwise be an abnormally high risk of undesirable progeny resulting.[394] The large-scale sterilisation programme had effectively been concluded. Perceived genetic problems would now be resolved by more radical measures.

In July 1933, Niek de Rooy, a Dutch journalist with the *'Noordbrabantsch Dagblad Het Huisgezin'* newspaper was one of a party of 22 invited to inspect the 'New Order' in Nazi Germany. On 13 July he reported that "the people have to get used to the new situation, but once they discover the new regime also does good things, they will certainly decide to support it with heart and soul." De Rooy, who was either a fellow-traveller or else easily duped, continued that on being questioned that rumours were spreading to the effect that Hitler intended "to kill 800,000 weak people", his unnamed German interlocutor explained that it was intended to strengthen the German race, but killing? No, "weak and less valuable people" would be housed in institutions, and only be prevented from marrying.[395] Little more than a year later, in autumn 1934 a Nazi doctor was writing:

> It must be made clear to anyone suffering from an incurable disease that the useless dissipation of costly medications drawn from the public store cannot be justified. Parents who have seen the difficult life of a crippled or feeble-minded child must be convinced that, though they may have a moral obligation to care

---

391 Burleigh and Wipperman, *The Racial State*, p. 250.

392 de Mildt, *In the Name of the People*, p. 224.

393 David Pilgrim, *The Eugenic Legacy in Psychology and Psychiatry* (International Journal of Social Psychiatry, Vol.54, No.3, 2008), p. 276.

394 Proctor, *Racial Hygiene*, p. 117.

395 Aart Vos, *"Wegens bijzondere omstandigheden..."—s'Hertogenbosch in bezettingstijd 1940–1945* (Alphen aan de Maas: Uitgeverij Veerhuis, 2008), p. 207. My thanks to the late Martin van Liempt for providing this information.

> for the unfortunate creature, the broader public should not be obligated...to assume the enormous costs that long-term institutionalization might entail.[396]

Fritz Bartels, Gerhard Wagner's lackey in the Nazi Physician's League, was reflecting the views of many, and not just the Party, when he commented:

> From time immemorial, the nation has always eliminated the weak to make way for the healthy. A hard, but healthy and effective law to which we must once again give credence. The primary task of the physician is to discover for whom health care at government expense will be worth the cost.[397]

In neighbouring Austria, the eminent physician, Julius Wagner-Jauregg, renowned for being, in 1927, the first neuropsychiatrist to receive the Nobel Prize, was another fervent racial hygienist and latent National Socialist. In fact, shortly before his death in 1940 he applied to join the Nazi party, an application denied because his first wife had been Jewish. He supported the German sterilisation law, criticizing the then Austrian government for rejecting policies of racial hygiene. Wagner-Jauregg was of the opinion that people with mental diseases or "criminal genes" were "individuals who, because of lasting genetic mental defects, are a danger to the community and unable to fit in."[398] It was becoming increasingly apparent which way the eugenic wind was blowing.

By 1935 Hitler was already proposing that he would use the cover of war to murder psychiatric patients in fulfilment of a long-held belief he had expressed in *Mein Kampf*. The war, Hitler reasoned, would provide both a distraction and an excuse for officially killing those deemed undesirable. If war should break out, he informed Gerhard Wagner that he would take up the "euthanasia" question and implement it.[399] Wagner had not attained his eminent position without reason. In the wake of the Nuremberg Laws, he spoke at the 1935 Party Rally of "the inherently different value" of different lives. The concept of equality was not only wrong, but pernicious, since it resulted in valuing "the sick, the dying, and the unfit on a par with the healthy and the strong." At last Germany was recognizing "the natural and God-given inequality of men." At the following year's rally he expanded on this theme.:

> We must oppose the three great dangers of racial and biological decline that have repeatedly destroyed states, peoples, and cultures in the past if they did not succeed in resisting them in good time. We must therefore contend with these three issues: the decline in the birth rate, the increase in sick and unfit genes in our people, and the mixing of the blood of our people with that of foreign and unrelated peoples, in particular with Jewish blood...The millions and billions that we have spent and the past, and the about one billion marks that we sacrifice today for the care of the genetically ill, is a squandering of our national resources that we National Socialists cannot justify when we consider the needs of the healthy population... The National Socialist state cannot repair the failings of the past, but through the "Law for the Prevention of Genetically Ill Offspring," it has seen to it that in the future the inferior will not be able to pro-

---

396 Proctor, *Racial Hygiene*, p. 183.

397 Ibid., p. 185.

398 Gabriel M Ronen, Brandon Meaney, Bernard Dan, Fritz Zimprich, Walter Stögmann, Wolfgang Neugebauer, *From Eugenic Euthanasia to Habilitation of "Disabled" Children: Andreas Rett's Contribution* (Journal of Child Neurology, Vol.24, No.1, 2009), p. 118.

399 Eugen Kogon, Hermann Langbein, Adalbert Rückerl (eds.), *Nazi Mass Murder: A Documentary History of the Use of Poison Gas* (New Haven: Yale University Press, 1993), p. 14. Hitler commented that in time of war it would be easier to "free the people from the burden of the mentally ill." He is also reported as stating that "the food supply does not allow for the incurably ill to be dragged through the war." (McFarland-Icke, *Nurses in Nazi Germany*, p. 211).

duce more inferior children, saving the German people from a steady stream of new moral and economic burdens resulting from genetic illnesses.[400]

Yet in 1936, the Ministerial Committee for Penal Law still felt able to declare: "There can be no question of an authorization of the extermination of so-called unworthy life...the strength of the moral standard of the prohibition on killing should not be allowed to weaken through the exemption for victims of severe illnesses or accidents because of considerations of mere expediency..." Compulsory sterilisation was to mark the limit of attempts to enhance the nation's biological inheritance.[401] Or so many assumed; others had different ideas. Commenting on the proposed introduction of voluntary assisted suicide in England that same year (a measure which was never implemented), one psychiatrist felt the drafting of the English law was seriously at fault because "the possibility of being permitted to kill incurable mental patients...has been entirely unconsidered; precisely this question demands a solution."[402]

In 1937, the result of legal proceedings against a farmer who had shot his sleeping adolescent son was reported in a Frankfurt newspaper. Accused of murder, the father defended his action on the grounds that the boy had been "mentally ill in a manner that threatened society". At the ensuing trial, the father's attorneys and attending Nazi Party officials contended that, amongst other things, the son had been a heavy financial burden on his family—the kind of economic reasoning that was increasingly being applied to cases of mental illness. The father received a sentence of only three years' imprisonment, of which he served just one.[403] The climate was ripe for the kind of radical action Hitler had promised.

Albert Hartl, a former Catholic priest who joined the SD in 1934, claimed that in the second half of 1938 he had been ordered by Reinhard Heydrich to report to Viktor Brack at the KdF, where Hartl was informed that the commencement of the "euthanasia" programme was under consideration. However, before a final decision was made, it was necessary to establish the position of the Catholic Church on the matter. Therefore, in early 1939 Hartl had met with Joseph Mayer, a Catholic professor of moral theology, in an attempt to gauge the position of the Church with regard to "euthanasia." Mayer, by no means an opponent of compulsory sterilisation, then produced a one hundred page long document arguing the pros and cons of such a policy, without coming to any firm conclusions. Hartl claimed to have relayed the contents of Mayer's memorandum to Brack, who thereafter informed him that since "a unanimous and unequivocal opposition from the...Churches was not to be expected", the "euthanasia" programme would begin. Hitler felt sufficiently emboldened by Mayer's position, so Hartl stated, to pass on to members of the Catholic hierarchy his intentions regarding the subject. Consequently, it is claimed that the Catholic Church was forewarned of the intention to commence "euthanasia", and by not objecting, gave their tacit approval to it.[404]

---

400 http://www.calvin.edu/academic/cas/gpa/pt36rasse.htm (Accessed 1 September 2008).

401 de Mildt, *In the Name of the People*, p. 54.

402 McFarland-Icke, *Nurses in Nazi Germany*, p. 131.

403 Mostert *Useless Eaters*, p. 160.

404 Burleigh, *Death and Deliverance*, pp. 169–170; Burleigh, *Ethics and extermination*, pp. 138–139; Gitta Sereny, *Into That Darkness—From Mercy Killing to Mass Murder* (London: Pimlico, 1995, pp. 64–73)

Whether Hartl's account can be regarded as entirely reliable is a matter for conjecture. There is no doubt that meetings did take place between Herbert Linden and Hans Hefelmann of the KdF on the one hand and Bishop Heinrich Wienken of Berlin on the other, with the purpose of seeking at least a degree of approval of the "euthanasia" policy on the Church's behalf. But the discussions collapsed after Pius XII issued an unmistakable condemnation of the destruction of "life unworthy of life"—although he was careful to refer to "killing" rather than "murder"—on 2 December 1940.[405]

What is incontrovertible is that by the end of 1938, the regime was receiving requests from the families of newborn or very young children with severe deformities and brain damage for the grant of a "mercy killing" (*Gnadentod*). The extent to which such appeals were at least encouraged, if not initiated, by local Nazi party officials is unknown, but seems to have been likely. These petitions were directed to the KdF, an agency headed by Philipp Bouhler and created by Hitler in 1934, ostensibly to keep him in direct touch with the concerns of the population, but acting in practice as Hitler's private office.[406] In particular, a petition was received (probably from the child's father) in respect of an infant named Gerhard Herbert Kretschmar, the so-called "Knauer" child, who had been born on 20 February 1939 in Pomssen, a small village in Saxony, blind, with one leg and part of one arm missing, and who was described as an "idiot."[407] It was Hans Hefelmann, head of department IIb of *Amt* (Central Office) II at the KdF who dealt with petitions to Hitler, and he passed this one on without comment to Albert Bormann (brother of Martin), head of *Amt* II. Bormann in turn forwarded it to Hitler, who instinctively understood that this was an opportune moment to put into practice what he had for so long been preaching.[408] As an agency of the Nazi party the KdF was outside the ambit of the state bureaucracy, and was thus perfectly placed to undertake any command of Hitler's without supervision, still less interference, on the part of others.[409]

Hitler ordered Karl Brandt, his escort physician, to visit the child at a hospital in Leipzig. Brandt testified at his post-war Nuremberg trial he had been instructed that if the facts provided by the child's father proved to be correct, he was to inform the attending physicians in Hitler's name that "euthanasia" could be carried out—which it was, on 25 July 1939.[410] Brandt's testimony also clarified two important points. Firstly,

---

405 Burleigh, *Death and Deliverance*, p. 170–171; Burleigh, *Ethics and extermination*, pp. 138–139.

406 Friedlander, *The Origins of Nazi Genocide*, pp. 39–40. Hitler maintained four separate chancelleries. In addition to the KdF there was the Nazi Party Chancellery, based in Munich and headed by Martin Bormann, the Presidential Chancellery under Otto Meissner, and the Reich Chancellery headed by Hans Heinrich Lammers. (Ibid, p. 40).

407 Christopher R Browning, *The Origins of the Final Solution: The Evolution of Nazi Jewish Policy, September 1939–March 1942* (London: William Heinemann, 2004) p. 185. The child's father had first approached Werner Catel to perform a *Gnadentod*, but Catel had refused on legal grounds—or so he claimed. [Michael S Bryant, *Confronting the "Good Death": Nazi Euthanasia on Trial 1945–1953* (Boulder: University Press of Colorado, 2005). p. 28].

408 Schmidt, *Karl Brandt*, p. 118.

409 http://www.memorialgugging.at/pdf/Czech_MedizinverbrechenGugging (Accessed 17 September 2009.)

410 Burleigh, *Death and Deliverance*, pp. 97–99; Friedlander, *The Origins of Nazi Genocide*, pp. 39–40. It has been suggested that Brandt personally carried out the killing of the 'Knauer' child, which was indeed possible [Friedländer, *The Years of Persecution*, p. 331]. He certainly had no compunction about admin-

the parents were not to be made to feel that they were responsible for the child's demise. For that reason a fictitious cause of death was to be registered—in the case of Gerhard Herbert Kretschmar, it was "heart weakness." Brandt stated: "...members of the family were told that the child had other weaknesses, that it had some sort of heart disease too..." This provided the template for the mass murder to come.

Secondly, Brandt testified that he had been instructed to state that "if these physicians were to become involved in any legal proceedings because of this measure, these proceedings would be quashed by order of Hitler." Although in fact Hitler did not possess the authority to issue such a dispensation, Brandt's words provided the medical staff concerned with the comfort they needed. And it would have been a brave man or woman in Nazi Germany who would have contested a *Führer-Befehl* (Führer order). It also suggests that everybody involved, from Hitler down to the parents, was aware of the illegality of the matter in which they were involved.[411]

It is arguable that the "Knauer" case was the catalyst for all that followed, although it could equally be argued that Hitler's dedication to "euthanasia" was such that its introduction was inevitable at some point in the brief history of National Socialism. There is some evidence that "euthanasia" was being unofficially practised even before the killing of the "Knauer" child, since asylum death rates began to rise unnaturally in 1938, and a 1942 planning commission report referred to "...certain active psychiatrists with positive attitudes [who] were already practising medical euthanasia extensively in their asylums in completely inconspicuous form even before the Operation" (a code word for the official "euthanasia" programme).[412] In his post-war testimony, Brandt commented that under certain circumstances, it was the practice for doctors to perform "euthanasia" on newly born children in cases such as that of the "Knauer" child; "No precise instructions were given in that respect."[413]

In May 1939, Hitler had instructed Brandt to pave the way for the killing of children by setting up a body with the impressive, if somewhat cumbersome title of "The *Reich* Committee for the Scientific Registration of Severe Hereditary Ailments" (*Reichsausschuss zur wissenschaftlichen Erfassung von erb- und anlagebedingter schwerer Leiden*).[414] This organization, usually referred to simply as "The *Reich* Committee", was completely fictitious, and was created simply as a cover for the KdF.[415] By a decree dated 18 August 1939, in return for a payment of 2 *Reichsmarks* per case, doctors and midwives were ordered to communicate all cases of "malformed" newborn to the *Reich* Committee. "The earliest possible registration" of all children under 3 years of age in whom "serious hereditary diseases" were "suspected" was required. Included on the initial list of diseases were certain physical conditions; mental retardation, microcephaly, hydrocephaly, paralysis, spastic conditions, and malformations of all kinds, but especially of the limbs, head and spinal column. Midwives were required to make these re-

istering, together with Conti, the first lethal injections of adult "euthanasia" [Schmidt, *Karl Brandt*, p. 122].

411 Schmidt, *Karl Brandt*, pp. 117–122.

412 Aly, Chroust, Pross, *Cleansing the Fatherland*, p. 35.

413 Schmidt, *Karl Brandt*, p. 120.

414 Members of the committee included Karl Brandt, Helmut Unger, Carl Schneider, Ernst Wentzler, Hans Heinze, and Werner Catel. (Proctor, *Racial Hygiene*, p. 186).

415 Friedlander, *The Origins of Nazi Genocide*, p. 44. The Committee's mailing address was a post office box.

ports at the time of a child's birth, while doctors were to report all cases of impaired children known to them. District medical officers were responsible for the accuracy of the reports, and the chief physicians of maternity clinics and wards were notified that the reports—which took the form of questionnaires from the *Reich* Health Ministry—were required. Subsequently, additional conditions such as learning problems, behavioural difficulties, and even bedwetting were added to the list of reportable "defects".[416] By June 1940, the questionnaires had been expanded to include not only a child's specific illness or condition, but details of their family history as well, including particulars of hereditary illnesses and any cases of alcohol, nicotine or drug abuse within the child's family. The revised reports also required a more detailed evaluation of the child's condition by a physician, indicating possibilities for improvement, life expectancy, prior institutional observation and treatment, details of physical and mental development, and descriptions of convulsions and related phenomena. The parameters for inclusion in children's "euthanasia" were gradually extended; by July 1941, not only infants but all minors suffering from any of the listed ailments were required to be registered. Teachers were required to report handicapped pupils. Within a few months thereafter, the programme had been progressively broadened to include children as old as teenagers of sixteen or seventeen.[417]

Even before war came in September 1939, the Nazis had thus established a government sanctioned process for murder. Two laymen (Hans Hefelmann and Richard von Hegener) made a preliminary selection of cases, which was then reviewed by three medical professors, Werner Catel, Hans Heinze, and Ernst Wentzler (all enthusiastic supporters of "euthanasia"), who determined the fate of the child.[418] The adjudicators, masters of life and death, never saw those they condemned or consulted their medical histories. Their "expert" decision was based solely on the contents of a form, the *Meldebogen*, prepared by the reporting medical authority. A child's fate was decided by a simple mark on the form—a red plus sign for death, a blue minus sign for life. Should the expert be undecided, "observation" was recorded on the document, postponing a final decision. If selected for "euthanasia", after approval by Brandt the child was transferred to one of a list of "*Children's Wards for Special Care*" (*Kinderfachabteilungen*), where, their parents were assured, the children would receive the most advanced available treatment for their condition.[419] After admission the child was either killed immediately, or after a period of further observation would be reassessed, ostensibly in order to prevent mistakes. If there had been none made, and the child's condition was confirmed, the *Reich* Committee issued authorization for the child to be killed. In reality, few children escaped once committed to a killing ward.[420]

In summer 1939 another of Hitler's personal physicians, Dr Theodore Morell, reviewed a survey of the views of parents of severely handicapped children, conducted in

416 Florian P Thomas, Alana Beres, Michael I Sheveli, "*A Cold Wind Coming*": *Heinrich Gross and Child Euthanasia in Vienna* (Journal of Child Neurology, Vol.21, 2006), p. 343.

417 Proctor, *Racial Hygiene*, p. 188.

418 Hefelmann's doctorate was in agronomy, not medicine. (Burleigh, *Death and Deliverance*, p. 269). Von Hegener was a former bank clerk (de Mildt, *In the Name of the People*, p. 67).

419 Burleigh, *Death and Deliverance*, pp. 103–104; Aly, Chroust, Pross, *Cleansing the Fatherland*, p. 190.

420 Friedlander, *The Origins of Nazi Genocide*, pp. 39–61.

the 1920s by Ewald Meltzer, an asylum director.[421] Morell reported to Hitler on the results of Meltzer's investigation:

> A number of parents expressed the view: 'If only you had done it [that is, "euthanasia"] and then told us that our child had died from an illness.' There is a lesson for us there. We need not suppose that we cannot carry out any salutary measure without the consent of the sovereign people.[422]

In Morell's opinion, therefore, the conclusion to be drawn was that the regime could anticipate no great negative reaction to the "euthanasia" programme from the general populace. This view was confirmed by another survey conducted in April 1941 which revealed that 80 percent of the relatives of those murdered in the "euthanasia" programme were in agreement with the decision, 10 percent spoke out against it, and 10 percent were indifferent. In fact, the programme was far from unpopular; in many cases parents of handicapped children welcomed, and in some cases requested euthanasia for their afflicted offspring. By 1939, propaganda concerning the removal of mental patients of all ages had been so intensive and effective, that their eradication seemed both natural and desirable.[423] It has been suggested that this policy of "official secrecy", where people knew while pretending not to know, and only a very few protested, was an invitation to denial and moral indifference on the part of both the German establishment and the German nation as a whole. It laid the foundation for a similar reaction to the "Final Solution." If people did not protest at the murder of their own relatives, they were hardly likely to do so when Jews, Gypsies, and foreigners were slaughtered.[424] By 1945, institutionalized murder had become so much the norm that it no longer remained an issue in Germany.[425]

Hitler must have taken comfort too from the almost complete dearth of public reaction to the raft of anti-Semitic legislation rapidly introduced after his attainment of power in 1933. Whilst not necessarily approving of such discrimination, protest by the clergy, academics, teachers, and the general populace was notable by its absence. In truth, other problems preoccupied them. This was not symbolic of a national eliminationalist anti-Semitism—merely of indifference and an overriding concern for self-preservation. Public acceptance of these early discriminatory measures was indicative of the extent to which questions of morality could safely be ignored in the Third *Reich*. The lack of concern shown by the great majority of Germans regarding the fate of their

---

421 Meltzer's questionnaire, distributed to the male parents or guardians of 200 children residing in the Katharinenhof institution, posed four questions; the first three asked whether the recipient would give their consent to a "painless shortening of your child's life" in certain hypothetical circumstances. The fourth question asked for the recipients' wife's response to the first three questions. Of the 162 replies received, 73 percent contained at least one "yes" response. (Aly, Chroust, Pross, *Cleansing the Fatherland*, pp. 29–30).

422 Götz Aly and Susanne Heim, *Architects of Annihilation: Auschwitz and the Logic of Destruction* (London: Phoenix, 2003), p. 171. Meltzer concluded that people were glad to release both themselves and the child from the burden whilst not wishing to have their consciences troubled. [Ibid., p. 330 note 29]. If his research contributed to the implementation of children's "euthanasia" (to which he was opposed), Meltzer lived to regret his efforts. The Katharinenhof institution was taken under state supervision, many children were killed with or without their parents' or guardians' consent, and Meltzer was forcibly retired. [Aly, Chroust, Pross, *Cleansing the Fatherland*, pp. 30–31 footnote].

423 Proctor, *Racial Hygiene*, p. 194.

424 Aly and Heim, *Architects of Annihilation*, p. 172–173.

425 Aly, Chroust, Pross, *Cleansing the Fatherland*, p. 92.

neighbours enabled Hitler to gradually expand the boundaries of what was acceptable. What began with the removal of Jews from the civil service was eventually to end in genocide.[426]

As early as July 1939, Werner Heyde, who was to play a prominent role in the "euthanasia" programme, attended a meeting at which he learned of the forthcoming killing of adult mental patients.[427] At this meeting, Bouhler reiterated that by eliminating some of the mentally ill, hospital space and medical personnel would be released for the forthcoming war. There was no compulsion to participate. But all of those present agreed to do so.[428] One month later the committee responsible for drafting a new criminal code came up with a clause proposing that a person suffering from an incurable or terminal somatic illness be permitted to request euthanasia. Furthermore, it was suggested, the legislation should include provision for the life an individual suffering from incurable mental sickness to be involuntarily terminated in appropriate circumstances. The suggestion was quietly shelved, overtaken by events, but it was evident that some kind of policy for dealing with "life unworthy of life" was imminent.[429]

Sometime immediately prior to the summer of 1939, it is suggested, Hitler had summoned Martin Bormann (head of the Party Chancellery), Hans Heinrich Lammers (head of the *Reich* Chancellery) and State Secretary Leonardo Conti (*Reich* Health Leader) to a meeting, at which Conti was instructed to implement the "euthanasia" programme.[430] According to Lammers, Hitler stated that that seriously ill mental patients should be "got rid of." The ending of the "worthless lives of such creatures" would result in savings of hospital and other medical costs.[431] Lammers testified that Conti insisted upon having his position legalised by way of receipt of a written confirmation, whereupon a furious Hitler replaced Conti with Bouhler.[432] But at Nuremberg, Brack suggested that in the never-ending struggle for supremacy among the Nazi hierarchy, the true reason for Bouhler's appointment was as the result of scheming on behalf of a coalition of Himmler, Frick, and Göring, who jointly considered that Conti was Bormann's puppet. Since Hitler was subsequently prepared to issue written authorization to Bouhler and Karl Brandt, this seems a conceivable scenario, although it should be born in mind that Brack's testimony was in many cases far from reliable.

In another variation on the origins of adult "euthanasia", it is said that Helmut Unger had been informed by Gerhard Wagner sometime in 1939 that Hitler had issued an authorization to Conti for the commencement of the programme. Unger so informed Hefelmann, who in turn told Brack on 18 July 1939. Conti, Bouhler and Herbert Linden

---

426 Friedländer, *The Years of Persecution*, pp. 59–60.

427 Brandt, Bouhler, Conti, and Linden were also present at this meeting. (Browning, *The Origins of the Final Solution*, p. 186.)

428 Padfield, *Himmler*, p. 262.

429 Robert Gellately, *Backing Hitler: Consent and Coercion in Nazi Germany* (Oxford: Oxford University Press, 2002), p. 101.

430 Peter Longerich, *The Unwritten Order: Hitler's Role in the Final Solution* (Stroud: Tempus Publishing Limited, 2003), p. 80.

431 Burleigh, *Death and Deliverance*, pp. 114–115.

432 Lucy S Dawidowicz, *The War Against the Jews 1933–1945* (New York: Bantam Books, 1979), p. 177.

then conceived a plan for the elimination of all German mental patients.[433] Karl Brandt's version of events was somewhat different. He claimed that so far as "euthanasia" was concerned, neither he nor Bouhler had attempted to manipulate events to their advantage out of ambition, or lust for power. Rather, the duo were of a mind; altruistic visionaries who had become involved in the affair for purely unselfish, compassionate reasons, Bouhler because he and the office he headed provided an essential link between Hitler and the general population, Brandt in order to ease the suffering of medically hopeless cases, both to prevent Bormann gaining control. Brandt claimed that shortly after the surrender of Poland, Hitler had summoned him and stated:

> ...that because of a document he [Hitler] had received from Reichsleiter Bouhler, he wanted to bring about a definite solution in the euthanasia question. He gave me general directives on how he imagined it, and the fundamental instruction was that any insane persons who were in such a condition that they could no longer take any conscious part in life were to be given relief through death. General instructions followed about petitions which he himself had received, and he told me to contact Bouhler himself about the matter.

Hitler wanted the KdF to be in charge of "euthanasia"; Brandt was to become involved only so that he could report directly to Hitler regarding developments in the programme.[434] Thus this was not a position Brandt had sought, he claimed, but one that had been thrust upon him. Furthermore, it was only after this conversation that Brandt discovered that Bouhler had advised Hitler that unauthorised "euthanasia" was *already* being carried out by some *Gau* leaders, and that to ensure that there was some kind of management in place, the KdF should be placed in charge of the operation. This conveniently left the responsibility for the scrambling to gain control of "euthanasia" at Bouhler's door, which was probably largely true. However, that Brandt was more extensively involved in all of these Machiavellian dealings than he was prepared to admit seems certain.[435] From a medical point of view he was quite unsuited to be responsible for a programme that was in any way dedicated to issues of mental health, since he was not a trained psychiatrist, had little knowledge of the subject, and may never have even visited an asylum prior to assuming joint control of the "euthanasia" programme.[436]

Whatever the precise timetable of events, it would appear that Bouhler and Brandt somehow got wind of Hitler's decision, and anxious to keep "euthanasia" within the ambit of the KdF, pressed him to confirm his instructions.[437] Ultimately, it is considered that Bouhler and Brandt were chosen by Hitler to head the programme because while Bouhler was considered to be in a position to eliminate the influence of Lammers and Franz Gürtner (the Minister of Justice), Brandt, as Hitler's escort physician since 1934 and the Party representative of the medical profession, was assumed to know doctors who would be prepared to cooperate with the proposed scheme. But Bormann proved too wise in the ways of Nazi government for Bouhler, Brandt, and their supporters. The

---

433 Proctor, *Racial Hygiene*, p. 188.

434 It has been suggested that the KdF was chosen by Hitler to administer the "euthanasia" programme simply because of the office's low profile, something that made it the perfect vehicle for such an important top secret project. [Donald Bloxham, *The Final Solution: A Genocide* (Oxford: Oxford University Press, 2009), p. 171].

435 Schmidt, *Karl Brandt*, pp.;128–130.

436 Ibid., p. 371.

437 Burleigh, *Death and Deliverance*, p. 115.

two leaders of the newly established organisation were forced to consult with Conti, who is known to have signed the first "euthanasia" orders.[438]

As with the "Final Solution", "euthanasia" proved zygotic regarding the two essential elements of National Socialist ideology—the biological and the economic. For example, the director of Stetten mental hospital, Pastor Ludwig Schlaich reported:

> When it became increasingly apparent, in the final years before 1939, that war was imminent, we learned that there had been discussion in the Reich Ministry of the Interior about putting the inmates of hospitals and nursing homes for the mentally ill, the mentally deficient and epileptics on drastically reduced rations in the event of war. When we objected that this would mean condemning our patients to a slow but certain death by starvation, they carefully began to sound out how the Home Mission would react if the state were to contemplate the elimination of certain categories of sick person in wartime, given that food supplies, once imports dried up, would no longer suffice to feed the entire population.[439]

Notwithstanding Morell's view and the verbal authorization that had been given to the "euthanasia" enthusiasts sometime prior to July 1939, Hitler wished to avoid the passing of an official law. In a dictatorship, no debate was necessary, no act of government required. It was enough for Hitler to state his intentions in the most general terms. Thereafter it was up to the recipient to interpret those objectives to Hitler's satisfaction. Commands were frequently implied, rarely written. If the listener or reader so chose, Hitler's wishes could effectively attain the force of law, although as will be seen, in reality what is often incorrectly referred to as the "euthanasia decree" was no more than a dispensation. Hitler, indolent and often indecisive, bored by the minutiae of government, ruled not by edict, but by implication and inference. One needed sensitive antennae to interpret his often imprecisely expressed intentions and opinions, for in the National Socialist state the most important attribute for any aspiring courtier was to understand the importance of "working towards the Führer."[440] The "euthanasia" authorization was a classic example of this principle. So in October 1939, a brief unofficial sanction was issued on Hitler's private stationery and signed by him:

> Reich Leader Bouhler and Dr Brandt are charged with the responsibility for expanding the authority of physicians, to be designated by name, to the end that patients considered incurable according to the best available human judgement of their state of health, can be granted a mercy death. (*Reichsleiter Bouhler und Dr. med. Brandt sind unter Verantwortung beauftragt, die Befugnisse namentlich zu bestimmender Ärzte so zu erweitern, dass nach menschlichen Ermessen unheilbar Kranken bei kritischster Beurteilung ihres Krankheitszustandes der Gnadentod gewährt werden kann*)[441]

---

438 Gerald Reitlinger, *The SS, Alibi of a Nation: 1922–1945* (New York: Viking Press, 1957), pp. 272–273.

439 Aly and Heim, *Architects of Annihilation*, pp. 169–170. On 6 September 1940, Schlaich wrote to the *Reich* Minister of Justice, protesting against the "euthanasia" programme (Kogon, Langbein, Rückerl (eds.), *Nazi Mass Murder*, p. 33.)

440 "Very often, and in many places, it has been the case that individuals, already in previous years, have waited for commands and orders. Unfortunately, that will probably also be so in future. Rather, however, it is the duty of every single person to attempt, in the spirit of the *Führer*, to work towards him. Anyone making mistakes will come to notice it soon enough. But the one who works correctly towards the *Führer* along his lines and towards his aim will in future as previously have the finest reward of one day suddenly attaining the legal confirmation of his work." Werner Willikens, quoted in Ian Kershaw, *Hitler 1889–36: Hubris* (London: Allen Lane, The Penguin Press, 1998), p. 529.]

441 Lifton, *The Nazi Doctors*, p. 63. It is possible that the document was created as late as winter 1939. (Schmidt, *Karl Brandt*, p. 131). The authorization is frequently (and mistakenly) referred to as a decree, which implies a spurious legality.

This sanction made no mention of mental patients *per se*, but the inference was clear. Asylums were expensive, ate up resources, and allowed patients to live who had no hope of being cured. Would it not be a relief to the nation and to the inmates themselves if they could painlessly be relieved of the burden of life? Such had been the arguments advanced by physicians and party *apparatchiks*, and which Hitler willingly endorsed.[442]

Although Hitler supposedly dictated the text, given the ambition of Bouhler and Brandt it seems more likely that it was they who determined the format and wording of the document in order to ensure that the programme remained in their hands.[443] For the reasons already stated Hitler was doubtless happy to do so. Deliberately drafted in the vaguest terms, the authorization allowed enormous leeway to the future perpetrators, a position they were more than ready to exploit. It should also be born in mind that even by the standards of Nazi government, this dispensation was not a law. It is feasible that some of those subsequently involved in "euthanasia" may have believed that such a law existed, or that a *Führer* decision (*Führerentscheid*) carried the weight of law, but it is simply a fact that it was not until 1942 that the *Reichstag* accorded Hitler the right to promulgate laws himself. The authorization was typed on Hitler's personal notepaper and not that of the *Führer und Reichskanzler*, was not countersigned by Lammers as head of the *Reich* Chancellery, and had not been published in the *Reich* legal gazetteer. These were points completely misunderstood by the Allies in presenting evidence at post-war trials.[444]

So furtive was the dispensation that Franz Gürtner, the Minister of Justice, was only informed of its existence and shown a photocopy of the document almost one year after adult "euthanasia" had commenced.[445] Gürtner had been informed of the *Aktion* by Pastor Paul Braun of the Inner Mission of German Evangelical Churches in June 1940. So shocked was Gürtner that he confronted Lammers, demanding that either the killings stop, or that they be given some kind of legal basis. Lammers referred Gürtner to Bouhler, who one month later, in August 1940, produced a copy of the authorization for the Minister of Justice (the only copy to have survived). On 5 September 1940, Bouhler informed Gürtner: "Based upon the authorization of the Führer as the only one responsible for the implementation of the requisite measures, I have issued directives that seemed necessary to me. Any additional written regulations for implementation are no longer required".[446] That was good enough for Gürtner; he offered no further resistance.

Brandt claimed never to have doubted the legality of the dispensation:

---

442 Raul Hilberg, *Perpetrators Victims Bystanders: The Jewish Catastrophe 1933–1945* (New York: Harper Collins, 1993), pp. 15–16.

443 Schmidt, *Karl Brandt*, pp. 131–132. Conti was also almost certainly involved in drafting the authorization, but was prepared to be the *eminence grise* in this endeavour. (Ibid). Others see the hand of de Crinis in the wording. [Lifton, *The Nazi Doctors*, p. 63 (footnote].

444 Bryant, *Confronting the "Good Death"*, p. 78–85. No West German court ever accepted the authorization as a valid law (Ibid, p. 38).

445 de Mildt, *In the Name of the People*, note 6.

446 de Mildt, *In the Name of the People*, p. 338, note 73; Schmidt, *Karl Brandt*, pp. 145–146; Friedlander, *Nazi Genocide*,p. 120–121.

For us, who were involved, there was no doubt that (the authorization) had the force of law, and there was therefore no question that our responsible actions were also formally legitimate...For those who were involved in euthanasia there was no doubt...that their action was morally and ethically legitimate and legally secure. Without this assurance and the conviction of this fact, none of the doctors would have acted in the way they did.[447]

Brandt recorded these comments in 1946, when he was facing a trial that he knew might very well end in his death sentence. His remarks may be viewed as an attempt to retrospectively rationalise the behaviour of himself and others, to justify his actions for the courtroom and for posterity. Whether in fact he gave the question of legality any thought at the time that "euthanasia" was implemented is another matter.

Those responsible for subsequent events later drafted several bills that would have made ""euthanasia"" legal, but all of these were rejected by Hitler.[448] He believed (almost certainly correctly) that this would present an unacceptable propaganda opportunity to the enemy. It was necessary to wait until final victory had been achieved before passing such a law.[449] In the meantime, this kind of *ersatz* legitimatization provided a sop to the consciences of those most intimately involved, whilst maintaining the programme's secrecy.[450] The authorization was backdated to 1 September 1939 to coincide with the date of commencement of the Second World War.[451] In the words of Alfred Rosenberg, "The new German style [...] is the style of a marching column, no matter where, or to what end, this marching column may be directed."[452]

Yet another example of a *Führer-Befehl* and eugenics can be seen in the state's attitude regarding abortions. Although abortions performed upon 'Aryan' women could be punished by up to two year's imprisonment, and in wartime by death,[453] in another secret decree of 19 November 1940, Hitler orally authorized "abortions [in cases] where it can be assumed with great probability that the birth of additional children would be

---

447 Schmidt, *Karl Brandt*, pp. 132–133.

448 The 1940 'Law on Euthanasia for the Incurably Ill' was drafted, but never implemented. It included provision for both voluntary and involuntary euthanasia. (Aly, Chroust, Pross, *Cleansing the Fatherland*, p. 26). The draft read in part:

1. Anyone suffering from an incurable illness that leads to strong debilitation of either oneself or others can, upon explicit request of the patient and with permission of a specially appointed physician, receive dying help from a physician.
2. A patient who, as a consequence of incurable mental illness requiring lifelong care, can, through medical intervention and without his knowledge, have his life terminated. (Proctor, *Racial Hygiene*, p. 193).

The planned euthanasia law contained provision for the creation of a central state office, which would take control of the various existing institutions. As part of the proposed eventual legislation, the new office of *Reich* Commissioner for Mental Hospitals was created by a statute of 23 October 1941; its first and only occupant was Herbert Linden. (Aly, Chroust, Pross, *Cleansing the Fatherland*, pp. 165–166).

449 Friedlander, *Nazi Genocide*, p. 154; Burleigh, *Death and Deliverance*, p. 166. It is probable that there were also other considerations. Hitler may have believed such a law would be unacceptable to the German people. He may also have felt any such law might restrict his options by defining potential victims too specifically. (Gellately, *Backing Hitler*, p. 102).

450 McFarland-Icke, *Nurses in Nazi Germany*, p. 9.

451 Burleigh, *Death and Deliverance*, p. 115.

452 http://www.phil.gu.se/sffp/reports/16.%20von%20Cranach.pdf (Accessed 25 February 2008.)

453 Burleigh, *The Racial State*, p. 250.

undesirable." In the hands of Brandt and Brack, this became a blank cheque—abortions up until the sixth month of pregnancy became permissible for "eugenic, racial, and ethnic reasons." As with the written "euthanasia" sanction, this authorization was completely illegal, but utilising it enabled T4 to organize the elimination of "inferior" children, even before their birth. However, should such a child slip through the net, "alternatives are available through the Reich Committee," Brack noted.[454] There had been an earlier 'Law for the Alteration of the Law for the Prevention of Genetically Diseased Offspring'[455] introduced in 1935, which legalised similar compulsory abortion, but only in cases where the prospective mother had been ruled "hereditarily ill" by the Health Courts.[456] And as previously described, from 1936 Himmler had his own personal anti-abortion programme. This new, illegal authorization significantly widened the scope for feticide.

At governmental level, Lammers was constantly bemoaning the lack of a legally constituted mandate for "euthanasia", but Hitler was not to be persuaded on that issue. In late May 1940, when the German campaign in France was about to end in triumph, Hitler instructed Bouhler to contact Lammers and other officials with a view to finding "a solution to the euthanasia problem". As Lammers described it, Bouhler was pushing for total control of the programme. However, Lammers was not to be moved: "In discussing this problem I drew Bouhler's attention to the serious concerns which I had and said that the matter could only be carried out on the basis of a published law." Bouhler dismissed Lammers' proposals; it would take far too long to introduce a formal law, and anyway, although Lammers later claimed to be unaware of the fact, the killings had already started. In the event, these discussions made no difference. Shortly afterwards Hitler simply informed Lammers that he did not want a law "for political reasons".[457]

It is impossible to be certain about the identity of all of those involved and the precise chronology of the gestation of "euthanasia". By the end of the war, many of those who made up what might be termed the "executive" of the programme were either dead (Hitler, Bouhler, Wagner, Conti, Linden, de Crinis), living under false identities or in hiding (Blankenburg, Bernotat, Heyde), or were simply at liberty, their participation in the murder of the "worthless" largely unknown—or ignored. Apart from Brandt and Brack in Nuremberg, and Nitsche in Dresden, none of the major figures making up the executive appeared in a courtroom in the immediate post-war years.[458] It was not until 1958 with the creation of *Zentrale Stelle der Landesjustizverwaltungen zur Aufklärung nationalsozialistischer Verbrechen* (*The Central Agency of the National Justice System for the Investigation of Nazi Crimes*) in Ludwigsburg that the importance of investigating these crimes was recognized in the Federal Republic, and a comprehensive prosecution policy introduced.[459] Moreover, whether in the immediate aftermath of war or in later proceedings, the evidence provided by those involved in "euthanasia" affairs, either in

454 Aly, Chroust, Pross, *Cleansing the Fatherland*, p. 55.

455 This rather strange title described legislation amending the 1933 'Law for the Prevention of Genetically Diseased Offspring' (Burleigh, *The Racial State*, p. 140).

456 Ibid., p. 49.

457 Schmidt, *Karl Brandt*, pp. 143–145.

458 de Mildt, *In the Name of the People*, pp. 80–81.

459 Ibid., p. 26.

their interrogation or under cross-examination, should be viewed with the utmost caution. Almost without exception they sought to present themselves and their living colleagues in the best possible light, and their own participation as insignificant. It is also true that few individuals can recall with certainty the exact wording of a conversation that occurred ten or twenty years earlier, where and when it happened, and who was or was not present. All of these factors need to be born in mind when considering testimony about the probable course of events.

There is also another consideration in evaluating the available evidence concerning the development of the "euthanasia" programme—the form of Nazi government. As described, Hitler ruled, not by debate and statute, but by hint and innuendo. He who had the ear of the *Führer* wielded power and influence. Hitler frequently conducted meetings with his cohorts "*unter vier Augen*", ("under four eyes"), that is to say *tête-à-tête*.[460] At such an audience, no secretaries were present, no minutes taken. If any "suggestions" were made, only Hitler and whichever other individual was present knew of them. The recipient of such a suggestion stood or fell by his ability to understand and interpret Hitler's intentions. This is what "working towards the Führer" meant in practice, and it was probably how the "euthanasia" policy developed. This anarchic method of government also provided Hitler with another important asset, what today would be termed "plausible deniability"—things done in his name could be blamed on a misinterpretation of his pronouncements, or on the misplaced initiative of others. It is an argument beloved of that peculiar breed, Holocaust "revisionists". By apparently distancing himself from the excesses of the regime, Hitler could appear to be innocent of their instigation. "If only the *Führer* knew!" was a familiar cry in Germany. In fact he knew very well.

This technique of protecting oneself against accusations of criminal activity is familiar from the world of organised crime. Layers of management are created, so that the ultimate authority, responsible for ordering or endorsing the illegal act, is distanced as far as is possible from the perpetrator. The two, overlord and executor, rarely, if ever meet. Everything can subsequently be defended or denied by the leader. Among the Nazi hierarchy, Hitler was certainly not alone in practising this strategy. It reappears time and again in the defence offered at countless Nazi war crimes trials. Perhaps one of the best examples was Albert Speer, clearly a very intelligent individual, and equally clearly a very slippery customer. To the Nuremberg IMT his denial of any knowledge of and involvement in certain aspects of the Holocaust appeared plausible, even if we know today that it was a barefaced lie. But Speer was much cleverer than many others appearing before the post-war courts. No, he didn't know, he claimed, but as a responsible member of a government, he *should* have known. This *mea culpa*, coupled with his obvious distinction in comparison to the other gangsters in the Nuremberg courtroom, saved him from the hangman's noose. Nonetheless, had the proof accessible today been available to the International Military Tribunal (IMT), Speer's sentence of 20 years' imprisonment would instead almost certainly have been the same as Fritz Sauckel's, a man whose activities were directly comparable with Speer's. Indeed, Sauckel had simp-

460 Dick van der Vat, *The Good Nazi: The Life and Lies of Albert Speer* (London: Weidenfeld & Nicolson, 1997), p. 108.

ly been acceding to Speer's demands regarding the provision of slave labour.[461] Instead of imprisonment, Sauckel's sentence was execution. So far as the implementation of eugenic policies in the Third *Reich* was concerned, there were a significant number of distinguished members of the German medical profession who also simply said, like Speer, "I didn't know" in answer to any charges. To prove otherwise was not always easy.

Of course 'plausible deniability' in the wrong hands can become 'implausible deniability', as for example in the case of Viktor Brack. He presented himself as a great humanitarian who had supported "euthanasia" out of ethical beliefs. At one point he even denied any knowledge of T4, the organisation responsible for carrying out the "euthanasia" programme, as well as supplying the majority of the killers of *Aktion Reinhard*, and an agency of which he was arguably the most important and influential member.[462] In such cases, where an overwhelming amount of documentary and eyewitness evidence was available, denial became not just implausible, but little more than farcical. Yet it is also true to say that there was a significant degree of secrecy among the Nazi leadership concerning policies. Essentially, matters were discussed and determined on a "need to know" basis. As Karl Brandt stated: "It was so with many measures...there was no reason for more men—other than the very smallest circle of people—to know about such measures. In the circles around Hitler, no one discussed his tasks and assignments with others unless the Führer himself felt it necessary or spoke about it."[463] Albert Speer was not always necessarily the most reliable chronicler of conversations or events, yet his record of comments Hitler made in 1938 has the ring of truth about it. To a party which included Philip Bouhler and his wife, Hitler stressed "the need to keep to oneself things not meant for the ears of the public." When Bouhler's wife protested that surely such restrictions did not apply to this group of the *Führer's* intimates, all of whom knew how to keep a secret, Hitler supposedly replied, pointing to Speer: "Nobody here knows how to keep his mouth shut, except for one person."[464]

There is an argument that the style of government adopted by Hitler was itself one example of the much abused theory of the "survival of the fittest". This suggests that the endless rivalry and conflict amongst his henchmen was deliberately cultivated by Hitler. Doubtless the post-war statement of Otto Dietrich, Hitler's press secretary, that the political leadership of the *Reich* operated in a state of the greatest confusion, was entirely accurate. It could not be otherwise given the level of inefficiency, jurisdictional disputes, intrigue, plotting, alliances, counter-alliances, and enmities that existed at the highest levels of Nazi government, and which Hitler seemingly encouraged. Whether this state of affairs was regarded by Hitler as a practical demonstration of Darwinism in action is feasible, but perhaps credits him with too subtle a mindset.[465] It is just as likely

---

461 Ibid., pp. 129–130, *passim*. Sauckel, who joined the SA in 1922 and the Nazi party the following year, was appointed Commissioner-General for the Deployment of Labour in March 1942. [Ernst Klee, *Das Personenlexikon zum Dritten Reich–Wer war was vor und nach 1945* (Frankfurt am Main: S.Fischer, 2005), p. 520].

462 Paul Julian Weindling, *Nazi Medicine and the Nuremberg Trials—From Medical War Crimes to Informed Consent* (Basingstoke: Palgrave Macmillan, 2006), p. 256.

463 Schmidt, *Karl Brandt*, p. 250.

464 Albert Speer, *Inside the Third Reich* (London: Sphere Books Limited, London,1971), p. 164.

465 Roderick Stackelberg, *The Routledge Companion to Nazi Germany* (London: Routledge, 2007), pp. 122–124.

that Hitler functioned in the way he did out of a fundamental disinterest in the niceties of a rational system of leadership. And, of course, for so long as confusion reigned, so did Hitler. Divide and rule, like so much of Nazism, was not an original concept. Hitler was not remotely concerned with the day-to-day administration of the state; the Great Dictator was content to beget a succession of little dictators and simply leave the mundane details of management of the nation to them. As Ian Kershaw observed: "...it is...misleading to regard the Third Reich as a dictatorship with a coherent, unitary command structure providing for the regulated and centrally directed consistent implementation of Hitler's will."[466] A degree of power may have flowed downward from him to his minions, often by implication, but the demarcation lines of authority were usually blurred and frequently overlapped, so that it is often impossible to decipher exactly who was delegated to be responsible for what.[467]

Despite this incoherent system of government, any suggestion that Hitler was somehow unaware of major policy decisions is simply inconceivable. Maintaining confidentiality may have been of paramount importance to Hitler, but he was the one man who knew all of the secrets. In a speech delivered in 1938, Hans Frank, head of the Nazi Association of Lawyers, made Hitler's omniscience quite clear. Not only was the *Führer* Head of State and leader of the government (in effect both President and Prime Minister), he was also Commander-in Chief of all armed forces and "supreme judge of the nation." Whether he governed through a formal written constitution was irrelevant; "The legal question is only whether through his activity the Führer guarantees the existence of his people."[468]

At his Nuremberg trial, Brandt quite correctly insisted that the authorization of 1 September 1939 was not an order to kill. Rather it was a dispensation for killing to occur if, in the judgement of expert medical opinion, the patient was incurably ill. It was consequently a licence, not a directive.[469] Given that so many defendants were to plead the exact opposite at future trials, that is that they were led to believe the authorization had the force of law, the Nuremberg court was wrong in rejecting Brandt's definition.

By 1938, Dr Walter Ott, head of the Austrian People's Health Office, was writing: "National Socialism does not focus on the mortal individual but on the immortal bloodstream of the race, the people. Thus eugenics and racial hygiene enter the life of our nation as a new task and model of its future.[470] Hermann Pfannmüller felt able to comment in November 1939: "The problem of whether to maintain this patient material under the most primitive conditions or to eradicate it has now become a subject for serious discussion once more."[471] As one doctor confided to her diary: "We were told nothing about the operation officially; unofficially it was assumed that everyone had somehow already been sufficiently informed." However, Werner Catel for one was under no illusions; "What we are doing here is murder," he stated.[472]

---

466 Ian Kershaw, *The Nazi Dictatorship: Problems and Perspectives of Interpretation* (London: Arnold, 1996), p. 107.

467 David Welch, *Hitler: Profile of a Dictator* (London: Routledge, 2001), p. 98.

468 Ibid., p. 46.

469 Samuel Totten, William S Parsons, Israel W Charny (eds.), *Century of Genocide: Eyewitness accounts and Critical Views* (New York: Routledge, 2004), p. 206.

470 Thomas, Beres, Sheveli, "*A Cold Wind Coming*", p. 343.

471 Burleigh, *Ethics and extermination*, p. 120.

472 Aly, Chroust, Pross, *Cleansing the Fatherland*, p. 31.

CHAPTER 5: T4

When euthanasia was introduced, we welcomed it, because it was based on the ethical principle of sympathy and had humane considerations in its favour, of the same sort that the opponents of euthanasia claim for their own ideas. - Victor Brack[473]

The handicapped and the mentally ill were not killed for mercy. They were killed because they could no longer manufacture guns in return for the food they consumed; because beds in the German hospitals were needed for wounded soldiers; because their death was the ultimate logic of the National Socialist doctrine of promoting racial superiority and the survival of the physically fit. - Michael Straight[474]

The tragedy is that the psychiatrists did not have to have an order. They acted on their own. They were not carrying out a death sentence pronounced by someone else. They were the legislators who laid down the rules for deciding who was to die; they were the administrators who worked out the procedures, provided the patients and places, and decided the methods of killing; they pronounced a sentence of life or death in every individual case; they were the executioners who carried out or—without being coerced to do so—surrendered their patients to be killed in other institutions; they supervised and often watched the slow deaths. - Fredric Wertham[475]

The formal commencement of the killing programme is generally considered to have been a meeting at the KdF on 9 October 1939 to initiate the registration of the inmates of all mental hospitals. Those present included Viktor Brack, Werner Blankenburg, Hans Hefelmannn, and Reinhold Vorberg.[476] After operating from a number of different addresses, including the Columbus-Haus on Potsdamer Platz, a permanent headquarters for the new organization was established at Tiergartenstrasse 4 in Berlin in April 1940.[477] The organization's code name, T4, was taken from this address. The premises, having been confiscated from their Jewish owners, were rented by the KdF under the name of the "*Reich* Work Group of Sanatoria and Nursing Homes" (*Reichsarbeitsgemeinschaft Heil- und Pflegeanstalten:* aka *RAG*). Other dummy organisations were set up under seemingly innocuous titles—the "Charitable Foundation for Institutional Care" (*Gemeinnützige Stiftung für Anstaltspflege:* aka *Stiftung*), the "Charitable Society for the Transportation of the Sick Limited" (*Gemeinnütziger Krankentransport GmbH:* aka *Gekrat*), and the "Central Compensation Office for Sanatoria and Nursing Homes" (*Zentralverrechnungsstelle Heil- und Pflegeanstalten*: aka *ZVST*). These names were all intended to camouflage the true purpose of the plan.[478] Despite the supposed secrecy, it should not be thought that T4 operated in a vacuum. The active co-operation of a num-

---

473 Michael S Bryant, *Confronting the "Good Death": Nazi Euthanasia on Trial 1945–1953* (Boulder: University Press of Colorado, 2005), p. 103.

474 Robert N Proctor, *Racial Hygiene: Medicine Under the Nazis* (Cambridge Massachusetts: Harvard University Press, 1989), p. 182.

475 Fredric Wertham, *A Sign for Cain: An Exploration of Human Violence*, (New York: Paperback Library, 1969, p. 161.

476 Götz Aly, Peter Chroust, Christian Pross, *Cleansing the Fatherland: Nazi Medicine and Racial Hygiene* (Baltimore: The Johns Hopkins University Press, Baltimore, 1994), p. 36.

477 Michael Burleigh, and Wolfgang Wipperman, *The Racial State: Germany 1933–1945* (Cambridge: Cambridge University Press, 1991), p. 148.

478 Eugen Kogon, Hermann Langbein, Adalbert Rückerl (eds.), *Nazi Mass Murder: A Documentary History of the Use of Poison Gas* (New Haven: Yale University Press, 1993), p. 17; Henry Friedlander, *The Origins of Nazi Genocide: From Euthanasia to the Final Solution* (Chapel Hill: University of North Carolina Press, 1995), p. 73.

ber of different government departments and agencies was necessary for the new organization to function smoothly, in particular the Ministry of the Interior and its Department for the People's Health.[479]

There was never an agreed division of authority between Bouhler and Brandt as the joint heads of T4. In fact, the "euthanasia" *Aktion* never had a clear cut *modus operandi*, as such. Not surprisingly, since the murder of the incapacitated had never been envisaged on this industrialized scale before, much was decided "on the hoof" as the programme progressed.[480] In theory, Bouhler's principal role was the management and organisation of T4's affairs, but in practice the day-to-day administration of the operation was delegated to *Hauptamt* II (Main Office II) of the KdF, headed by Viktor Brack. It was the responsible agency for all aspects of the "euthanasia" and subsequent *Aktion Reinhard* policies.[481] Brack's deputy was Werner Blankenburg; Hans Hefelmannn and Richard von Hegener, both suitably unqualified to occupy their positions, were the T4 executives responsible for children's "euthanasia". Bouhler only participated when his authority was needed to intercede or negotiate with other branches of government. Apart from functioning as a link between T4 and the medical profession, it was Brandt's job to act as a conduit between Bouhler and Hitler and to keep Hitler informed "about medical matters which arose, or advise him (on medical matters) if these remained unclear to him."[482] Although nominally joint head of T4, Brandt was far too busy to become closely involved with the humdrum details of the organisation's activities. He was still Hitler's escort physician, accompanying the *Führer* on his travels, and there was his own surgical activity to occupy him. If Brandt was interested in matters such as the financing of T4 or the choice of killing centres, he disguised it well. It appears that he was happy to rely upon Bouhler to keep him informed of developments.[483] For his part, Bouhler was prepared to leave the actual management of T4's affairs to Brack and his cohorts. Much like their master, it could be said that Bouhler and Brack only involved themselves in the "big" decisions; everything else could be left to others, although this in no way absolved either of them (or their master) from responsibility for T4's actions.

Martin Bormann was never directly involved in "euthanasia" matters. At the KdF it was considered that if he were, "euthanasia would not stop at mental patients." Bormann was less than happy at being kept outside the inner circle of managers responsible for the programme, especially since his brother Albert worked for the KdF. In fact, all the KdF head office staff involved in "euthanasia" had other work commitments; they took the job on in addition to these other employments. Despite these distractions, the T4 personnel managed to remain thoroughly devoted to the business of murder.[484] After they had been vetted by Brandt, Bouhler would arrange for new doctors to be signed up to carry out the actual killings. Since Brandt was responsible for all medical

---

479 http://www.memorialgugging.at/pdf/Czech_MedizinverbrechenGugging_engl.pdf (Accessed 18 September 2009.)

480 Ulf Schmidt, *Karl Brandt: The Nazi Doctor—Medicine and Power in the Third Reich* (London: Hambledon Continuum, 2007), pp. 133–134.

481 *Aktion Reinhard* was the code name given to the plan to exterminate the Jews of Poland, later extended to include the Jews of other European countries.

482 Schmidt, *Karl Brandt*, p. 134.

483 Ibid., p. 140.

484 Robin O'Neil, *Belzec: Stepping Stone to Genocide* (New York: JewishGen Inc, 2008), p. 43.

matters, it was he who also decided upon the appointment of the expert referees the programme required, as well as the killing method to be used.[485] Most senior personnel were between thirty and forty years of age.[486] There was no coercion so far as recruitment for the project was concerned. Individuals were simply asked if they were prepared to participate. Doctors were never ordered to kill psychiatric patients and handicapped children. Even assuming the broadest interpretation of Hitler's "euthanasia" dispensation, it was not a directive to kill, but simply empowered doctors to do so in "appropriate" circumstances.[487] This was a question of choice, not command. However, all participants were required to confirm their understanding of the need to maintain absolute confidentiality. A number were informed that a "euthanasia" law existed, but could not be shown to them for the same reasons of secrecy. How many believed in the existence of such legislation (or wanted to believe it) is a matter of conjecture. Two documents discovered after the war at the Eglfing-Haar Hospital clearly indicate the nature of the oath to which accomplices (in this case nurses) were required to swear:

> Obligation: To the Director of the Cure and Care Institute, Eglfing-Haar, *Obermedizinalrat*, Dr. Pfannmüller.
>
> I have been informed about the nature of my activity and my duties in the special department of the children's house of the Cure and Care Institute, Eglfing-Haar, where children of the [*Reich*] Committee for Scientific Approach to severe illness due to heredity and constitution are housed. I undertake to carry out my duties in this department according to the directions of my chief and I confirm that my attention has been called to the fact that the treatment of these children in this department is a matter of the [*Reich*] which has to be kept absolutely secret. I have been instructed that I have to keep strictest silence concerning all happenings of which I should become aware during the treatment of these children, and that any breach of this silence on my part will warrant the death penalty. I have given my word to maintain strictest silence and I shall adhere to this at all times and toward all people.
>
> Eglfing 26 April 1941.

The second, similar document required the signature of a clerk, who was presumably not directly involved in the killing:

> Obligation: I, the undersigned, have been obligated by handshake instead of by an oath, on the part of the director, to receive and to copy matters concerning the [*Reich*] which have to be kept secret. Such papers are of a special confidential nature. I herewith undertake to keep all papers which should become known to me under the heading "secret [*Reich*] matter" strictly secret, and never to give anyone knowledge of them without [a] specific order from the director of the institution, Dr. Pfannmüller. My attention has been called to the fact that if I should not keep this oath of secrecy, I will face penal prosecution by the *Gestapo*, and that I will have to count with the possibility of the death penalty if I should either carelessly or deliberately divulge matters which have become known to me as "secret [*Reich*] matters.[488]

To a large degree, nepotism and personal contacts dictated recruitment to T4, with few of those approached declining an invitation to become involved. It was possible to refuse to participate, or to end one's participation, as some did. So far as is known, nobody was executed or sent to a concentration camp for declining to become a partici-

---

485 Schmidt, *Karl Brandt*, pp. 134–135.
486 Aly, Chroust, Pross, *Cleansing the Fatherland*, p. 28.
487 Proctor, *Racial Hygiene*, p. 193.
488 http://www.ess.uwe.ac.uk/genocide/alexrep4.htm#personnel (Accessed 24 February 2008).

pant, or as a result of ceasing their involvement. Careers were not blighted as a consequence of such action.

All the non-medical T4 staff were initially interviewed by Brack or Blankenburg.[489] The recruitment of Dieter Allers was typical of many. Conscripted by the *Wehrmacht* in 1939, Allers, a young lawyer, was posted to Poland. Blankenburg (an old SA comrade of Allers) met Allers' mother in the street one day in November 1940.[490] On being informed by Aller's mother that her son was in the army, Blankenburg immediately offered him a job at the KdF and arranged for his discharge. In January 1941 Brack appointed Allers (in effect) managing director of T4.[491] In a speech made by Brack to a conference on 23 April 1941, the parameters for participation in "euthanasia" were set out. "Find men with courage to implement", and "nerves to endure", Brack pronounced.[492] The purpose of this meeting, organized by Brack and Werner Heyde, was to confirm to the leaders of the legal profession the existence of the "euthanasia" programme and to persuade them of its legality. There was no problem in convincing the assembly of lawyers, who willingly accepted that "every existing legal norm" was to be executed "according to the wishes of the Führer."[493] This unqualified endorsement of despotism simply confirmed that those functioning at any level within the programme shared an enthusiastic, sometimes fanatical belief in Nazi ideology, and a complete lack of any moral or ethical principles. It might be suggested that the two go hand in hand.

The majority of lower echelon staff recruited to the programme appear to have been of average intelligence. They had received a basic education, and lacked the talent, ambition, or desire to become members of the SS or the police. Their personal records and, where available, their post-war testimony, seem to indicate that they were surprised at being appointed to T4, although care should be taken in evaluating the evidence supplied personally by any of these functionaries. None had applied to join; they were ordered to do so and had no idea why, nor what they would be doing. They did not possess any immediately obvious murderous attributes; however, it may be supposed that those who chose them had excellent grounds for doing so. Most of these "ordinary men" were initially employed by T4 at jobs such as doormen, telephone operators or general handymen, and were gradually introduced into the actual killing process. There were, of course, exceptions to whom that killing came naturally, but in general the KdF, through Christian Wirth, bred a team of specialist murderers who could kill without any misgivings or, if they were concerned, managed to overcome their scruples and continued to kill anyway. The later dedicated slaughterers of European Jewry came predominantly from this "euthanasia" team of nondescript individuals in commonplace pre-T4 employment: chefs, carpenters, drivers, plumbers, photographers, and nurses.[494] T4, and subsequently *Aktion Reinhard*, were the catalysts that transformed them into a collection of brutal and insensitive murderers and provided them with a status

489 "Brack...saw everybody, including the chars [cleaners]." Allers to Sereny—[Gitta Sereny, *Into That Darkness—From Mercy Killing to Mass Murder* (London: Pimlico, 1995), p. 81].

490 SA = *Sturmabteilung*, or Storm Trooper.

491 Sereny, *Into That Darkness*, p. 79.

492 Aly, Chroust, Pross, *Cleansing the Fatherland*, p. 43.

493 Ibid., p. 62.

494 http://www.jewishgen.org/Yizkor/belzec1/bel020.html (Accessed 5 November 2006).

and a degree of power they could never have otherwise have hoped to attain. Neither should the financial benefits that they enjoyed, both in terms of salary and of theft, be overlooked, nor the considerable advantage of avoiding front-line military duty.

As already mentioned, whilst given its self-serving nature it is always wise to approach any perpetrator testimony with a healthy degree of scepticism, Franz Stangl, later to become commandant of the death camps at Sobibor and Treblinka, provided some ostensibly plausible details concerning his recruitment to T4. He claimed to have received an order in November 1940, signed by Himmler, to report to the *Stiftung* (T4) in Berlin. Stangl was then a member of the Linz *Gestapo.* The motive he provided for his eventual willingness to transfer (conflict with a senior officer), appears improbable, and it seems reasonable to assume that he was selected by his superiors for good, if unknown, reasons. He had probably been an undercover member of the illegal pre-*Anschluss* Austrian Nazi party, and had rendered useful service; he was certainly not a man who was likely to disobey an order. In Berlin he was interviewed by *Kriminalrat* Paul Werner (an SS member who was also a fully certified detective), who explained that both Russia and America had a law permitting euthanasia for people who were "hopelessly insane or monstrously deformed." It was now proposed that such a law was to be passed in Germany, but only very slowly after a great deal of "psychological preparation." In the meantime, the "euthanasia" programme had already secretly begun. Although Stangl claimed to be horrified at this revelation, his sensibilities were soothed by the assertion that merely being asked to become involved in this operation was proof of the exceptional trust others had in him. Besides, all the killing would be done by doctors and nurses; he would only be responsible for law and order. The purportedly humanitarian Stangl—he claimed to have been reassured by Werner's comments concerning the intended careful medical vetting of patients, as well as the State's general concern for the wellbeing of the population and the fact that there was already a Russian and American euthanasia law in existence—then reported to Brack at Tiergartenstrasse 4. There he was instructed as to his specific duties and eventually posted to Hartheim.[495] Allers confirmed that Stangl's job at Hartheim was the issuing of death certificates.[496]

Even before the issue of the edict, the killing of Polish patients had begun in what were to become the incorporated territories of Danzig-West Prussia and the *Warthegau*, as well as that of German patients in Pomerania[497] and East Prussia. When the Lauenberg mental hospital was dissolved in January 1940, the patients were transferred to another institution at Kosten, where they were housed overnight in an empty room. The following morning, as the nurses from Lauenberg who had accompanied the patients on their journey left, they saw soldiers in grey uniforms entering the hospital

495 Sereny, *Into That Darkness*, pp. 48–53.

496 Ibid., p. 57.

497 The *Warthegau* was a part of western Poland incorporated into the *Reich* in October 1939. In a trade-off the same month, the *Gauleiter* of Pomerania, Franz Schwede-Coburg (nicknamed 'Nero' by his constituents) agreed to allow the SS and *Wehrmacht* to use local asylums as barracks in return for first shooting the inmates. (Burleigh and Wipperman, *The Racial State*, p. 148).

grounds. On waving to the soldiers, one nurse was invited to stay behind. She declined, replying that if she did she would be shot. Because they were so close to the Russian front, the nurse had been jokingly referring to the possibility of being caught up in any fighting, but when the party reached the railway station a military official asked to see the nurse who had spoken of shootings. She was only freed with great difficulty from what was in all probability a life-threatening situation by the intervention of the accompanying doctor and the senior nurse.[498] The fate of the patients may be imagined.

The units involved in these early killings were the *Wachsturmbann Eimann* (Eimann commando), *Einsatzkommando* 16 and the *Selbstschutz* (a paramilitary organization of *Volksdeutsche*). On 22 September 1939, the Eimann commando began killing Polish patients in the mental hospital at Kocborowo, south of Danzig (Gdansk). In the following months, thousands of Polish mental patients from a number of mental hospitals were killed.[499] Initially all executions had been by shooting, but in October, Dr Albert Widmann, chief chemist of the *Kripo* (Criminal Police) suggested the use of bottled carbon monoxide as a killing agent. In his post-war testimony, Widmann stated that he had been approached by Arthur Nebe, head of the *Kripo*, and asked if Widmann's department could supply large quantities of poison. When Widmann enquired: "To kill people?" Nebe replied, "No, to kill animals in human form, that is to say the mentally ill."[500] If true, it was a typically callous and cynical Nazi response. The first gas chambers were already being constructed at Grafeneck and Brandenburg in late November, by which time gassing experiments on mental patients utilizing both carbon monoxide and an agent similar to *Zyklon B* had already taken place at Fort VII in Posen (Poznan.)

By January 1940, Polish patients were being gassed, not in stationary gas chambers, but in mobile gas vans; bottled carbon monoxide was introduced into a sealed compartment at the rear of the vehicle. The first of these vans was operated by a commando headed by Herbert Lange, formerly the chief of staff of Naumann's *Einsatzgruppe* VI. Lange graduated from this relatively small-scale murder operation to become the first commandant of the Chelmno death camp.[501] Throughout the last months of 1939 and the first months of 1940, Lange's commando travelled around the *Warthegau* in a van bearing a sign advertising *Kaiser's Kaffeegeschäft* ("Kaiser's Coffee Company"). The commando also journeyed to East Prussia, where in a period of less than three weeks more than 1,800 German and Polish patients were gassed, among them 1,558 victims from the surrounding region killed at Soldau between 21 May and 6 June 1940.[502] It has been estimated that a total of 7,700 sick and handicapped patients fell victim to this early, experimental killing phase, in which up to seventy patients received a tranquiliz-

498 Bronwyn Rebekah McFarland-Icke, *Nurses in Nazi Germany* (Chichester: Princeton University Press, 1999), p. 222.

499 Eimann, an unskilled labourer, joined the SS in 1932, rising to the rank of *Sturmbannführer* by 1939. He personally shot an unknown number of patients. In January 1941, the Eimann unit reported the killing of more than 3,000 victims to date. For these and other crimes, in 1968 Eimann received a sentence of four years imprisonment from a West German court. [Michael Burleigh, *Death and Deliverance : 'Euthanasia' in Germany c. 1900–1945* (London: Pan Books, 2002), pp. 128–129.]

500 Peter Padfield, *Himmler: Reichsführer-SS* (London: Papermac, 1990), p. 262.

501 Israel Gutman (ed), *Encyclopedia of the Holocaust* (New York: Macmillan Publishing Company, 1990), p. 283.

502 Christopher R Browning, *Fateful Months: Essays on the Emergence of the Final Solution* (New York: Holmes & Meier, 1991), p. 3.

ing injection of morphine-scopolamine before being loaded into the gassing van.[503] In total, between 1939 and 1944, a total of approximately 12,850 Polish psychiatric patients were killed.[504] Following the invasion of the former Soviet Union in 1941, thousands of mentally ill patients were murdered in places such as Riga, Dvinsk, Aglona, Poltava, Mogilev, Kiev, and elsewhere. Both members of *Einsatzgruppen* and individual physicians participated in the killing.[505] To quote a single example, *Einsatzgruppe* A shot 748 asylum inmates in Lithuania and Northern Russia since the patients had neither guards, nurses, nor food, and were considered a "danger to security."[506]

Lodz, now renamed Litzmannstadt after a First World War German general who had won a 1914 victory in the region, had been incorporated into the *Reich* as the principal city of the *Warthegau.* In March 1940, 40 Jewish patients were removed from the Lodz ghetto mental hospital and murdered in a nearby forest.[507] A little more than a year later, a commission consisting of two German doctors and the head of the Lodz *Judenrat*, Chaim Rumkowski, visited the hospital again and compiled a list of some 72 patients, 12 of whom, it was claimed, had been cured. In fact it was agreed that five of the patients could be released. The remaining 67 men and women were first sedated with scopolamine injections, then on 29 July 1941 transferred in two batches to the forest for execution.[508] It seems clear that these patients were murdered under the umbrella of the "euthanasia" programme, although by July 1941 the fact that they were Jewish was enough of itself to condemn them. Curious then, that it was felt necessary to go through the process of a medical inspection, and even more curious that five potential victims were spared and allowed to return to their home.

During the first two years of the war, the "euthanasia" programme in the *Reich* entered four distinctive but overlapping phases, relating sequentially to the killing of infants, adults, institutionalised Jews and concentration camp prisoners. Although during

---

503 Kogon, Langbein, Rückerl, *Nazi Mass Murder*, p. 37ff. The first gassings of patients from the Owinska Mental Hospital at Fort VII in Poznan were terminated no later than 15 November 1939.

504 Burleigh, *Death and Deliverance*, p. 130.

505 Dr. Wilhelm Gustav Schueppe was ordered to the Kiev Pathological Institute in September 1941 to be, first deputy head, then head of a commando of approximately twenty persons, including about ten physicians, augmented by members of the SD dressed as medical personnel. At the Institute Schueppe's commando conducted a special operation he described as the "destruction of life unworthy of life." This "unworthy life" included the handicapped as well as members of "inferior Races", that is, "Jews, Gypsies, and so forth". Schueppe estimated that during his service at the institute 110,000–140,000 victims were killed by means of lethal injection within the period of nine months between his arrival and March 1942. This was almost certainly an over-estimate. Subsequent research suggests a figure of 80,000 victims during Schueppe's stay in Kiev. Schueppe testified: "I believe in this system. It is comparable to pruning a tree, thereby removing the old undesirable branches in order to produce the highest yield. In a nation this system must be carried out to prevent decadence" (Friedlander, *The Origins of Nazi Genocide*, p. 142, p. 347, note 33); http://library2.lawschool.cornell.edu/Donovan/show.asp?id=481&query= (Accessed 20 December 2007).

506 Raul Hilberg, *The Destruction of the European Jews* (New Haven: Yale University Press, 2003), p. 341.

507 Lucjan Dobroszycki, *The Chronicle of the Lodz Ghetto* (New Haven: Yale University Press, 1984), p. 67, note 79.

508 Ibid., pp. 67–69. Another account of this incident differs somewhat in detail. This source states that it was the *Gestapo* who ordered the compilation of the list of patients; these numbered 65, rather than 72, of whom 58 rather than 67 were killed. There is no mention of any medical commission, this being entirely a *Gestapo* inspired operation. [Alan Adelson and Robert Lapides, *Lodz Ghetto: Inside a Community Under Siege* (New York: Viking, 1989),pp. 155–156]. Such confusion of detail is common in contemporary accounts. However, there can be no doubting the essentials of this matter.

the first phase young mentally ill patients were gassed together with adults,[509] 31 special wards were very quickly established in hospitals throughout the country[510] for the murder of children.[511] In time, this programme was extended from infants to older children, and in some cases even to teenagers. As is true of so many aspects of Nazi mass murder, it is impossible to be completely accurate regarding numbers of victims unless the perpetrators chose to record details of their activities and those records are still extant. Not all bodies of children were immediately cremated. As was to occur with adult victims, autopsies were sometimes conducted on children's corpses in the interests of "science" by doctors attempting to find causes for a child's disability. Autopsy activities of this kind on both adults and children provided experience for novice surgeons, who often received academic credit for their efforts.[512]

It has been estimated that by the end of the war at least 6,000 children had been killed. The actual figure is almost certainly higher.[513] In general, children were killed either by overdoses of medication, starvation, or a combination of the two. In the so-called "Luminal treatment" the sedative drug, sometimes mixed with Veronal, was administered orally, by injection, or rectally by enema. In some hospitals morphine-scopolamine was the killing agent of choice.[514] The consequences of the "Luminal treatment " were described in the following somewhat understated terms at the trial of Dr Fritz Kühnke:

> During the deep sleep brought on by Luminal, the respiration of the patients became weaker and weaker. As a consequence of the continuous insufficient breathing, regularly, after two to five days of unconsciousness, an acute pneumonia developed itself, which again no longer caused any discomfort (or even pain) to the patients, who were fast asleep. For carefully considered reasons, this pneumonia, caused intentionally, was not medically treated, as it was not intended to be cured. On the contrary: according to the previously calculated plan, it should develop even further. For this reason it was deliberately not treated. Thus, the pneumonia developed in individual cases as desired and soon caused the pre-calculated and intended death.

Methods were frequently not quite so humane. At the Kaufbeuren asylum, a 14 year-old boy suffering no major mental or physical handicap, but knowing he had been singled out for killing, handed a photograph of himself inscribed "In Remembrance" to a nurse. The boy refused to drink his Luminal laced coffee. The drug was added to the boy's food; he refused to eat it. Then he was told he was suspected of having typhoid fe-

509 Dick de Mildt, *In the Name of the People: Perpetrators of Genocide in the Reflection of Their Post-War Prosecution in West Germany. The 'Euthanasia' and 'Aktion Reinhard' Trial Cases* (The Hague: Martinus Nijhoff Publishers, 1996)., p. 67.

510 This is the number of hospitals with children's killing wards that have been identified to date (April 2008).

511 Unlike adults, children were sometimes kept under observation in these hospitals for several weeks before being murdered. (Sereny, *Into That Darkness*, p. 55 footnote).

512 Burleigh, *Death and Deliverance*, p. 146; Friedlander, *The Origins of Nazi Genocide*, p. 97–98; Kogon, Langbein, Rückerl, *Nazi Mass Murder, Mass Murder*, pp. 27–28; http://www.irren-offensive.de/kle espeech.htm (Accessed 10 January 2007).

513 Burleigh, *Death and Deliverance* , p. 114. Von Hegener estimated that approximately 100,000 children's *Meldebogen* had been submitted by the war's end. About 20,000 of these had been forwarded to Heinze, Catel, and Wentzler, of which an estimated 5,200 had resulted in the murder of the child in question. Von Hegener, of course, had good reason to minimise the number of deaths in which he had been involved. (Aly, Chroust, Pross, *Cleansing the Fatherland*, p. 190).

514 Bryant, *Confronting the "Good Death"*, p. 35.

ver, and must receive an injection. When Luminal tablets were forced upon him, he spat them out. Finally the boy was held down by two men while a nurse injected him with morphia-scopolamine. The next day the boy was dead.[515]

Ludwig Lehner, a visitor to the *Reich* Committee mental hospital at Eglfing-Haar, near Munich described the cynicism with which the director, Dr Hermann Pfannmüller, boasted of his system, openly admitting that among the children to be murdered were some who were not mentally ill. They were simply the children of Jewish parents.[516] "No Jews are allowed in my institution!" Pfannmüller proudly declared. His technique was simple. Rations were gradually reduced and consisted of nothing other than vegetables boiled in water—no bread, fats, meat or carbohydrates. For adults, a slow lingering death would occur in about three months. The starvation process could be accelerated by mixing overdoses of sedatives in the inadequate food, or by lethal injections. It required much less time to kill children. Sometimes through starvation, latterly through overdoses of luminal or injections of morphine and scopolamine, they usually died in anything from three days to three weeks.[517] Nor was Pfannmüller alone in practising such methods, for the directors of other asylums were quick to take up what its creator, Valentin Faltlhauser of Kaufbeuren-Irsee, dubbed the *Sonderkost* or *'E-Kost'* diet.[518]

Perhaps nothing illustrates the callous inhumanity of the children's "euthanasia" programme better than the case described by Götz Aly. Charlotte P., the wife of an SD officer found guilty of murder and rape by an SS court (for which he received a sentence of ten years imprisonment!), communicated with Himmler in order to thank him for his help in mitigating her husband's sentence. At the same time she took the opportunity to write of her four-year-old son: "He suffers from the effects of a severe cerebral infection. The doctors, especially Professor Bamberger at the university clinic, whom I visited at the recommendation of *Gauleiter* Erich Koch, are of the opinion that his illness is incurable. I hope, however, that I will be able to give my seriously-ill child significant relief one day." Nine days later Koch wrote to Himmler: "I spoke quite openly with Frau P. about the case. She would be grateful if the child could be brought to one of your clinics and freed from his suffering." Himmler responded immediately and passed the request to the KdF. Condemned by his own mother, Rudiger P. became one of the many victims of children's "euthanasia".[519]

The strategy for adult "euthanasia" was much more comprehensive than that which had been in place for children. The children's "euthanasia" three-man team of referees was expanded to encompass the new criteria through the willing recruitment of a number of eminent academics and asylum directors (*Gutachter*). The Interior Ministry had issued another decree on 21 September 1939, requiring medical facilities to complete

515 de Mildt, *In the Name of the People*, pp. 68–69.

516 Robert Jay Lifton, *The Nazi Doctors: Medical Killing and the Psychology of Genocide* (London: Papermac, 1990), p. 62.

517 Bryant, *Confronting the "Good Death"*, p. 195.

518 Burleigh, *Death and Deliverance*, p. 229.

519 http://www1.yadvashem.org/download/about_holocaust/studies/aly_full.pdf (Accessed 14 May 2010).

questionnaires (*Meldebogen*) in respect of all patients who (a) suffered from an extensive list of medical conditions and were unable to work; or (b) had been institutionalized for more than five years; or (c) had been committed as criminally insane: or (d) were non-"Aryan". As had been the case with children, whether or not an individual was to be condemned was initially decided by a panel of physicians.[520] To overcome suspected "sabotage" of the system and a reluctance on the part of some doctors to complete the *Meldebogen*, T4 also employed roving "Doctors' Committees" (*Ärztekommissionen*) whose job it was to visit institutions and carry out instant selections.[521] These committees were often hardly scientific. Dr Rudolph Boeckh, head physician of the asylum at Neuendettelsau, protested at the manner in which the medical commission had behaved at his establishment: "The commission did not examine one single patient out of the 1,800 concerned...They merely questioned the nursing staff, whose opinions were recorded on the forms...Cases were observed in which the opposite of the staff member's true statement was recorded on the questionnaire."[522] Others were prouder of their contribution. As a referee, Dr Josef Artur Schreck boasted of having conscientiously completed 15,000 forms within little more than a nine month period.[523] Schreck sentenced about 8,000 of these patients to death.[524] Paul Nitsche's "Doctor's Committee" registered over 900 patients in a single day, 843 of whom were subsequently gassed at Hadamar.[525]

Not all doctors were prepared to become part of the nefarious plan, although there was often a disquieting acceptance of the principle involved. When approached to become head of the children's "euthanasia" programme at Eglfing-Haar by the aforementioned Pfannmüller, Dr Friedrich Hölzel declined, writing: "It is one thing to approve of measures of the state with full conviction, and another to carry them out oneself in their final consequences. I am reminded of the difference which exists between a judge and an executioner."[526] But in general, the medical establishment was deeply implicated in the "euthanasia" programme, and many were indeed content to become willing executioners. Heinrich Bunke gave the following explanation for accepting the invitation to

---

520 Friedlander, *The Origins of Nazi Genocide*, pp. 75–77. A list of men who served as a T4 *Gutachter* at various times includes the names of forty physicians, of whom nine were university medical professors. The list also indicates the names of other doctors who were either serving at T4 head office, the killing centres, or were engaged in research at Görden or Heidelberg. Some names are included more than once under different headings [Ernst Klee, *"Euthanasie" im NS-Staat: Die "Vernichtung lebensunwerten Lebens"* (Frankfurt am Main: Fischer Taschenbuch Verlag, 1991), p. 228–229]. The forms were submitted to 54 selected psychiatrists, the elite of German psychiatry, university professors and hospital directors. [http://www.phil.gu.se/sffp/reports/16.%20von%20Cranach.pdf (Accessed 25 February 2008.)] A red plus or blue minus notation, similar to that employed in assessing children, was utilized on the forms to indicate death or life respectively. Doubtful cases were indicated with a question mark for review by senior evaluators (*Obergutachter*) [Burleigh and Wipperman, *The Racial State*, p. 144].

521 de Mildt, *In the Name of the People*, p. 59.

522 Kogon, Langbein, Rückerl, *Nazi Mass Murder*, p. 21–22.

523 Burleigh and Wipperman, *The Racial State*, p. 146.

524 de Mildt, *In the Name of the People*, p. 100.

525 Ibid., p. 59.

526 Burleigh, *Death and Deliverance*, pp. 107–108.

join T4 as a physician : "It provided the opportunity to collaborate with experienced professors, to do scientific work, and to complete my education [*Ausbildung*]."[527]

One notable exception to this overwhelming complicity in murder should be noted. Heinrich Hermann, the Swiss born warden of the Wilhelmsdorf asylum for deaf mutes, wrote to his superior in August 1940:

> I know the aim of this planned registration...I am simply convinced that the authorities are doing wrong in killing certain patients...by exterminating such a patient, or a member of the family or the asylum who is simply abnormal, we are acting against God's will...I am sorry, but we must obey God more than human beings. I am prepared to accept the consequences of my disobedience.[528]

Men of principle, like Hermann, were fighting a losing battle. By the late 1930s, both the regional health departments as well as the office of general administrator at most psychiatric institutions had been taken over by Party officials, who used their political affiliation to browbeat those who remained non-political. The non-medical middle managers, often recruited from the police, were motivated by personal considerations of advancement and ideological commitment; they were the arbiters of life and death.[529]

In total, some 200,000 *Meldebogen* were returned to T4 from various institutions. Tragically, some doctors interpreted the forms as an attempt to register suitably qualified patients for employment in the armaments industry. Hoping to save relatively healthy individuals from labour for which they were mentally unsuited, these doctors exaggerated the patient's disabilities, thereby unknowingly condemning them to death.[530] Asylum patients with criminal records became doubly at risk as medical and sociological criteria were combined, with inmates of mental institutions held in preventive detention among the earliest victims of "euthanasia."[531] In late 1939, Dr Wilhelm Bender of the asylum at Berlin-Buch complained to the Main Health Office (*Hauptgesundheitsamt* – HGA) concerning the 100 criminals who were in his institution; they were both uncooperative and expensive to care for. "I request that you consider whether the institution could be freed of the unnecessary burden of these people," he wrote. The Health Office procrastinated, but Bender found another way of resolving his problem. In June 1940 he wrote again to the HGA; there was no need to pursue the matter further, "since we can assume that it will be taken care of in the course of measures arranged by the Minister of the Interior for the methodical economization of the psychiatric institutions."[532] "Methodical economization" was coded language for "euthanasia".

Upon receipt, the forms were distributed to the *Gutachter*, who were instructed by Brack to assess the cases in the harshest terms. If there was any doubt, the patient was

---

527 Friedlander, *The Origins of Nazi Genocide*, p. 127. After a tortuous legal process lasting almost thirty years, on 28 May 1987 Bunke was convicted of having been an accessory to the murder of 11,000 individuals. For this he received a sentence of four years imprisonment. In December 1988, an appeal court reduced the assumed number of victims to 9,200, and cut the sentence proportionately to three years. [Burleigh, *Death and Deliverance*, pp. 279–280.] See the appendix to this manuscript for a brief biography of Bunke.

528 Aly, Chroust, Pross, *Cleansing the Fatherland*, p. 34.

529 http://www.jewishgen.org/Yizkor/belzec1/bel020.html (Accessed 5 November 2006).

530 de Mildt, *In the Name of the People*, p. 57

531 Aly, Chroust, Pross, *Cleansing the Fatherland*, pp. 60–61.

532 McFarland-Icke, *Nurses in Nazi Germany*, pp. 177–178.

to be registered for killing.[533] When returned to Berlin, the *Meldebogen* were dispatched to the senior evaluators, the *Obergutachter.* Initially, there was only one such higher-ranking expert, Herbert Linden, head of the health department in the Ministry of the Interior. Although a doctor of medicine Linden was no psychiatrist, but that did not prevent him from assuming the responsibility for deciding matters of survival or quietus. He was only too eager to ensure that Brack's instructions were followed to the letter, even when he was himself no longer an *Obergutachter*. In a letter dated 26 July 1940 to the *Oberpräsident* of Hannover, who had queried the parameters to be applied when determining a patient's capacity for labour (those only capable of "mechanical work" were much more likely to be killed), Linden stated:

> Regarding the general instructions as to how the questionnaires are to be filled in...I cannot agree to your proposed limitations on the notion of mechanical work. Too many rather than too few patients should be registered. Selection is to be made according to standard criteria. In case of doubt, the extent of the incapacity to work should be noted, this being a factor of great importance.[534]

Werner Heyde quickly joined Linden, who after six months was succeeded by Paul Nitsche. There were thus never more than two *Obergutachter* to sit in judgement of more than 200,000 patients, over 70,000 of whom were gassed.[535]

Unlike the children, who had been murdered in the main by lethal doses of medication or starvation, the much greater number of adults to be liquidated demanded a more efficient killing system. At Christmas 1939, Brack met with *SS-Oberscharführer* August Becker of the RSHA[536] and explained to him that it was proposed to institute an adult "euthanasia" programme.[537] After a variety of options had been considered, those responsible arrived at a suitably cheap and efficient method. Dr Richard von Hegener, chairman of the "Real Committee for the Scientific Approach to Severe Illness caused by Heredity and Constitution" and a member of the KdF testified:

> Originally the killing of single incurable persons ... by injections or an overdose of sleeping pills was proposed. But this possibility ... was considered impracticable from the technical point of view ..., by the majority of the consulted doctors. Therefore it was proposed ... to kill the selected persons by some method simultaneously in a bigger group. After many consultations ... and having heard the chemist of the *Reichskriminalpolizeiamt*, it was decided to install a room in a number of conveniently located sanatoriums. The room would be filled with carbon monoxide gas...[538]

Karl Brandt, the man charged with the responsibility for deciding such matters, was not initially enthusiastic about the use of carbon monoxide gas. It was not the "medical" solution he was looking for:

> In my mind coal oxide [carbon monoxide] had never been an instrument of medicine. I myself had learned about it a few years ago and had experienced a little coal oxide poisoning when I was in a room where there was something wrong with the stove. Then I remembered that at that time when I went to

---

533 de Mildt, *In the Name of the People* p. 337, note 45.

534 Kogon, Langbein, Rückerl, *Nazi Mass Murder*, p. 20.

535 de Mildt, *In the Name of the People*, pp. 58–59.

536 RSHA = *Reichssucherheitshauptamt*, the Central Office for Reich Security, created by Himmler in 1939 to unify the SIPO (*Sicherheitspolizei* – Security Police) with the SD (*Sicherheitsdienst* – Security Service of the SS).

537 Burleigh, *Death and Deliverance*, p. 131.

538 http://www.deathcamps.org/euthanasia/t4intro.html (Accessed 14 December 2007)

> the door, I had fallen out of the room without feeling anything and had simply passed out. So on reflection I thought that coal oxide death might be the most humane form of death. I did not come to this conclusion quickly, because it was clear to me that the moment I accepted this idea I would be bringing into being a wholly new medical concept. I thought very hard and earnestly about all this, in order to put my own conscience at rest.[539]

Despite his earlier reservations, in time Brandt became inordinately proud of his contribution to mass murder: "This is just one case of a major leap being made in medical history," he told his post-war interrogators.[540]

With Brandt's conscience salved, and utilising Albert Widmann's expert advice as well as the experience gained at Posen, it was decided to employ bottled carbon monoxide in stationary gas chambers situated in six isolated killing centres—Grafeneck (operational January 1940–December 1940), Brandenburg (January 1940–September 1940), Bernburg (November 1940–April 1943), Hadamar (January 1941–August 1941), Hartheim (May 1940–December 1944) and Pirna-Sonnenstein (June 1940–September 1942).[541] Only four of these sites were functioning at any given time. Each killing centre was assigned a code that was to be used in all telephone conversations and written communications with the T4 central offices: "A" was assigned to Grafeneck, "B" to Brandenburg, "C" to Hartheim, "D" to Sonnenstein, "Be" to Bernburg and "E" to Hadamar.

A report discovered after the war at Hartheim provided a total number of 70,273 victims for the six killing centres from inception to the issue of the so-called "stop" order in August1941.[542] This figure may be conservative. After a meeting with Bouhler on 31 January 1941, Goebbels noted in his diary: "Discussed with Bouhler the quiet liquidation of the mentally ill. 80,000 are gone. 60,000 still have to disappear." It is certain

539 Schmidt, *Karl Brandt*, p. 137. Compare this incident with Arthur Nebe's remarkably similar alleged experience: "[Nebe] had come home drunk from a party one night and passed out in his garage with his car still running. The carbon monoxide from the exhaust nearly killed him...This near miss convinced him that gassing could be used effectively against the Jews and other Nazi enemies. Gas would be cheaper than bullets, and no Nazi would directly take a life." [http://www.pbs.org/auschwitz/40-45/killing/ (Accessed 19 January 2009)].

540 Schmidt, *Karl Brandt*, p. 138. In Widmann's version, it was Conti who had first proposed using gas (ibid). It is also possible that Brandt consulted individuals such as Ferdinand Flury, a professor of pharmacology at the University of Würzburg, and an expert on poison gas, before arriving at his decision. [Paul Julian Weindling, *Nazi Medicine and the Nuremberg Trials—From Medical War Crimes to Informed Consent* (Basingstoke: Palgrave Macmillan, 2006), p. 183)].

541 Burleigh, *Death and Deliverance*, pp. 142; Friedlander, *The Origins of Nazi Genocide*, p. 89. On 14 October 1939, Grafeneck Castle near Marbach in the Münsingen district was confiscated specifically for use as a "euthanasia" centre. At Brandenburg, close to Berlin, some of the buildings of the former prison/concentration camp were acquired by T4 in 1939 for the same purpose. A mental home was founded in 1875 at Bernburg, near Magdeburg and served as a T4 replacement for Brandenburg when the latter was closed in 1940. Originally a kind of half-way house for released prisoners, the *Korrigenden-Anstalt* at Hadamar, near Limburg-an-der-Lahn in Hesse, became a mental institution in 1906. Between November 1940 and January 1941 it was converted into a killing centre, superseding Grafeneck. Hartheim castle, situated in the village of Alkoven near Linz in Austria, and formerly a home for physically and mentally handicapped children, was similarly adapted between late 1939 and early 1940. *Schloss* Sonnenstein, located at Pirna near Dresden, had been used as a mental home since 1811. Between early 1940 and April of that year, the part of the castle located in buildings 1–3 at the front of the property was converted into a "euthanasia" killing centre. See chapter six of this monograph for more comprehensive details of the six principal T4 killing centres.

542 Grafeneck 9,839, Brandenburg 9,772, Hartheim 18,269, Sonnenstein 13,720, Bernburg 8,601, Hadamar 10,072. (Friedlander, *The Origins of Nazi Genocide*, p. 109).

that by then the programme had become more ambitious in scope. Following consultations between Brandt and Bouhler, it seems likely that the targeted number of victims had now been doubled.[543] When the "stop" order was issued there were still hundreds of patients waiting in transit institutions for transportation to killing centres, which appears to indicate that the programme's objectives had by no means been achieved.[544]

Within each of the killing centres two positions were paramount. The first, the physician-in-chief, was the individual responsible for the actual gassing of victims, the maintenance of their medical records and the decision regarding their supposed cause of death. He also served to maintain the pretence that the institution was just another hospital so far as outsiders were concerned. The second, the administrative supervisor, had responsibilities which included the security and documentation of the operation, and the efficient functioning of an incident-free killing procedure. At different times, and in different centres, either one of these individuals could be the dominant figure in the establishment's activities.[545] Both types will be encountered in due course.

August Becker, who was a chemist by profession, collected the material necessary to conduct an experimental gassing at the former Brandenburg-Havel prison in December 1939 or January 1940, when 18–20 handicapped patients were gassed in a test which was decisive in arriving at the decision to use carbon monoxide gas as the killing agent for the "euthanasia" programme. It is impossible to state with certainty who attended the gassing demonstration, but there is testimony that the following persons were present: Hitler's two plenipotentiaries for "euthanasia", Karl Brandt and Philipp Bouhler, as well as Leonardo Conti, the State Secretary for Health in the Ministry of the Interior. Also in attendance were Herbert Linden from the Ministry of the Interior and the KdF representatives Viktor Brack, Werner Blankenburg, Hans Hefelmannn, Reinhold Vorberg, Richard von Hegener, and Gerhard Bohne. Further, the T4 physicians Werner Heyde, Paul Nitsche, Irmfried Eberl, Horst Schumann, and Ernst Baumhard participated in the demonstration, as well as the KTI (Criminal Technology Institute) chemists Albert Widmann and August Becker.[546] Finally, the Stuttgart police officer Christian Wirth, who was to play an essential role in Nazi killing operations, was also present during this first demonstration of systematic mass murder.

Watching through a peephole, Becker observed how very quickly "people toppled over, or lay on benches"– all without "scenes or commotions."[547] Becker noticed that Eberl had opened the gas container too quickly and the escaping gas made a hissing sound. Fearful that this noise "would make the victims uneasy," Becker demonstrated how to open the valve slowly and quietly, and "thereafter the killing of the mental patients progressed without further incidents."[548] The gas chamber was quickly ventilated, then SS-men used special stretchers to mechanically push the corpses into the

543 Schmidt, *Karl Brandt*, p. 155.

544 Ibid., p. 164.

545 Friedlander, *The Origins of Nazi Genocide*, p. 99.

546 Ibid., pp. 87–.88.

547 Lifton, *The Nazi Doctors*, p. 71.

548 Friedlander, *The Origins of Nazi Genocide* , p. 210. Brandt and Brack had explicitly ordered that a doctor should operate the gas-tap, since the *Führer* authorization referred only to doctors. [de Mildt, *In the Name of the People*, p. 61]. However, this instruction was often ignored. Doctors frequently delegated the task to others, and merely supervised the killing. [Friedlander, *The Origins of Nazi Genocide*, p. 96].

crematory ovens. Brack and Brandt both expressed their satisfaction with the test. 4–8 further inmates were administered lethal injections of curare and scopolamine, but when the poison had little immediate effect, these inmates were also gassed. The experiment was then repeated and considered a success. It was Becker who then delivered gas cylinders to the functioning killing centres and demonstrated how to operate the equipment.[549]

Gekrat was created to ship the victims to the killing centres, with a fleet of buses (usually three at each killing centre) hired from the *Reich* Post Office.[550] The *Gekrat* bus drivers were all SS-NCOs recruited from the concentration camps.[551] The steel tanks or pressurised containers containing the gas were ordered by T4 through the Chemical Department of the KTI[552] from Mannesmann *Röhrenwerke*, a manufacturer of tubes and piping in Buss-an-der-Saar. Ordered in batches of one hundred, each tank had a capacity of approximately six cubic metres of carbon monoxide. The gas itself was supplied by the Ludwigshafen plant of I G Farben (today *Badische Anilinund Sodafabriken: BASF*) to the KdF, which traded under the name "Jennerwein (Brack) and Brenner (Blankenburg)".[553] The use of code names was common for senior T4 staff.

Many of the deceptive practices that were later used in the death camps in Poland were first in evidence in the "euthanasia" programme. For example, at Hadamar, only minutes before they were taken to the gas chamber, the victims were given a stamped postcard on which to report to relatives a safe arrival at their destination.[554] Apparently, it was Bouhler who suggested that the gas chamber at Brandenburg should be disguised as a shower room, a deception that was to become a common feature of Nazi killing centres. To enhance the chicanery some victims were even provided with soap and towels to take with them into the gas chamber.[555] The man responsible for the construction of the gas chambers and crematoria at four of the killing centres[556], Erwin Lambert, subsequently put his knowledge and experience to more extensive use at Sobibor and Treblinka.[557]

With minor exceptions, a similar killing method applied at all of the centres. At his trial at Nuremberg, Viktor Brack had described the simple design of the gas chambers :

---

549 Burleigh, *Death and Deliverance*, pp. 131–132.

550 It has been suggested that postal vans were chosen because they would be relatively inconspicuous in more remote areas, and would also be easier to service and maintain nationwide. (Burleigh and Wipperman, *The Racial State*, p. 148.)

551 http://www.jewishgen.org/Yizkor/belzec1/bel020.html (Accessed 5 November 2006).

552 When Walter Heess, head of the KTI, was asked how one could justify using gas to kill human beings, he replied: "What are you talking about; after all, it works." (Friedlander, *The Origins of Nazi Genocide*, p. 86.)

553 Kogon, Langbein, Rückerl, *Nazi Mass Murder*, pp. 30–31. Some asylums, for example Hadamar, obtained supplies via local pharmacies. [Aly, Chroust, Pross, *Cleansing the Fatherland*, p. 83].

554 http://www.jewishgen.org/Yizkor/belzec1/bel020.html (Accessed 5 November 2006).

555 Michael Burleigh, *The Third Reich: A New History* (London: Pan Books, 2001), p. 393.

556 Friedlander, *The Origins of Nazi Genocide*, p. 215. Having first undertaken the partial rebuilding and renovation of Tiergartenstrasse 4, the four killing centre gas chambers Lambert constructed were at Hartheim, Sonnenstein, Bernburg and Hadamar. He was also responsible for the building works and renovation of the T4 rest home, Haus Schoberstein, at Weissenbach am Attersee.

557 de Mildt, *In the Name of the People*, pp. 266–267.

No special gas chambers were built. A room suitable in the planning of the hospital was used, a room attached to the reception ward.... That was made into a gas chamber. It was sealed, it was given special doors and windows, and then a few metres of gas pipe were laid, some kind of pipe with holes in it. Outside of this room there was a bottle, a compressed bottle, with the necessary apparatus, necessary instruments, a pressure gauge, etc.[558]

The gas chambers were of a similar size—approximately 3 metres x 5 metres x 3 metres in height. Benches lined the walls, hiding a 3 centimetre perforated pipe that encircled the chamber.

On arrival, the patients undressed and were presented to a doctor, who briefly glanced at them without conducting any actual medical examination. This was considered necessary in order to establish that the prospective victim was indeed the person registered in the accompanying documentation, and that the appropriate authorization to murder the patient had been given by the *Obergutachter.* It was also required to confirm the individual's nationality, since only German citizens were to be victims. Any exceptions to the established criteria could be singled out at this stage, as indeed a very few individuals were. The immediate murder of these fortunate souls was then cancelled. The inspection also served to reassure the patients, whilst at the same time directing the staff towards finding possible fictitious causes of death. The patients were then marked on their backs with strips of adhesive paper or coloured pencil for future identification and photographed.[559]

Usually in batches of anything of from 25–50 at a time, the patients were told they were to enter an "inhalation room" for therapeutic reasons.[560] Troublesome or resistant patients were sedated or manhandled into the chamber by brute force. Initially, in accordance with orders from the KdF, the gassing apparatus was operated by the medical director at each "euthanasia" institution, who became known as the *Vergasungsarzt* ("gassing doctor"). In less than five minutes the victims were unconscious; within ten to fifteen minutes all were dead. A hiatus of about one hour ensued, following which the chamber was ventilated and the bodies removed to either the autopsy room or to the crematorium. Before cremation occurred, where considered appropriate the victim's body was looted. In a precursor of action that was to be practised on an industrial scale at the extermination camps, any gold-filled teeth or dental bridges were extracted from the corpse's mouth for smelting. In the autopsy room, organs, particularly brains, were removed from the corpses of previously noted patients considered to be of special interest. The brains were processed by the Kaiser-Wilhelm-Institute of Neurological Research in Berlin and the Kaiser-Wilhelm-Institute of Psychiatry in Munich (today the Max-Planck-Institutes).[561]

---

558 Friedlander, *The Origins of Nazi Genocide*, p. 88.

559 de Mildt, *In the Name of the People*, p. 60.

560 Some gas chambers were later enlarged to accommodate as many as 75 victims. Sometimes the numbers gassed together were even greater. A Hartheim staff member stated: "Once 150 persons were gassed at one time. The gas chamber was so full that the people in it could scarcely fall down, and the corpses were therefore so jammed together that we could pry them apart only with great difficulty." (Friedlander, *The Origins of Nazi Genocide*, pp. 96–97.)

561 It is probable that a department for the murder of children was established at the Kaiser Wilhelm Institute for Brain Research in Berlin itself. (Aly, Chroust, Pross, *Cleansing the Fatherland*, p. 220).

As in the death camps which were to follow, disposing of the remains of the victims exceeded by far any problems arising in connection with the comparatively straightforward killing method employed. The killing centre crematoria had a restricted capacity. Following the burning of between two and eight bodies at any one time, the remaining bone fragments were crushed, and the ashes either forwarded to the deceased's next-of-kin, or simply discarded .[562] Each killing centre had a so-called special registry office, from where a standard falsified death certificate and letter of condolence were sent to victim's next-of-kin. A list of causes that could explain a sudden natural death was available. Relatives were informed that it had been necessary to cremate the body for public health reasons; the ashes of the deceased were available upon request. Nazi functionaries and German bureaucrats were stereotypically precise and efficient when it came to keeping records about mass murder.[563] Originals or copies of all paperwork generated by the killing programme were retained, including record books and lists for internal T4 use, correspondence with outsiders, and the medical records of the killed patients. To organize the paperwork, each killing centre kept a death book (*Sterbebuch*), sometimes known as the book of patients (*Krankenbuch*).[564] However, in order to enhance the secrecy of the operation, files were exchanged between the various killing centres so that for example, the death certificates of many of those who had been murdered at Hartheim were issued by Brandenburg. Consequently even today a victim's place of death is often attributed to the incorrect killing centre.[565]

Thinking that their lucky fellow inmates were being taken on an excursion, if at first some patients who were not on the list of transferees were disappointed at being left behind, they quickly realised that these were trips from which nobody returned. Consequently, subsequent transfers often resulted in traumatic scenes, as distressed patients clung to nursing staff they trusted and relied upon. On occasion patients were placed in strait-jackets or handcuffs before being loaded into the vans.[566]

In March 1941, Bouhler and Brack issued guidelines concerning the assessment of those to be considered "hereditarily ill." These instructions stated, *inter alia*, that (a) all those who were unproductive were to be liquidated, not simply those considered to be "life unworthy of life" ; (b) war veterans who had distinguished themselves at the front or had been decorated were to be excluded (which category was in the event frequently disregarded.) Other ex-servicemen were to be included in the programme, with doubtful cases referred to Brack for a decision; (c) geriatric cases were to be evaluated with the greatest care, and action taken only in extreme circumstances (for example cases of

---

562 Mostert *Useless Eaters*, p. 165.

563 Notes taken at a meeting on 23 April 1941 of the leaders of the legal profession addressed by Werner Heyde make the methodology clear, and the cynicism even clearer: "Patient dies of fabricated causes; reason: *Führer's* call for secrecy. Death certificate. Date and cause of death incorrect. In addition, however, a true registry will be kept. Now the estate is carefully processed, which is generally most important to the relatives..." (Aly, Chroust, Pross, *Cleansing the Fatherland*, p. 44).

564 Friedlander, *The Origins of Nazi Genocide*, p. 100 ff.

565 http://www.memorialgugging.at/pdf/Czech_MedizinverbrechenGugging_engl.pdf (Accessed 18 September 2009.)

566 Burleigh and Wipperman, *The Racial State*, p. 148.

criminal or "asocial" behaviour.) Again, Brack was to be the final arbiter; and (d) only *Reich* Germans were to be considered for inclusion in the programme.[567]

In the early phase of T4, victims were taken by *Gekrat* from the institutions to which they had been committed directly to a designated killing centre. However, in order to improve efficiency and at the same time disguise the murderous activity by making it more difficult for family and friends to track the whereabouts of inmates, from April-May 1940 patients were first transferred to so-called *Zwischenanstalten*, (intermediary institutions), assembly points where for several weeks they awaited onward transportation to their final destination. For example, in the province of Hessen-Nassau there were five such dedicated mental homes serving as feeder stations for Hadamar-Scheuern, Weilmünster, Herborn, Kalmenhof-Idstein, and Eichberg.[568] A typical example of the kind of notification sent by the director of an intermediate institution, in this case Kalmenhof, to the next-of-kin following the transfer of the victim to Hadamar for gassing, read:

> By Order of the Reich Defence Commissioner, your sister...was transferred on 25 July 1941 by the Charitable Society for the Transportation of the Sick [that is *Gekrat*]...to another institution whose name and address are not known to me. The receiving institution will send you a corresponding communication. I would ask you to abstain from further inquiries until this notice is received.
>
> If, however, you receive no notification from the receiving institution within fourteen days, I would recommend that you make inquiries with the Charitable Society for the Transportation of the Sick..."[569]

This delay between initial transportation and ultimate murder also provided a lacuna sufficient to enable any administrative errors to be corrected. At the same time, schedules sent to individual mental institutions containing details of the number of patients to be killed were inflated by about 25 percent, so that a list might contain 94 names, but only 65 individuals were to actually be transported. Thus the director of the asylum became not just an executor of the orders of others; he became part of the decision making process, able to remove patients from the list at his own discretion.[570] In certain respects, this can be seen as a precursor of the 'selections' which were to become so commonplace with the expansion of genocide.

Although the programme had officially been concluded in August 1941, in reality "euthanasia" never ceased, but simply continued by alternative means. Over time, the process was polished and refined and a degree of sophistication added, so that by 1944 Hans-Joachim Becker[571] of the Central Clearing Office for Mental institutions, the administrative and financial heart of T4, was able to issue precise instructions to hospitals regarding the manner in which relatives were to be notified of a patient's admission, as well as the procedure to be followed if a patient was to be discharged. Concerned relatives were to be notified by mail of a significant deterioration in a patient's condition; should a patient die, a telegram was to be immediately sent to the closest relatives stat-

567 Ibid., p. 147.

568 de Mildt, *In the Name of the People*, p. 146.

569 Horst Biesold, *Crying Hands: Eugenics and Deaf People in Nazi Germany* (Washington: Gallaudet University Press, 1999), p. 102.

570 Aly, Chroust, Pross, *Cleansing the Fatherland*, pp. 40–41.

571 Not to be confused with August Becker, mentioned above.

ing the deceased was to be buried in either the institution's or a local cemetery. In order to allow relatives to attend, the funeral should occur no earlier than four days after the date of death. The property of the deceased, including clothing and underclothing, was to be forwarded to their relatives. Objects of minor value were also permitted to be sent to the next-of-kin. However, "where objects of real value are included in the estate, I am to be consulted as to whether they should be claimed to partially cover costs."[572]

In 1939 there had been approximately 300,000 mental patients in greater Germany. By 1946 there were 40,000. Not all of these individuals fell victim to "euthanasia"; hundreds of thousands of civilians died from a variety of other causes during the Second World War. But there can be no doubt that a significant number of those who perished did so at the hands of the medical killers.[573] Evidence of this exists in both the minutes and the private notes of a meeting of the German Association of Cities held in April 1940, a convocation of local authority representatives including the mayors of all major cities, where the principal speaker was Viktor Brack. Having estimated that 30–40 percent of sanatoria and nursing home inmates were "asocial elements or unworthy of life", he was explicit about the putative benefits of "euthanasia":

> In the many mental institutions in the Reich there are an infinite number of incurably ill patients of all kinds who are completely useless to humanity; in fact they are nothing but a burden, their care creates endless expense, and there is no possibility that these people will ever become healthy or useful members of human society. They vegetate like animals, and are antisocial people unworthy of living...they only take nourishment away from other, healthy people, and often need two to three times as much care. Other people must be protected from these people.
>
> If, however, we must already make preparations for healthy people, then it is all the more necessary to first eliminate these beings, even if only to better maintain curable patients in mental hospitals. The space that would thus become free is needed for all sorts of things important to the war effort: military hospitals, regular hospitals, and auxiliary hospitals.

And how was this to be achieved? Brack continued:

> Thus...incurable patients...must be packed into very primitive special asylums...nothing must be done to maintain these seriously ill patients; on the contrary, everything must be done in order to have them die as quickly as possible...It would be best if those involved were to be placed in very bad barracks where they could contract pneumonia; in other words, accelerate their death rather than artificially maintaining them.

In these "primitive lodgings...mortality will naturally be substantially greater, especially in time of war." Brack then went on to detail the methodology to be followed regarding the disposal of corpses, the notification to the next-of-kin of the deceased, and who was to be responsible for the costs involved. In addition to the possibility of protest by the churches, concern was expressed regarding the reaction of the populace if this proposal became common knowledge. It was for these reasons that the programme would be conducted under conditions of the utmost secrecy, with cremation the preferred method of disposal of the victim's bodies.[574] Although the notes and minutes

---

572 Aly, Chroust, Pross, *Cleansing the Fatherland*, pp. 182–183.

573 Hugh Gallagher, *What the Nazi "Euthanasia Program" Can Tell Us About Disability Oppression* (Journal of Disability Policy Studies, Vol.12, No.2, 2001), p. 97.

574 Aly, Chroust, Pross, *Cleansing the Fatherland*, pp. 49–50; Biesold *Crying Hands*, pp. 161–162.

were circumspect concerning method with no mention of gas chambers, neither document was a verbatim transcript, and it is a moot point whether such matters were discussed. In any event it may be regarded as certain that those present knew exactly what any coded message about desired "increases in mortality" meant.

CHAPTER 6: THE KILLING CENTRES

Only one idea is humane, the promotion of what is good and the elimination of what is bad. The will of Nature is God's will. Just look around...Nature sides with the strong, the good and the fit and separates the chaff from the wheat. We fulfil her commandment. No more. No less. – Achim Gercke (1933)[575]

Believe me, in every case it is possible to distinguish these creatures without a soul from those who will become human. – Werner Catel[576]

It is a wise man who refrains attempting a comprehensive description of all of the killing centres associated with Nazi "euthanasia". Apart from their sheer number, there would be an unacceptable amount of repetition, for as has already been described, the extermination process was very similar at many of the locations.[577] Instead, what follows is brief information concerning the six establishments associated with the first phase of "euthanasia" (in all of which gas chambers were installed), followed by similar details of just a few of the better-known other killing institutions where there were no gas chambers. In the latter death was as a result of lethal injection or medication, starvation, disease, neglect, or a combination of any of these factors. What follows should therefore be regarded as representative rather than exhaustive.

575 Claudia Koonz, *The Nazi Conscience* (Cambridge Massachusetts: Harvard University Press, 2003), p. 166.

576 Michael Burleigh, *Death and Deliverance : 'Euthanasia' in Germany c. 1900–1945* (London: Pan Books, 2002), p. 273

577 The reader will find an encyclopaedic list of the institutions from where victims were selected at http://www.irren-offensive.de/institutions.htm (Accessed 19 August 2008).

Brandenburg
Source: © 2017 by Robert Priseman

Zuchthaus Brandenburg-Görden mainbuilding
Source: Bundesarchiv / Signature 102-11695

Brandenburg

In 1790, a home for paupers was founded in Brandenburg, near Berlin. In 1820, the complex of buildings was extended and subsequently used as a prison, until in 1932, the prison was closed following the building of a penitentiary in Brandenburg-Görden. From 24 August 1933 until 2 February 1934 the Nazis used the city-centre buildings of the former prison at 90–C Neuendorfer Strasse as a concentration camp and police barracks. Up to 1,200 prisoners were incarcerated in the cells. In 1939 some of the buildings were converted into a euthanasia killing centre named "Brandenburg State Hospital and Nursing Home" (*Heil- und Pflegeanstalt Brandenburg*).

The site was probably chosen for its convenient location, since not the least of Brandenburg's advantages was its proximity to Berlin; rapid communication with the KdF posed no problem. A gas chamber (3 x 5 metres) was created in a former brick barn, although the exact location of the gas chamber within the barn is unknown, since only the foundations of the building remain. The gas chamber was disguised as a shower room, but as initially no showerheads had been installed, patients were informed that they were entering an "inhalation room" for therapeutic reasons. Later showerheads were added to enhance the illusion that patients were to be bathed prior to their admission to their new home. Only the floor of the adjacent former storage building is now visible. The prison cells no longer exist; it was not until 1996 that their foundations were uncovered.

The killings took place in similar fashion at all of the "euthanasia" centres; buses brought the victims to the site, where they were registered, undressed, examined, photographed, and gassed. At Brandenburg there were seven SS stokers and ten SS guards who were also sometimes employed as "burners" (*Brenner*).[578] Bodies were cremated at night in two oil-fired mobile cremation ovens attached to the chimney of the building. Flames often escaped from the chimney, which added to the unpleasant smell of burning flesh, caused the crematory ovens to be moved in July 1940 to an isolated house surrounded by a high wooden fence and located about three miles outside the town. This site was camouflaged as *Chemisch-Technische Versuchsanstalt* (Chemical and Technical Research Institute); the corpses were driven there in a post office van at night.[579] As already described, the first killings at Brandenburg took place in December 1939 or January 1940, although it would appear that "formal" T4 functions commenced no earlier than February 1940.[580] The final killings took place on 29 October 1940.

A large hospital complex, Brandenburg-Görden, provided the first killing ward for children, many of whom were among the recorded victims of the newly created Brandenburg killing centre. Moreover, Görden was headed by Hans Heinze, who served as one of the three *Gutachter* for children's euthanasia. The Görden ward and the methods introduced there by Heinze served as the template for other children's wards; Görden became a training centre for physicians assigned to direct the killing of children. In addition, unlike almost all other children's wards, Görden possessed facilities sufficient to

578 Henry Friedlander, *The Origins of Nazi Genocide: From Euthanasia to the Final Solution* (Chapel Hill: University of North Carolina Press, 1995), p. 233.

579 Ibid., p. 90.

580 Ibid., p. 328, note19.

enable Heinze and his students to perform medical research on children before and after they were killed.[581] According to the testimony of Dr Heinrich Bunke, during the three months he worked there about 100 children between approximately eight and twelve to fourteen years of age were transferred from Görden and gassed at Brandenburg. Some of the bodies were dissected by Professor Julius Hallervorden, who was specifically interested in the victim's brains.[582] However, children were not the only patients observed, tested, killed, and then dissected at Görden. During the first nine months of 1942, 97 male and female patients suffering from a variety of ailments were so treated.[583]

Concern that knowledge of events at both Brandenburg and Grafeneck was arousing disquiet among the local population probably led to the eventual closure of both killing centres, but since there was still a need for such facilities, it was decided that Bernburg and Hadamar would replace them.[584] In all, at least 9,772 individuals lost their lives in the Brandenburg "euthanasia" centre within a period of less than nine months. Among the victims were more than 400 Jews. The T4 electrician, Herbert Kalisch, witnessed the murder of a transport of Jewish patients at Brandenburg in June 1940:

> We drove in six large buses of the Reich railroads to the mental hospital Buch near Berlin, and there collected about 100 women with children and about 100 men, all members of the Jewish race.... The transport went to the city of Brandenburg on the Havel, to the old prison in the centre of the city, which, being empty, had been remodelled into a crematorium. After arrival at the prison, the persons were put in cells, separated by gender. Still on the same day, immediately after arrival, about twenty persons at a time were taken from the cells. The persons were undressed completely, as they were told that they would be taken to another building for bathing and delousing. First they took women and children for gassing. To pacify these patients, physicians gave them a cursory examination. Thereafter, they were placed in a room with wooden benches, which looked, more or less, like a shower room. But before they entered the room, they were marked with consecutive numbers. The doors were locked as soon as the prescribed number of persons had entered the "shower room." At the ceiling were shower heads through which gas entered the room.[585] The gas was ventilated after fifteen to twenty minutes, as soon as one had discovered by looking through the peep-hole that all people inside were no longer alive. As the earlier examination had noted which persons had gold teeth, these persons could now be discovered by their marked number. The gold teeth were pulled from the dead people. Thereupon SS-men stationed at the prison carried the dead people from the "shower room" and took them to the crematorium. On that very day, the entire transport was eliminated in this fashion.[586]

In his 1940 pocket diary, Irmfried Eberl, as physician-in-charge at Brandenburg, recorded the details of transports arriving at the centre for gassing, frequently indicating the number, sex, and racial identity of the victims by using the capital letters "M" for men (*Männer*), "F" for women (*Frauen*), and "J" for Jews (*Juden*). It is evident from the diary that the transports carried a disproportionately large number of Jews.[587] A report

---

581 Ibid., p. 59.

582 Götz Aly, Peter Chroust, Christian Pross, *Cleansing the Fatherland: Nazi Medicine and Racial Hygiene* (Baltimore: The Johns Hopkins University Press, Baltimore, 1994), p. 225.

583 Friedlander, *The Origins of Nazi Genocide*, p. 130.

584 Ibid., p. 106

585 This statement was inaccurate—"About 4 inches above the floor, a pipe with a circumference of about one inch ran along the wall; in the pipe there were small holes through which the gas could enter the chamber." (Friedlander, *The Origins of Nazi Genocide*, p. 87). The shower heads here and in other gas chambers were solely for the purpose of camouflage.

586 Friedlander, *The Origins of Nazi Genocide*, p. 278.

587 Ibid.

produced in the same year for the Württemberg Ministry of the Interior stated that some nurses at Brandenburg had been provided with guns to prevent patients from escaping.[588]

After the cessation of the euthanasia programme the buildings served as a prison for forced labourers and a police barracks. During the war some buildings were destroyed; others were demolished after 1945. Further traces of the crimes were lost when new buildings were constructed on the site.

588 Bronwyn Rebekah McFarland-Icke, *Nurses in Nazi Germany* (Chichester: Princeton University Press, 1999)., p. 212.

Bernburg
Source: © 2017 by Robert Priseman

Bernburg killing building
Source: Wikimedia Commons / Public Domain

Bernburg

A mental home with a capacity of 132 beds was founded in 1875 in Bernburg, near Magdeburg. As a replacement for Brandenburg, in September 1940 a section of the complex was confiscated for use by *Aktion T4*. From that time onwards, Bernburg mental home was divided into both a normal institution (renamed the "Anhalt Psychiatric Clinic" under Dr Willi Enke), and a euthanasia killing site directed by Dr Irmfied Eberl.[589] Within a period of four weeks during October/November 1940, two months later than originally intended, 80 square metres of the cellar in the former *Männerhaus 2* (Men's House No.2) were converted into a killing facility. The Brandenburg staff, including the physician-in–chief, Eberl (pseudonym: "Schneider" or "Meyer"), moved as a group to the new killing centre. A small room measuring 14 square metres served as a gas chamber, and was disguised as a shower room. It was equipped with a peephole and the walls and floor were tiled. Another room was converted into a crematorium by installing two crematory ovens, with a third room utilised as a dissecting room, and a fourth as a mortuary.

The killing began on 21 November 1940 with 25 persons from the mental home at Neuruppin, which served as a *Zwischenanstalt* (Intermediate Home). As at the other killing centres, the victims were first registered, then required to undress and hand over any valuables in their possession. A superficial inspection of the victims occurred next, in order to see which plausible cause of death the Bernburg administration could pass on to the deceased's relatives. The victims were then photographed and led to the gas chamber. After gassing, the corpses were cremated and urns filled with ashes. No attempt was made to identify specific remains. Finally, if so requested, the urn containing a quantity of ashes together with a falsified death certificate was forwarded to the next-of-kin.

Victims were killed immediately. In general large grey buses were used to transport patients to Bernburg, although here and at Hartheim they sometimes arrived by train.[590] Following examination in one of several ground floor rooms, they were ushered to the basement in groups of 60–75, accompanied by nurses. Doctors marked those having distinctive physical features with a red cross on their back. After gassing, the two gas chamber doors remained closed for one hour, until the room had been ventilated. The corpses bearing the red cross were separated and were subjected to an autopsy in the adjoining dissection room. The other bodies were cremated immediately by the *Brenner*, T4 usually picking members of the SS to serve in this capacity. In Bernburg there were normally seven *Brenner*, although for a short time their number was increased to nine.[591]

A minimum of 8,601 persons were gassed in first-phase T4 operations at Bernburg, although other estimates place the number of victims somewhat higher.[592] Under the aegis of *Sonderbehandlung* 14f13, about 5,000 further individuals were killed there between 1941 and April 1943. In the main these were Jews from the concentration camps

589 Friedlander, *The Origins of Nazi Genocide*, p. 92.
590 Ibid., p. 94
591 Ibid., p. 233
592 Ibid., p. 110.

at Buchenwald, Flossenbürg, Gross-Rosen, Neuengamme, Ravensbrück and Sachsenhausen. As an example of how the system functioned, on 19 and 20 January 1942, about 214 prisoners were selected at Gross-Rosen and their questionnaires sent to T4 headquarters in Berlin, which then transmitted the list of selected prisoners to Bernburg. On 3 March, Bernburg requested the transfer of the 214 prisoners; three days later Gross-Rosen replied that the 125 prisoners who remained would be transferred on 23 March (the other prisoners originally selected by then either having died or been deemed capable of working). As a single example of the 14f13 victims, the Jewish social scientist, Dr. Käthe Leichter, gained a distinguished academic and political reputation with her research on the life conditions of maids and home workers and through her union activities. As an activist member of the Revolutionary Socialist (that is the Communist) Party, she was arrested on 30 May 1938 by the Gestapo and transferred to Ravensbrück concentration camp in January 1940. Leichter was gassed in March 1942 at Bernburg.[593]

In October 1941, Franz Stangl arrived in Bernburg from Hartheim with orders to wind up the operation. His duties consisted of "(looking) after property rights, insurance and that sort of thing. After all, some of those who died left children who had to be properly provided for. Bernburg was a mess." Stangl claimed to have no knowledge of *Sonderbehandlung* 14f13, which given his position and associates seems extremely unlikely.[594] Six months later Stangl was in Lublin, en route to taking up his position as commandant of first Sobibor and then Treblinka. Less than a year after that, a report from Bernburg dated 15 January 1943 stated: "The Public Foundation for the Maintenance of Asylums, and therefore all asylums, have been idle since 24 August 1941 [the date of the 'stop' order]. Since then, only a very small number of disinfections [that is, gassings] have occurred. A very limited number will continue to occur."[595]

The Bernburg buildings not occupied by T4 continued to operate as a normal mental institution throughout the entire T4 and 14f13 periods. Following an order from the WVHA on 27 April 1943, the crematory ovens were dismantled and the resident T4 personnel transferred to Poland. In 1949, the nurse Anna Maria L. stated that it was still possible to see benches in the basement on which the victims had waited for their "shower", as well as the sprinklers in the gas chamber and the dissecting table. During the course of works undertaken in 1988/89 in connection with the construction of a memorial to the victims, the peephole in the gas chamber wall and the gas chamber door were uncovered.

Eberl was arrested, but evaded trial by committing suicide in February 1948. He had been succeeded by Dr. Heinrich Bunke, whose locum at Bernburg was Dr. Kurt Borm. An estimated 42 personnel were directly involved in the killing process at Bernburg, with at least 20 others concerned solely with office work[596] Some members of the Bernburg staff were tried and sentenced, but most were never brought to justice.[597]

---

593 http://www.doew.at/projekte/holocaust/shoahengl/euth/euth.html (Accessed 15 January 2008).

594 Gitta Sereny, *Into That Darkness—From Mercy Killing to Mass Murder* (London: Pimlico, 1995), p. 77. Emphasis in original.

595 Aly, Chroust, Pross, *Cleansing the Fatherland*, p. 39.

596 Friedlander, *The Origins of Nazi Genocide*, p. 101

597 http://www.bernburg.meyersch.de/ (Accessed 16 January 2008).

Grafeneck
Source: © 2017 by Robert Priseman

Schloss Grafeneck
Source: Wikimedia Commons / CC-BY-SA 3.0
https://creativecommons.org/licenses/by-sa/3.0/deed.en

Grafeneck

Situated on a hill near Marbach in the Münsingen district, some 60 kilometres from Stuttgart, Grafeneck Castle was built between 1556 and 1560. Between 1762–1772, Duke Karl Eugen converted the castle to a more contemporary style. In 1929 a Samaritan foundation in Stuttgart took over the building to serve as a hospice for invalids. Early in October 1939, Dr Herbert Linden of the Health Ministry visited Dr Eugen Stähle, the official in the Württemberg Ministry of Interior responsible for health care, and asked for Stähle's cooperation in finding a relatively small institution "to implement euthanasia". Stähle offered Grafeneck, and on 14 October 1939 Grafeneck Castle was duly confiscated. 10–15 manual labourers from nearby villages began to convert the castle into a killing centre. Stähle had been a member of the Nazi party since 1927,[598] and had a boundless enthusiasm for racial hygiene and anti-Semitism, writing as he did that blood

> [was] not merely symbolic, but has also a physical, material meaning...Think what it might mean, if we could identify non-Aryans in the test tube! Then neither deception, nor baptism, nor name change, nor citizenship, and not even nasal surgery could help [the Jew escape detection]. One cannot change one's blood.[599]

A short distance from the castle several barracks were built, fenced in with a hoarding up to a height of 4 metres. An additional 5 metre screen hid the gas chamber and crematorium. On the first floor of the castle the required amenities were installed; accommodation and offices for doctors, the registrar, the police, and others. Quarters for personnel were on the second floor. The main building of the killing facility was a barrack (68 metres long and 7 metres wide), which included several rooms. In one of them 100 beds were placed, covered with straw mattresses. A wooden barrack served as the reception centre for arriving patients. An adjoining old coach house contained the gas chamber, disguised in the usual fashion as a communal bath room complete with shower heads and wooden benches. Initially the gas chamber held 40–50 persons, but was later enlarged to hold 75 victims at any one time. Next door to the gas chamber was a small room containing the equipment necessary to carry out the gassing process. From this room, a window permitted the operator to view the interior of the chamber as he released the gas.[600]

Three buses for the transportation of the victims and an ambulance car stood ready in a wooden garage, whilst two mobile oil-fuelled cremation ovens were located in another wooden barrack. Because of the immense heat generated by the round-the-clock cremations, the roof of this barrack was removed; consequently after a short time the surrounding trees were blackened from the noxious smoke and fumes emitted by the ovens. A former horse stable (circular and 15 metres in diameter) probably served as a mortuary. At the bottom of the hill, at the access road, a high hoarding and a guardhouse were built. Fences with barbed wire surrounded the whole castle, whilst armed

598 Ernst Klee, *Das Personenlexikon zum Dritten Reich–Wer war was vor und nach 1945* (Frankfurt am Main: S.Fischer, 2005), p. 594.

599 Robert N Proctor, *Racial Hygiene: Medicine Under the Nazis* (Cambridge Massachusetts: Harvard University Press, 1989), pp. 78–79.

600 Friedlander, *The Origins of Nazi Genocide*, pp. 90–91.

guards with dogs patrolled the perimeter, which bore notices carrying the warning "Danger of Infection." Grafeneck was now officially designated as a "Reich Care Institute" or "State Care Facility".

In mid-November 1939, SS-men, typists and other personnel arrived and were supplemented during early January 1940 by approximately 25 male and female nurses. In mid-January the cremation ovens were delivered, and on 18 January 1940 the first transport of 25 handicapped men arrived from Eglfing-Haar, near Munich, accompanied by the Grafeneck physician-in-chief, Dr Horst Schumann, whose stay was to be brief. In early summer 1940 Schumann was ordered to the Sonnenstein killing centre. His successor at Grafeneck was Dr Ernst Baumhard, who was succeeded in turn by Dr Günther Hennecke. The chief administrator at Grafeneck was the monstrous Christian Wirth, a detective superintendent and SS-*Obersturmführer*, who supervised the first gassings there and later was to become the principal organizer of *Aktion Reinhard*. A former nurse described a transport from her hospital to Grafeneck:

> The evening before the transport was due, we received a list with the names of the patients who were to be picked up. Early in the morning, the buses drove up; the windows were painted gray up to the top. The patients received a slip of paper with a number. Then they filed by one by one, and we wrote the number on their bare back in ink. Because they thought that we were going to transfer them to another institution, they were generally quite calm. Indeed, they did not know what was going to happen to them. Then they were led into the bus... A few weeks later their clothes were sent back from Grafeneck.[601]

Another male nurse testified about the gassing procedure:

> I don't know how long the physician let the gas flow and take effect. I only know that it was according to a precise rule which I am not acquainted with. The doors were opened and the ventilation turned on by one of the male nurses who had seen to locking up the interior. After half an hour or an hour perhaps, I can't say exactly, these male nurses received the order from the physician to open the doors and turn on the ventilation.... In the beginning it was the physician, wearing a gas mask, who opened the doors... The gas chamber would be left open for a certain time to allow fresh air to circulate. I don't know now if it was one or two hours. The oven crew was also responsible for transporting the corpses from the gas chamber to the oven.[602]

Yet another nurse provided a graphic description of the operational method at Grafeneck:

> ...On the arrival of the patients at Grafeneck, they were taken to the huts there and briefly examined by Drs. Schumann and Baumhardt on the lines of the questionnaires. These two doctors gave the final decision whether a patient was to be gassed or not. In certain cases gassing was postponed. But the majority of the patients were killed within twenty-four hours of arriving at Grafeneck. I was there nearly a year and know of only a few cases in which patients were not gassed. As a rule they were given, before gassing, an injection of 2 c.c. of morphine and scopolamine. These injections were given by the doctor. The gassing was undertaken by certain picked men. Some of the corpses were dissected by Dr. Hennecke. Some idiotic children between 6 and 13 years old were also included in the programme...[603]

Theoretically, veterans of the Great War were exempt from "euthanasia", particularly if they had been awarded medals, been wounded, or had shown exceptional bravery

601 McFarland-Icke, *Nurses in Nazi Germany*, p. 219

602 Eugen Kogon, Hermann Langbein, Adalbert Rückerl (eds.), *Nazi Mass Murder: A Documentary History of the Use of Poison Gas* (New Haven: Yale University Press, 1993), p. 28.

603 http://home.earthlink.net/~thetabus/eugenics/eutt-6.htm (Accessed 22 February 2009).

while serving. In practice, status as a disabled veteran did not always exclude the possibility of being murdered. 58 year-old Karl Rueff, who had been awarded the Iron Cross First Class, was institutionalized in southern Germany as the result of a head wound he had suffered in the Great War. He was a burden neither financially nor medically, since his disability pension paid for his institutional care, and he suffered from only occasional epileptic seizures. However, none of this prevented his being sent to Grafeneck for gassing in 1940.[604]

Despite the bloodcurdling threats concerning breaches of security, when Baumhard was head of Grafeneck he invited Dr Otto Gutekunst, director of the Winnenthal hospital, to visit Grafeneck and see for himself how the patients Gutekunst had selected for transport to Grafeneck were being treated. Gutekunst later testified:

> Of course I was interested in what was happening over there. I just could not imagine how the killing of so many people was accomplished. So I answered that I would be glad to come.
>
> The doctor (Baumhard) showed me a barrack with beds which probably were never slept in, since they were all freshly made, the gas chamber with false shower heads, and the crematory oven; apart from that I saw, in a side-room, a high mountain of ashes and pieces of bones. I remember how an employee of the institution was hammering these pieces of bones into smaller bits.
>
> After my return, I told my priest Flachsland in Winnenden that he should not use the usual formula, "I give your ashes to the ashes" at the internment of urns from Grafeneck, but rather "I give these ashes to the ashes", because they would not be the ashes of any particular deceased.[605]

The killing continued until December 1940, when Grafeneck ceased to operate, probably because of the hostility of local residents.[606] Although situated in southern Germany, Grafeneck's operational area had actually extended beyond the Austrian border into Italy. Supposedly sick ethnic German patients from the Italian provinces of Bolzano and Trentino were brought to Grafeneck from the Pergine Institute in Italy with the cooperation of the Italian authorities, and gassed.[607] Rumours that wounded members of the *Wehrmacht* were killed there seem substantiated by the inordinate number of death notices relating to members of the military which arrived in Stuttgart from the special registry office at Grafeneck. Similar rumours were also circulating about Hadamar, adding to the concerns of some members of the judiciary.[608] The supposedly secrecy of the programme was thus not much in evidence.

Following the closure of Grafeneck, some of the personnel were sent on leave, whilst others were ordered to Hadamar.[609] A few remained at the castle to cover up all traces of the murders committed there. At least 10,654 victims were gassed and cremated at this facility from more than 40 so-called "Care Facilities" in Baden, Würtemberg, and

---

604 Friedlander, *The Origins of Nazi Genocide*, p. 174

605 Gutekunst's final paragraph is a variation of 'ashes to ashes, dust to dust', difficult to translate precisely, but the general meaning is clear.

606 There is an alternative view that Grafeneck, like Brandenburg, was closed because both had completed their intended purpose and were scheduled to be shut down. (Friedlander, *The Origins of Nazi Genocide*, p. 336, note 140.)

607 Kogon, Langbein, Rückerl, *Nazi Mass Murder*, p. 18.

608 Friedlander, *The Origins of Nazi Genocide*, pp. 119–120.

609 Kogon, Langbein, Rückerl, *Nazi Mass Murder*, p. 19

Bavaria.[610] Of the 80–100 persons who carried out the "euthanasia" programme in Grafeneck, only eight were called to account for their actions; all others were untraceable. A trial was held at the Tübingen Jury Court from 8 June until 5 July 1949. At its conclusion three men were sentenced to terms of imprisonment ranging from 18 months to 5 years.[611]

610 The number of victims for Grafeneck as recorded by Edmund Brandt—(see chapter 7) was 9,839. However, the Tübingen court established that a further 815 individuals had been murdered there. (Kogon, Langbein, Rückerl, *Nazi Mass Murder*, p. 37.)

611 http://www.gedenkstaette-grafeneck.de/304.htm (Accessed 12 December 2007).

Hadamar
Source: © 2017 by Robert Priseman

Hadamar undated
Source: Chris Webb Private Archive

Hadamar

Another killing centre located on a hill, the Möchsberg, in this case overlooking the town of Hadamar, near Limburg-an-der-Lahn in Hesse, was a former *Korrigendenanstalt* (Correctional Institution for Released Prisoners) founded in 1883.[612] As the number of mentally disturbed persons in Germany increased in the early years of the twentieth century, additional medical facilities were considered necessary. Thus in 1906 the *Korrigendenanstalt* in Hadamar was converted into a mental institution, and by 1930 contained 320 patients.[613] As decreed by Nazi law, from 1934 nursing costs for mental patients were reduced, so that by 1936 all mental homes had become overcrowded, and conditions significantly worsened as a result of the reduced quantity and quality of food supplies. The building at Hadamar had been designed to accommodate 250 patients, but by 1939 about 600 inmates were crowded into its cramped quarters.

During late August 1939, following Hadamar's designation as a military hospital, patients were distributed to surrounding mental homes. However, between November 1940 and January 1941 the hospital's sickrooms were converted into quarters for the medical and administrative staff of the new T4 killing centre. The staff had been transferred from Grafeneck following the closure of that facility. Rooms to receive arriving patients were located on the first floor. A gas chamber disguised as a shower room was installed in the basement of the building, together with a crematorium with two ovens which were attached to a chimney. Gas entered the chamber from an adjacent room through pipes with holes punched in them, as in the other killing centres. Large grey buses arrived daily carrying victims from nearby mental homes at Herborn, Weilmünster, Kiedrich, Idstein, Nassau, Langenfeld, Andernach, Wiesloch, and Weinsberg (the *Zwischenanstalten* / intermediate mental homes). From the garage the victims were conducted through the so called "sluice" (a narrow fenced-in path) to the extermination building. The Hadamar "sluice" became the model for the "sluices" or "tubes", later used in the extermination camps of *Aktion Reinhard.*

Potential visitors were warned off by signs forbidding entry to Hadamar because of the danger of epidemics, but the chimney's smoke and the nauseating odour did nothing to disguise the nature of the operation from local inhabitants.[614] From the commencement of gassing operations on 13 January 1941 until August of that year, approximately 100 victims were killed every day. Walter Schmidt, physician-in-chief at the Eichberg sanatorium, testified that his superior, Friedrich Mennecke, told him that patients transferred from Eichberg to Hadamar in the morning were dead by that same night.[615] A witness testified about the killing sequence at Hadamar:

> After the doors were closed, the air was sucked out of the gas chamber through a ventilator by the same doctor who carried out the earlier "examination". Then for about ten minutes, carbon monoxide was let in [by that doctor] and its effect observed through a small window. As soon as he thought that those shut in had died, he had the gas chamber emptied. First fresh air was introduced through the ventilator, and

612 Friedlander, *The Origins of Nazi Genocide*, pp. 92–93.

613 Burleigh, *Death and Deliverance*, pp. 142–149.

614 Friedlander, *The Origins of Nazi Genocide*, pp. 92–.93.

615 Alexander Mitscherlich, and Fred Mielke, *Doctors of Infamy: The Story of the Nazi Medical Crimes* (New York: Henry Schuman, 1949), p. 102.

the gas was forced out. From the beginning of the gassing until the reopening of the gas chamber took about one hour. The corpses that were to be dissected were removed to a special room. However, the great majority of corpses were immediately taken to the ovens and burned there.[616]

A further witness described observing the victims through the peephole during a gassing:

> Through it I saw 40–45 men who were pressed together in the next room and were now slowly dying. Some lay on the ground, others had slumped down, many had their mouths open as if they could not get any more air. The form of death was so painful that one cannot talk of a humane killing, especially since many of the dead men may have had moments of clarity. I watched the process for about 2–3 minutes and then left because I could no longer bear to look and felt sick.[617]

A third witness, Maximilian Friedrich Lindner, stated:

> Did I ever watch a gassing? Dear God, unfortunately, yes. And it was all due to my curiosity.... Downstairs on the left was a short pathway, and there I looked through the window.... In the chamber there were patients, naked people, some semi-collapsed, others with their mouths terribly wide open, their chests heaving. I saw that, I have never seen anything more gruesome. I turned away, went up the steps, upstairs was a toilet. I vomited everything I had eaten. This pursued me days on end.... Looking into the chamber, I could not imagine that this was completely without pain. Of course, I am a layman and this is just my opinion. A few were lying on the ground. The spines of all the naked people protruded. Some sat on the bench with their mouth wide open, their eyes wide open, and breathing with difficulty.[618]

Some time in August 1941, a bizarre party was held in Hadamar to celebrate the gassing and cremation of the ten thousandth patient:

> ... Dr Berner declared that the ten thousandth corpse would be burnt today and that all personnel should attend. Toward evening we gathered in the hall of the right wing, where everyone was given a bottle of beer, and from there we went down into the cellar. The naked body of a dead man with hydrocephalus lay on a stretcher. In answer to a question raised, I declare with certainty that it was a real corpse and not a paper one. The "cremators" put the body in a sort of trough and pushed it into the oven. Mr. Märkle, who was dressed as a sort of clergyman, pronounced a "funeral oration"[619]

Another witness reported that the celebration, which included music, continued with a drunken procession through the grounds of the institution.[620]

Despite the half-hearted attempts the warning signs represented, in reality there appears to have been little effort to maintain secrecy at Hadamar. A contemporary account of activities there written by a member of the Frankfurt court was, by and large, remarkably accurate:

> There is constant discussion of the question of the destruction of socially unfit life—in the places where there are mental institutions, in neighbouring towns, sometimes over a large area, throughout the Rhineland, for example. The people have come to recognize the vehicles in which the patients are taken from their original institution to the intermediate institution and from there to the liquidation institution. I am told that when they see these buses even the children call out: "They're taking some more people to be gassed." From Limburg it is reported that every day from one to three buses with shades drawn pass

---

616 Robert Jay Lifton, *The Nazi Doctors: Medical Killing and the Psychology of Genocide* (London: Papermac, 1990), pp. 73–74.

617 Doris L Bergen, *War & Genocide: A Concise History of the Holocaust* (Lanham: Rowman & Littlefield, 2003), p. 127.

618 Friedlander, *The Origins of Nazi Genocide*, p. 97.

619 Kogon, Langbein, Rückerl, *Nazi Mass Murder*, p. 37

620 Lifton, *The Nazi Doctors*, p. 100 (footnote).

through on the way from Weilmünster to Hadamar, delivering inmates to the liquidation institution there. According to the stories the arrivals are immediately stripped to the skin, dressed in paper shirts, and forthwith taken to a gas chamber, where they are liquidated with hydro-cyanic acid gas and an added anaesthetic. The bodies are reported to be moved to a combustion chamber by means of a conveyor belt, six bodies to a furnace. The resulting ashes are then distributed into six urns which are shipped to the families. The heavy smoke from the crematory building is said to be visible over Hadamar every day. There is talk, furthermore, that in some cases heads and other portions of the body are removed for anatomical examination. The people working at this liquidation job in the institutions are said to be assigned from other areas and are shunned completely by the populace. These personnel are described as frequenting the bars at night and drinking heavily. Quite apart from these overt incidents that exercise the imagination of the people, they are disquieted by the question of whether old folk who have worked hard all their lives and may merely have come into their dotage are also being liquidated. There is talk that the homes for the aged are to be cleaned out too. The people are said to be waiting for legislative regulation providing some orderly method that will insure especially that the aged feeble-minded are not included in the programme.[621]

On 24 August 1941, the killing at Hadamar was suspended on Hitler's orders. By that time the bishop of Limburg was also well aware of what was happening at this particular killing centre. On 13 August 1941, he had written to the Reich Minister of Justice:

About 8 kilometres from Limburg in the little town of Hadamar, on a hill overlooking the town, there is an institution which had formerly served various purposes and of late had been used as a nursing home. This institution was renovated and furnished as a place in which, by consensus of opinion, the above-mentioned euthanasia has been systematically practiced for months—approximately since February 1941.This fact is, of course, known beyond the administrative district of Wiesbaden because death certificates from the Hadamar-Moenchberg Registry are sent to the home communities.[622]

The bishop protested that pupils of schools in Hadamar called the *Gekrat* buses "killing crates" and threatened each other with the words: "You'll end up in the Hadamar ovens!"[623] Older people were saying, "Don't send me to a state hospital! When the feeble-minded have been finished off, the next useless eaters whose turn will come are the old people." The bishop continued:

And if anybody says that Germany cannot win the war, if there is yet a just God, these expressions are not the result of a lack of love for the Fatherland but of a deep concern for our people. The population cannot grasp the fact that systematic actions are carried out which in accordance with paragraph 211 of the German Penal Code are punishable with death!"[624]

In the summer of 1942 the installations in the cellar were dismantled and other rooms converted back to sickrooms.[625] By then, many of the staff had been ordered eastward to use the experience they had acquired as members of T4 in the newly established extermination camps of *Aktion Reinhard*. It is strongly suspected, although never conclusively proven, that in winter 1941/1942, a group of 20–30 doctors, nurses, and

---

621 Leo Alexander, *Medical Science Under Dictatorship* (The New England—Journal of Medicine 241, 1949), p. 39. Alexander dates this document to December 1939, which is clearly not possible since Hadamar only functioned as a killing centre between January and August 1941. The suggested use of hydro-cyanic acid gas (*Zyklon B*) is, of course, also incorrect.

622 Samuel Totten, William S Parsons, Israel W Charny (eds.), *Century of Genocide: Eyewitness accounts and Critical Views* (New York: Routledge, 2004), p. 186.

623 Kogon, Langbein, Rückerl, *Nazi Mass Murder*, p. 32

624 Roderick,Stackelberg and Sally Anne Winkle, *The Nazi Germany Sourcebook: An Anthology of Texts* (London: Routledge, 2002), p. 333.

625 Friedlander, *The Origins of Nazi Genocide*, p. 149.

other personnel, drawn mainly from Hadamar and Sonnenstein, had conducted "euthanasia" on severely wounded German soldiers on the Russian front as part of the so-called *Organisation Todt* mission. If that had indeed been the case, some were now to use their murderous expertise on an even greater scale at Belzec, Sobibor, and Treblinka.[626]

Killing at Hadamar recommenced later in 1942 as part of the second phase of the "euthanasia" programme, the so-called "wild euthanasia." Now the victims were murdered by administering lethal doses of barbiturates or morphine-scopolamine injections. While initially there had been some pretence of medical deliberation before deciding on a patient's fate, with the arrival of Polish and Russian workers in 1944, mostly diagnosed as "tubercular" despite a complete lack of medical examination until *after* their death, the killing became automatic. When pressed at a post-war trial, the chief female nurse finally admitted that what had occurred could only be described as murder.[627]

On 15 April 1943, a ward for Jewish children of mixed marriages, so-called *Mischlingskinder*,[628] was created, camouflaged as an "educational home." This ward would eventually receive children from all parts of the *Reich*, but started by receiving children from the immediate region. The precise number of Jewish children transferred to Hadamar during 1943 and 1944 is not known. What is known is that most of these children were not unhealthy, and that they were sent to Hadamar solely because they were partly Jewish. As the regulations exempted these children from deportation, the Ministry of the Interior, not the SS or police agencies, decided to destroy those children within its control as wards of the state who were defined by the regime as Jewish. These healthy children were thereafter killed at Hadamar with lethal injections.[629]

Until late August 1942, bodies had been buried at the town's cemetery, but from September of that year mass graves located behind the building were used. During October 1944, more than 700 inmates were crowded together at the home. On 26 March 1945, US Forces captured the town of Hadamar. In the institute's pharmacy 10 kilograms of the barbiturates "Veronal" and "Luminal", used to sedate or kill the victims, were found.

Some members of Hadamar's personnel were put on trial in 1945 in Wiesbaden, others in 1947 in Frankfurt.[630] They were found to be responsible for killing approximately 10,000 persons. At the Hadamar trial of October 1945, one of the first American war crimes trials, Colonel Leon Jaworski described the murder of patients by their physicians in the following terms:

> ...They [the patients] were brought into the death halls. They were given the promise that medication would be administered to them in their ailment. Oh, what a vicious falsehood, what a terrible lie, what an

---

626 Lifton, *The Nazi Doctors*, p. 143. Dieter Allers estimated the total number of T4 personnel sent to the Soviet Union in this operation at 400 (Sereny, *Into That Darkness*, p. 84).

627 Burleigh, *Death and Deliverance*, pp. 239–238

628 The German term for people of mixed race is *Mischlinge*, inevitably interpreted by the Nazis in a pejorative sense to mean 'mongrel.'

629 Friedlander, *The Origins of Nazi Genocide*, p. 295.

630 For convenience usually referred to as the first and second Hadamar trials. The first trial was conducted by the US military, the second by the West German government.

evil and wicked thing to do to a person who is already suffering and already carrying burdens, to build up false hope that a ray of sunshine was to enter their hearts. Yes, they were given medications, of poison that gripped their heart and closed their eyelids still...[631]

At least 10,072 victims were gassed at Hadamar. To this must be added a minimum of 4,000[632] further individuals killed as a result of "wild euthanasia." To this day the building is still in use as a mental home.[633][634][635][636]

---

631 Burleigh, *Death and Deliverance*, p. 260.

632 Between August 1942 and March 1945 there were 4,422 recorded deaths at Hadamar. It is reasonable to assume that a minority of these were from natural causes. (Burleigh, *Death and Deliverance*, p. 231.)

633 Burleigh, *Death and Deliverance*, p. 143.

634 http://chgs.umn.edu/histories/documentary/hadamar/index.html (Accessed 21 January 2008).

635 http://www.hlz.hessen.de/irj/HLZ_Internet?rid=HStK_15/HLZ_Internet/nav/d60/d60172d4-d87b-1126-6841-44e9169fccd5,4f87084f-0918-1b11-2668-4144e9169fcc,,,11111111-2222-3333-4444-100000005003%26overview=true.htm&uid=d60172d4-d87b-1126-6841-44e9169fccd5 (Accessed 22 January 2008).

636 Friedlander, *The Origins of Nazi Genocide*, p. 91.

Hartheim
Source: © 2017 by Robert Priseman

Hartheim postcard from 1929
Source: Chris Webb Private Archive

Hartheim

Hartheim castle, situated in the village of Alkoven near Linz in Austria, and close to the Mauthausen concentration camp, originally dated from the ninth century.[637] The present castle, built at the end of the sixteenth century, and by 1793 the property of the princes of Starhemberg, was donated to the *Landes wohltätigkeitsverein in Oberösterreich* (Provincial Charity Club of Upper Austria) in 1898 for the establishment of a home for physically and mentally handicapped children.

After the *Anschluss*, the building was confiscated, and between late 1939 and early 1940 the castle was converted into an *Aktion* T4 killing centre with a gas chamber 5.8metres long, 3.8metres wide, and a maximum height of 2.7metres.[638]There were also at least two crematoria, one located in a room beside of the interior courtyard and one in the courtyard itself. Offices and staff quarters were located on the upper floors of the castle, while the killing installation occupied the ground floor premises surrounding the inner courtyard. As in the other killing centres, the gas chamber was disguised as a shower room and had a maximum capacity of 150 persons.[639] Senior stoker (*Oberbrenner*) Josef Vallasta often administered the gas and oversaw the gassing in place of a physician.[640]

At the western side of the castle a wooden annex was built for the reception of the victims, constructed so that their arrival remained unseen by others. During May 1940 the first 633 patients were gassed.[641] Four drivers operated the large grey buses of the *Gekrat* organisation which carried the victims to the castle. These same buses were also used for staff excursions; every two months they would transport the personnel to the cinema in Linz.[642]

Any gold teeth were extracted from the corpses prior to cremation, following which bones still visible among the ashes were pulverized into dust with a bone mill.[643] If relatives had not requested the victim's ashes, a lorry left the castle and drove to either the River Donau (Danube) or the River Traun at least once a week to scatter the remains on the water. The Austrian Vinzenz Nohel was assigned work as a *Brenner* at Hartheim, receiving wages of 170 *Reichsmarks* (RM) per month, plus a 50 RM family separation allowance, a 35 RM *Brenner* allowance, and a 35 RM bonus for keeping his mouth shut. Moreover, "because the work [as a Brenner] was very strenuous and nerve-shattering, we also received ¼ litre schnapps every day."[644] Nohel's evidence at his post-war trial

637 Kogon, Langbein, Rückerl, *Nazi Mass Murder*, p. 49.

638 Ibid.

639 Friedlander, *The Origins of Nazi Genocide*, p. 91.

640 Ibid., p. 96. This was because George Renno, assistant physician-in-charge, initially refused to operate the gas taps, considering such tasks beneath his dignity as a trained doctor. On one occasion when Renno was absent, Vallaster did not administer sufficient gas; when the chamber was opened, not all of the victims were dead. On his return Renno was furious, and for the next two months operated the gas taps himself before permitting Vallaster to perform the task again, now always under supervision. [Mireille Horsinga-Renno, *Cher oncle Georg: La bouleversante enquête d'une femme sur un médecin de la mort impuni* (Strasbourg : La Nuée Bleue, 2006), p. 59].

641 Friedlander, *The Origins of Nazi Genocide*, p. 328, note 31.

642 Horsinga, *Cher oncle Georg*, pp. 104–105 includes a photograph taken on just such an outing.

643 An electric bone crusher was in use at Hartheim. (Horsinga, *Cher oncle Georg*, p. 66).

644 Friedlander, *The Origins of Nazi Genocide*, p. 233–234.

provides perhaps the most comprehensive available description of the entire killing process:

> The mentally ill were, as far as I know, brought to Hartheim from the various asylums by rail and by car. The transports arrived at Hartheim at irregular intervals and at no specific hours....The number of arrivals varied from between forty and one hundred and fifty. First of all, they were sent to the changing room. There the men and women were made to undress or were undressed in two separate areas. Their clothes and any luggage which they had brought with them was placed in a pile, labelled, recorded and numbered. The unclothed patients then went across a corridor to the so-called registry office. There was a large table in this room and a doctor with a staff of three or four helpers. The doctor on duty there was either Dr. Lonauer or Dr. Renno. As far as I can judge as a layperson, the doctors did not examine the patients; instead they only looked at the files of those who were brought past them. Someone stamped the patients. A nurse had to stamp the individual patients on the shoulder or chest with a consecutive number...Those who had gold teeth or a gold dental bridge were marked with a cross on the back. Following this procedure the people were taken to an adjacent room where they were photographed. From the photography room they were taken through a second exit back to the registry office and from there through a steel door into the gas chamber. Initially, the gas chamber was very simply furnished ...so that one could assume it was a bathroom. There were three shower heads mounted on the ceiling...Once the entire transport had been processed, i.e. the admissions recorded, the stamping carried out and those persons marked who had gold teeth, everyone was taken to the gas chamber that was disguised as a shower room. The steel doors were closed and the doctor on duty released the gas into the chamber. Within a short time, the people in the gas chamber were dead...
>
> Taking the dead from the gas chamber to the mortuary was a very difficult and nerve-racking process. It was not easy to separate the corpses, which were stiff and tangled up in one another, and to drag them into the mortuary...In the mortuary the corpses were stacked up. The crematorium was located next to the mortuary. The ovens were fitted with a corpse caddy which could be taken out of the oven. The bodies of the dead were placed on these caddies and then pushed into the crematorium as if into a bakery oven. Depending on the number of dead we incinerated two to eight corpses at a time. The ovens were fuelled with coke. Work continued night and day as necessary. Before the dead were incinerated the stokers extracted the gold teeth of those who had previously been marked with a cross...Once the corpses had been incinerated, the remains of the bones which had fallen through the oven grid were placed in a bone crusher and ground to powder. The bone powder obtained in this fashion was then sent to the bereaved relatives as the mortal remains. Approximately 3 kilograms of such ash was calculated for each deceased person."[645]

There was an unusual association between Hartheim and the state hospital of Niedernhart, near Linz. In 1938, Rudolf Lonauer was appointed director at Niedernhart, a position he retained when he became physician-in-charge at Hartheim. Niedernhart in effect became an annexe of Hartheim, with Lonauer and his assistant Georg Renno deciding which of the arriving patients were to be gassed immediately at Hartheim and which were to be temporarily accommodated at Niedernhart.[646] A note discovered after the war graphically illustrates Niedernhart's principal function:

> The mother of the patient Maria Mader, born 16.9.1917, Mrs Anna Mader, born Rein Munich, Amalienstr 95/I, appeared today before the director of the Institution and declared:
>
> "I request immediate release of my child Maria Mader to my care. In spite of all difficulties I want to take her home. My reason is that I have learned that people who are transferred from institutions die soon and that their bodies are then cremated. Only the ashes are sent to the relatives. Since I find it unbearable to

---

645 http://www.memorialgugging.at/pdf/Czech_MedizinverbrechenGugging_engl.pdf (Accessed 17 September 2009.)

646 Friedlander, *The Origins of Nazi Genocide*, p. 109. The transit period at Niedernhart was usually of 8–10 days duration. Lonauer also killed a number of patients at Niedernhart by means of lethal injection. (Horsinga, *Cher oncle Georg*, p. 83).

have my child cremated, which is against my principles, I want to take my daughter home with me. At any rate I would like to take possession of the corpse of my daughter if anything should happen so that the child could be buried in a decent ritual manner in a cemetery. According to your advice I shall submit a petition so that I can take my daughter away with me."

Since the patient had already been scheduled for the transport of 25 April 1941, the director declined to return the patient immediately to her mother. I put it at the discretion of the direction of the Cure and Care Institution in Niedernhart-Linz, as transition institution, to possibly return the patient to the mother upon request. The mother threatened immediate suicide if she learned that her child, Maria, was cremated.

Eglfing 23 April 1941.

Signed: Direktion her Heil- und Pflegeanstalt des Bezirksverbandes Oberbayern Eglfing-Haar. [Hermann Pfannmüller]

As a consequence of her mother's intervention, Maria Mader was held back from the transport scheduled for 25 April 1941, but she was not saved for long. Her name was added in handwriting to the bottom of a list of victims sent to Hartheim four days later, on 29 April 1941.[647]

Although it is impossible to calculate a definitive figure for the number of victims in respect of any of the euthanasia centres, it has been estimated that by August 1941 the death toll at Hartheim had risen to at least 18,269; later, during the *Sonderbehandlung* 14f13 operation, a minimum of 12,000 further victims were murdered,[648] among them more than 8,000 inmates of the concentration camps Dachau and Mauthausen / Gusen. These included Russian prisoners-of-war, Hungarian Jews, Jehovah's Witnesses and other *Muselmänner*.[649]

The first prisoners selected from Dachau (some of whom had voluntarily registered for transfer to a "convalescent camp" or "sanatorium") arrived at Hartheim no later than January 1942. When the personal effects of transferees were returned to the camp with the address of the sender marked as "Hartheim", the voluntary registrations ceased. Prisoners were also transferred to Hartheim from the Gross-Rosen, Sachsenhausen, and Ravensbrück concentration camps.[650] They all shared a common fate. After late August 1941, when Hitler ordered a cessation to the gassing of the mentally and physically disabled, only concentration camp prisoners were gassed in T4 killing centres.

During the summer of 1943 Hartheim became an administrative department of *Aktion T4*, since Berlin was no longer considered safe in view of the increasing frequency and intensity of bombing raids. Another administrative department was also installed at the *Aktion* T4 recreation facility "*Haus Schoberstein*" in Weissenbach, at Lake Attersee in Austria.[651] This villa had initially been designated as a recreation facility for T4

---

647 http://www.ess.uwe.ac.uk/genocide/alexrep2.htm (Accessed 24 August 2008).

648 Estimates of the total number of victims at Hartheim range from 20,000 (Georg Renno, sometime director of the institution) to 400,000 (Franz Ziereis, former commandant of Mauthausen concentration camp.) The latter figure is clearly impossible. The generally accepted figure is 30,000 fatalities. (Lifton, *The Nazi Doctors*, p. 142.)

649 Lifton, *The Nazi Doctors*, p. 138.

650 Kogon, Langbein, Rückerl, *Nazi Mass Murder*, p. 47.

651 http://www.chgs.umn.edu/histories/documentary/hadamar/sterilization3.html (Accessed 1 February 2008).

personnel, including the staff of *Aktion Reinhard*, who were often to be found there when on leave.

As seemingly occurred at all of the killing centres, the local population soon became aware of the nature of the activities being carried on there. After the arrival of one of the grey buses at Hartheim, a dark cloud always appeared over the castle. On days when there was a low cloud base, smoke smelling of burnt flesh and hair would spread over the town of Hartheim, causing the townsfolk to be physically sick. The fact that many people were brought to the castle, but nobody left, resulted in rumours spreading. To reassure the citizens, the Hartheim authorities arranged communal meetings. The townspeople were told that the vile odours had been caused by the use of contaminated oil. A warning was issued against persons speculating to the contrary, but it was impossible to completely hide the evidence of the killing. Hartheim's function quickly became common local knowledge.

The last 14f13 gassing at Hartheim occurred on 11 December 1944.[652] During December 1944 and January 1945 inmates of the Mauthausen concentration camp dismantled and removed the gassing installations,[653] and Hartheim again became a normal nursing-home for mentally ill patients.[654] Although most of the T4 documentation was destroyed toward the end of the war, a U.S. War Crimes Investigating Team discovered a box in Hartheim containing statistical reports on the number of patients killed. This data has formed the basis of many subsequent attempts to establish the number of T4 victims.[655]

In 1954 the rooms of the castle were temporarily converted into apartments for rent, and in 1969 a memorial was installed in the part of the building used for killing purposes. The "*Verein Schloss Hartheim*" was founded in 1995 with the objective of perpetuating the history of Nazi "euthanasia" at Hartheim, and establishing an appropriate museum at the castle. In 2001, further significant traces of the "euthanasia" process were uncovered.[656]

---

652 Kogon, Langbein, Rückerl, *Nazi Mass Murder*, p. 49.

653 Friedlander, *The Origins of Nazi Genocide*, p. 150.

654 Dick de Mildt, *In the Name of the People: Perpetrators of Genocide in the Reflection of Their Post-War Prosecution in West Germany. The 'Euthanasia' and 'Aktion Reinhard' Trial Cases* (The Hague: Martinus Nijhoff Publishers, 1996), p. 78.

655 Burleigh, *Death and Deliverance*, p. 154. This was the Brandt report referred to in chapter 7.

656 http://www.schloss-hartheim.at/index.asp?Seite=441&Lg=1 (Accessed 12 January 2008).

Sonnenstein
Source: © 2017 by Robert Priseman

Pirna Schloss Sonnenstein postcard from 1938
Source: Tall Trees Archive UK

Pirna Haus C 16
Source: Tall Trees Archive UK

Sonnenstein

Located at Pirna near Dresden, above the river Elbe, *Schloss* Sonnenstein was built on the site of a former medieval castle and had been used as a mental home since 1811. It was the first major state institution in Germany to be explicitly dedicated to treating, rather than just interning, mentally ill patients. In the nineteenth century, the reform-oriented hospital was one of the most renowned institutions of its kind in Europe. In 1922 as a *Heil und Pflegeananstalt* (Care and Cure institution) it had housed 672 psychiatric patients. By 1939, the highly respected psychiatrist, Paul Nitsche, had been in charge of the Sonnenstein state hospital for more than ten years.

In October 1939 the mental home was officially closed. Part of the institution, situated behind the main buildings and excluded from the area allocated to the killing centre, was used first as a military hospital and then to house *Volksdeutsche* refugees from Bessarabia. Another section, also at the rear of the property, remained a regular mental hospital and assumed the name "Mariaheim". Between early 1940 and June of that year, that part of the castle located in buildings 1–3 at the front of the property was converted into the "euthanasia" centre. Building 3 served as a reception area for the arriving victims and buildings 1–2 provided office space and staff quarters. Building 2 (now numbered C 16) contained the actual killing facility, including the gas chamber and a crematorium with two stationary coke-fired ovens located in the basement.

A high brick-wall on two sides of the complex shielded it from outside observation, while a tall hoarding was erected to serve a similar purpose on the other two sides. Sleeping quarters for the *Brenner* were provided in the attic of building C 16. The physician-in-charge, Horst Schumann, who transferred from Grafeneck, arrived in late April to supervise the final renovations in time for killing to commence in June 1940. The Sonnenstein "euthanasia" facility was to differ from most of the other killing centres because it did not occupy the entire hospital, which made secrecy even more difficult than it was to prove elsewhere.[657]

From the end of June 1940 until September 1942, at least 15,000 persons, probably many more, were killed within the scope of the "euthanasia" programme and *Sonderbehandlung* 14f13 at Sonnenstein; a minimum of 13,720 patients from various institutions alone had fallen victim to T4 during the first phase of the killing.[658] One nurse testified that during his three week stay in July 1940 about 1,000 people had been gassed, including youngsters aged fifteen and sixteen.[659] Well before gassing experiments had started at Auschwitz in late summer 1941, 575 prisoners from that camp were transferred to Sonnenstein for extermination on 28 July 1941.[660] Other prisoners also arrived in 1941 and 1942 from Ravensbrück, Sachsenhausen, and Buchenwald concentration camps. A prisoner working as a physician's secretary in Buchenwald was witness to the fate of 190 prisoners gassed at Sonnenstein in July 1941:

> ... A long list of prisoners, identified by their numbers, was ordered over the loudspeakers to report to the camp gate... The next day and the day after that, these prisoners were taken away from the camp. When they left we didn't know where they were being taken. Some days later the hospital orderly, Wilhelm, came in with some personal effects... false teeth, spectacles, crutches. We knew then for certain the prisoners had been killed... Shortly afterward the SS camp physician appeared... He gave me a list of prisoners who had been transferred and killed and told me to cross them off the hospital register and to remove the file cards. Hoven [the SS doctor] confirmed that the prisoners of both convoys were dead and that I should prepare the official death certificates. I asked what cause of death I should mention. He told me I should go and find something in the medical dictionary...[661]

On one occasion, the problem of flames escaping from the chimney required the expert advice of the KTI. It was concluded that the chimney had not been built correctly, and that too many corpses were being cremated at the same time. This kind of hitch was not uncommon; similar incidents occurred at Hartheim and Brandenburg, in the latter case requiring the crematorium to be removed to the outskirts of town.[662]

The Sonnenstein staff consisted of approximately 100 persons. Based upon their expertise in the deception, gassing, and burning of innocent victims. about one third of them were transferred to the extermination camps in occupied Poland. During August/September 1942, the Sonnenstein killing centre was liquidated and incriminating material such as the gas chamber installations and crematorium ovens dismantled.

657 Friedlander, *The Origins of Nazi Genocide*, pp. 91–92.
658 Kogon, Langbein, Rückerl, *Nazi Mass Murder*, p. 37.
659 Ibid., p. 19.
660 Ibid., p. 45, p. 51.
661 Ibid., p. 46.
662 Friedlander, *The Origins of Nazi Genocide*, p. 98.

From October 1942 until 1945 the buildings were used as a military hospital, and in the post-war years served a number of different local government and commercial functions. Out of public sight a large commercial plant was installed which utilised the buildings that had once housed the killing facilities.

In the summer of 1947, some T4 staff were among those accused in the *Dresdner Ärzteprozess* ("Doctors' Trial" in Dresden). Paul Nitsche, T4 *Obergutachter*, was sentenced to death, together with two male nurses from Sonnenstein. 40 years of silence followed, before the part Sonnenstein had played in the euthanasia programme was recognized. It was not until 1989 that the public commemoration of the history of this "euthanasia" centre occurred.

After reunification in 1989, an employee of the Tolkewitz municipal cemetery in Dresden discovered three mass graves containing the urns of "euthanasia" victims. They were exhumed by the German war graves agency, and laid to rest in a more dignified tomb in the Dresden-Tolkewitz cemetery in January, 2004. Since 9 June 2000 a permanent exhibition has been on display in the rooms of building C 16. The fate of twenty-two representative victims murdered at Sonnenstein are documented in a memorial tract in the basement of the building.[663][664]

Other Killing Centres

There were eventually to be more than one hundred hospitals, asylums, and medical facilities across the *Reich* where "euthanasia" was practiced.[665] In more than thirty of these institutions there was a ward dedicated to the killing of children.[666]

It is beyond the scope of this work to describe all of the centres where state-sanctioned murder occurred. Instead, the following is intended to provide brief descriptions of a few of the major establishments. In general terms, operational methods were common to all. Obviously, staffing levels were dependant upon the size of the institution.

Meseritz-Obrawalde

Previously situated in Posen/West Prussia, in 1939, the town of Meseritz was brought within the Prussian province of Pomerania. Today the town bears the name

---

663 http://www.stsg.de/main/pirna/ueberblick/einfuehrung/ (Accessed 21 January 2008).

664 http://www.geschichte-pirna.de/euthanasie.htm (Accessed 22 January 2008).

665 Mark P Mostert, *Useless Eaters: Disability as Genocidal Marker in Nazi Germany* (Journal of Special Education. Vol. 36, No. 3, 2002, p. 165.

666 Marius Turda and Paul J Weindling (eds.), "*Blood and Homeland*". *Eugenics and Racial Nationalism in Central and Southeast Europe, 1900–1940*, (Budapest: Central European University Press, 2007), p. 329.

To date, killing wards for children have been identified in the following 31 locations: Ansbach, Berlin-Wiesengrund, Brandenburg-Görden, Breslau, Dortmund Applerbeck, Egelfing-Haar, Eichberg bei Eltville, Graz, Grossschweidnitz bei Löbau, Hamburg-Langenhorn, Hamburg-Rotenburgsort, Kalmenhof-Idstein, Kaufbeuren, Königsberg, Konradstein, Leipzig-Dösen, Leipzig-Universitätskinderklinik, Loben/Oberschlesien, Lüneburg, Niedermarsberg, Sachsenberg bei Schwerin, Schleswig, Stadtrhoda in Türingen, Stuttgart, Tiegenhof/Gnesen, Uchtspringe, Ueckermünde, Waldniel, Wien, Wiesengrund bei Pilsnen, Wiesloch bei Heidelberg. [http://www.bundesarchiv.de/geschichte_euthanasie/Inventar_euth_doe.pdf (Accessed 27 February 2008.)]

Miedzyrzecz and is situated in Poland. The hospital at Obrawalde (now Obrzyce)[667] completed in 1904 and usually referred to as Meseritz-Obrawalde, is today considered one of the most notorious killing centres of so-called "wild euthanasia".[668]

In the early 1930s, Meseritz-Obrawalde was a general hospital enjoying a good reputation. However, the hospital's non-psychiatric facilities, which included internal medicine, neurology and obstetrics departments, were closed in 1938.[669] Meseritz-Obrawalde had been built to accommodate 700 mental patients and held 900 such cases in 1939, but within a year the institution was more than filled to capacity with 2,000 inmates under the care of just three physicians—Drs Theophil Mootz, Hermann Vollheim, and Hilde Wernicke.[670] According to the evidence of a former employee, who left Meseritz-Obrawalde in 1939, the hospital was

> the second or third newest such institution in all of Germany at that time, had modern equipment, and was run according to correct medical principles. Nurses and also other staff members very much enjoyed carrying out their duties in Obrawalde back then because Obrawalde was considered more or less a model institution.

In contrast, a former patient claimed that absolutely no medical treatment occurred at Meseritz-Obrawalde during his stay there in the last year of the war.[671] Killing had replaced therapy.

During the period preceding the suspension of the "euthanasia" programme in August 1941, large numbers of patients had been transferred from Meseritz-Obrawalde "to the east" and had, like patients from other institutions in the region, simply disappeared. At the beginning of 1942, at which time "euthanasia" had entered its so-called "wild" phase, the first trains, each containing about 700 handicapped patients, arrived at the hospital. They were eventually to be transported to Meseritz-Obrawalde from at least twenty-six German cities, frequently arriving in the middle of the night.[672] At the end of that year, and more particularly during 1943, these trains arrived with increasing regularity. All the nurses and orderlies had to assist in the unloading of the sick patients, who were in a deplorable condition; many were emaciated and very dirty. As Wernicke testified:

> Only one who has witnessed once the arrival of such a transport can have an idea of what we were dealing with. The patients were half-starved, often totally covered in their own dirt, exhausted. Many were crippled and helpless after journeys that sometimes took several days. Some died during the travel and others soon thereafter. We very often had to be on our feet day and night with only a few hours of rest in order to take care of the arrivals. Because of the crowded conditions, it was almost impossible to fight problems like lice, scabies, and dysentery.[673]

---

667 Susan Benedict, Arthur Caplan, Traute Lafrenz Page, *Duty and 'Euthanasia': the Nurses of Meseritz-Obrawalde* (Nursing Ethics, Vol. 14(6), 2007), p. 783

668 McFarland-Icke, *Nurses in Nazi Germany*, p. 17.

669 Susan Benedict, and Tessa Chelouche, *Meseritz-Obrawalde: a 'wild euthanasia' hospital of Nazi Germany* (History of Psychiatry, Vol. 19 (1), 2008), p. 70.

670 Benedict, Caplan, Lafrenz, *Duty and 'Euthanasia'*, p. 783

671 McFarland-Icke, *Nurses in Nazi Germany*, p. 213.

672 Friedlander, *The Origins of Nazi Genocide*, p. 160.

673 Benedict, Chelouche, *Meseritz-Obrawalde*, pp. 70–71.

It would appear that these abysmal conditions contributed to the nursing personnel's ability to distance themselves emotionally from those entrusted to their care. The patients were in such an undignified condition that the staff could be convinced to kill thousands of them without compunction. To coin a phrase, they were less than human.[674]

They were helped in this attitude by the appearance and mind-set of their superiors. Appointed in November 1941, Walter Grabowski, the first non-medical director at Meseritz-Obrawalde, patrolled the wards dressed in hunting attire, accompanied by a large dog.[675] Conditions at the hospital rapidly deteriorated under his administration. Apart from chronic overcrowding, there was a shortage of medications and cleaning materials. Even the well supplying the hospital's water began to run dry.[676] It was Grabowski who instituted "euthanasia" at the hospital in late 1942[677] or early 1943,[678] reputedly terrifying the institution's staff by frequently holding roll-calls at which he impressed upon them their duties and obligations to everyone except their patients. Soldiers were protecting them by doing their duty at the front, he proclaimed; they were equally required to do their duty at Meseritz-Obrawalde.[679] Any infraction of his orders would result in severe punishment, he threatened. In fact, for all of his bluster, as with so many bullies, it seems that when faced with determined opposition, Grabowski simply backed down. But the threats were often enough in themselves for most employees—or so they later said. One nurse explained: "If I had not followed official orders back then, I was certain that I would have been subject to reprisals from Grabowski. What kind of reprisals these might have been, I could not imagine, but without a doubt I had something to be afraid of."[680] This may or may not have been true. In any event, it certainly provided the necessary degree of self-justification required to follow unpalatable commands.

Mootz and Wernicke were responsible for selecting the condemned, sometimes after reviewing a patient's file or a cursory medical examination,[681] although in practice those selected for killing included "patients who caused extra work for the nurses, those who were deaf-mute, ill, obstructive, or undisciplined, and anyone else who was simply annoying." The choice of victim was completely arbitrary, and also included some "who had fled and were recaptured, and those engaging in undesirable sexual liaisons."[682] The actual killing was done by nurses and 'carers', rather than any of the three doctors.[683] For those not selected for death, conditions in the hospital were appalling. A former inmate testified:

---

674 Benedict, Caplan, Lafrenz, *Duty and 'Euthanasia'*, p. 783

675 McFarland-Icke, *Nurses in Nazi Germany*, p. 215.

676 Benedict, Chelouche, *Meseritz-Obrawalde*, p. 71.

677 Benedict, Caplan, Lafrenz, *Duty and 'Euthanasia'*, p. 783

678 Michael S Bryant, *Confronting the "Good Death": Nazi Euthanasia on Trial 1945–1953* (Boulder: University Press of Colorado, 2005), p. 118. A former salesman, Grabowski, described as an "unbearable person and an absolute despot", was rumoured to have been involved in the murder of Jews in Kalisch before his appointment at Meseritz-Obrawalde. (McFarland-Icke, *Nurses*, p. 214).

679 McFarland-Icke, *Nurses in Nazi Germany*, p. 249.

680 Ibid., pp. 242–245.

681 Benedict, Caplan, Lafrenz, *Duty and 'Euthanasia'*, p. 783 .

682 Friedlander, *The Origins of Nazi Genocide*, p. 160.

683 Benedict, Caplan, Lafrenz, *Duty and 'Euthanasia'*, p. 783.

> The people lay in metal frame beds on top of each other, on straw sacks. The privy lay open between the rows of beds...Here there was only a minimum to eat. In the morning there was bread and coffee. At midday there was watery soup with a few globules of fat for the workers. If one was lucky, there were a few more globules, but it was all basically water, so that one could not be nourished by it.

For non-workers, rations were even worse. Their "soup" consisted of potato peelings in water. Relatives were prohibited from sending food to patients, who, so the relatives were informed, had enough to eat.[684]

The starving patients selected daily for liquidation were taken to so-called "killing rooms" (in Houses 1,3,6,8,9,10, and18),[685] where they were murdered by means of an orally administered drug overdose or a lethal injection. Normally ten tablets of Veronal or Luminal were dissolved in water. If the patient was unable or unwilling to drink the medication, they were injected with morphine and scopolamine instead. Sometimes a simple injection of air into a vein was sufficient to kill within minutes.[686] There is evidence that not all patients were so benumbed that they were prepared to be murdered without a struggle.[687] Because of this, and for the psychological relief of the perpetrators, the killing of patients was not usually undertaken by a single nurse; experience had shown that it was more efficient and less stressful (for the perpetrators, that is) that it be done by two or more nurses.[688] As one nurse described the process:

> In general either the ward caregiver or I would sit the patient up in her bed, put an arm around her, and talk to her consolingly. So one of us would hold the patient in an upright position and the other caregiver would hold onto the glass with the medication. Then the patient either was able to swallow the liquid on her own, or it was given to her with a spoon. If the patient was extremely restless, which also happened quite frequently, then three caregivers were needed for the procedure.

The patients would quickly fall asleep and usually be dead within half a day. Those receiving injections of morphine and scopolamine died within a few hours.[689] After the patient had been killed by the male and female nurses, a fraudulent death certificate was prepared and sent to the victim's family. Most of the naked corpses were buried in mass graves in Obrawalde's own cemetery, but some were cremated in Frankfurt an der Oder. In the early days of the killing programme the staff had to remove the corpses themselves, but as the number of murders increased a group of male patients, the so-called "cemetery gang", was made responsible for the disposal of the bodies.[690] Construction of a gas chamber to accelerate the killing and a crematorium to handle the large number of corpses was begun, but the project was not yet completed when Soviet troops liberated the hospital on 29 January 1945. Russian investigators interviewed some surviving inmates, who testified that 30–50 patients had been killed at Meseritz-Obrawalde each day for several years.[691] Some 1,000 mentally ill patients were liberat-

---

684 McFarland-Icke, *Nurses in Nazi Germany*, p. 213.

685 Benedict, Caplan, Lafrenz, *Duty and 'Euthanasia'*, p. 783.

686 Ibid., p. 784.

687 McFarland-Icke, *Nurses in Nazi Germany*, p. 214.

688 Susan Benedict, and Jochen Kuhla, *Nurses' Participation in the Euthanasia Programs of Nazi Germany* (Western Journal of Nursing Research, Vol. 21(2), 1999), p. 253.

689 Benedict, Caplan, Lafrenz, *Duty and 'Euthanasia'*, p. 784

690 Benedict, Chelouche, *Meseritz-Obrawalde*, p. 72.

691 Burleigh, *Death and Deliverance*, pp. 257–258. This may have been an over-estimate. 30 patients murdered daily would amount to nearly 11,000 victims per annum, which in the two years that "euthana-

ed by the Red Army, who found more than adequate evidence that the hospital "was actually a national facility for the extermination of the German population."[692]

The precise number of victims at Meseritz-Obrawalde remains uncertain, and is likely to remain so, as most of the institute's records had been destroyed by the end of the war, but even the most cautious estimate of 6,991 fatalities is more than three times the hospital's maximum occupancy. An alternative post-war evaluation could only arrive at the probable number of handicapped patients killed at Meseritz-Obrawalde as being in excess of 10,000.[693] Other sources suggest as many as 18,000 victims.[694] There is evidence that physically or mentally impaired German soldiers may have been included among that number.[695] It is also known that following the bombing raids on Hamburg in late July 1943, a number of inmates from old-age homes were transported to Meseritz-Obrawalde for extermination under the aegis of *Aktion Brandt*.[696] In 1944, the death rate among incoming patients reached a staggering 97 percent,[697] so it was hardly surprising that the institution's deadly reputation spread far and wide. When one nurse travelled home by train, a fellow passenger, noticing the address on her suitcase asked: "Isn't that where people are given lethal injections?"[698]

Several thousand unused urns found at the time of liberation indicate that many further murders were planned. The dreadful conditions at Meseritz-Obrawalde were graphically illustrated by the post-war testimony of former patients, who described a concentration camp-like regime at the hospital, including roll call, selections, forced labour, and the appointment of trusted inmates to positions analogous to that of *Kapos* in the concentration camps.[699] Nurse Amanda Ratajczak, captured by the Russians in March 1945, admitted to killing more than 1,500 patients herself, the last of them one day before the arrival of the Soviet army. After a brief trial, along with caregiver Hermann Guhlke, Ratajczak was shot on 10 May 1945.

After the war Meseritz-Obrawalde continued to function as a psychiatric hospital, and has remained one to this day.[700]

Eichberg

Eichberg, near Eltville am Rhein, is an idyllically situated mental hospital founded in 1849 in the Wiesbaden district of Hessen-Nassau. Originally designed for 220 patients, by the mid-1880s there were 450 cases in residence.[701] As further buildings were added, the hospital's capacity was increased to 900 beds by 1937; after 1939 and the ad-

---

sia" was practised at Meseritz-Obrawalde would give a total of about 22,000 fatalities. Either the estimated size of the batches was incorrect, or killing did not occur every day. But this is really to split hairs.

692 Aly, Chroust, Pross, *Cleansing the Fatherland*, p. 91.

693 Friedlander, *The Origins of Nazi Genocide*, pp. 160–161.

694 A second Soviet investigation discovered part of the hospital's death register in which the final entry (dated 30 January 1945) listed death number 18,232 (de Mildt, *In the Name of the People*, p. 65).

695 Lifton, *The Nazi Doctors*, p. 78.

696 Bryant, *Confronting the "Good Death"*, p. 52. .

697 McFarland-Icke, *Nurses in Nazi Germany*, p. 214.

698 Ibid., p. 234.

699 Friedlander, *The Origins of Nazi Genocide*, p. 166.

700 Benedict, Benedict, Caplan, Lafrenz, *Duty and 'Euthanasia'*, p. 783.

701 McFarland-Icke, *Nurses in Nazi Germany*, p. 16.

vent of "euthanasia", the maximum number of patients the hospital could hold was eventually doubled to 1,800.

Hans Hefelmannn and his deputy, Richard von Hegener, visited Eichberg in 1941 to instruct Dr Friedrich Mennecke, already involved in adult euthanasia, to open a killing ward for children. Later Mennecke claimed that he could no longer recall the precise date the Eichberg children's ward opened, but assumed that it was some time in April 1941. After Grafeneck closed in December 1940, Eichberg also became a transit centre for Hadamar,[702] and from the time of the "stop order" in August 1941, commenced operations as a killing centre for adults in its own right.[703]

When rumours of the killings at Grafeneck began to circulate at Eichberg, Mennecke informed his staff that anyone spreading such stories without providing supporting evidence faced the prospect of a prison sentence; naturally enough this was to have precisely the opposite effect to that intended, and confirm the speculation in the minds of the more astute.[704] Whilst generally claiming to be unaware of the fate of the patients who had been sent to Hadamar, some nurses felt uncomfortable about the subject. One claimed:

> Although I did not personally know the purpose of the transports, I had an unpleasant feeling about them from the beginning, because I was aware from the speeches of the National Socialist leader and of influential personalities that the excessively high number of psychiatric patients was complained about again and again.

Others were more perceptive: "When the clothes first came back, we could imagine nothing else than that the patients had been done away with there."[705] Between January and August 1941, 2,262 patients were transferred from Eichberg to Hadamar, where virtually every one was gassed on arrival. Not all were seriously ill; some were only mildly disabled and capable of work.[706]

Although director of the institution (he had been appointed to this position in 1938)[707] and supervisor of the children's ward, Mennecke left the daily operation of the killing to his deputy, Dr Walter Eugen Schmidt. Mennecke completed the paperwork and ordered the inmate's death; Schmidt supervised the killings, sometimes personally carrying them out, and even went so far as to claim that a few severely malformed adults with *Reich* Committee documentation (that is authorizing them to be killed) arrived with a family request for their homicide.[708] A fanatical Nazi bureaucrat, Schmidt strutted around in SS uniform with pistol at his side, supervising the daily murder of patients. Described by one male nurse as a "hothead and psychopath," Schmidt gave orders to his nurses to shoot dead any patient who attempted to escape. At least one of them, nurse Friedrich Golsmann, may have been prepared to do so, for he received a gun of his own for use when on duty.[709] Such was the character of Schmidt that some

702 Friedlander, *The Origins of Nazi Genocide*, p. 109.
703 Bryant, *Confronting the "Good Death"*, p. 51.
704 McFarland-Icke, *Nurses in Nazi Germany*, p. 226.
705 Ibid., p. 224.
706 Bryant, *Confronting the "Good Death"*, p. 123.
707 Ibid., p. 121.
708 Lifton, *The Nazi Doctors*, p. 97 (footnote).
709 McFarland-Icke, *Nurses in Nazi Germany*, p. 217

nurses felt it was quite conceivable that he was killing of his own volition: "the measures were completely in the spirit of Dr Schmidt [and] corresponded with his views," stated one.[710] Another, who admitted to assisting Schmidt in killing patients, claimed that he could not refuse to carry out Schmidt's orders: "Dr Schmidt was a boor who would not have stood for refusal and undoubtedly would have ordered the most far-reaching consequences for me if I had refused."[711]

When Mennecke rejoined the *Wehrmacht* in 1942 (he had also been summoned for a brief period of army service between the outbreak of war and January 1940), Schmidt became acting director, in charge of the children's killing ward.[712] Even under the supervision of Mennecke's predecessor, Wilhelm Hinsen, conditions at Eichberg had been appalling. In 1938, an amount of 0.49 *Reichsmarks* was allocated for each patient's daily food allowance, and straw bedding was in widespread use.[713] Rations were described as follows:

> In the morning there was turnip on bread. We usually smeared that on the bread the night before so that it had time to sink in. There was also coffee in a tin mug, with no milk or sugar. Lunch was at noon. Turnips or swedes once more, cooked without any fat. Always the same thing. There was usually three potatoes, boiled in their skins. Once a week we had beetroot. At three o'clock we had coffee again and bread with syrup.[714]

Wartime conditions became even worse. One doctor testified:

> Most of [the patients] were physically reduced to a state bordering on the skeletal. This was a consequence of the fact that the food was quantitatively and qualitatively totally inadequate...The then asylum director, Dr Mennecke, forbade the sort of forced feeding which was normal in the case of catatonic or depressed patients who refused food. During mealtimes, the patients were left to their own devices, which meant that weak, apathetic and helpless patients were subject to the sick and asocial thieving compulsions of certain individuals, thus losing their own food.[715]

Apart from overcrowding and starvation, patients had to contend with lack of heating and proper clothing. Of still greater concern was the prospect of being confined in the so-called "bunker", a series of underground cells where any patient could be incarcerated on the whim of a staff member. Nurses testified that some patients were kept in the dark, filthy, rat-infested, unheated "bunker" for six months or more, with no supplement to the detainee's already inadequate rations.[716]

At a meeting of T4 physicians at the KdF in early 1942, plans were made for a new-style children's ward to be opened, where innovative kinds of treatment would be explored and tested. Unresponsive children were to be sent to Eichberg for killing. In the interests of efficiency, some of those present even wished to combine research, treatment, and murder in a single institution[717] Eichberg had already become a vital source

---

710 Ibid., p. 228.
711 Ibid., p. 229.
712 Friedlander, *The Origins of Nazi Genocide,* p. 51.
713 Burleigh, *Death and Deliverance,* pp. 54–55.
714 Ibid., p. 75. It is sickening to contrast Mennecke's well recorded gluttony with the starvation of his patients at Eichberg. See Aly, Chroust, Pross, *Cleansing the Fatherland,* p. 238ff.
715 Burleigh, *Death and Deliverance,* pp. 248–251.
716 McFarland-Icke, *Nurses in Nazi Germany,* p. 215.
717 Friedlander, *The Origins of Nazi Genocide,* p. 156–157.

of brains for Carl Schneider's Heidelberg research into the mentally retarded. Children from various hospitals, some as far away as Hamburg, were transferred to Eichberg. There they were observed, killed, dissected, and their brains sent to Schneider. In late 1942, Schneider urgently requested T4 to provide a list of institutions holding retarded patients, and in the summer of 1943, exhorted Paul Nitsche to transfer 10–12 such patients to Heidelberg every month. However, facilities for Schneider's research became steadily worse as the war progressed. In August 1944 he was complaining that Eichberg was delivering only one or two retarded subjects monthly; by the final year of the war, Schneider's research had been completely thwarted, since autopsies were not being performed on the children killed at Eichberg, leaving him without the crucial material he required.[718]

At least 430 children were murdered at Eichberg in the "Children's Speciality Department", lodged in a separate single-storied building close to the main hospital. If, after a period of observation, the child showed no evidence of improvement, Schmidt applied to T4 in Berlin for "authorization for treatment" (*Behandlung*)—that is, the killing of the child.[719] In a minimum of thirty documented cases, lethal doses of medication were administered to children personally by Schmidt.[720] Four-year-old Friedrich S. was admitted to Eichberg, where his parents visited him on his birthday, 21 October 1941. He was underfed and covered in bruises. Two weeks later Friedrich's father wrote to the asylum and received a reply dated 14 November 1941 to the effect that his son was dead. A number of parents began to speculate how their child could appear healthy at the time of their visit, yet be dead a few days later.[721]

Eichberg also served as a training facility for other physicians dedicated to the theory and practice of children's "euthanasia". One such was Dr. Magdalene Schütte, who having been suitably tutored in the methods to be employed, went on to become head of the 'special children's ward' at the *Städtische Kinderkrankenhäuser und Kinderheime* in Stuttgart, where, so recent research indicates, between January 1943 and April 1945, at least 46 suspicious deaths of children occurred, with some death certificates signed with a false name. Post-war, Schütte worked as a paediatrician in private practice and was also the head physician of the children's department of the regional hospital in Aalen, before retiring in 1967. Investigations into Schütte's activities at the Stuttgart children's hospital commenced in 1963, but were terminated in the same year. Schütte died in 1980, one of many to never be brought to account for her "euthanasia" related actions.[722]

Sick foreign workers were also transferred to Eichberg, where within a few weeks at most they were dead, the great majority almost certainly murdered. In a letter to the Labour Office at Frankfurt am Main, Schmidt outlined his cynical abuse of his position as a "healer": "The above-named is a mentally ill Russian female who is no longer capable of deployment in Germany. Nor can we anticipate that she will be capable of work-

718 Ibid., p. 131.
719 http://www.chgs.umn.edu/histories/documentary/hadamar/assassination.html (Accessed 22 February 2008).
720 Bryant, *Confronting the "Good Death"*, p. 125
721 Burleigh, *Death and Deliverance*, p. 110.
722 http://www.uvm.edu/~lkaelber/children/stuttgart/stuttgart.html (Accessed 29 October 2016)

ing again in the foreseeable future; she is lying here simply as a burden upon our authorities." One month later the woman was dead.[723]

At the post-war trial of Mennecke, Schmidt, and others, Dr Elisabeth V. testified about conditions at Eichberg in 1942 after her return there to serve as a physician. Amongst other horrors, patients were being directed by non-medical staff to the rat-infested bunker mentioned earlier. When she complained to Schmidt about the treatment of foreign workers, he replied that she was just a woman, and would never be able to understand him regarding these matters.[724]

Eglfing-Haar

Prior to the commencement of "euthanasia", a surgical department had been established as part of the sterilisation programme at the large hospital complex of Eglfing-Haar, near Munich in Bavaria.[725] The removal of adult patients from the hospital to the killing centres at Grafeneck and Hartheim was initiated by a letter from Dr. Walter Schultze of the Health Department of the Bavarian Ministry of the Interior in Munich, dated 8 January 1940. The first transport, consisting of 25 male patients, left on 18 January 1940. In total, from that time until 20 June 1941, when the last shipment in this initial phase of "euthanasia" was dispatched, 20 transports were sent from Eglfing-Haar at fairly regular intervals. The number of deportees varied between12–149 per trip.

At first it was believed that a total of 1,857 patients had been taken away to be killed, all but one of whom were non-Jews. However, when the surviving lists of those deported were checked, it was discovered that this figure did not include a transport of 191 Jewish inmates sent to Lublin for extermination on 20 September 1940. Eglfing-Haar served as an assembly point for Jewish patients from all Bavarian hospitals, who were housed in two isolated buildings. One male nurse later described an arrival of such Jewish inmates in September 1940: "There were many elderly gentlemen, including a city councilman from the Rhineland, business people, lawyers, an acquaintance of Thomas Mann," as well as "frail, elderly ladies" and "a boy from Gmund whose parents had escaped to England." On 20 September, *Gekrat* collected the191 handicapped Jewish patients. 33 were from Eglfing-Haar itself.[726] After several weeks had passed, relatives were informed that the patients had died in the hospital at Cholm (Chelm) in Poland.[727] On the same day that they were deported, Dr Hermann Pfannmüller, the man in charge of Eglfing-Haar, sent a list of the 191 Jews destined to be killed in the Lublin region to the Ministry of the Interior in Munich, smugly declaring: "I herewith report to the State Ministry that henceforth my institution will accommodate only Aryan mental patients."[728]

The medical complex at Eglfing-Haar treated both adults and children, with the killing ward for children, established in September 1940, located separately from the regular children's section. The aforementioned Pfannmüller was an early participant in both

---

723 Burleigh, *Death and Deliverance*, p. 245, and p. 337 (notes 85 and 86.)

724 Ibid., pp. 248–251.

725 Ibid., p. 62.

726 Friedlander, *The Origins of Nazi Genocide*, pp. 273–274.

727 http://www.phil.gu.se/sffp/reports/16.%20von%20Cranach.pdf (Accessed 25 February 2008.)

728 Friedlander, *The Origins of Nazi Genocide*, pp. 273–274.

adult and children's "euthanasia". In Eglfing-Haar, he introduced a brutal system intended to murder his charges in a heartless manner; he also conducted tours through his institution in order to educate the public about the mental and physical deficiencies of his patients.[729] That this tutoring apparently achieved its objective can be gauged from the fact that some visitors questioned why the patients were still being kept alive.[730] They had a sympathetic listener in Pfannmüller, for even before "euthanasia" had officially commenced, Pfannmüller was murdering children at Eglfing-Haar.[731] In 1946, Ludwig Lehner, a Bavarian schoolteacher, provided testimony concerning his observations during the course of one such visit, undertaken in the autumn of 1939:

> During my tour, I was eyewitness to the following events: After visiting a few other wards, the institution's director himself, as far as I remember he was called Pfannmüller, led us into a children's ward. This hall impressed me as clean and well-kept. About 15 to 25 cribs contained that number of children, aged approximately one to five years. In this ward Pfannmüller explicated his opinions in particular detail. I remember pretty accurately the sense of his speech, because it was, either due to cynicism or clumsiness, surprisingly frank: "For me as a National Socialist, these creatures" (meaning these children) "obviously represent only a burden for our healthy national body. We do not kill" (he might also have used a euphemism instead of the word 'kill') "with poison, injections, etc., because that would only provide new slanderous campaign material for the foreign press and certain gentlemen in Switzerland. No, our method is, as you can see, much simpler and far more natural." As he spoke these words, [Pfannmüller] and a nurse from the ward pulled a child from its crib. Displaying the child like a dead rabbit, he pontificated with the air of a connoisseur and a cynical smirk something like this: "With this one, for example, it will still take two to three days." I can still clearly visualize the spectacle of this fat and smirking man with the whimpering skeleton in his fleshy hand, surrounded by other starving children. Furthermore, the murderer then pointed out that they did not suddenly withdraw food, but instead slowly reduced rations."[732]

On 30 November 1940, Pfannmüller submitted his first report on admissions and discharges to the new children's department. Of the eleven children admitted during the preceding month, five had already died.[733]

In January 1993, at the suggestion of the Archbishop of Munich, four nuns accused Dr. Hans-Joachim Sewering of participating in the transfer of more than 900 German Catholic children from Schönbrunn Sanatorium in the city of Dachau to Eglfing-Haar, where they died. Schönbrunn Sanatorium, which was nearly 100 years old by the time that Sewering worked there, was home to a large number of institutionalized children. Sewering, a member of the Nazi Party and the SS since 1933, rose from Staff Physician at Schönbrunn to Chief Medical Officer. Between 1942 and 1945, he participated in the "wild euthanasia" programme. About 900 children were sent from Schönbrunn to Eglfing-Haar for killing.

There is clear documentary evidence that Sewering signed an order on 23 October 1943 authorizing the transfer of fourteen-year-old Babette Fröwis from Schönbrunn to Eglfing-Haar, an order equivalent to a death sentence. Two weeks after her transfer from Schönbrunn, Babette Fröwis was dead, probably as a result of an intentional overdose of phenobarbital, despite the fact that the notation on her chart upon admission

729 Between 1933 and 1939, over 21,000 people were treated to a tour of Eglfing-Haar. Almost 6,000 of them were members of the SS. (Burleigh, *Death and Deliverance*, p. 47.)

730 Burleigh, *Death and Deliverance*, p. 49.

731 de Mildt, *In the Name of the People*, p. 53.

732 Friedlander,*The Origins of Nazi Genocide*, pp. 49–50.

733 http://www.ess.uwe.ac.uk/genocide/alexrep3.htm (Accessed 23 February 2008).

indicated that she was physically healthy. Sewering, her doctor, signed the transfer order, proof of his active role in the extermination process.

Sewering denied all knowledge of the killing of his patient. He continued to practice medicine in post-war Dachau, until public outcry about his involvement in "euthanasia" forced his resignation as President-Elect of the World Medical Association in 1993 (blamed, in true Nazi fashion, on "a world Jewish conspiracy").[734] Twenty years earlier he had been appointed president of the German Medical Association. A man apparently devoid of any moral compass, he even attempted to shift the blame for the death of the murdered children onto the Franciscan nuns who had cared for them throughout the years 1933–1945. Sewering's statements so outraged the nuns that they issued a press release refuting his claims, and indicating that over 900 children had been taken away from them to be murdered as part of the "euthanasia" programme. The nuns stated that everyone at Schönbrunn, including the children themselves, knew that transfer to Eglfing-Haar meant death.[735] Quite unbelievably, despite all of the evidence of his participation in children's "euthanasia", in May 2008 the German Federation of Internal Medicine awarded Sewering its highest honour, the Gunther-Budelmann medal "for services to the nation's health system".[736]

Pfannmüller was proud of his achievements, telling a post-war German court that "putting the children to sleep was the cleanest form of euthanasia."[737] At Pfannmüller's trial, testimony was offered concerning the death of a boy named Karl Memmel, who had consistently complained about the Eglfing-Haar starvation diet to his parents. They therefore supplied food for Karl, and were assured that it had been given to him. When they received notification of Karl's death, the boy's stepfather and his wife travelled to Eglfing-Haar. Eventually they were allowed to see Karl's emaciated body. Karl's stepfather told the court that "he can only have starved to death." Karl Memmel's death at Eglfing-Haar, in the final moments of the war, was not exceptional. Pfannmüller's successor wrote about the killing of two Silesian refugee girls — Ruth and Marie — on 30 April and 1 May 1945, only days before German surrendered and after Hitler had al-

---

734 The World Medical Association (WMA) was created in late 1946. In September 1947, shortly after the final judgment at the Nuremberg Medical Trial, the first official meeting of the WMA was held in Paris. The WMA formulated a new physician's oath to promote and serve the health of humanity. One of the first issues discussed by the 1947 general assembly was the German "betrayal of the traditions of medicine." The assembly asked, why did these doctors lack moral or professional conscience and forget or ignore the humanitarian motives and ideals of medical service, and how can a repetition of such crimes be averted? The "widespread criminal conduct of the German medical profession since 1933" was acknowledged. The WMA endorsed "the judicial action taken to punish those members of the medical profession who shared in the crimes" and it "solemnly condemned the crimes and inhumanity committed by doctors in Germany and elsewhere against human beings." The assembly continued: "We undertake to expel from our organization those members who have been personally guilty of the crimes.... We will exact from all our members a standard of conduct that recognizes the sanctity, moral liberty and personal dignity of every human being." Sewering had been a German delegate to the WMA since 1959 and its treasurer for twenty years. [John J Michalczyk (ed), *Medicine, Ethics, and the Third Reich: Historical and Contemporary Issues* (Lanham: Rowman & Littlefield, 1994), pp. 202–206];

735 Michael A Grodin, George J Annas, Leonard H Glantz, *Medicine and Human Rights: A Proposal for International Action* (The Hastings Center Report, Vol. 23, Issue 4, 1993), pp. 8–12.

736 In March 2001, Sewering was still practicing medicine.

737 Friedlander, *The Origins of Nazi Genocide*, p. 51.

ready committed suicide. He concluded that both girls "had been...intentionally killed in a German hospital."[738]

So successful were Pfannmüller's methods that in 1943 he established two "starvation houses" (*Hungerhaüser*) for adults. Often patients who were not considered sick enough for the gas chambers elsewhere were starved to death at Eglfing-Haar, where the motto was: "We give them no fat, then they go on their own." Patients often suffered from hallucinations, and endlessly fantasized about food, or suffered from feelings of guilt over having done something wrong for which they were now being punished. From available records, it is estimated that at least 444 patients died at Eglfing-Haar from the effects of malnutrition.[739] The actual figure was probably greater.

Kaufbeuren-Irsee

The Bavarian state hospital at Kaufbeuren and its branch at Irsee, affiliated in 1876, served as a transfer institution prior to the "stop order" and thereafter as a centre for "wild euthanasia." During the First World War, the institution had served as a military hospital for shell-shocked and severely disabled soldiers. Since 1929 it had been headed by Dr Valentin Faltlhauser, from 1940 a T4 expert, and an enthusiastic supporter of both adult and children's "euthanasia."[740] Clearly a dedicated eugenicist, he had been responsible for initiating the sterilisation of patients from Kaufbeuren-Irsee at the municipal hospital. Since he was also a judge in the local Hereditary Health Court, some might have viewed this as a conflict of interests, as well as being ethically indefensible, but neither issue apparently bothered Faltlhauser. Nor did it seem to concern members of the medical profession elsewhere, who found themselves in the similar position of being, in effect, prosecutor and adjudicator.[741]

In 1948, nurse R., a member of a religious order, reported the following concerning the conduct of affairs at Kaufbeuren:

> Until August 1940, the patients were respected. They were well taken care of, and the director tried to improve their physical and mental health as best he could. But all of a sudden this changed. When I returned from my vacation in August 1940, eleven of the patients on my ward F 3 b, where primarily calm patients stayed, were gone. No one knew at that time where they had been taken. We believed they had been transferred to an asylum where they would be well cared for. But when, on 8 November 1940, the second transport of women disappeared, and when later their clothes and underwear were returned in an incredible state — it appeared as if the underwear and clothes had been ripped off the patients — we became suspicious. The third transport of female patients occurred on 9 December 1940. It was especially difficult for us nurses to deliver these patients, for whom we had cared for many years, like cattle to an almost certain death. The personnel of the buses from Berlin were rough and frightening characters, some women, some men. They grabbed the patients roughly and tied them down in the cars, sometimes even with chains. I had the impression that they were disguised SS people. The ambulance cars did not arrive at the main entrance, but came before dawn, collected the patients in the inner yard of the so-called country house, and left the hospital before daybreak.
>
> The patients gradually suspected what was going on, got terribly frightened and cried and screamed at times. The selection of patients took place according to lists that were on hand in the office of the inspector. Many patients suspected their fate in advance. One female patient who was transferred from ward F

738 Ibid., pp. 166–167

739 Lifton, *The Nazi Doctors*, pp. 98–99.

740 Ibid., p. 51.

741 Burleigh, *Death and Deliverance*, pp. 61–62.

3 b to the so-called country house, from where the transports departed, said: "Now I know what is ahead of me." Prior to her transfer, she asked for a good-bye pancake and went to confession. During confession, she cried bitterly. Some time after her deportation, her sister was notified that the patient had died as a result of dysentery.[742]

A number of nurses attempted to persuade relatives to remove patients from the hospital and take them home, something which did occur in certain cases. If it did not, the patient's next-of-kin would receive a letter similar to the following:

> This is to inform you that your son, in connection with economically necessary measures for the clearance of patients from some facilities, has been transferred to another institution unknown to us. The patient transfers have been ordered centrally according to instructions from the National Secretary of Defence. The hospital has absolutely no influence concerning the transfer or non-transfer of its patients. You will be notified about the condition of your son by the receiving institution at an appropriate time.[743]

This deliberation obfuscation of the patient's whereabouts all formed part of the camouflaging of the killing operation. In fact, as another nurse recalled, patients "were picked up, on the next day they were dead and the relatives got news, dead of typhus or whatever, and it was all lies."[744]

A conference, knowledge of which was immediately declared to be a state secret, was held by Walter Schultze of the Bavarian Ministry of the Interior on 17 November 1942, at which the directors of mental hospitals in the region were asked to provide a "special diet " (*Sonderkost,* or *E-Kost,* as Faltlhauser termed it) for inmates. Since far too *few* patients were dying in the asylums, it was said, it was unnecessary to treat many of the diseases that occurred. Moreover, the death rate could be cheaply and efficiently "improved", as Faltlhauser illustrated. Like Pfannmüller in Eglfing-Haar and others elsewhere, Faltlhauser was a devotee of the deprivation of food as a killing method, stating at the meeting that following the cancellation of the formal "euthanasia" programme, patients could now be killed by gradual starvation. Initially he had been opposed to "euthanasia", but then he had become informed about the official programme, and now regretted its abolition. He produced the Kaufbeuren-Irsee condemned patients' menu for the meeting's edification. Initially introduced at Irsee in August and at Kaufbeuren in October 1942,[745] it was totally fat free, consisting of potatoes, yellow turnips, and green or red cabbage cooked in water. He boasted that this diet would ensure a slow death within about three months.[746] Following the meeting, Schulze produced a directive on 30 November 1942, stating:

> With regard to wartime food supplies and the state of health of those asylum patients who work, we can no longer justify the fact that all inmates of asylums are receiving the same rations, without regard on one hand to whether they perform productive labour or are in therapy, or on the other hand to whether they are simply being kept in the asylum without performing any labour worthy of the name.
>
> It is therefore decreed, with immediate effect, that those inmates of asylums who do productive work or who are receiving therapy, and in addition children who are capable of being educated, war casualties,

742 http://www.phil.gu.se/sffp/reports/16.%20von%20Cranach.pdf (Accessed 25 February 2008.)
743 Ibid.
744 McFarland-Icke, *Nurses in Nazi Germany*, p. 220.
745 Burleigh, *Death and Deliverance*, pp. 229–230.
746 Lifton, *The Nazi Doctors*, p. 98.

and those suffering from geriatric psychoses, shall be better fed—in both a quantitative and qualitative respect—than the remaining inmates.[747]

Different diets were produced for carefully categorized patients. The best fed ate twice daily, with an additional ration of bread; then followed those on the standard diet, and finally the condemned, provided with the "basic diet" or *E-Kost*. A male nurse stated:

With regard to the E-Kost diet, I can repeat the following, which was told to me by the kitchen nurse: at one time, there were two pots of meat broth in the kitchen. The nurse begged the administrative inspector to be allowed to serve the broth to the patients on the E-Kost diet, since they were almost assaulting each other due to hunger. He started to shout and curse — one could even say scream — that he would rather spill the broth into the trash dump than give it to the patients on the E-Kost diet.[748]

Until a crematorium was built, a number of the working patients had to bring corpses to the cemetery and for some time had to dig the required graves. The priest conducting burial ceremonies (as many six or seven every afternoon) reported that the ringing of church bells was prohibited lest people in the vicinity became suspicious about the number of deaths occurring at the asylum. In the event, he was permitted to ring the bells for just the first of the daily funerals.[749]

Kaufbeuren also served as one of eleven collection centres for adult foreign psychiatric patients who did not recover from their sickness within the six-week period allowed for so doing. Typically, their fate was as already described for other foreign workers at Eichberg.[750] More than 2,000 adult and child patients from Kaufbeuren-Irsee were either deported to extermination facilities, starved to death, or died as a result of lethal injections and/or overdoses of medication.[751] Condemned patients were given Luminal or Veronal and sometimes Trional in pill form, as well as Luminal and Morphine-Scopolamine in liquid form. Morphine-Scopolamine was given when Luminal or Veronal alone did not yield the desired result.[752] The final component in the extermination process was provided in July 1944 with the building of a crematorium in Kaufbeuren.[753]

Faltlhauser continued to kill children even after the war had ended. In late April 1945, American troops occupied Kaufbeuren but, placing the state hospital off limits, did not interfere with its operation. For more than two months, the institution was able to function without change. It was only on 2 July, after rumours had reached the mili-

747 Burleigh, *Death and Deliverance*, p. 230.

748 http://www.phil.gu.se/sffp/reports/16.%20von%20Cranach.pdf (Accessed 25 February 2008).

749 Burleigh, *Death and Deliverance*, p. 231.

750 Ibid., p. 244.

751 http://www.phil.gu.se/sffp/reports/16.%20von%20Cranach.pdf (Accessed 25 February 2008). This authoritative source states that in 1943, 1944 and 1945, a total of 1,808 patients died in Kaufbeuren: The free beds were immediately filled by patients from other psychiatric clinics which were cleared to be used for different purposes, and also by so-called "eastern workers" (*Ostarbeiter*), Russian, Polish and Baltic forced labourers, who had become mentally ill in the camps where they had been interned. A directive from Berlin expected... the directors to stop any treatment if the patient was unable to return to work within four weeks. This meant death. [Ibid]. The life or death decision was thus no longer in the hands of a doctor, but was made by the Central Clearing Office for Mental institutions on the basis a report submitted by the hospital. [Aly, Chroust, Pross, *Cleansing the Fatherland*, p. 184].

752 http://www.phil.gu.se/sffp/reports/16.%20von%20Cranach.pdf (Accessed 24 February 2008).

753 Aly, Chroust, Pross, *Cleansing the Fatherland*, p. 91.

tary in Munich, that American soldiers entered the hospital.[754] A German physician had returned home from the war to discover that mental patients were still being exterminated at Kaufbeuren; doctors, nurses, and other staff had continued to murder patients unmolested.[755] What the Americans found when they entered Kaifbeuren was described by an officer:

> When asked to see the second doctor in charge, investigators were nonchalantly informed that he had hanged himself the night before. No one seemed to be aroused or emotionally upset at his violent end. Such was the callous attitude the doctors and nurses had for violent death. Observers found, in an uncooled morgue, stinking bodies of men and women who had died days before. Their weight was between 26 and 33 Kilograms. Among the children still living was a 10 year-old boy whose weight was less than 10 Kilograms and whose legs at the calf had a diameter of 2 ½ inches. An informant stated that tuberculosis and other disease are rampant. Scabies, lice and other vermin were encountered throughout, linen was dirty and quarantine measures non-existent upon [the] investigators' arrival.[756]

On 29 May 1945, Sister Mina Wörle, head nurse of the children's ward at Kaufbeuren-Irsee, murdered four-year-old Richard Jenne with a lethal injection. At 1:10 P.M, Faltlhauser recorded the death of the child from "typhus." Richard Jenne was the last known victim of the "euthanasia" programme to be directly targeted in this way, although the inmates of many other institutions were to die in the days to come as a result of the treatment they had received in establishments like Kaufbeuren.[757]

Wörle and her fellow nurses Olga Rittler and Paul Heichele joined Faltlhauser in the dock at Augsburg in 1949. The three nurses were accused of killing either adults or children at Kaufbeuren or Irsee, the minimum numbers of victims involved varying from 20 to 100 in each case. "The principles of justice...required retrospective atonement," the court ruled, but the degree of expiation imposed on the quartet was hardly excessive. After noting their alleged dedication to healing, their "more or less" regret for their actions, and the other mitigating circumstances applying in each case, the court imposed sentences of 18 months imprisonment on Wörle, 21 months on Rittler and 12 months on Heichele. Arriving at an appropriate punishment for Rittler in particular was no easy task, since she had been involved in the first (gassing) and second (medication) phases of "euthanasia", had been married to a man who served at Bernburg and Treblinka, and had initially retired from then returned to the practice of state sponsored killing entirely of her own volition, in the full knowledge of exactly what was involved. However, the court decreed, she had no previous convictions, a fact which apparently weighed heavily in her favour.[758] As for Faltlhauser, this mass murderer of children and adults received a jail sentence of just three years.

Wiesloch

The first transport of patients to be sent to a killing centre from Wiesloch, a mental hospital near Heidelberg, left on 29 February 1940 for Grafeneck.[759] In his capacity as

---

754 Friedlander, *The Origins of Nazi Genocide*, pp. 51–52.
755 Proctor, *Racial Hygiene*, pp. 192–193.
756 http://www.phil.gu.se/sffp/reports/16.%20von%20Cranach.pdf (Accessed 25 February 2008.)
757 Friedlander, *The Origins of Nazi Genocide*, pp. 162–163.
758 de Mildt, *In the Name of the People*, pp. 189–190.
759 http://www.ess.uwe.ac.uk/genocide/alexrep5.htm (Accessed 27 February 2008).

director of the institution, Dr Josef Artur Schreck committed 580 of his patients to certain death, having already personally witnessed the Grafeneck gas chamber in operation.[760] Between March and May 1940, Dr Ludwig Sprauer, in charge of the health department in the Baden Ministry of Interior, issued orders that a number of patients (usually in batches of 120) were to be prepared for transport. *Gekrat* removed 800 patients in total from Wiesloch, the last of the transports leaving in the middle of 1941. The nature of the transports eventually became the subject of discussion within the local community because of the stereotyped nature of the letters received by relatives. These letters usually read "Your mother/brother/sister/son," respectively, "has died. Unfortunately all medical skill has failed to keep him/her alive. His/her ashes can be obtained by you by writing to ....."[761]

A children's killing ward was established at Wiesloch early in 1941, supervised by Schreck. Having personally murdered three children and conducted autopsies on their bodies,[762] Schreck declined to perform further homicides, stating that "a hospital is not the appropriate place" for such killings; he did, however, continue to supervise the killing ward. During visits to Wiesloch, Dr. Fritz Kühnke, a young physician from Eglfing-Haar,[763] performed the actual killing of a further eight children,.[764] The last of these children was killed in May 1941. Aged between three and five, the children were killed by means of injections of Luminal, which induced terminal pneumonia. Schreck claimed to have been motivated by purely humanitarian considerations—it was "inhumane" to prolong the life of a handicapped child.[765]

Schreck, director of the Rastatt hospital since 1934,[766] was recruited by Sprauer in 1939. Although he had not been informed of their purpose, Schreck shrewdly suspected that the *Meldebogen* he was asked to process in late 1939 were not unconnected with a programme "along the lines of Binding/Hoche."[767] In October 1939, Sprauer had been summoned to a meeting in Berlin by Herbert Linden, where, having first been sworn to secrecy, he was informed of the plan to kill incurable mental patients under the guise of "euthanasia". By doing so, beds for wartime casualties would be made available. A legal basis for such action existed, Linden stated, although no formal law had yet been enacted.

Sprauer controlled twelve hospitals in the Baden region. However, he only succeeded in convincing two of the directors of those hospitals to cooperate fully with the "euthanasia" programme—Schreck and a Dr Gercke who was in charge of the mental home at Hub. All of the other leading medical personnel of the region resisted implementation of the proposal.[768] Some were more successful than others. For example, in December 1939, when Sprauer informed Dr Hans Römer, director of the Illenau mental home that

760 de Mildt, *In the Name of the People*, p. 100.
761 http://www.ess.uwe.ac.uk/genocide/alexrep5.htm (Accessed 26 February 2008.)
762 Bryant, *Confronting the "Good Death"*, p. 173.
763 Friedlander, *The Origins of Nazi Genocide*, p. 52.
764 http://www.ess.uwe.ac.uk/genocide/alexrep5.htm (Accessed 28 February 2008.) One other child allegedly died of natural causes.
765 Bryant, *Confronting the "Good Death"*, p. 173
766 de Mildt, *In the Name of the People*, p. 99.
767 Bryant, *Confronting the "Good Death"*, p. 173
768 Ibid., p. 170

Römer's patients were to be killed in accordance with a *Führer* order, Römer simply refused to cooperate. He managed to delay the removal of the victims for a few months, but after Römer had been forcibly retired, Schreck was temporarily appointed in Römer's place long enough to have the patients gassed. Illenau was then closed down.[769]

On the other hand, August Schilli, head of the Fussbach hospital and a member of the SS, was rather more successful in his opposition to the planned killings. Having discovered that his patients were being murdered he began discharging others he felt were threatened, or alternatively made sure that they were hidden or otherwise unavailable when transports when scheduled. On one occasion he arrived in his full SS uniform, and having argued fiercely with the *Gekrat* leader, managed to withhold sixty of the ninety patients slated for extermination. The next day Schilli visited Sprauer to file an official protest, and to inform him that no further patients would be handed over. Despite making threatening noises, Sprauer was unable, or unwilling to press the matter, and Schilli succeeded in saving the lives of the majority of the Fussbach inmates.[770] These actions, and those of others, indicate that it was possible to resist the "euthanasia" programme, at least to some degree, and to retain a modicum of morality. But in general, Sprauer's bullying tactics, coupled with threats of imprisonment or worse,[771] ensured that the project eventually attained its objectives.[772]

In an interrogation conducted on 9 June 1945 with an understandably nervous Dr. Wilhelm Möckel, then director at Wiesloch, the latter stated that in 1940, a delegation came from Berlin which was headed by Professor Carl Schneider of Heidelberg. This delegation arranged for the removal of patients by *Gekrat*, with killings to be carried out at Grafeneck and Hadamar. Möckel claimed: "I never wanted to hear about what they did at Grafeneck. It always nauseated me. Later I could not stand it any longer. Whenever a transport was to leave at night, I usually left the institution at noon and did not come back until the next day."

At first the patients did not know where they were being taken, but later when they became aware of what was happening there were terrible scenes as inmates refused to be transported. Eventually a storm of protest arose among the local populace because of the formulaic death notices which were being received from Grafeneck, Hadamar, and Brandenburg. At some time during that period Sprauer held a meeting and said that patients who had been transferred would no longer be "put away", but 8 or 14 days later another batch of death notices were dispatched. Because of the wave of public protest, Möckel stated, the first phase of the mass killing of patients in special killing centres was discontinued in the middle of 1941, and a policy of extermination by starvation in the state institutions themselves was introduced. Starvation had the advantage of spreading deaths out over a longer period of time, so that large numbers of patients were not dying on the same day.

---

769 de Mildt, *In the Name of the People*, p. 104

770 Ibid., p. 105.

771 Sprauer threatened Dr Viktor Mathes of the Emmendingen hospital that "one could be sent to Grafeneck oneself." (de Mildt, *In the Name of the People*, p. 347, note 40.)

772 Bryant, *Confronting the "Good Death"*, pp. 170–171.

600 inmates from Wiesloch were transferred to Hadamar for killing in the three years to June 1944, when the last of these transports occurred. This was in addition to the 800 patients who were transferred to Grafeneck and Hadamar for killing between January 1940 and June 1941. Möckel also transferred patients for death by starvation to various other institutions, including Eichberg and Kaufbeuren. 150 cases, 60 men and 90 women, constituted the last transport to Kaufbeuren on 4 June 1944. A starvation centre named a "research house" was set up on Möckel's own hospital grounds. It was taken over and staffed by Carl Schneider, and operated throughout 1942 and 1943. Möckel claimed that he did not know how many patients were killed in this centre, and stated that he had not been aware until shortly before his interrogation that any patients were killed there at all. Schneider had his own staff functioning at the "research house", separate from the Wiesloch employees. "Our personnel were indignant over the way these people lived. They received wine and beer by the car-load directly from Berlin." It may come as no surprise that the Wiesloch staff were more concerned about supplies of alcohol than the fate of those supposedly in their care.

The inmates transported to killing centres as well as the patients killed at Wiesloch itself from 1940 to 1945 did not include any Jewish victims, since as early as 1935 the Jewish patients of Wiesloch had been dispatched elsewhere, together with healthy Jews living in Wiesloch and the surrounding area. They had all been taken away to an unknown destination which was supposed to be "somewhere in the Pyrenees". Jewish patients were written off the books as "transferred to an institution outside of the province of Baden".[773]

Möckel stated that one of his nurses, Amalie Widmann, once set out to have a look at the killing centre in Grafeneck. This nurse had apparently become depressed from seeing so many of her patients transported to their death, and developed some kind of an obsession about seeing the institution for herself. She was arrested soon after she arrived at Grafeneck. Möckel had unexpectedly received a telephone call from Grafeneck informing him that one of his nurses had been arrested on a charge of spying on the hospital. She was arrested there by the S.S. At Möckel's request she was finally released, but not before she was threatened with a concentration camp if she tried a thing like that again "and that means death".

Widmann personally testified that among the patients on the first transport leaving Wiesloch for a killing centre on 29 February 1940, there were a good many that had become endeared and attached to her. After they had been taken away, she found herself constantly thinking about their sad fate, unable to rest day or night. Finally, she felt that it might give her peace of mind if she could actually see what happened at Grafeneck, and she therefore decided to visit there for herself. So she asked for a leave of absence without telling anybody what she planned to do, and went to Grafeneck on 22 July

---

773 Whether Möckel 's memory failed him, or he deliberately offered incorrect dating, the fact was that 6,300 Jews from Baden were dumped in unoccupied France in October 1940. They were subsequently interned in camps at Gurs, Rivesaltes, Le Vernet, Les Milles and elsewhere; many died, either in these camps or after transportation to Auschwitz or Sobibor. Included in this expulsion were the Jews of the Saar-Palatinate. Both provinces had been earmarked for incorporation into the *Reich*, which were thus to be first rendered *'Judenrein'* ['Jew-free']. [Hilberg, *Destruction of European Jews*, pp. 651 and 665–666; Saul Friedländer, *The Years of Extermination: Nazi Germany and the Jews, 1939–1945* (New York: HarperCollins, 2007), pp. 93–94.]

1940. When she got off the train at Marbach a.d. Lauter bei Münzingen, which was the station for Grafeneck, the people who she asked for directions to the hospital looked at her in a peculiar way, as if there was something strange or funny about her. When she finally arrived in front of the institution at Grafeneck, she found a sign reading: "Entry strictly prohibited because of danger of infection" ("*Zutritt wegen Seuchengefahr strengstens verboten*"). There were heavily armed men in green uniforms about the area, obviously policemen. Suddenly Widmann felt gripped by an overwhelming feeling of anxiety and she ran away over an open field, crying bitterly. After a while she saw that she was on the grounds of a stud farm. The farmer came and asked her whether he could do anything for her, and she told him that she wanted to go and see the institution in Grafeneck. The farmer then told her: "Do not go there. One must not say anything". Shortly afterwards SS-men with hounds appeared. She was brought before an official who asked her what she was doing at Grafeneck. She said that she wanted to see some of her old patients and find out how they were. The official then stated that the patients liked it so much there that they would never want to leave again. He interrogated her roughly about her antecedents and any connections she may have had with other groups. Then he telephoned Möckel. Widmann added that she felt she owed her life to Möckel, because if he had not spoken on her behalf, they would have killed her. The sole reason for her going to Grafeneck was because of her close relationship with her patients.[774]

When also interviewed in the summer of 1945, Dr. Alfred Schwenninger was rather less complimentary about the good Samaritan Möckel, who, so Schwenninger claimed, was an old National Socialist of many years standing, and was fully behind the extermination of the mentally ill. Möckel was closely tied by bonds of personal friendship to the originators and main activators of the "euthanasia" programme, said Schwenninger, who was himself director of the military section at Wiesloch. In Schwenninger's opinion the principal players in the implementation of the policy in Baden were Carl Schneider and his associates at the psychiatric clinic of the University of Heidelberg, Werner Heyde in Würzburg, Paul Nitsche, a Dr. Packheiser, the principal expert in the Public Health Office of Baden, and Möckel himself. "I once met the whole gang here. They used to come to Wiesloch regularly for discussions," reported Schwenninger, who felt that Packheiser, originally a physician who took up a career in the Health Ministry, was a particularly evil individual. It was Packheiser who had first informed Möckel about the coming arrangements concerning the killing of the mentally ill in the summer of 1939, passing on the information that it would commence as soon as war broke out.

After the 'stop order' in August 1941, the operation at the State Institution in Wiesloch headed by Schneider killed patients locally in small groups, continued Schwenninger. One day, shortly after Stalingrad, that arrangement ceased.. Then the third method began—the systematic starvation at special institutions to which patients were transferred in large numbers. From these institutions relatives received the standard letter announcing the death of the patient. Schwenninger claimed that he had always protested against these activities. "I used to stick my neck out almost to the

774 http://www.ess.uwe.ac.uk/genocide/alexrep5.htm (Accessed 28 February 2008).

point of being sent to a concentration camp. Here I was always surrounded by spies, Möckel's spies".

Schwenninger added that psychopaths were sometimes transferred from state institutions to concentration camps. He knew of ninety-four such patients who were taken to a concentration camp which supplied slave manpower to the radium mines in Johannesthal. Of these ninety-four, eight were still alive. Since 1934 many criminals had been shunted into state institutions for the insane under paragraph 42b of the Penal Code, and these individuals were later supposedly transferred to concentration camps.[775]

---

775 Ibid.

CHAPTER 7: REORGANIZATION

Everything that was considered until now as the holiest obligations of medicine—to care for the sick without paying attention to their race, to deal in the same way with all diseases, to help ill men everywhere and to ease their pain—all this is viewed by the National Socialists as sheer sentimental stuff which should be thrown away. The only matter of importance in their eyes is leading a war of annihilation against the less worthy—the incurable patients...If this line of thought will win the upper hand the German medical profession will lose its ethical norms...the physician will act as a killer, the doctor will become a murderer
- Julius Moses, February 1932.[776]

It is said of these patients: They are like an old machine which no longer runs, like an old horse which is hopelessly paralyzed, like a cow which no longer gives milk...We are not talking here about a machine, a horse, nor a cow...No, we are talking about men and women, our compatriots, our brothers and sisters. Poor unproductive people if you wish, but does this mean that they have lost their right to live?
- Bishop Clemens August Graf von Galen[777]

On 9 October 1939, a meeting of the organisers of the "euthanasia" programme was held, at which a formula was presented by which the number of future potential *Reich* "euthanasia" victims could be calculated. The basis of the calculation was the ratio 1,000:10:5:1, meaning that for every 1,000 of the population, ten would require psychiatric treatment. Of those requiring treatment, five would receive this as in-patients, and one of those in-patients would fall within the scope of the programme. That is to say, one in every 1,000 of the population would be subject to "euthanasia". Applying this ratio to the population of the *Reich* as a whole resulted in an aggregate figure of 65,000–75,000 potential victims.[778] This figure proved to be an under-estimate. In 1942, the statistician Edmund Brandt (one of many bureaucrats to find employment with the post-war West German government) prepared an analysis entitled "*What has so far been accomplished by the various institutions in terms of disinfection?*" The "institutions" were the six principle killing centres; "disinfection" was a code word for murder. By Brandt's calculations, between January 1940 and August 1941, when the killings were officially, although not actually suspended, precisely 70,273 patients had been liquidated by gassing. Brandt went on to calculate the savings made in food, clothing and accommodation on the assumption that those murdered had been allowed to survive for a further ten years. Based on the average daily needs for an institutional inmate, he estimated that the killing of these sick persons, incapable of labour—"useless mouths" in Nazi terminology—would yield precisely 885,439,800 *Reichsmarks* in savings by 1951.[779] Setting purely fiscal considerations to one side, put another way it has been calculated that as many as 33 percent of the beds occupied by patients of mental institutions in 1939 had been made available through "euthanasia" for other use by the end of 1941.[780]

---

776 John J Michalczyk (ed), *Medicine, Ethics, and the Third Reich: Historical and Contemporary Issues* (Lanham: Rowman & Littlefield, 1994), p. 42.

777 Robert Jay Lifton, *The Nazi Doctors: Medical Killing and the Psychology of Genocide* (London: Papermac, 1990), pp. 93–94.

778 Michael Burleigh, *The Third Reich: A New History* (London: Pan Books, 2001), p. 385.

779 Götz Aly, and Susanne Heim, *Architects of Annihilation: Auschwitz and the Logic of Destruction* (London: Phoenix, 2003), p. 168.

780 Lifton, *The Nazi Doctors*, p. 142.

In addition to those gassed, at least 20,000 further victims had been killed by the time of the "stop order" by means of lethal injection. In truth, despite a pause, the killing in various guises continued until the end of the war. Ludwig Sprauer, senior medical officer in Baden, and thus the official responsible for "euthanasia" in that state, testified that it continued uninterrupted from 1941, stating: "The persons killed in the course of this programme included not only those who were mentally sick but also those who suffered from arteriosclerosis, tuberculosis, cancer, and other ailments."[781] Had Sprauer been completely honest, he would have admitted that, if anything, following the pause, the killing was actually expanded to include the *Gemeinschaftsfremde* ("aliens to the community") as the Nazis labelled them—the maladjusted, sick foreign slave labourers, civilians suffering breakdown as a result of Allied air raids, and anybody else considered to be "useless."[782] A drafted but never implemented law of 1945 defined an "alien to the community" as "one whose personality and way of life renders him incapable of fulfilling the minimum demands of the national community through his own efforts, especially as a result of extraordinary defects in judgment or character." Apart from those who were obviously criminal, encompassed within this definition were the rowdy, the drunken, beggars—in fact, anybody deemed by the state to be "anti-social."[783]

So far as children were concerned, the killing never stopped at all. At the Kelsterbach hospital alone, 68 children under the age of three died between 1943 and 1945. One Polish woman described how a man she assumed was a German physician, together with his assistants, had murdered children at Kelsterbach:

> ...They took the child of someone I knew, undressed it, and laid it on the table which stood in our room...As the child lay on the table he [the assumed physician] gave it an injection in the lower part of the spinal column, and then sucked out a white, transparent fluid, which looked like water...This process took about half an hour. During the process the child began to scream, but it was then held by the orderlies until it was quiet and had died.

The woman's own child was then killed. She was instructed to take her child's body to the washroom, where she witnessed the corpses of approximately ten children. The oldest was no more than eighteen months of age.[784]

Sources suggest that, taken together, the original estimate of a total of 160,000 victims of "euthanasia" now seems conservative. A figure of 200,000 appears to be minimal, with even double that number quoted by some historians.[785] The more recent the research, the greater the evidence is uncovered of the true extent of the killing. In short, what had originally been conceived by the eugenicists as the extermination of the med-

---

781 Joseph Tenenbaum, *Race and Reich: The Story of an Epoch* (Westport: Greenwood Press, 1976), p. 95.

782 Götz Aly, Peter Chroust, Christian Pross, *Cleansing the Fatherland: Nazi Medicine and Racial Hygiene* (Baltimore: The Johns Hopkins University Press, Baltimore, 1994), p. 15.

783 Ibid., p. 52.

784 Michael Burleigh and Wolfgang Wipperman, *The Racial State: Germany 1933–1945* (Cambridge: Cambridge University Press, 1991), p. 165.

785 Ian Kershaw, *Hitler 1936–45: Nemesis* (London: Allen Lane, The Penguin Press, 2000), p. 261; Israel Gutman (ed), *Encyclopedia of the Holocaust* (New York: Macmillan Publishing Company, 1990), p. 453. Aly, Chroust, Pross, *Cleansing the Fatherland*, p. 22 claims: "...the murder of more than two hundred thousand psychiatric patients, camp inmates who had fallen ill, people suffering from major depression and nonconformists." Paul Julian Weindling, [*Nazi Medicine and the Nuremberg Trials—From Medical War Crimes to Informed Consent* (Basingstoke: Palgrave Macmillan, 2006) p. 188] puts the "currently estimated" number of victims of "euthanasia" at 400,000.

ically "incurable", concluded with the murder of those who, for whatever reason, were either unable to contribute to the economic well being of the *Reich*, or were considered politically or racially undesirable. By 1942, "euthanasia" was considered an essential element in the modernization of mental institutions to the point where a planning review was able to predict that, in future, "with few exceptions, death by euthanasia will hardly be distinguishable from natural death. That is the goal to be striven for."[786]

It is dangerous to simply rely upon statistics without giving further consideration to their composition, but the decrease in the number of psychiatric patients in Berlin from 9,204 in 1939 to 1,807 in 1945 is surely indicative of the scale of the killing. Some Berlin patients had certainly been transferred to other institutions (often for liquidation), since by 1943 Wittenau was the only remaining public psychiatric institution in the city; others had undoubtedly died of natural causes, but there is significant evidence to indicate that an unknown number of patients were killed in Wittenau itself either by lethal doses of medication, starvation, or a combination of the two.[787]

There were some in a position of authority who had the courage to object to this policy of state murder. Lothar Kreyssig was a judge in the province of Brandenburg who discovered that patients from the Brandenburg asylum were being murdered. In July 1940 he protested to Franz Gürtner, the Minister of Justice, deploring what were unquestionably illegal killings. After a confrontation with Roland Freisler, at that time a State Secretary, but later to become the notorious head of the *Volksgericht*, the People's Court, Kreyssig was informed by Gürtner that he would be compulsorily retired that December—which he was.[788] Others were treated more harshly. The Protestant Pastor, Paul Gerhard Braune, submitted a memorandum to Lammers in 1940, in which he questioned how 2,019 people had died within a period of 43 days at Grafeneck,[789] when the asylum only had a capacity of 100 beds. Braune was arrested by the *Gestapo* two months later for his troubles, but was released after three months imprisonment at Prinz-Albrecht-Strasse, the Berlin headquarters of the *Gestapo*, and survived the war.[790]

The Nazi leadership remained nervous about public reaction to the programme. On 25 November 1940, Else von Löwis of Menar, a member of an aristocratic family, a

---

786 Aly, Chroust, Pross, *Cleansing the Fatherland*, p. 35.

787 Bronwyn Rebekah McFarland-Icke, *Nurses in Nazi Germany* (Chichester: Princeton University Press, 1999), p. 212.

788 When Kreyssig, protested about the illegality of the "euthanasia" action, Gürtner replied: "If you cannot recognize the will of the Führer as a source of law, as a basis of law, then you cannot remain a judge." (Kershaw, *Hitler 1936–45: Nemesis*, p. 253–254) This was a complete reversal of Gürtner's position in 1934, when in rejecting proposals for the legalisation of euthanasia he had stated: "If we start out in this direction, it would touch on the very foundations of Christianity's teachings to humanity, it would be the fulfilment of Nietzschean thoughts." [Dick de Mildt, *In the Name of the People: Perpetrators of Genocide in the Reflection of Their Post-War Prosecution in West Germany. The 'Euthanasia' and 'Aktion Reinhard' Trial Cases* (The Hague: Martinus Nijhoff Publishers, 1996), p. 55)].

789 Braune had calculated the total of fatalities from the numbers printed on the urns containing the ashes of victims which had been sent to relatives [Michael Burleigh, *Ethics and extermination: Reflections on Nazi genocide*, (Cambridge: Cambridge University Press, 1997), p. 138].

790 Michael Burleigh, *Death and Deliverance : 'Euthanasia' in Germany c. 1900–1945* (London: Pan Books, 2002), p. 160–166; Michael Burleigh, *Political Religion and Social Evil.* (Totalitarian Movements and Political Religions: Vol. 3, No. 2, 2002) 1–60*l*, p. 32. Reinhard Rürup (ed), *Topography of Terror: Gestapo, SS and Reichssicherheitshauptamt on the "Prinz-Albrecht-Terrain." A Documentation* (Berlin: Verlag Willmuth Arenhövel, 2000), p. 161.

committed Nazi, and a leader of the party's women's movement, sent a letter to her friend, the wife of Walter Buch, the presiding judge of the Nazi party court. Von Löwis asked the Buchs to inform Hitler of her concern that the killing of patients was straining the allegiance of the citizenry to National Socialism, writing:

> Surely you know about the measures currently used by us to dispose of incurable mental patients. Still, you may not fully realize how it is accomplished and the vast scope of the undertaking, nor the terrible impression it leaves with the population! Here in Württemberg the tragedy takes place in Grafeneck, and this place has thus acquired an ominous reputation.

Events in Grafeneck and in other places had become common knowledge and thus were now a "public secret," resulting in a "terrible feeling of insecurity"; people were asking, "What can one still believe? Where will this lead us and what will be its limits?" The population remained convinced that "the *Führer* obviously does not know about this," but the party could lose the confidence of the people if it continued to deceive in this manner. A legal euthanasia law based upon the existing sterilisation law would be acceptable to the *Volksgemeinschaft*, but the present unlawful system was simply unsustainable.

Walter Buch forwarded this letter to Heinrich Himmler on 7 December 1940, pointing out that if the activities in Grafeneck could not remain secret and were causing such unrest, an alternative *modus operandi* must be put in place. On 19 December, Himmler replied to Buch:

> Many thanks for your letter of 7 December 1940. I can inform you in confidence that the events that take place there are authorized by the *Führer* and are carried out by a panel of physicians.... The SS only assists with trucks, cars, and the like. I agree with you on one point. The process must be faulty if the matter has become as public as it appears.... I will immediately contact the office that has jurisdiction to point out these errors, and advise them to deactivate Grafeneck.

Himmler did exactly that. On the same day he received Buch's letter, he wrote to Viktor Brack, advising him to close Grafeneck:

> As I have heard, there is great excitement in the Swabian Jura due to the institution Grafeneck. The populace recognizes the gray automobile of the SS, and thinks it knows what is happening under the constant smoke of the crematorium. What takes place there is a secret, and yet is no longer a secret. Thus the worst public mood has taken hold there, and in my opinion there remains only one option: discontinue the operation of the institution in this locality.[791]

Brack heeded Himmler's suggestion. Grafeneck ceased operations on 13 December 1940.

The concept of "Jewish disease" had a long pre-Nazi history. According to some so-called "experts", a disproportionate Jewish tendency to suffer from certain sicknesses extended to anything from mental illness to myopia, glaucoma, diabetes, tuberculosis, gall, bladder, and kidney stones, rheumatism, and much else. As an indication of how

791 Henry Friedlander, *The Origins of Nazi Genocide: From Euthanasia to the Final Solution* (Chapel Hill: University of North Carolina Press, 1995), pp. 107–108.

ludicrous many of these judgements were, in a contrary opinion, Otmar Freiherr von Verschuer, head of the Frankfurt Institute for Racial Hygiene, actually considered Jews *less* susceptible to tuberculosis than non-Jews.[792] However, for the Nazis, the image of the Jew as the bearer of disease was potent, indeed totemic. It provided the justification, if one were needed, for the ghettoisation of the Jews in the occupied countries of eastern Europe, ostensibly on eugenic rather than racial grounds, for as a Nazi newspaper in the *Generalgouvernement*[793] stated: "The separation of the Germans from the Poles—and particularly from the Jews—is not merely a question of principle; it is also, at least as far as Warsaw is concerned, a hygienic necessity."[794] It was to become "a hygienic necessity" in many places other than Warsaw. Thus the reason given for the establishment of the Lodz ghetto was the alleged danger of the outbreak of epidemics.[795]

The districts chosen to house the ghettos were inevitably situated in the most impoverished parts of cities and towns. The housing was dilapidated, often with no piped water or electricity. The number of people packed into a ghetto produced staggering levels of population density. In Warsaw, 30 percent of the population were forced to live in 2.4 percent of the city's area; the ghetto district occupied about 425 acres, of which 375 acres (approximately 152 hectares) was residential space. The Germans calculated a density of 6–7 people per room in the Warsaw Ghetto. According to calculations made after the war the density actually reached 9.2 people per room, while the population density of the ghetto as a whole rose to 128,000 per square kilometre.[796] The allocated living space of the ghetto in the town of Checiny was fixed at 2–2.5 square metres per person.[797] In the small ghetto of Odrzywol, 700 people lived in an area previously occupied by 5 families, so that between 12 and 30 people had to share a single small room. The intolerable population density, inadequate hygienic and sanitary facilities—in the Lodz Ghetto 95 percent of apartments had no sanitation, piped water or sewerage—almost complete lack of medical supplies, absence of fuel for heating, and starvation rations, combined to produce conditions in which sickness and epidemics were inevitable.[798]

Lice plagued the ghetto population. In the Kutno Ghetto, which the Germans nicknamed *Krepierlager* ("Pegging out Camp"), between March and December 1941, 42 percent of all deaths were of typhus patients.[799] The overall mortality rate during that period in Kutno was almost ten times the pre-war rate, for other contagious diseases were also commonplace.[800] On 16 December 1941, Wilhelm Kube, *Generalkommissar* of Byelorussia pointed out that there were 22 epidemics prevalent in the country at that time.

---

792 Robert N Proctor, *Racial Hygiene: Medicine Under the Nazis* (Cambridge Massachusetts: Harvard University Press, 1989), pp. 196–197.

793 That part of occupied Poland not annexed to the *Reich* or initially ceded to the Soviet Union.

794 Proctor, *Racial Hygiene*, p. 199.

795 Ibid., p. 200.

796 Yisrael Gutman, *The Jews of Warsaw 1939–1943* (Bloomington: Indiana University Press, 1989), p. 60.

797 Raul Hilberg, *The Destruction of the European Jews* (New Haven: Yale University Press, 2003), p. 229.

798 Isaiah Trunk, *Judenrat: The Jewish Councils in Eastern Europe under Nazi Occupation* (Lincoln: University of Nebraska Press, 1996), p. 145.

799 Ibid., pp. 148–149.

800 Ibid., p. 153.

No serum was available for their treatment.[801] In Nazi ideology the Jews had always been regarded as the bearers of disease. Now, because of conditions the Nazis had themselves created, this took on the nature of a self-fulfilling prophecy. Moreover, the deliberate creation of this appalling environment provided part of the rationale for the extermination of the ghettoised Jews. However, the precedent and method for doing so had been set much earlier.

Ludwig "Israel" Alexander was the only Jew listed on the Eglfing-Haar transport of twenty-five men destined for Grafeneck on 18 January 1940. After 1 January 1939, all male Jews were forced to take the second name "Israel", and all Jewish women the second name "Sarah". It is thus evident that Alexander was considered Jewish, as defined by the Nuremberg laws, and was therefore probably the first handicapped Jewish patient murdered in the gas chamber of a "euthanasia" killing centre.[802] Initially no distinction had been made between Jewish and non-Jewish adult patients in the "euthanasia" programme, even though as early as 24 October 1939, a mere fifteen days after T4 had effectively commenced operations, *Reich* Health Leader Leonardo Conti ordered mental institutions to report to his office details of all Jewish inmates in their care.[803] In a follow-up dated 15 April 1940, Herbert Linden required local health authorities to provide him with details of Jewish patients within their jurisdiction.[804] On 30 August 1940, the Ministry of the Interior ordered that Jewish patients be separated from German inmates and placed in specified asylums, from where some were selected for gassing. At the end of that year, all remaining Jewish patients were incarcerated in a single asylum at Bendorf-Sayn prior to their murder. The director of the Munich hospital replied to a complaining relative in a letter: "The transfer (of the Jewish patients) has not only administrative reasons, but it makes sense in so far as many Aryan patients and nurses repeatedly refused to share the same hospital with Jewish patients."[805] By November 1942, Bendorf-Sayn could be closed. There were no longer any Jewish mental patients.[806]

In response to enquiries from the families of a number of Jewish patients who had actually been gassed in Brandenburg, *Gekrat* replied that the victims had been sent to an asylum in Chelm (sometimes referred to as either "Chelm II", "Cholm", or Cholm II), in the Lublin District of Poland. Notices containing fictitious dates of death were mailed from Chelm in order to add to the deception. *Aktion Cholm,* as it was known, proved particularly lucrative for T4, since it was not only possible to collect the gold teeth and jewellery of the deceased, but also their hospital fees, which continued to be received for some months after the actual date of death.[807]

---

801 Yitzhak Arad, Israel Gutman and Abraham Margaliot (eds.), *Documents On The Holocaust* (Lincoln: University of Nebraska Press, 1999)., p. 408.

802 Friedlander, *The Origins of Nazi Genocide*, p. 271.

803 Hilberg, *The Destruction of the European Jews*, p. 467.

804 Eugen Kogon, Hermann Langbein, Adalbert Rückerl (eds.), *Nazi Mass Murder: A Documentary History of the Use of Poison Gas* (New Haven: Yale University Press, 1993), p. 31.

805 http://www.phil.gu.se/sffp/reports/16.%20von%20Cranach.pdf (Accessed 25 February 2008.)

806 Hilberg, *The Destruction of the European Jews,* pp. 467–468.

807 Kogon, Langbein, Rückerl, *Nazi Mass Murder*, pp. 31–32.

Following a decision of the Ministry of the Interior in early 1941, all Jews in German hospitals were ordered to be exterminated.[808] From that year German Jewish patients were sent to the *Generalgouvernement*, and there either shot or killed in gas vans, although from the time of Hitler's decision in the autumn of that year to deport German Jews *en masse*, these killings did not form part of *Aktion* T4, but rather of the so-called "Final Solution." By the summer of 1940, the medical condition of Jewish patients had already become irrelevant. They were being killed simply because they were both Jewish and a patient.[809] Very shortly, one of those qualifications also disappeared. Selected members of the staff of T4 began killing Jews of all nationalities on a gigantic scale in the Polish death camps solely because they were Jews.

By the early summer of 1941 the "euthanasia" genie was out of the bottle. An undated letter to the *Reich* Minister of Justice quoted at the Nuremberg Medical Trial read:

> I have a schizophren[i]c son in [a] Württembergian mental institution. I am shocked about the following absolutely reliable information.
>
> Since some weeks, insane persons are being taken from the institutions allegedly on the grounds of military evacuation. The directors of the institutions are enjoined to absolute secrecy. Shortly afterwards the relatives are informed that the sick person has died of encephalitis. The ashes are available if so desired. This is plain murder just as in the concentration camps. This measure uniformly emanates from the SS in Berlin. The institutions dare not inform the authorities. Inquire at once at Rottenmuenster, Schassenried, Winzertal, all in Würtemberg. Have the lists of two months ago examined and submitted to you, check upon the inmates who are there now and ask where the missing persons went to. For seven years now this gang of murderers defiled the German name. If my son is murdered, woe! I shall take care that these crimes will be published in all foreign newspapers. The SS may deny it as they always do. I shall demand prosecution by the public prosecutor.
>
> I cannot give my name nor the institution where my son is, otherwise I, too, won't live much longer. Heil Hitler Oberregierungsrat N.[810]

Knowledge of the "euthanasia" programme had become widespread, in part because the killings were taking place on German or Austrian soil. Visiting a lady doctor friend who had founded a private clinic in Pirna in 1937, Victor Klemperer noted her remarks in a diary entry of 21 May 1941:

> Sonnenstein has long ceased to be a regional mental asylum. The SS is in charge. They have built a special crematorium. Those who are not wanted are taken up in a kind of police van. People here call it "the whispering coach". Afterwards the relatives receive the urn. Recently one family here received two urns at once.[811]

---

808 Proctor, *Racial Hygiene*, p. 207.

809 Aly, Chroust, Pross, *Cleansing the Fatherland*, p. 44. It had never been a requirement for Jewish patients to meet the usual medical criteria established by T4 prior to their being murdered. No consultations or discussions were appropriate or necessary: "The total extermination of this group of asylum inmates was the logical consequence of the 'radical solution' of the Jewish problem being embarked upon." (Lifton, *The Nazi Doctors*, p. 77).

810 http://140.247.226.138/NurTranscript/TranscriptPages/99_60.html (Accessed 26 June 2008).

811 Victor Klemperer, *I Shall Bear Witness : The Diaries of Victor Klemperer 1933–41* (London: Weidenfield & Nicolson, 1998), p. 368.

In the town of Absberg, the SD reported, citizens had actually resisted the removal of the disabled inmates from the home at Ottilien.[812] Still, not everybody objected. The murder of handicapped children appears to have received broad public support across the country, and it is suggested that the liquidation of mentally sick adults was not entirely unpopular.[813] Where there were objections, the Nazis were quick to learn the lesson. Future killing centres would be established far away, where possible in isolated areas. Meanwhile, awareness by any sector of the general population, however small, of what had been from the beginning a *Staatsgeheimnis* (state secret) rang alarm bells, as Himmler's reaction over Grafeneck illustrated. Aware of such growing public disquiet, and in the wake of explicit sermons preached by a number of churchmen, in particular the Roman Catholic Bishop Clemens August Graf von Galen of Münster, on 24 August 1941 Hitler ordered the suspension of adult "euthanasia", at the same time threatening post-war revenge on the bishop for his audacity.[814] Von Galen had certainly pulled no punches, saying:

> Do you, do I have a right to live only as long as we are productive, as long as long as others recognize us as being productive? If the principle is established and applied that "unproductive fellow citizens" can be killed, woe to all of us when we grow old and feeble! If it becomes permissible to kill unproductive people, woe to the invalids who invested and sacrificed and lost their energies and sound bones during their working careers. If unproductive fellow citizens can be eliminated by force, woe to our brave soldiers who return to their homeland severely injured, as cripples, as invalids. Once it becomes legal for people to kill "unproductive" fellow citizens—even if for at presently only our poor defenceless mentally ill are concerned—then the basis is laid for murder of all unproductive people, the incurable, the invalids of war and work, of all of us when we become old and feeble.[815]

It has been suggested that this ostensible cessation was not so much caused by the protests of a few churchmen and others,[816] but rather because the T4 personnel were now required to practice their killing technique on a vastly greater scale as part of *Aktion Reinhard.*[817] Considered by his contemporaries to be a man of no great intellect, nor prior to his revelatory sermon one overly endowed with moral courage, Von Galen's condemnation of "euthanasia" occurred in the last of three sermons delivered by him in summer 1941 on the subject of Nazi outrages. Following a conversation in early June of that year with the Dominican priest, Odilo Braun, in the course of which Braun revealed the extent of state seizure of church assets, von Galen's principal motivation for his condemnation of the regime appears to have been not so much the murder of the incapacitated, but rather his concern that Catholic institutions in Germany were threatened

---

812 Hugh Gallagher, *What the Nazi "Euthanasia Program" Can Tell Us About Disability Oppression* (Journal of Disability Policy Studies, Vol.12, No.2, 2001), p. 97.

813 Robert Gellately, *Backing Hitler: Consent and Coercion in Nazi Germany* (Oxford: Oxford University Press, 2002), p. 104.

814 Von Galen had first been informed about the "euthanasia" programme more than a year earlier. (Burleigh, *Ethics and extermination*, p. 126).

815 Samuel Totten, William S Parsons, Israel W Charny (eds.), *Century of Genocide: Eyewitness accounts and Critical Views* (New York: Routledge, 2004), p. 189.

816 This was certainly the view of Dieter Allers: "Nobody cared about what those fellows said in church...Hardly anyone went to church anyway. All we cared about was our crust of bread and getting the war over and done with." [Gitta Sereny, *Into That Darkness—From Mercy Killing to Mass Murder* (London: Pimlico, 1995), p. 75)]. Which in Aller's case meant continuing to commit genocide.

817 Burleigh, *Death and Deliverance*, p. 174.

with the prospect of excision.[818] The *Gestapo* were seizing church properties and hounding priests in Münster, and it was these acts that convinced von Galen to voice his alarm.[819]

An even simpler reason for the official halt may have been that the envisaged target figure of victims had been reached, although that seems unlikely in the light of the increase in the number of potential victims believed to have been agreed by Bouhler and Brandt in January 1941.[820] It seems likely that there was no single motivation for the "stop order", but simply a combination of all of the factors mentioned. Whatever the reasoning, children's "euthanasia", *Sonderbehandlung* 14f13, and the so-called "wild euthanasia" of adults in hospitals continued. In fact, the suspension should more accurately be called a pause for reorganization.[821] Indeed, it is certain that more victims of "euthanasia" perished after the official order to cease than had been killed before it was issued.[822] Thereafter, gas chambers at the killing centres functioned on a greatly reduced scale, but thousands of patients were murdered by means of lethal drugs or starvation at mental hospitals like Eichberg, Meserlitz-Oberwald, Kaufbeuren and many others.[823] What occurred was in reality not a cessation, but rather a change of direction and method. That the "stop order" came as a surprise to the "euthanasia" perpetrators is indicated by the report of a Bavarian planning commission in October 1942, which indicated that the previous year's target figures for patient mortality had not been

818 http://www.foreignpolicy.com/articles/2010/05/06/why_did_the_pope_keep_quiet_about_the_holocaust?page=0,0 (Accessed 14 May 2010)

819 Beth Griech-Polelle, *Image of a Churchman-Resister: Bishop von Galen, the Euthanasia Project and the Sermons of Summer 1941* (Journal of Contemporary History, Vol 36, No.1, 2001), p. 42. This revisionist assessment of von Galen proposes that he was not the heroic resister as so often portrayed, but rather "an example of a high-ranking German clergyman who offered selective opposition to certain Nazi policies". (Ibid., p. 43.) There is particular criticism of a complete absence of any specific mention of the persecution of the Jews in von Galen's pronouncements: In keeping with traditional Catholic teaching, von Galen could blame the Jews for their own misfortunes since they had failed to recognize the "true Messiah". As St Augustine had written centuries earlier, Jews could be mistreated in order to be reminded of their "crime"; they did not need to be loved. Following this line of reasoning, turning a blind eye to the misfortunes of the Jews would not necessarily make one a bad person. (Ibid., p. 52). It seems evident that if von Galen may be considered to some degree anti-Semitic, this was on traditional religious rather than racial grounds. In his defence, it has been pointed out that von Galen was highly critical of Nazi racist anti-Semitism, and maintained good relations with the Münster rabbi Fritz Leopold Steinthal, indeed going so far as to enquire after the rabbi's well-being following *Kristallnacht*. [http://www.foreignpolicy.com/articles/2010/05/06/why_did_the_pope_keep_quiet_about_the_holocaust?page=0,1 (Accessed 14 May 2010)].

820 Burleigh, Wipperman, *The Racial State*, p. 153. Allers quoted Brack as stating that it was expected to complete the "euthanasia" programme by July 1941 at the latest (Sereny, *Into That Darkness*, p. 76). If that is true it is clear that Brack was referring only to the first phase of the operation.

821 It was assumed by the leading lights in T4 that the cessation was only temporary, and that it would soon be business as usual. Nitsche was insistent that registration of potential victims continue in readiness for a resumption of the killing. (de Mildt, *In the Name of the People*, p. 339, note 79). In March 1943, Hans and Margot Räder-Grossmann, in charge of gassing at Bernburg, wrote to Friedrich Mennecke: "It is said we will soon get a lot of work, and by the beginning of May we should know whether yes or no." (Aly, Chroust, Pross, *Cleansing the Fatherland*, p. 79). In the event, the answer was "no". Bernburg ceased gassing operations in April 1943.

822 Friedlander, *The Origins of Nazi Genocide*, p. 151.

823 In addition to patients with psychiatric conditions considered incurable, from August 1941 the criteria for those slated for extermination included being a "public menace", as well as "criminality", a "psychopathic condition", "antisociality" and racial and social "inferiority." (Aly, Chroust, Pross, *Cleansing the Fatherland*, p. 46.) This could, of course, mean almost anybody.

achieved because they had been suggested "on the assumption that the Operation [that is "euthanasia"] would continue". Such assumptions had not been limited to Bavaria.[824]

With Paul Nitsche and Herbert Linden the driving forces behind a resumption of "euthanasia", albeit in an amended form, in September 1942 Linden's superior, Fritz Cropp, issued a directive to the institutions under his control. In view of the overcrowding in certain hospitals because of bombing raids, he asked for details of the mentally ill who were "eligible for transfer" to asylums in less dangerous regions. The coded message was clearly understood by those to whom it was directed; "euthanasia" was back on the agenda.[825] In this second phase of "euthanasia", murder became decentralised. No longer were there to be just a handful of killing centres, each equipped with a gas chamber in which a number of individuals could simultaneously be annihilated. For the purpose of *Sonderbehandlung* 14f13, the gas chambers at Bernburg and Sonnenstein remained operational for a short time, that at Hartheim for somewhat longer. Elsewhere, killing of the undesirable was now to be on an individual basis, be simpler in method, and occur at institutions throughout the *Reich*.

With the reorganization, it was no longer just those deemed medically "inferior" who became the target of T4. Linden busily compiled the registration of those considered socially unacceptable.[826] Thus the list of those earmarked for elimination was actually expanded. In a "*Manual of Hereditary Diseases*" published in 1942, Hans Heinze listed those unfit for marriage and therefore reproduction: prostitutes, vagrants, professional criminals, pimps, paupers, swindlers, cheats, frauds, confidence tricksters, "hysterical scoundrels" (whatever that may have meant), "incorrigible and active confirmed homosexuals", and "incorrigible shirkers"—amongst others.[827] Compulsory sterilisation was urgently required for "antisocial and recidivist criminal psychopaths whose hereditary deviance can be inferred from their genealogy." Within the context of Nazi racial hygiene, "psychopath" meant not just those who fell within the psychiatric definition of the term, but individuals who were simply nonconformists—"aliens to the community", as described earlier. Ominously, Heinze added: "We hope that after the war, the fight against or extermination of subhumanity through purposeful measures will take its honoured place as a further great deed beside those already accomplished."[828]

Following a meeting held in September 1942, Otto Thierack, the Minister of Justice, noted:

> With regard to extermination of antisocial lives, Dr Goebbels' point of view is that all Jews and Gypsies, Poles who have served prison sentences of around 3–4 years, and Czechs or Germans who have been sen-

824 Aly, Chroust, Pross, *Cleansing the Fatherland*, pp. 166–167.

825 Ulf Schmidt, *Karl Brandt: The Nazi Doctor—Medicine and Power in the Third Reich* (London: Hambledon Continuum, 2007), pp. 211–212.

826 Aly, Chroust, Pross, *Cleansing the Fatherland*, p. 64.

827 A 1938 article written for the Office of Racial Policy drew comparisons between two of the pariah groups in Nazi Germany—Jews and homosexuals. Like the Jews, the latter built a state within a state. They were not sick people to be pitied, or "treated", but enemies of the state to be eliminated. (Proctor, *Racial Hygiene*, p. 213).

828 Aly, Chroust, Pross, *Cleansing the Fatherland*, p. 53.

tenced to death, life in prison, or preventive detention should be exterminated. The best idea is extermination through work.[829]

A secretary confirmed the new categories of potential victims when she testified in 1946: "I was assigned by the Foundation [*Stiftung*] to duty in the Rummelsburg workhouse. Here files were prepared that would process the so-called antisocial elements for extermination."[830]

By October 1942, two doctors were already visiting German mental hospitals for the purpose of evaluating inmates with criminal records, with particular reference to the patient's capacity for work. These found "no longer [in] need [of] treatment in mental institutions and [who] are capable of working", had "KZ" marked on their records, meaning recommendation for the transfer of the individual to a concentration camp (KZ). Such patients were listed by the Ministry of Justice and placed "at the disposal of the police for accommodation in police labour and disciplinary camps", in reality meaning "for liquidation through labour". This led to a rather odd conflict of interests, for while Himmler and the SS intended to use those capable of labour by working them to death, the directors of mental institutions wished to retain the very same individuals alive and resident for the purpose of the asylum's own productivity. After much negotiation, the Ministry of Justice were forced to retract; they issued a new directive, excepting "from delivery to the police...those employed in important work either within or outside the mental hospitals, whom it is either impossible or inexpedient to replace with other workers." Shortly thereafter, Herbert Linden was able to inform asylum directors that the decision now rested with them whether to send a patient to a concentration camp for extermination through labour, to a killing centre for death by other means, or to retain them for work at the institution.[831]

Karl Brandt, the pioneer of "euthanasia" and T4, was appointed plenipotentiary for medicine and health (*Bevollmächtigter für das Sanitäts- und Gesundheitswesen*) in July 1942, and granted additional powers by Hitler in September 1943. Brandt thus became the central medical authority of the *Reich*, responsible in particular for the allocation of hospital beds and other medical resources.[832] There is some evidence that Brandt received Hitler's approval in summer 1943 to begin a second phase of "euthanasia". In post-war testimony, Nitsche stated that in August 1943 he had "managed to obtain Hitler's expressed approval, through the mediation of Professor Brand[t], to authorise certain responsible physicians to perform euthanasia in individual cases by means of drugs." Although by this stage Brandt had become a major player in his own right in the chaotic Nazi government, and had "the unlimited trust of the Führer", it is highly unlikely that he would have issued such an authorization without first obtaining Hitler's consent, or even perhaps his specific instructions.[833] Albert Widmann, the chief chemist of the Kripo,[834] who had recommended the use of bottled carbon monoxide in the gas chambers of the killing centres, now became responsible for the delivery of lethal

829 Ibid., p. 70.

830 Ibid., p. 64.

831 Ibid., pp. 71–74.

832 Friedlander, *The Origins of Nazi Genocide*, p. 157.

833 Schmidt, *Karl Brandt*, pp. 235–236.

834 Kripo = *Kriminalpolizei*, the Criminal Police.

pharmaceutical substances to various T4 related institutions, which in turn distributed them to an unknown number of mental hospitals. As the end of the war neared and it became increasingly difficult to obtain drugs, some physicians relied upon simple injections of air or gasoline to kill their victims.[835]

Together with Linden, Brandt created a programme codenamed *Aktion Brandt*, under the auspices of which mentally disabled patients were evacuated from mental institutions in areas vulnerable to Allied air raids, purportedly for the patients' protection. In fact, in many cases the evacuees were eventually transported to killing centres for elimination.[836] Under the umbrella of *Aktion Brandt*, in early 1943 several hundred patients were deported from hospitals in the Rhineland to asylums in the *Generalgouvernement* for killing. Within eighteen months all were dead.[837] In April 1943, Brandt informed Conti that he had visited north-west Germany. He considered that a "large scale plan for the supply of hospital beds, etc is necessary." And who did Brandt intend to turn to for aid in instituting this plan? "The organisation which has helped us in the registration of the cure and nursing homes through the Chancellery of the Führer [that is, T4]...Herr Blankenburg will also collaborate in this."[838]

The reason given for the removal of patients was to allegedly create room in hospitals for civilian casualties of the Allied bombing offensive and wounded *Wehrmacht* soldiers.[839] The reality was that this was simply another manifestation of "euthanasia", one which extended the concept of killing the unwanted "useless eaters", originally limited to specific categories of the community, into an instrument of unrestricted population control.[840] For example, in addition to the inmates of old-age homes in Hamburg sent to Meseritz-Obrawalde following the air raids of July 1943, between 22 June and 7 August, 349 mentally ill women of that city were transported to Hadamar for killing. About 20 percent of them had literally lost their minds as a result of the firestorm that had raged through Hamburg as a result of "Operation Gomorrah."[841] Ample evidence of their fate is provided by the correspondence passing between the victims' relatives and the killing centre. One son enquiring after his 77 year-old mother was informed by Hadamar: "Frau K. was so frail she had to be kept in bed constantly. Rapid degeneration began several days before her death. A weak heart in addition to this caused her death."[842]

The truth about the end of these deportees was revealed by the testimony of a surviving ex-patient of Meseritz-Obrawalde:

> I recall a transport from Berlin of patients who had suffered severe shock as a result of air raids on Berlin. These patients were not even distributed among the various houses, but instead were brought directly from the train to the hospital, where they were killed the same day.[843]

---

835 Aly, Chroust, Pross, *Cleansing the Fatherland*, p. 83.

836 Burleigh, *Death and Deliverance*, p. 247.

837 Aly, Chroust, Pross, *Cleansing the Fatherland*, p. 84.

838 Schmidt, *Karl Brandt*, p. 222.

839 Michael S Bryant, *Confronting the "Good Death": Nazi Euthanasia on Trial 1945–1953* (Boulder: University Press of Colorado, 2005), p. 89.

840 Ibid., p. 52.

841 Burleigh, *Death and Deliverance*, p. 247.

842 Aly, Chroust, Pross, *Cleansing the Fatherland*, pp. 86–87

843 Ibid., pp. 88.

Other mentally impaired evacuees were sent to hospitals such as Eichberg and Kaufbeuren, where they were either starved to death or killed with lethal injections or overdoses of medication.[844] It was not now quotas to be fulfilled, as had been the case in the first phase of T4, but rather individuals who were to be killed on the basis of "need." As the bombing raids increased in intensity, space had to be created for the treatment of those considered to be essential for the prosecution of the ongoing conflict. Total war precluded the use of hospital beds for any considered unlikely to again become economically productive, namely the aged and the infirm,[845] so that following an air raid on Bremen on 26 November 1943, most of the patients of a bombed hospital in that city were transferred to Meseritz-Obrawalde. By 2 February 1944 the great majority of them were dead.[846]

It has been estimated that between 1941 and 1944, some 35,000 patients were deported from various mental institutions to other more remote asylums, where many thousands were murdered.[847] Some died because of questions of explication. Nazism had so corrupted the German language that it was no longer possible to be certain of the meaning of certain words. If patients were to be "evacuated", was the expression to be taken literally or was some more sinister objective implied? Those responsible for interpreting such orders were often uncertain of what was required of them, and sometimes chose the more extreme option, even when this had not been the original intention.[848]

The initial phase of the killing system had been largely devoted to the elimination of supposedly "worthless life". The official conclusion of the application of "euthanasia" to the handicapped in August 1941 shifted the emphasis of the programme to the marginalized members of society—those "antisocial" elements listed so extensively by Heinze in the "Manual of Hereditary Diseases" mentioned earlier.[849] From 1943, all of those considered to be unproductive were exterminated in order to make room for others who could be nursed back to health and work, or whose lives were judged to be of greater worth. The sick and "antisocial" of all kinds were murdered wherever and whenever their removal provided essential hospital space and medical supplies for the healing of those deemed likely to be of future value to the state.[850] Thus, following the destruction of the University of Kiel in 1944, 700 patients of the mental hospital in Schleswig were transferred to Meseritz-Obrawalde for killing, in order to make room for the Kiel university students.[851] As their friends and family disappeared, elderly individuals were terrified by the prospect of becoming inmates of geriatric homes or sanatoria. Even a visit to the doctor became a feared ordeal.[852]

---

844 Bryant, *Confronting the "Good Death"*, p. 89.
845 http://motlc.wiesenthal.com/site/pp.asp?c=gvKVLcMVIuG&b=395193 (Accessed 20 February 2008).
846 Aly, Chroust, Pross, *Cleansing the Fatherland*, p. 85.
847 Schmidt, *Karl Brandt*, p. 223.
848 Ibid., p. 233.
849 Aly, Chroust, Pross, *Cleansing the Fatherland*, p. 76.
850 Ibid., pp. 81–82.
851 Ibid., pp. 85–86.
852 Burleigh, *Ethics and extermination*, p. 128.

In August 1943, the majority of the administrative staff of T4 was moved from Berlin to the new central office which had been established at Hartheim.[853] At the same time, Blankenburg, who by that time had succeeded Brack, reorganised T4 into three departments. Department I under Nitsche was responsible for the identification and examination of patients; department II headed by Gerhardt Siebert oversaw the place and time of the actual killings; and Friedrich Lorent was in charge of department III, which took care of the organisation's finances.[854]

Starting one month later, barrack-like structures either were, or were intended to be, erected within the grounds of at least 32 mental institutions. It was proposed that defined categories of patients would be held in these barracks for treatment, thereby making room in the hospitals themselves for others more likely to respond to medical care. The barracks, with their three tiered bunk-beds, bore an obvious similarity to the structures evident in concentration camps. The standard of treatment meted out to patients confined within them can be imagined. Only one addition was necessary to complete the parallel, and in the summer of 1944, Linden, who had been responsible for the planning and financing of this entire scheme, authorised the construction of crematoria in German mental hospitals.[855]

Future intentions were made abundantly clear in a planning report of July 1942 by the T4 *Gutachter*, Robert Müller, following an inspection of mental institutions in Baden:

> ...In implementing the contemplated euthanasia law, we cannot allow asylums to attain the reputation of institutions of death—in other words, asylums in which death awaits those transferred there. One of the basic conditions for implementing euthanasia is that it be as inconspicuous as possible. This means first of all an inconspicuous milieu...There we can accommodate incurable, chronic, or far advanced cases for which euthanasia has been or will be decided upon...These establishments and the general conditions for euthanasia cases must be indistinguishable from traditional nursing institutions. Euthanasia directives and their implementation must remain completely integrated into normal asylum routine. Thus, with few exceptions, euthanasia deaths will be all but indistinguishable from natural deaths. That is what we strive for...[856]

It is only recently that evidence has emerged of human experiments being conducted on psychiatric patients, for example the Schaltenbrand experiments in 1940 at the Werneck psychiatric hospital, in which Professor Georg Schaltenbrand injected chronically mentally ill patients intradernally and cisternally with spinal fluid from apes, the latter having been previously injected with spinal fluid from multiple sclerosis patients.[857] There is also an increasing body of evidence emerging of the involvement of the pharmaceutical industry (I.G. Farben) in the testing of drugs on patients.[858]

As conditions of wartime life in Germany inexorably deteriorated, the potential victims of Nazi eugenic and racial policies (to all intents and purposes the two had long since ceased to be distinguishable) were no longer solely Jews and the inmates of asy-

853 Aly, Chroust, Pross, *Cleansing the Fatherland*, pp. 183–184.

854 Schmidt, *Karl Brandt*, p. 233–234.

855 Aly, Chroust, Pross, *Cleansing the Fatherland*, pp. 89–91. Ludwig Trieb, administrative director of the Günzburg mental hospital was in charge of the "barracks operation." (Ibid, p. 169).

856 Ibid., p. 176.

857 http://www1.uni-hamburg.de/rz3a035//psychiatry.html (Accessed 25 February 2008.)

858 http://www.phil.gu.se/sffp/reports/16.%20von%20Cranach.pdf (Accessed 25 February 2008.)

lums. In November 1942, Kurt Blome, Deputy *Reich* Physician's *Führer* wrote to Arthur Greiser, *Gauleiter* of the *Warthegau*, concerning tubercular Poles of the region governed by the latter. Earlier Greiser had written to Himmler advocating a "radical procedure" for the resolution of the problem. Everybody understood what the coded "radical procedure" implied, but Blome was rather more circumspect than others in his response:

> Since some time ago the Führer halted the programme in the mental institutions, it occurs to me that he may not regard "special treatment" of the hopelessly sick as politically feasible at the moment. In the case of the euthanasia programme, German citizens afflicted with hereditary diseases were involved. This time it would be infected members of a subjugated nation. There can be no question that the proposed method represents the simplest and most radical solution. If there were assurances of complete secrecy, all reservations, regardless of the reason, could be withdrawn. But I regard such secrecy as downright impossible.

Rather than gassing, Blome proposed "strict quarantine and institutionalization of all infectious, hopeless tubercular patients. This solution would tend to make the patients die off fairly rapidly. The necessary inclusion of Polish physicians and nurses would to some extent rob such institutions of the character of death camps." In other words, murder by alternative means. The following month Himmler instructed Greiser "to select a suitable area into which incurable tubercular patients can be sent." Whether such a scheme was implemented is unknown.[859] Blome was one of those acquitted at the Nuremberg Medical Trial, allegedly in return for providing the Americans with details of his research into biological and chemical warfare, regarding which subject he had probably had practical experience. Whilst admitting that he was a convinced Nazi, and approving of "euthanasia" in principle, Blome claimed to have been opposed to *Aktion* T4 on the grounds that Hitler's 1939 authorisation was of dubious legality.[860] It was an assertion impossible to refute, and one that appeared to sit comfortably with the Nuremberg Tribunal, who in any event had only the haziest notion of the legality or otherwise of Hitler's dicta.[861]

Presumably having exhausted the list of other helpless victims and with the war already clearly lost, in November 1944 the Ministry of Justice were considering killing the unattractive:

> During various visits to the penitentiaries, prisoners have always been observed who—because of their bodily characteristics—hardly deserve the designation human; they look like miscarriages of hell. Such prisoners should be photographed. It is planned that they too shall be eliminated. Crime and punishment are irrelevant. Only such photographs should be submitted that clearly show the deformity.[862]

---

859 Eugen Kogon, *The Theory and Practice of Hell* (New York: Berkley Books, 1998), p. 227.

860 Weindling, *Nazi Medicine and the Nuremberg Trials*, p. 254.

861 Having served in the army throughout the First World War, Blome became active in extremist right-wing political movements during the Weimar years. He joined the NSDAP and SA in 1931, and advanced steadily upwards through the Nazi medical hierarchy following the *Machtergreifung*. (Wolfgang Uwe Eckart (ed), *Man, Medicine, and the State: The Human Body as an Object of Government Sponsored Medical Research in the 20th Century* (Stuttgart: Franz Steiner Verlag, 2006), pp. 199–200). Another source dates Blome's membership of the NSDAP to 1922, which given his background seems more likely. (Ernst Klee, *Das Personenlexikon zum Dritten Reich–Wer war was vor und nach 1945* (Frankfurt am Main: S.Fischer, 2005), p. 54).

862 Hilberg, *The Destruction of the European Jews*, pp. 1066–1067. Emphasis added.

It would be reassuring to think that the killing of patients ceased with the end of the war. Regrettably, that was not the case, for thousands perished from the effects of starvation and neglect in the months that followed the termination of hostilities. Heinz Faulstich was one of the first psychiatrists to document the results of murder by starvation. He proposed a minimum number of 20,000 deaths due to malnourishment in the post-war period alone. It is impossible to obtain precise figures, as so many of the asylums and institutions destroyed data and relevant documentation, but some statistics are indicative. On 28 April 1945, the day of liberation, the mental institution at Teupitz accommodated 600 inmates. By the end of October 1945, their number had been reduced to 54. At Grossschweidnitz, 1,012 inmates died in May 1945 alone. The mortality rate at a variety of hospitals is also revealing. At Altscherbitz it was 36·5 percent in 1945, rising to 38 percent in 1947. At Zwiefalten it was 46·5 percent in 1945, double that of 1944. At Ueckermünde, the rate rose to 55 percent in 1945. At Bernburg/Saale, it doubled in 1945. At the Düsseldorf-Grafenberg institution, between 1946 and 1947 the rate was 55 percent, and even as late as 1948/49 was still 30 percent.[863]

So much for a "stop order".

863 http://www1.uni-hamburg.de/rz3a035//psychiatry.html (Accessed 25 August 2008).

CHAPTER 8: 14F13

People lose the sense of being one species and try to make other kinds of people into a different and mortally dangerous species, one that doesn't count, one that isn't human...You can kill them without feeling that you have killed your own kind. - Erik Erikson[864]

They brought people there from Mauthausen; I don't know whether from other places too. But I have even heard tell that they were still gassing at 'C' (Hartheim) when the 'Amis' (Americans) were already on the Rhine. - Franz Suchomel[865]

The fourth manifestation of "euthanasia" was the so-called "*Aktion* 14f13,"[866] usually referred to in official SS terminology as *Sonderbehandlung* ("special treatment"—a favourite synonym for killing) 14f13.[867] This was the code name given to the extension of T4 operations to the concentration camps of the Greater *Reich*. In early 1941, Himmler and Bouhler reached an agreement whereby the T4 killing centres were to be utilised for the purpose of gassing "sick", that is non-productive camp inmates.[868] Until that time, the involvement of the SS in T4 had been limited to the provision of certain of the personnel (doctors, transport, guards), and gassing expertise.[869]

A distinction should be born in mind between the concentration camps of the *Reich* (Dachau, Buchenwald, Sachsenhausen, Mauthausen, Ravensbrück, etc) together with their associated sub-camps, and the extermination camps such as Chelmno, Belzec, Sobibor, and Treblinka, situated in occupied Poland. In addition to the Polish *Vernichtungslager* (death camps), there were further extermination camps operating on a somewhat smaller scale in other countries under Nazi occupation. Whilst many thousands died in concentration camps from a variety of causes, unlike the extermination camps, where the overwhelming majority of deportees were killed within hours of their arrival, usually by gassing, the prime function of the concentration camps was not necessarily the immediate liquidation of prisoners. The extermination camps existed for the sole purpose of theft and murder, almost exclusively of Jews. By contrast, as initially conceived, the concentration camps were considered to be a kind of extension to the

864 James Waller, *Becoming Evil. How Ordinary People Commit Genocide and Mass Killing* (Oxford: Oxford University Press, 2002), p. 244.

865 Gitta Sereny, *Into That Darkness—From Mercy Killing to Mass Murder* (London: Pimlico, 1995), p. 77.

866 Henry Friedlander, *The Origins of Nazi Genocide: From Euthanasia to the Final Solution* (Chapel Hill: University of North Carolina Press, 1995), pp. 142–150. At the Inspectorate of Concentration Camps, the category 14f included all files involving the death of prisoners. Thus,14f7 files concerned death through natural causes, 14f8 applied to suicides, 14f14 involved executions, and so on. From April 1941, 14f13 was the coded file number for the killing of prisoners in T4 centres. There would appear to be some confusion regarding 14f categories. For example, Eugen Kogon, Hermann Langbein, Adalbert Rückerl (eds.), *Nazi Mass Murder: A Documentary History of the Use of Poison Gas* (New Haven: Yale University Press, 1993) p. 40, cites 14f1 as death from natural causes, 14f2 as suicides or accidental deaths,14f3 as shot while trying to escape, and 14fI as execution. But there is no uncertainty regarding the meaning attributed to 14f13.

867 Dick de Mildt, *In the Name of the People: Perpetrators of Genocide in the Reflection of Their Post-War Prosecution in West Germany. The 'Euthanasia' and 'Aktion Reinhard' Trial Cases* (The Hague: Martinus Nijhoff Publishers, 1996), p. 73.

868 Michael Burleigh, and Wipperman, Wolfgang, *The Racial State: Germany 1933–1945* (Cambridge: Cambridge University Press, 1991), p. 161.

869 de Mildt, *In the Name of the People,* p. 73.

penal system, even if in many cases no recognizable criminal offence had been committed by those interned. By 1937, in addition to "professional and habitual criminals", camps held homosexuals, the "work shy", beggars, "vagrants", prostitutes, even "fault finders" and "traffic offenders". It was not difficult to fall foul of the authorities in pre-war Nazi Germany, who even came up with a word for these adversaries—*Volksschädling* ("pest harmful to the people"). However, having started life as means of holding perceived political opponents of the regime as well as those deemed eugenically worthless, this function decreased with the advent of war and the incarceration of foreign nationals. Although camp populations increased rapidly in wartime, German prisoners came to represent no more than 5–10 per cent of inmates.[870]

A further distinction should be made between the concentration camps and the hundreds of labour camps, and even between the nominal categories of camp, for they varied both in the severity of their regime (ranging from the terrible to the indescribable), and in the composition and nationality of their prisoner inmates.[871] In essence, those incarcerated in a labour or concentration camp generally had at least a hope of survival, however remote that may have seemed and to have actually been; in an extermination camp there was no such hope. Those not killed on arrival had merely postponed their death for a few days or weeks. The Auschwitz complex, and to a lesser extent Majdanek, were untypical in that they combined elements of both extermination and labour camps, as well as being orthodox concentration camps, and for a time at least, holding camps for Soviet prisoners-of war.

In late 1940, Oswald Pohl, at that time in charge of the SS Administration and Economy Main Office (*Hauptamt Verwaltung und Wirtschaft*) ordered the formation of a Labour Action Office, appointing as its head Wilhelm Burböck. The sole function of this new department was to assess the productivity of concentration and labour camp inmates. Burböck in turn created a new office within each camp's *kommandantur* called the "Detention Camp *Führer* Labour Office" (*Schutzhaftlagerführer "Einsatz"*), the principal function of which was to collate a card index containing relevant details of all working prisoners. In time, as the appalling conditions within the camps inevitably led to a shortage of labour within SS industries and elsewhere, these cards came to include information concerning inmates considered unfit for work, thereby providing the source material necessary for the selection of the victims of *Sonderbehandlung* 14f13.

As head of the Concentration Camps Inspectorate, earlier in 1940 Richard Glücks had determined that Dachau was to serve as a collection point for sick and dying prisoners. It was a disastrous decision, for the camp was soon overflowing with *Muselmänner*—the living dead. When Heinrich Himmler visited the camp in January 1941 he was appalled at what he found. Thousands of terminally diseased *Häftlinge* overwhelmed the camp's limited facilities. For Himmler the solution was obvious, the Nazis' standard

870 Wolfgang Sofsky, *The Order of Terror: The Concentration Camp*, (Princeton: Princeton University Press, 1997), pp. 28–43.

871 So far as prisoners were concerned, the difference between concentration, labour and a variety of alternatively categorized camps was often one of terminology. Conditions could be just as bad whatever the classification of the camp, and often were. The fundamental characteristics of the various categories of camp were frequently not understood by the Allies in the immediate post-war years, and remain vague to many today.

answer to problems of this kind. Having created the charnel house that Dachau had become, the issue would be resolved by the simple expedient of killing the invalids rather than waiting for them to die and tieing up valuable resources while they did so. But how was this to be achieved? For reasons that are unclear, Himmler was not yet prepared to authorise mass murder by the SS in the concentration camps themselves. The creation of the extermination camps in Poland was a year or more away, but there existed an organisation dedicated to killing on just the scale required—T4. And so meetings with Bouhler and Brack ensued, and within a few days *Aktion* 14f13 was underway.[872]

As originally conceived, 14f13 had been intended to eliminate the mentally and physically disabled held within the camp system, in essence a natural extension of T4, but the ability to categorize a prisoner as merely "unfit to work" significantly extended the power of the camp hierarchy to simply dispose of any prisoner considered undesirable, an opportunity that was now grasped with enthusiasm.[873]

Following the "stop order" of August 1941 and the availability of both facilities and personnel by now thoroughly well versed in the practice of murder, T4 assigned at least twelve physicians to visit the camps.[874] *Sonderbehandlung* 14f13 introduced the concept of the extermination camp (in the early days of 14f13 probably then not even in the planning stage) to the concentration camps. It can thus be seen as the bridge between T4—"euthanasia"—and *Aktion Reinhard*—genocide. Doctors visited the concentration camps, but did not conduct medical examinations. Prisoners were supposedly assessed utilising similar criteria to those applied to the handicapped. In practice inmates who had artificial limbs, wore spectacles or had "unsatisfactory" personal histories, were routinely selected for extermination. Again, simply being Jewish was enough to incur a death sentence. As was made clear in post-war trial testimony, when reaching the decision to kill an individual, it was largely immaterial to doctors whether this was on medical, racial or political grounds.[875] Requests by the camp SS to dispose of "undesirables" were fulfilled without question. Prisoners were deliberately deceived into volunteering for transfer to a "rest camp"—which in reality meant transfer to a killing centre. Those prisoners selected were transported to either Hartheim, Sonnenstein, or Bernburg for gassing.[876]

With his colleague Theodor Steinmeyer, director of the mental institutions in Warstein and Mühlhausen, Friedrich Mennecke, head of the Eichberg hospital and its children's ward, visited Sachsenhausen concentration camp in early April 1941,[877] where

---

872 Nikolaus Wachsmann, *KL: A History of the Nazi Concentration Camps* (Little, Brown, London, 2015), pp. 243–245.

873 Miriam R Levin (ed), *Cultures of Control* (Amsterdam: Harwood Academic Publishers, 2000), pp. 181–187.

874 Friedlander, *The Origins of Nazi Genocide*, p. 145. Those known to have visited the camps at least once as part of *Aktion* 14f13 included Doctors Heyde, Nitsche, Mennecke, Steinmeyer, Wischer, Lonauer, Renno, Robert Müller, Schmalenbach, Ratka, Gorgass, and Hebold. (Wachsmann, *KL: A History of the Nazi Concentration Camp*, p. 693, n.30). Brief biographies of most of these individuals can be found in the appendix to this book.

875 Robert N Proctor, *Racial Hygiene: Medicine Under the Nazis* (Cambridge Massachusetts: Harvard University Press, 1989), p. 209.

876 Sofsky, *The Order of Terror*, pp. 241–243.

877 de Mildt, *In the Name of the People*, p. 74 suggests this visit was to Orainienburg concentration camp. There is frequent confusion regarding the camps at Oranienburg (established 1933, closed 1934) and

together they selected 400 prisoners. This visit marked the actual start of 14f13 killings, although preparatory work had commenced earlier. The *Aktion* was only gradually introduced at individual camps over the next 12 months, with as many as seven doctors enlisted for processing potential victims at Dachau in September 1941.[878] Mennecke was an obsessive letter writer, particularly so far as his wife was concerned,[879] in the case of Sachsenhausen informing her that the prisoners had not been chosen on the grounds of their sickness, but because they were "'anti-socials' of the highest degree."[880]

He was also an enthusiastic collector of photographs of Jewish male and female camp inmates, some of whom are known to have been victims of 14f13. A cache of these photographs was discovered after the war. The comments Mennecke made on the back of many of these photographs, particularly of Jewish women, provide a clear indication of both his racism and apparent obsession with matters sexual. The prisoners photographed were guilty of "race defilement with German soldiers, like on a conveyor belt". Many were designated prostitutes, for example a prisoner described as "[a] pure bred Jewish whore with venereal disease". Another was a "sexually impulsive and insatiable Jewess". He was equally intolerant of homosexuality, for at Buchenwald he condemned virtually all Jewish men deemed homosexual out of hand.[881]

In fact, although the inspection of prisoners was conducted by doctors, they could have been carried out by anybody, since the medical diagnosis of the victims was perfunctory at best.

The process was similar at each of the concentration camps the physicians visited. A form similar to the *Meldebogen* developed for the "euthanasia" programme was completed. And also as with "euthanasia", the judgement of life (a minus sign), or death (a plus sign) was speedily made and entered in a box on the form. In many cases that decision had already been taken before the prisoner even appeared before the assessor. The completed forms were forwarded to T4 in Berlin, from where, after review, a list of victims was sent to one of the three available killing centres, which in turn would arrange with the relevant camp for the transport of the victims.[882]

In August 1941, the Dachau commandant, Alex Piorkowski, together with his camp doctors, had registered 7,000 inmates as potential victims of 14f13. Julius Muthig, the camp doctor of the Dachau concentration camp, testified that in the autumn of 1941, Werner Heyde, accompanied by Paul Nitsche and three or four other representatives of

---

Sachsenhausen (established 1936), a confusion compounded by the fact that Sachsenhausen was situated on the outskirts of Oranienburg, and is sometimes referred to as Sachsenhausen-Oranienburg. From the timing of Mennecke and Steinmeyer's visit, it is clear that the reference is to Sachsenhausen. [see http://www.stiftung-bg.de/gums/en/geschichte/gesch_neu_en.html (Accessed 10 April 2008)].

878 Wachsmann, *KL: A History of the Nazi Concentration Camp*, p. 246.

879 Approximately 2,500 pages of Mennecke's letters had been located by 1987, representing about one-third of their total number. (Götz Aly, Peter Chroust, Christian Pross, *Cleansing the Fatherland: Nazi Medicine and Racial Hygiene* (Baltimore: The Johns Hopkins University Press, Baltimore, 1994), pp. 251 and 296). A representative sample of extracts from Mennecke's letters to his wife can be found at http://nuremberg.law.harvard.edu/documents/157-letters-to-mrs-mennecke?q=issue:%22Euthanasia+program+%28Action+14+f+13%29%22(Accessed 26 December 2016).

880 de Mildt, *In the Name of the People*, p. 74.

881 Wachsmann, *KL: A History of the Nazi Concentration Camp*, p. 254

882 Ibid., p. 247.

T4 (including Mennecke),[883] arrived at the camp and began to select the unemployable among the inmates for gassing.[884] The records of certain prisoners were removed from the political department of the camp to Block 29, where four men in civilian clothing sat. The prisoners to whom these records referred were then produced and asked a few desultory questions, following which a number of them were separated on the left side of the room. The leader of the group (Heyde) informed the chosen prisoners that they would be sent to a camp where conditions were better. These inmates were subsequently transferred, and shortly afterwards their personal possessions were returned to Dachau.[885]

The ubiquitous Mennecke wrote to his wife from Dachau on 3 September 1941: "There are only two thousand men, who will be quickly done, as they can be examined only in assembly-line fashion."[886] Another inmate recalled Mennecke and his colleagues sitting behind some tables following the evening roll-call. Sick prisoners, or those considered unfit for work were instructed to register for transfer to a sanatorium. When only two to three hundred did so, other physically broken prisoners were added to their number. Subsequently 700 of these inmates were loaded onto lorries, to be followed a few days later by a further 650 victims.[887] Following this initial visit and several subsequent similar assessments by putative "experts", transfers to Hartheim for gassing commenced in January 1942; over the next 12 months, 2,593 prisoners in 32 transports were murdered as part of this programme, at least some of whom were quite healthy.[888] Among the victims were 324 members of the clergy, including many Polish priests. In a later letter, Mennecke would write from Ravensbrück, "In Berlin (Jennerwein! [Viktor Brack]) they simply state: 2,000 are to be dealt with. Whether or not so many qualify in view of the fundamental guiding principles, nobody cares!"[889]

Hans-Bodo Gorgass arrived at the Buchenwald concentration camp together with Horst Schumann for the specific purpose of familiarising themselves with the 14f13 process. The visit was made on the instructions of Brandt, and conveyed by Brack. Among those assembled for examination were Jews in protective custody, other Jews who were categorized as "shirkers," Poles, Czechs, and also half-Jewish children and Gypsies. Waldemar Hoven, later chief physician at Buchenwald, testified:

> Camp commandant Koch called all important SS officials together and informed them... the higher authorities from Berlin had ordered that all the Jewish inmates of the Buchenwald concentration camp be

---

883 de Mildt, *In the Name of the People,* p. 77 Another of those probably present was Rudolf Lonauer, chief physician at Hartheim. [Mireille Horsinga-Renno, *Cher oncle Georg: La bouleversante enquête d'une femme sur un médecin de la mort impuni* (Strasbourg : La Nuée Bleue, 2006), p. 100].

884 The gas chamber building in Dachau was not erected until 1942. [Harold Marcuse, *Legacies of Dachau. The Uses and Abuses of a Concentration Camp, 1933–2001* (Cambridge: Cambridge University Press, 2001), p. 11]. Trial gassings of at least two groups of prisoners were conducted there between summer 1942 and spring 1945, but Dachau was never an extermination camp in the sense attributed to the *Aktion Reinhard* and certain other camps. (Ibid., p. 46, p. 254).

885 Michael Burleigh, *Death and Deliverance : 'Euthanasia' in Germany c. 1900–1945* (London: Pan Books, 2002), p. 219.

886 Kogon, Langbein, Rückerl *Nazi Mass Murder*, p. 41.

887 Burleigh, *Death and Deliverance,* p. 219.

888 de Mildt, *In the Name of the People,* p. 77.

889 Ibid., p. 341, note139.

included in the euthanasia programme. This special operation was carried out under the code name 14f13.

Subsequently some 384 Jews were sent to Bernburg, with a further 187 transported to Sonnenstein.[890] As well as Jews, Koch informed his staff, Himmler had ordered that all feeble-minded and crippled inmates in German concentration camps were to be killed.[891] Hoven also recorded the assistance rendered by his colleagues at Bernburg in disposing of the bodies of victims of the Buchenwald sub-camp Schönebeck:

> Contractual obligations of physicians and negotiations with cemetery authorities have often led to insurmountable difficulties...for this reason I immediately communicated with Dr Eberl, Medical Director of the mental institution at Bernburg...He is the physician who carried out 14f13. Dr Eberl showed unusual understanding and courtesy. All current bodies of prisoners from Schönebeck-Wernigerode were shipped to Dr Eberl at Bernburg, where they were cremated even without a death certificate.[892]

In summer 1941, Horst Schumann, who would become notorious even among this company, chaired a committee that visited Auschwitz and selected 573 prisoners, who were transferred to his establishment at Sonnenstein for liquidation, thereby confirming that there were no suitable gassing facilities available at Auschwitz at that time.

Questionnaires were transmitted to the camps prior to the physicians' arrival with the SS filling in obligatory information in advance: name, date and place of birth, last residence, family status, citizenship, religion, race, and date of arrest. Initial selections were undertaken by camp commanders. At roll-calls, prisoners were informed that those among them who were sick or unable to work could request a transfer to an *Erholungslager* ("recovery camp"), where they would receive medical care and perform less arduous labour. A post-war court recorded the process that occurred at Gusen, a sub-camp of Mauthausen concentration camp:

> ...It was announced in the spring of 1941 that the "Dachau recovery camp" had 2,000 vacancies, and that prisoners feeling ill or incapable of work could report for transfer to this camp, which was also referred to as a "sanatorium". In order to make the operation credible, a few prisoners were indeed taken to Dachau, where they were ordered to write to their fellow prisoners on their arrival. As, nevertheless, many prisoners suspected a trap, only about 1,200 of them reported...After the camp doctors had selected more prisoners, so that the stipulated total number of 2,000 persons for *Sonderbehandlung* was reached, the doctors' committee from T4 arrived in the camp in the summer of 1941.

After inspection by the T4 doctors, the selected prisoners were moved to a special block. In most cases no medical examination was undertaken. Between August 1941 and February 1942, the isolated prisoners were transported to Hartheim and gassed on arrival. But not before they had been commercially evaluated: "On the eve of every transport the camp doctors entered the 'Invalids Block' and inspected the mouths of those prisoners who were to leave the camp the following morning. Afterwards they marked those with gold fillings with a cross on their chest or back."[893]

---

890 Joseph Tenenbaum, *Race and Reich: The Story of an Epoch* (Westport: Greenwood Press, 1976), p. 93; Harry Stein, *Buchenwald concentration camp. 1937–1945: a guide to the permanent historical exhibition* (Göttingen: Gedenkstätte Buchenwald, Wallstein Verlag, 2004), p. 124.

891 Proctor, *Racial Hygiene*, p. 208.

892 Eugen Kogon, *The Theory and Practice of Hell* (New York: Berkley Books, 1998), p. 226.

893 de Mildt, *In the Name of the People*, pp. 74–75.

Glücks issued a secret directive on 10 December 1941 which further simplified the 14f13 procedure. Apart from a prisoner's specific biographical details, only four questions were of consequence: "Since when in institution?; Incurable illness?; Crime?; Former punishable offences?"[894] The T4 physicians therefore only needed to check the completed information, enter the diagnosis, and make the final decision. Thereafter, the questionnaires were delivered to T4 headquarters in Berlin. Mennecke admitted: "This did not involve medical evaluations, because in the concentration camps I only had the assignment to fill out questionnaires." For Jews, the physicians did not even bother to any longer enact the pretence of a physical examination. In yet another of his letters to his wife, Mennecke described the procedure:

> As a second allotment there then followed altogether 1,200 Jews, who did not first have to be "examined", but where it is sufficient to extract from the files the reasons for their arrest (often very extensive!) and to record them on the questionnaires.

Unlike T4, the deaths were recorded by the registry office responsible for each concentration camp. The camp was noted as the place of death and a fictitious cause of death created, although from early 1942 death notices were no longer sent to the next-of-kin of the deceased.[895]

Within little more than a year, 14f13 as originally conceived was over. The last transport from Neuengamme arrived at Bernburg in June 1942. Ostensibly this curtailment was caused by increasing economic demands, with Himmler insisting upon significantly increased productivity from the concentration camps. The vast pool of slave labour available to the SS in the camps would be put to more gainful use, rather than simply being murdered in haphazard fashion, although extermination through labour would remain a constant in Himmler's world . For now, only those permanently unable to work would be killed, and medical commissions were not necessary for making that kind of assessment.

In truth, economics had little to do with the decision to end the pretence of supposedly'expert' medical opinion justifying homicide. In reality the expertise of T4's killing centre staff was now required for the purpose of perpetrating murder on a much grander scale in the newly established or proposed Polish extermination camps of Belzec, Sobibor, and Treblinka. Himmler's initial reluctance to use his SS to kill *Muselmänner* within the concentration camps had been dissipated. Outside medical 'expertise' was no longer required for this simplified programme. Camp doctors were authorized to select and kill their victims, usually within the camps themselves, although transports of prisoners selected for killing did continue.to those killing centres which still had a functioning gas chamber.[896]

---

894 Ibid., p. 74.

895 Robert Jay Lifton, *The Nazi Doctors: Medical Killing and the Psychology of Genocide* (London: Papermac, 1990), pp. 139–142; Burleigh, *Death and Deliverance*, pp. 216–220; Kogon, *Mass Murder*, p. 40 ff.

896 Wachsmann, *KL: A History of the Nazi Concentration Camp* pp. 255–257.

But It was due to the persistent shortage of manpower that in April 1943 the 14f13 programme was further modified. The WVHA (*Wirtschaftsverwaltungshauptamt*—the SS Economics and Administrative Department) issued a decree authorised by Himmler, to the effect that in future *only* mentally ill prisoners should be murdered, evidence, if such were needed, that 14f13 had never applied solely to prisoners suffering from mental disabilities.[897] Productivity was to be the new criterion. All other inmates who were unfit for work should nevertheless be given tasks commensurate with their physical capabilities. Bedridden prisoners should be allocated tasks that they could perform in bed. The decree notwithstanding, camp physicians continued to select and kill prisoners physically unfit for work, either in the concentration camp itself, or in another camp or killing centre equipped with a gas chamber. In any event, within a year the new decree was rescinded. By April 1944, 14f13 was fully functional again, except that now the façade of a medical commission had been entirely dispensed with; the camp doctors simply chose the victims, then if not killing them within the camp themselves, usually by lethal injection, dispatched the victims to Hartheim, which was now the only one of the original "euthanasia" killing centres with an operational gas chamber.[898] Mauthausen and Gusen again provide a typical example:

> One day before each transport one of the camp doctors accompanied by several SS officers appeared in the sick-bay or in the "Jewish Block". Often, he merely pointed at some prisoners, whose numbers were taken down by SS-men. Most of the time he only declared: "Tomorrow there will be another transport to the recovery camp. For this we need 80 men." The number of 80 was a minimum number; the maximum number lay somewhere around 150. It was the duty of the prisoners' doctor and the Block eldest to select as many prisoners as the camp doctor had requested. As both knew, or at least suspected, that the invalids were to be killed, they regularly chose newcomers or prisoners whom they disliked, in order to protect their friends against transportation.[899]

Although unable to provide precise details of dates or the number of victims, the camp physician of the women's concentration camp at Ravensbrück, Dr Percival Treite, testified that two transports had been sent from that camp to "a sanatorium in Thüringen" (probably Sonnenstein, if before August 1943), and to Linz (probably Hartheim). A third convoy, made up of female Jehovah's Witnesses, had been sent to "a sanatorium near Linz." None of this latter party was sick. The good doctor later learned that all of the women in this transport had been gassed, presumably because of their religious beliefs. It is known that it was common for convoys of prisoners to leave from a number of different concentration camps for unknown destinations. It is also known that none of the prisoners included in these transports survived.[900]

At their Nuremberg trial, both Brandt and Brack denied any knowledge of *Sonderbehandlung* 14f13, a claimed ignorance that was innately improbable, to say the least.[901] When asked if he thought Hitler was aware of the *Aktion*, Brandt replied that he did not believe that Himmler could have authorized the killings without Hitler's ap-

897 Tenenbaum, *Race and Reich*, p. 443, note 11.

898 Kogon, Langbein, Rückerl *Nazi Mass Murder*, p. 49. Hartheim was now known as "the Dachau sanatorium" (Ibid.)

899 de Mildt, *In the Name of the People*, p. 77.

900 Kogon, Langbein, Rückerl *Nazi Mass Murder*, pp. 50–51.

901 Paul Julian Weindling, *Nazi Medicine and the Nuremberg Trials—From Medical War Crimes to Informed Consent* (Basingstoke: Palgrave Macmillan, 2006), p. 252.

proval, however discrete and indirect that approval may have been: "It is my opinion that he (Hitler) knew about it in some form. The question is whether he gave orders or whether he gave hints so that Himmler carried them out in this form."[902] The same might be said, of course, of any of the Nazis' murderous policies.

As is so often the case, it is impossible to provide precise data concerning fatalities. By the war's end the victims of *Sonderbehandlung* 14f13 have been estimated to number about 20,000, although it is quite conceivable that the actual figure was considerably greater.[903] The 14f13 programme was relatively small in scope in comparison to the killings in the death camps, but it provided an important link between the "euthanasia" of mental patients in the T4 programme and the subsequent mass murder of *Aktion Reinhard*, Auschwitz-Birkenau, and elsewhere. 14f13 transferred the practices of the "euthanasia" programme to the concentration camps, thereby helping to create the dual purpose labour and extermination camps such as Auschwitz and Majdanek. Moreover, to some extent, 14f13 provided the model for future *modus operandi*, particularly insofar as the "selection" of victims was concerned.

---

902 Ulf Schmidt, *Karl Brandt: The Nazi Doctor—Medicine and Power in the Third Reich* (London: Hambledon Continuum, 2007), p. 236.

903 de Mildt, *In the Name of the People*, p. 78 states that the number of victims of 14f13 "must have run into tens of thousands." Weindling, *Nazi Medicine and the Nuremberg Trials*, p. 250 estimates "50,000 concentration camp prisoners killed" between April 1941 and 1944.

CHAPTER 9: *AKTION REINHARD*

'Aussiedlung'—the destruction of people and theft of their possessions.[904]

There is great unhappiness and fear among the Jews...They are... resettling the Jews from various towns and villages and taking them somewhere towards Belzec...Today they deported the Jews from Izbica—they were also taken to Belzec where there is supposed to be some monstrous camp. - Zygmunt Klukowski, 26 March 1942[905]

Following a series of meetings with Karl Brandt, Hitler ordered an official halt to the "euthanasia" programme on 24 August 1941. Brandt immediately telephoned Philip Bouhler to pass on the instructions, and Bouhler in turn contacted Viktor Brack. As the message passed down the list of apparatchiks, Hans Hefelmann queried whether all "euthanasia" was to cease, including that of children. Back came the reply—children were not included in the cessation. The "stop order" was only to apply in the case of adults.[906] "Euthanasia" had accounted for the lives of 4,000–5,000 Jewish patients, overwhelmingly killed for being Jews rather than for reasons of incapacity.[907] The time had now arrived to accelerate that rate of murder.

By the time of the suspension of adult "euthanasia", *Fall Barbarossa*, the invasion of the Soviet Union, and with it the shooting of Jews, communist party functionaries, and other *Untermenschen* by the *Einsatzgruppen*, had been underway for two months. Most of the personnel at the "euthanasia" killing centres were dismissed following the "stop order", only to be hurriedly recalled two weeks later. The exact chronology of events during the critical period following the issue of the "stop order" and the commencement of the construction of the Belzec extermination camp is imprecise, and such details as exist are often conflicting, but it seems that at some time soon after the notional cessation, Bouhler was asked, probably by Himmler, to provide the necessary manpower to Odilo Globocnik for the implementation of the planned extermination of the 3.3 million Polish Jews. Immediately thereafter Bouhler visited Globocnik in Lublin, doubtless to confer about the relevant details.[908]

So far as Himmler was concerned, Globocnik, a career gangster of Austrian-Slovene ancestry, was a natural choice to oversee the annihilation of Poland's Jews. Born in Trieste in 1904, and having made his reputation as an uscrupulous thug in pre-*Anschluss* Austria, Globocnik was appointed SS and Police Leader for the Lublin District in November 1939. A member of the then illegal Austrian Nazi party since 1930 and having joined the Austrian SS in 1933, his fanatical and ruthless devotion to the Nazi cause ensured early recognition and rapid promotion following the *Anschluss*. By May 1938 he

---

904 Rudolf Reder, *Belzec* (Oswiecim: Fundacja Judaica Panstwowe Muzeum Oswiecim-Brzezinka, 1999), p. 107.

905 Martin Gilbert, *The Holocaust: The Jewish Tragedy* (London: Collins, 1986), p. 308.

906 Ulf Schmidt, *Karl Brandt: The Nazi Doctor—Medicine and Power in the Third Reich* (London: Hambledon Continuum, 2007), p. 166.

907 Peter Longerich, *The Unwritten Order: Hitler's Role in the Final Solution* (Stroud: Tempus Publishing Limited, 2003), p. 82.

908 Christopher R Browning, *The Origins of the Final Solution: The Evolution of Nazi Jewish Policy, September 1939–March 1942* (London: William Heinemann, 2004), p. 115.

had been appointed *Gauleiter* of Vienna, but his incumbency was brief. In January 1939 he was relieved of his office and stripped of his party honours, guilty of illegal currency dealings. This might have proved the end of another man's burgeoning SS career, but Globocnik remained a protégé of Himmler, who recognized amorality when he saw it and was quick to appoint Globocnik to the crucial position in Lublin. There the venemous anti-semite, "Globus" as Himmler called him, instituted the anticipated regime of terror and murder, and surrounded by a coterie of like-minded killers was, from Himmler's perspective, the right man in the right place when the decision was taken to expedite the liquidation of the Jews of Poland.[909] Globocnik was delighted with the prospect. "The Reichsführer SS was just here and has given us so much new work," he enthused. "I am so very grateful to him, that he can be certain that these things that he wishes will come true in no time."[910]

It is evident that the need to utilise the services of men experienced in the process of mass murder had been recognized. According to Brack:

> In 1941, I received an oral order to discontinue the euthanasia programme. In order to retain the personnel that had been relieved of these duties, and in order to be able to start a new euthanasia programme after the war, Bouhler asked me—I think after a conference with Himmler—to send this personnel to Lublin and place it at the disposal of SS-*Brigadeführer* Globocnik.[911]

Brack had good reason to create spurious excuses for retaining the T4 personnel. There was certainly an intention to one day recommence the "euthanasia" programme, but it was not for this reason that the staff was to be sent to Poland. Brack had accompanied Bouhler on the trip to see Globocnik;[912] exactly what the trio of murderers discussed at their meeting in Lublin was either unrecorded, subsequently destroyed, or remains as yet undiscovered, but is not difficult to surmise. The SS had only been marginally involved in "euthanasia", which had essentially been an operation conducted through and by the KdF. Whilst some Jews had certainly been among the victims of T4, any continent-wide mass murder was to be very much an SS affair. By providing some personnel and expertise the SS had enabled T4 to function relatively smoothly. Now with the planned expansion of killing through the creation of dedicated extermination camps, the favour was to be returned. The SS-men who had served at the killing centres were easily recalled to duty, but they represented only a minority of T4 personnel. Globocnik needed the other, non-SS members who had been, and were to remain, employees of T4; although they were eventually to become nominally members of the SS whilst serving in the *Aktion Reinhard* camps, these men continued to be paid by and receive benefits from T4.[913]

Having been hastily ordered to report back to headquarters for a new assignment, the T4 staff found themselves at something of a loose end. Some were temporarily re-

---

909 http://www.deathcamps.org/reinhard/globocnik.html (Accessed 28 December 2016).

910 Nikolaus Wachsmann, *KL: A History of the Nazi Concentration Camps* (Little, Brown, London, 2015) p. 293.

911 Yitzhak Arad, *Belzec, Sobibor, Treblinka: The Operation Reinhard Death Camps* (Bloomington: Indiana University Press, 1987), p. 17.

912 Browning, *The Origins of the Final Solution*, p. 362.

913 Robin O'Neil, *Belzec: Stepping Stone to Genocide* (New York: JewishGen Inc, 2008), pp. 93–94; Arad, *Belzec, Sobibor, Treblinka*, p. 18.

turned to the "euthanasia" centres,[914] whist others dawdled in Berlin, killing time rather than patients. But all knew that there was something in the air. Hans-Bodo Gorgass testified: "I knew police captain Wirth, the administrative head of various euthanasia institutions, who told me in the late summer of 1941, that he...was being transferred to a euthanasia institute in the Lublin area." Rumours were rife that there was now to be a "euthanasia" programme exclusively devoted to Jews.[915]

At a moment when the collapse of the Soviet Union seemed to be just weeks away and mastery of Europe seemed assured, the rumours became reality. The total destruction of Polish Jewry, until then still largely a matter for contemplation and debate, was to be set in motion. But if the SS administrators of the planned extermination were in place, the operatives, the men who would man the camps and do the actual killing were not. However, it required little imagination to conclude that there was nobody better qualified to undertake the appointed task than the murderous staff of Grafeneck, Sonnestein, and the other killing centres. And so appropriate orders were issued, and the by now vastly experienced T4 killers were shipped eastwards to join Globocnik's entourage.

There is testimony extant that Christian Wirth was in fact among the earliest of the T4 staff to arrive in Lublin and gain admission to the small circle of those who were to plan and execute the probably as yet unnamed *Aktion*. The prime source for this evidence was Adolf Eichmann, who despite the many inconsistencies in his various depositions, provided one version that, at least in its essentials, provided a plausible account of some of the events of those pivotal weeks. Eichmann claimed that two or three months after *Barbarossa* had commenced, that is in August or September 1941, he was summoned by Reinhard Heydrich and informed, "The *Führer* has ordered the physical extermination of the Jews". Eichmann was ordered to visit Globocnik, who had already received his instructions concerning the intended genocide, and having done so, to report back to Heydrich on the progress of Globocnik's programme.[916]

On reporting to Lublin, Eichmann was sent on a tour of inspection, accompanied by Herman Höfle, Globocnik's chief-of-staff. During his interrogation by the Israeli police after his arrest and extradition in 1960, Eichmann had several convenient lapses of memory or moments of confusion, among the most unlikely of which was his inability to identify exactly where Höfle had taken him. He claimed that it might have been Treblinka, which was clearly impossible from his description of the geography of the region they visited, quite apart from the timing involved. It is unlikely that Treblinka, where initial inspections were made in late April or early May 1942, construction begun in late May or June,[917] and operations only commenced on 23 July 1942, was even under consideration as a potential extermination camp in the late summer or autumn of 1941. Eichmann did visit Treblinka, but from his description of the events he witnessed

---

914 Ibid., p. 93

915 Browning, *The Origins of the Final Solution*, p. 362

916 Eichmann stated that Heydrich's words were: "Go and see Globocnik, the *Führer* has already given him instructions." This suggests that Globocnik has received his orders direct from Hitler, which was quite possible, but given the chain of command and the need to disassociate Hitler from direct involvement in mass murder, it seems more likely that Globocnik, who was very much Himmler's man, received his orders from the *Reichsführer*.

917 Arad, *Belzec, Sobibor, Treblinka*, p. 37.

it was at a much later date, certainly sometime after the erection of the fake railway station buildings at the camp in late December 1942.

It is quite clear that it was Belzec, much closer to Lublin than Treblinka, that Eichmann saw: "There were patches of woods, sort of, and a road passed through—a Polish highway. On the right side of the road there was an ordinary house, that's where the men who worked there lived." This is an accurate description both of the Belzec camp site and the location of the staff living quarters. It does not remotely describe Treblinka. It is equally clear that it was Wirth who greeted Eichmann and Höfle on their arrival: "A captain of the regular police (*Ordnungspolizei*) welcomed us...He had, let's say, a vulgar, uncultivated voice. Maybe he drank. He spoke some dialect from the southwestern corner of Germany..."[918] In the course of further Israeli interrogation Eichmann's memories became somewhat clearer. The place he had visited had a more Polish-sounding name than Treblinka, and the police captain's name, he now recalled, was Wirth.

Eichmann's account of subsequent events varied in detail with each telling. Wirth led the men along a small forest path on the left side of the main highway, leading to two or three wooden houses still under construction. Wirth described how he had to hermitically seal all the windows and doors. It may be that Eichmann was describing this sealing-up when he refers to construction work. In an alternative version, Eichmann described how they came to "two small peasant huts standing under the deciduous trees". In a third variation, the wooden structures were "in a forest, a deciduous forest, a quite dense deciduous forest, large trees and in full colour...It was therefore 1941 in the fall." None of these descriptions fit the known layout of Belzec as a functioning extermination camp.

This testimony notwithstanding, there is a school of thought that, for whatever reason, Eichmann deliberately misstated the time of his visit to Belzec during these interrogations in Israel, and that in fact his visit took place not in the autumn of 1941, but rather sometime in early 1942, when the final touches to the gas chambers were being made (Belzec commenced operations on 17 March 1942).[919] The source of this version of events was Eichmann's self-serving autobiography, *Ich Adolph Eichmann*, a work which bears the imprimatur of no less an authority than David Irving, and was derived from some eighty tapes recorded by Eichmann in the 1950s.[920] In view of the known chronology of events and other witness testimony, this account is open to question. According to it, Eichmann dated Heydrich's informing him of the decision to physically exterminate the Jews to "the turn of the year 1941/42", following which, as in his other accounts, he journeyed to somewhere in the Lublin region ("I don't know what the place is called"), where Wirth showed him a small house, "completely secluded" and informed Eichmann: "Here the Jews are being gassed now."[921] No mention of deciduous trees "in full colour", which is hardly surprising if Eichmann's visit did occur in the midst of a Polish winter.

---

918 Jochen von Lang and Claus Sibyll, *Eichmann Interrogated: Transcripts from the Archives of the Israeli Police*, (New York: Da Capo Press, 1999), pp. 75–76.

919 O'Neil, *Belzec*, pp. 88–89.

920 http://www.fpp.co.uk/Auschwitz/Eichmann/Sudholt171079.html (Accessed 11 October 2009).

921 Arad, *Belzec, Sobibor, Treblinka*, p. 26.

No explanation has been proffered as to why, when in captivity, Eichmann would have found it necessary to construct such an elaborate lie about the timing of his visit to Belzec, a lie which was not in any way in his interests.[922] Like other criminals, Eichmann was only likely to be mendacious when it was to his perceived benefit. It is difficult to see any advantage accruing to Eichmann by his placing himself at the centre of the "Final Solution" at an earlier rather than a later date. It seems more feasible that events occurred in the autumn of 1941 largely as Eichmann described them, not least because they place Wirth in exactly the right place at the right time, doing the thing he knew best. The timing of Eichmann's visit is only important for this very reason. In fact, Christopher Browning has constructed a scenario which provides a logical pattern to at least some of Wirth's activities at that decisive juncture, as will be demonstrated.

Eichmann claimed not to have seen any actual gassings taking place at Belzec: "The motor was not yet there, the installation was not yet in operation." Whilst it is true that the fixed gassing motor was not acquired for the camp until much later, in the earliest phase of killing at Belzec gassing was carried out with bottled carbon monoxide.[923] There was also another possibility; a *Zyklon B* type of killing agent had been used to dispose of mental patients at Posen as early as autumn 1939. For such a chemical to be effective all that was required was an airtight room. Whether he actually did or did not do so, it would certainly have been possible for Eichmann to have seen Wirth in action. Moreover, as Arthur Nebe and Albert Widmann had discovered at Novinki and Mogilev, it did not require excessive ingenuity to divert the exhaust fumes of a stationary petrol-fuelled vehicle into a hermetically sealed room in order to kill the occupants,[924] although admittedly to do so in the middle of the kind of forest Eichmann described would have presented considerable problems. However, it is known that sometime after the arrival of the first SS-men at Belzec in December 1941, a Post Office delivery van was converted into a mobile gas chamber. Even at that late stage, was Wirth perhaps considering a Chelmno-style extermination camp based upon the methods he had witnessed there, or a combination of mobile and stationary gas chambers?

Then there is the testimony of Ferdinand Hahnzog, commander of the Gendamerie (rural police) in the Lublin district, who stated that experiments into methods of mass killing were being investigated in the Belzec region at an early date. His description of "a primitive facility near Belzec hidden deep in the forest bordering on Galicia ... consisting of a sealed shed into which the Security Police and the SD from Zamosc pumped exhaust fumes from the vehicles used to bring the 'morituri' there!" corroborates the idea that the use of carbon monoxide produced by a petrol-driven engine was quickly recognized as a potential method for the extermination of large numbers of people. Hahnzog, however, dated these experiments to "the spring of 1941, if not earlier, in the fall of 1940."[925] Hahnzog may have been mistaken regarding dates, so that perhaps it was Wirth's tests twelve months later to which he referred, although in general

---

922 Browning, *The Origins of the Final Solution*, p. 529, n.251.

923 Arad, *Belzec, Sobibor, Treblinka*, p. 26.

924 Eugen Kogon, Hermann Langbein, Adalbert Rückerl (eds.), *Nazi Mass Murder: A Documentary History of the Use of Poison Gas* (New Haven: Yale University Press, 1993), pp. 52–53.

925 http://www1.yadvashem.org/odot_pdf/Microsoft%20Word%20- %203222.pdf (Accessed 11 October 2009).

Hahnzog has been found to be a reliable witness so far as data is concerned. On the other hand, it is equally possible that research into different killing methods were being undertaken in that region of eastern Poland long before the decision or decisions implementing the "Final Solution" had been arrived at. It is also possible that Hahnzog and Eichmann were describing two separate activities conducted quite independently of each other.

Hahnzog also testified that in autumn 1941, he had been personally informed of the impending extermination programme by Globocnik himself, who needed Hahnzog's cooperation in providing guards for the unloading of transports at the proposed new camp at Sobibor; yet more evidence of the early planning for a second killing facility and a commensurate expansion of the *Aktion*. In early 1942 Hahnzog attempted to visit Sobibor, but found that even at that basic stage of construction the camp site was surrounded by barbed wire and the gates firmly shut.[926]

Browning's thesis concerning Wirth's experimental phase at Belzec posits that Wirth first arrived there in September 1941, and initially considered the conversion into gas chambers of two peasant huts that were situated deep in the surrounding forest (not on what was to become the extermination camp site). It was these hermetically sealed structures that Eichmann saw, and which were used to test the efficacy of a number of possible killing agents. This use of existing small buildings as gas chambers was in fact exactly what was to happen at Auschwitz-Birkenau, where the "little red house" and "little white house" were the first killing installations, functioning long before the much larger crematoria II-V were completed. Having satisfied himself concerning the method and technique to be employed, Wirth left Belzec to return as the camp's first commandant a few weeks later.

There is one further source which can be interpreted as providing confirmation of Browning's reconstruction of events. Although it should be approached with the caution necessary when evaluating the evidence provided by any ex-member of the SS, particularly those directly involved in mass murder, in his post-war interrogation, Josef Oberhauser, Wirth's right-hand man, made it clear that there had been not one, but two experimental phases at Belzec:

> The gassing of Jews which took place in Belzec camp up till 1 August 1942 can be divided into two phases. During the <u>first series of experiments</u> there were two to three transports consisting of four to six freight cars each holding twenty to forty persons. At that stage the gassings were not yet part of a systematic eradication action but were carried out to test and study closely the camp's capacity and the technical problems involved in carrying out a gassing. After these first gassings, Wirth and Schwarz along with the entire German personnel left Belzec... About a week after Brack had come to Globocnik [that is in mid-May 1942] Wirth and his staff returned to Belzec. The <u>second series of experiments</u> went on until 1 August 1942... During the first experiments and the first set of transports in the second series of experiments bottled gas was still used for gassing; however, for the last transports of the second series of experiments the Jews were killed with the exhaust gases from a tank or lorry engine which was operated by Hackenholt.[927]

Oberhauser was either confused, or more probably simply lying so far as dating was concerned. He claimed that "the large-scale extermination programme was due to start

---

926 Jules Schelvis, *Sobibor—A History of a Nazi Death Camp* (Oxford: Berg, 2007), p. 27.

927 Ernst Klee, Willi Dressen, Volker Riess (eds.), *The Good Old Days—The Holocaust as Seen by Its Perpetrators and Bystanders* (New York: Konecky & Konecky, 1991), pp. 228–230. Emphasis added.

[at Belzec] on 1 August 1942", which was clearly untrue. However, the proposition that Wirth conducted initial experiments in September 1941 in the manner suggested by Browning, then departed from Belzec only to return as commandant on 22 December 1941 and then carry out a second series of experimental gassings in the newly constructed gas chambers before the full-scale extermination of Jews began on 17 March 1942, seems entirely feasible.

Bogdan Musial has persuasively argued that by early October 1941, the decision to exterminate the Jews of the *Generalgouvernement* and the Warthegau had been taken. Later, probably following Hitler's declaration of war against the United States on 11 December 1941, the extermination order was expanded to encompass the Jews of all Europe, a decision ratified at the Wannsee conference (originally planned for 9 December, then postponed to 20 January 1942).[928] At some point in October 1941 Hahnzog had been informed by Walther Griphan, the new commandant of the regular police in the Lublin District:

> ...quite plainly that the moment had now arrived for settling accounts with all enemies of the Reich — Poles, Jews and even Germans! ... This first shock was soon followed by a second: probably in November 1941, once again completely unprecedented, I was ordered, just as suddenly and unexpectedly, to report to Globocnik himself. He introduced me to a young SS leader who had been given the job of setting up the Sobibor camp and wanted the support of the gendarmerie office in Wlodawa for that purpose.[929]

Following a meeting on 13 October 1941 of Himmler, Globocnik, and Friedrich Krüger, *HSSPF* ((Higher SS and Police Leader) of the *Generalgouvernement*, it had been determined to radically expand and accelerate the first phase of the killing. Two small huts would certainly not be sufficient to cope with the thousands who would be arriving daily by train for extermination. Since a facility which was both close to the railway line as well as the major ghettos was clearly necessary, a purpose-built camp, for which Belzec had been earmarked at an early stage, was the obvious solution.[930] There was more; by autumn 1941 plans were already being laid for a second extermination camp at Sobibor.[931] Two days after the Himmler-Globocnik–Krüger meeting the deportation of German Jews "to the East" began, yet another decisive factor in the developing "Final Solution of the Jewish Problem".[932]

After departing from T4 following the issue of the "stop order", it has been suggested that Wirth may also have spent some time at Chelmno, the first of the Polish extermination camps. Although an SS concern, Chelmno was an operation entirely independ-

---

928 It is interesting to note that the conference concluded with "a discussion of the various possible forms which the solution [meaning extermination] might take." [Yitzhak Arad, Israel Gutman and Abraham Margaliot (eds.), *Documents On The Holocaust* (Lincoln: University of Nebraska Press, 1999), p. 261.] Even at this late stage different killing options were still under consideration.

929 http://www1.yadvashem.org/odot_pdf/Microsoft%20Word%20-%203222.pdf (Accessed 11 October 2009).

930 Browning, *The Origins of the Final Solution*, pp. 362–365. There had been a Jewish labour camp at Belzec between early 1940 and October of that year, when the camp was disbanded. [Arad, *Belzec*, p. 23].

931 Schelvis, *Sobibor*, p. 26

932 Longerich, *The Unwritten Order*, p. 130.

ent of both T4 and *Aktion Reinhard*, being primarily concerned with the murder of Jews of the *Warthegau*. Wirth considered Herbert Lange, commandant of Chelmno, a "bungler",[933] but the killing method eventually determined by Wirth for use in *Aktion Reinhard* was the same as that used by Lange at Chelmno—carbon monoxide poisoning generated by petrol exhaust fumes—albeit in purpose built stationary gas chambers rather than the mobile gas vans Lange favoured. Chelmno was quite different in character from other extermination centres, in that it was not a purpose built camp, consisting as it did of the town's manor house and adjacent buildings and a quite separate *Waldlager* (Forest Camp), where the victims were buried.[934] Thus there was no need for any great preparatory work before commencing operations, and such work as was necessary was only undertaken in November 1941, with mass gassings initiated in early December. If Wirth did visit Chelmno, it seems unlikely that he would have done so before it had been chosen as a killing site, that is in November 1941. By that time, following the visit of an SS surveying party in late October 1941,[935] construction of the first of the *Aktion Reinhard* camps at Belzec had begun on 1 November 1941 initially using local labour, then Jews from nearby ghettos.[936]

There was yet another connection between T4 and Chelmno. The HSSPF for the Warthegau, Wilhelm Koppe, testified that "in 1940, or it may have been in 1941" the Jews of the region were to be "evacuated" by an SS commando led by a man "whose name I later discovered was Lange." This was a barefaced lie, as Koppe had been responsible for supervising Lange and his murder of mental patients since at least May 1940.[937] Koppe continued:

> This idea was based upon the fact that a certain Dr Brack, of Hitler's private chancellery, had already done some preparatory work with poison gases, and that these were to be tried out by *Sonderkommando* Lange...Dr [Rudolf] Brandt [Himmler's secretary] told me that Dr Brack had already carried out experiments with gas in Berlin, that these experiments had almost been completed, and that it was planned that he, Dr Brack, would be put in charge of testing these gases in the Warthegau. *Sonderkommando* Lange was the obvious choice for carrying out the gassings...

The remainder of Koppe's remarks may be dismissed in their entirety as the usual farrago of lies and distortions common to mass murderers. Among other things, this great humanitarian claimed to have suffered a disconcerting moral and ethical dilemma: "Day and night I considered possibilities of averting the planned operation by some clever tactic or other."[938] Sadly for the 150,000 dead of Chelmno he was unable to chance upon a suitable strategy.

---

933 Dick de Mildt, *In the Name of the People: Perpetrators of Genocide in the Reflection of Their Post-War Prosecution in West Germany. The 'Euthanasia' and 'Aktion Reinhard' Trial Cases* (The Hague: Martinus Nijhoff Publishers, 1996), p. 240.

934 Ibid., pp. 231–233.

935 O'Neil, *Belzec*, p. 115.

936 Arad, *Belzec, Sobibor, Treblinka*, p. 25.

937 de Mildt, *In the Name of the People*, p. 229, p. 382, n.23.

938 Kogon, Langbein, Rückerl, *Nazi Mass Murder*, pp. 74–75.

Despite all of the correspondence, meetings, and experimentation that occurred in the autumn of 1941, and which was so crucial to the evolution of genocide, the majority of T4 personnel seem to have been uninvolved in the planning of this unprecedented crime. In November 1941 a conference of senior members of the "euthanasia" staff was held at Sonnenstein. According to Gorgass, Brack informed them that, "the *Aktion* had not been completed in August 1941, but was to continue...in some other form."[939] A schedule of T4 staff was sent to Krüger, but nothing is known of the activities of those listed. It has been suggested that members of this group were sent to Chelmno to assist with the killing operation there;[940] if so, it would be logical to conclude that, assuming Wirth did visit Chelmno, it would have been at around this time.

There followed the enigmatic *Organisation Todt* (OT) mission. In his post-war testimony, Brack claimed that in late 1941, Fritz Todt, then Minister of Armaments, persuaded Hitler that medical and other facilities for the troops on the eastern front were in desperate need of improvement, whereupon Hitler ordered all health-related organizations to supply whatever aid they could. Bouhler lost no time in gathering several parties of "volunteers" from among the T4 personnel, and sent them eastwards in January 1942. Included in these contingents (one of which was commanded by Brack) were some of the most prominent of the T4 doctors—Irmfried Eberl, Horst Schumann, Kurt Schmalenbach, Aquilin Ulrich, Hans-Bodo Gorgass, and Kurt Borm. But there were also those with lesser medical qualifications, and some with none at all. It is this mix of medical and non-medical personnel that has given rise to the proposition that these men and women were really sent on this assignment for the primary purpose of administering "euthanasia" to a different type of victim—brain-damaged members of the *Wehrmacht*.[941] This was certainly the evidence submitted by the nurse Pauline Kneissler, a member of one of these units. It is true that some "normal" medical aid-posts for wounded soldiers were set up in the Minsk region and elsewhere, but it would certainly have been within the realms of possibility for such establishments to offer a combination of healing and, in the worst cases, killing. This would undoubtedly have been a logical extension of the principle of *lebensunwertes Leben* to the battlefield, and might explain why men such as Kurt Bolender, Werner Dubois, Otto Stadie, and Heinrich Unverhau, some of whom were members of the SS, but none of whom were in any way medically qualified, were among those sent on the OT mission.

Whatever Wirth had been doing since the issue of the "stop order", it is known that he arrived at Belzec on 22 December 1941 to take command of the partially completed camp. Included among the earliest members of his staff were Josef Oberhauser, Erich Fuchs, Lorenz Hackenholt, Werner Borowski, Johann Niemann and Siegfried Graetchus—all ex-T4 "euthanasia" centre employees.[942] More with a similar background were to follow, with the most significant injection of new faces occurring in April-July 1942 as first Sobibor and then Treblinka became operational. SS staff were routinely trans-

---

939 Ernst Klee, "*Euthanasie" im NS-Staat: Die "Vernichtung lebensunwerten Lebens*" (Frankfurt am Main: Fischer Taschenbuch Verlag, 1991), p. 418.

940 O'Neil, *Belzec*, p. 92.

941 Henry Friedlander, *The Origins of Nazi Genocide: From Euthanasia to the Final Solution* (Chapel Hill: University of North Carolina Press, 1995), pp. 296–297.

942 O'Neil, *Belzec*, p. 118.

ferred between the camps, but as it transpired, due to reconstruction of the railway line between Lublin and Chelm, Sobibor was forced to temporarily close in late July 1942. There was thus only a relatively brief period when all three of the *Aktion Reinhard* camps were functioning simultaneously—between early October and mid-December 1942.

On reporting to the Lublin headquarters, the men transferred by T4 to *Aktion Reinhard* had been required to sign a statement affirming their binding observance of the secret nature of the mission. Amongst other things they acknowledged that it was impermissible to "pass on any information, verbally or in writing, on the progress, procedure or incidents in the evacuation of Jews to any person outside the circle of the *Einsatz* Reinhard staff." Photography within the camps was absolutely prohibited. Moreover, "the obligation to maintain secrecy" continued even after they had left *Reinhard.*[943]

Since few of the T4 personnel had enjoyed any military experience, those who had not were first sent to the camp at Trawniki, where like the Ukrainians who were to provide the bulk of the guards at the camps, they went through a course of basic training prior to receiving their posting.[944] This could eventually be to one of a number of places, since the *Aktion* extended far beyond the three extermination sites. In Lublin itself there were three major camps for the sorting and packaging of the possessions of the murdered Jews and the exploitation of Jewish forced labour—the Airfield Camp (*Zwangsarbeitslager Flugplatz*), the Lipowa Street Camp, and the *Sportplatz* Camp. Another building, at 27 Chopin *Strasse*, evaluated the jewellery and currency of the victims. On the outskirts of Lublin, the enormous concentration camp at Majdanek provided an inexhaustible supply of labour for sorting and other work. Within the Lublin region were smaller camps like Trawniki, Poniatowa, Budzyn, and Dorohucza. All contributed in their own way towards the gigantic programme of oppression, murder, and theft.[945]

Debate continues concerning the expansion of the killing programme from the Jews of the Soviet Union and Poland to those of all Europe and, perhaps, beyond. The Wannsee Conference, held on 20 January 1942 is usually seen as critical to this process, but it may be that its importance was more symbolic than practical. By the time the conference was held, killings on a massive scale were already being carried out by the *Einsatzgruppen* in the Soviet Union, and by an *SS-Sonderkommando* at Chelmno. Moreover, construction was already well advanced on the extermination camp at Belzec. But, importantly, the endorsement of a genocide already in progress by those present at Wannsee provided Heydrich, organizer in chief of the planned extension of the killing, with the comfort of knowing he had a free hand to implement "The Final Solution" .[946]

---

943 Arad, *Belzec, Sobibor, Treblinka,* p. 18.

944 O'Neil, *Belzec,* p. 93.

945 http://www.deathcamps.org/reinhard/areconomics.html (Accessed 14 October 2009).

946 https://www.jewishvirtuallibrary.org/jsource/Holocaust/wannsee.html (Accessed 28 December 2016).

Whatever the purpose of the conference might have been, in little more than eighteen months, *Aktion Reinhard* was complete. Within days of the uprising by the inmates of Sobibor on 14 October 1943, Himmler ordered the closure of the last of the three principal Lublin extermination camps still functioning.[947] Auschwitz-Birkenau with its four operational crematoria was more than capable of handling any future large scale *Aktionen*, as the annihilation of the Hungarian Jews the following year was to prove.

The T4 killers were not disbanded, however. Instead they were transferred en masse to Italy, their base of operations an abandoned rice mill, La Risiera di San Sabba in Trieste (home territory for Globocnik and several of his cohorts). Under the unsubtle code name of *Aktion R*, ostensibly their duties consisted of combating partisans, but in reality San Sabba became a prison and extermination camp for hostages, partisans, political and Jewish prisoners, a transit point for many destined for Auschwitz and other camps. Few of those transported were to survive. The creation of a crematorium on the premises in April 1944 provides confirmation that an uknown number of prisoners were killed at San Sabba itself.

It was not until 1976 that a trial was held in Trieste in connection with the atrocities committed at San Sabba. The prosecutions were effectively only symbolic. Many of the perpetrators were dead, and even if some survived and could be found, the Italo-German extradition treaty did not cover crimes committed before 1948. Nevertheless, the trial shone a light on events that had been too quickly forgotten.[948]

---

947 http://www.deathcamps.org/sobibor/sobibor.html (Accessed 29 December 2016).

948 https://web.archive.org/web/20120907014927/http://www.retecivica.trieste.it/triestecultura/new/musei/risiera_san_sabba/brochure/Risiera%20inglese%20per%20e-brochure.pdf (Accessed 29 December 2016).

CHAPTER 10: HIMMLER AND EUGENICS

Should we succeed in establishing this Nordic race again from and around Germany...and from this seedbed producing a race of 200 million, then the world will belong to us. Should Bolshevism win, it will signify the extermination of the Nordic race...devastation, the end of the world...We are called, therefore, to create a basis on which the next generation can make history. - Heinrich Himmler[949]

The people which has many children has the candidature for world power and world dominion - Heinrich Himmler[950]

As "*Reich* Commissioner for the Strengthening of German Nationhood" (*Reichskommissar für die Festigung deutschen Volkstums*—RKF) Heinrich Himmler was responsible for overseeing *Generalplan Ost*, the scheme to completely reorganize the demographics of eastern (and in time probably western) Europe. This fantastic plan to either redistribute or alternatively to annihilate whole populations largely owed its origins to ideas propagated by German social Darwinists both under and prior to the advent of National Socialism, and was essentially eugenic in its aims. As Himmler described it, the prime function of the RKF was to "transplant entire ethnic contingents". The perceived importance of the RKF within the overall Nazi genocidal strategy is illustrated by the invitation extended by Heydrich to the head of the organisation, the economist and SS-*Gruppenführer* Ulrich Greifelt, for the latter's attendance at the Wannsee conference, although as matters transpired Greifelt was unable to be present when the reconvened conference met on 20 January 1942.[951]

Although he only had a limited involvement with T4, Himmler had obviously been fully aware of that programme and supported it enthusiastically with personnel and equipment when called upon to do so.[952] As early as September 1933 Himmler had been invited by Wilhelm Frick to become a member of the commission for population and racial policy, a body intended to promote the annihilation of those considered "inferior", "burdensome existences", or "life unworthy of life". It was a position he accepted with alacrity.[953] However, it was only with *Sonderbehandlung* 14f13 and subsequently with *Aktion Reinhard* that T4 and Himmler's SS merged into a single killing whole. Despite his apparent lack of direct participation in the "euthanasia" programme, in his long term pursuit of Nordic supremacy and the "Aryan" myth, and therefore always ready to embrace virtually any hare-brained notion, in 1935 Himmler was godfather to

---

949 Peter Padfield, *Himmler: Reichsführer-SS* (London: Papermac, 1990), p. 101.

950 Gerwin Strobl, *The Bard of Eugenics: Shakespeare and Racial Activism in the Third Reich* (Journal of Contemporary History, Vol. 34, no.3, 1999), p. 335.

951 Götz Aly, and Susanne Heim, *Architects of Annihilation: Auschwitz and the Logic of Destruction* (London: Phoenix, 2003), pp. 74–75.

952 Hitler's allocation of "euthanasia" activities to the KdF was apparently a source of some irritation to Himmler, writing as he did to Bouhler and Brack in December 1940 on receipt of the letter from von Löwis concerning the evidence of killing at Grafeneck: "If operation T4 had been entrusted to the SS, things would have happened differently...When the *Führer* entrusts us with a job, we know how to deal with it correctly, without causing useless uproar among the people." [Robert Jay Lifton, *The Nazi Doctors: Medical Killing and the Psychology of Genocide* (London: Papermac, 1990), p. 95].

953 Horst Biesold, *Crying Hands: Eugenics and Deaf People in Nazi Germany* (Washington: Gallaudet University Press, 1999), p. 160.

two other organisations, both of which contained elements of "positive" and "negative" eugenics.

The first of these entities was *Ahnenerbe* (Ancestral Heritage), a venture jointly founded by Richard Walter Darré, Minister of Food and Agriculture and a leading "blood and soil" ideologue, the Dutch-German quasi-historian, Herman Wirth, and Himmler. It was an attempt to "scientifically prove" that the "Aryan" race (that is the Germans) had arrived on earth fully-formed.[954] This pseudo-scientific research institution led by Wolfram Sievers engaged in archaeological digs and extensive research of early German history, as well as expeditions to places like Tibet, in a hopeless attempt to validate a ludicrous proposition.[955] This alleged study might have been regarded as relatively harmless nonsense in comparison to other of Himmler's more sinister preoccupations, were it not for the suggestion that the "Aryans" had been preserved in "eternal ice" before emerging with "superhuman" powers. The concept of an eternal struggle for supremacy between two opposing forces—the heat of the sun and immense quantities of ice—fitted perfectly with the Nazi's interpretation of Darwin's "struggle for existence". When it was pointed out to Himmler that the notion was ludicrous, he indignantly defended his right to promote research into the *Welteislehre* ("world ice doctrine"). He believed himself in good company, "since the *Führer*...Adolf Hitler, has also been a convinced adherent for a long time of this despised doctrine..."[956]

The *Welteislehre* hypothesis led directly to experiments on concentration camp inmates under the umbrella of *Ahnenerbe*'s "Institute for Functional Research in Military Science" (*Institut für Wehrwissenschaftliche Zweckforschung*) established in late summer 1942. Perhaps the most notorious of these experiments were the attempts to assess the ability of non-"Aryans" to withstand the effects of (a) lack of pressure or of oxygen at high altitude, and (b) lengthy immersion in freezing water or other exposure to intense cold. These particular tests, conducted by Dr Sigmund Rascher at Dachau, resulted in the death, often in great agony, of many of the several hundred prisoners selected for experimentation.[957] Himmler was unconcerned. In October 1942 he wrote to Rascher:

---

954 The formal name of the organisation was *Studiengesellschaft für Geistesurgeschichte, Deutsches Ahnenerbe e.V.* (*'Study Society for Primordial Intellectual History, German Ancestral Heritage [registered society]'*). In 1937 it was renamed *Forschungs- und Lehrgemeinschaft das Ahnenerbe e.V.* (*'Research and Teaching Community of the Ancestral Heritage [registered society]'*).

955 Wolfram Sievers was born in 1905 in Hildesheim; he joined the Nazi party in 1928/1929 and the SS in 1935. A defendant at the Nuremberg Medical Trial, accused of aiding in the acquisition of skeletons for August Hirt's Strasbourg collection, as well involvement in human medical experiments at Dachau and Natzweiler, Sievers was sentenced to death. He was executed on 2 June 1948. [Ernst Klee, *Das Personenlexikon zum Dritten Reich–Wer war was vor und nach 1945* (Frankfurt am Main: S.Fischer, 2005), p. 583].

956 Padfield, *Himmler*, p. 171.

957 Ibid., pp. 374–376. Equally fatal tests were also conducted at Dachau concerning the potability of sea water. All branches of the military had an interest in these medical experiments, which despite often resulting in the death of the prisoners selected, had little or no practical value. Among other objections, the physical condition of concentration camp inmates could in no way be compared to that of healthy young serving soldiers, sailors or airmen. [Alexander Mitscherlich and Fred Mielke, *Doctors of Infamy: The Story of the Nazi Medical Crimes* (New York: Henry Schuman, 1949), pp. 4–41].

> People who today still disapprove of experiments on human beings, but who prefer to let brave German soldiers die from the consequences of intense cold, are to me nothing but traitors to their country, and I shall not hesitate to supply the names of these people to the authorities who are in a position to take action against them.[958]

In a letter written in the same year to Erhard Milch, *Generalinspekteur der Luftwaffe*, Himmler stated that he "personally assumed the responsibility for supplying asocial individuals and criminals who deserve only to die from concentration camps for these experiments". Moreover, Himmler requested that "the low pressure chamber [be put] at our disposal again, together with step-up pumps, because the experiments should be extended to include even greater altitudes".[959] By this time *Ahnenerbe* and its 197 employees were part of Himmler's personal staff.[960]

Other medical experiments were sponsored by *Ahnenerbe* via the Institute for Military Science, including investigation into insect control, typhus vaccines and the use of mustard gas.[961] When the anatomist August Hirt requested the bodies of "Jewish Bolshevik criminals" for his anthropological skeleton collection at the University of Strasbourg, Himmler arranged for Jews from Auschwitz to be murdered and their corpses supplied to Hirt.[962] At the Nuremberg IMT, a letter dated 2 November 1942 from Sievers to Rudolf Brandt, Himmler's secretary, was produced which stated, *inter alia*:

> As you know, the *Reichsführer-SS* has directed that SS-*Hauptsturmführer* Professor Dr. Hirt be supplied with everything needed for his experiments. For certain anthropological experiments—I have already reported to the *Reichsführer-SS* on them—150 skeletons of Jewish prisoners are required, which are to be supplied by KZ Auschwitz. The only thing that remains to be done is that the RSHA receive an official directive from the *Reichsführer-SS*. This, however, can also be given by you, acting for the *Reichsführer-SS*.

On 6 November, Brandt responded to Sievers' request with a letter to Adolf Eichmann at the RSHA:

> The *Reichsführer*-SS has issued a directive to the effect that SS-*Hauptsturmführer* Professor Dr. Hirt, who is the director of the Anatomical Institute at Strasbourg and the head of a department of the Institute for Military Science Research in the *Ahnenerbe* Society, be furnished with everything he needs for his research work. By order of the *Reichsführer*-SS, therefore, I ask you to be of assistance in materialising the planned collection. SS-*Obersturmbannführer* Sievers will get in touch with you to discuss the details.

In due course, Sievers advised Eichmann that 115 prisoners, 109 of them Jews, had been selected at Auschwitz for Hirt's collection. The potential victims were being kept in quarantine before being transferred to Natzweiler (Struthof) concentration camp for "processing". Because there was the risk of an epidemic at Auschwitz, Eichmann was instructed to arrange for clean and disinfected prisoner uniforms to be sent from Natzweiler to Auschwitz in order to avoid the possibility of spreading any infection from one camp to the other at the time the prisoners were transferred.

958 Padfield, *Himmler*, p. 437.

959 http://www.ess.uwe.ac.uk/genocide/rascher3.htm (Accessed 29 May 2008).

960 Guenter Lewy, *Himmler and the 'Racially Pure Gypsies'* (Journal of Contemporary History, Vol 34(2), 1999),, p. 204.

961 Paul Julian Weindling, *Nazi Medicine and the Nuremberg Trials—From Medical War Crimes to Informed Consent* (Basingstoke: Palgrave Macmillan, 2006), p. 51.

962 Ibid., p. 187. Hirt was also extensively involved in the mustard gas experiments. (Mitscherlich and Mielke, *Doctors of Infamy*, p. 76).

Joseph Kramer, later the final commandant of Bergen-Belsen concentration camp but at that time commandant of Natzweiler, testified that in August 1943 he met with Hirt at Strasbourg, and was informed that a group of eighty prisoners was *en route* from Auschwitz to Natzweiler. On arrival they were to be gassed, and their corpses delivered to Hirt at his Anatomical Institute. Hirt passed a bottle to Kramer, probably containing *Zyklon* B, and instructed him on the approximate dosage to be used to kill the victims. Kramer estimated that he subsequently gassed 50–55 individuals at Natzweiler and dispatched their bodies to Hirt. Kramer claimed to have no knowledge of what Hirt intended to do with the bodies. A witness who worked at the Anatomical Institute stated that in fact a total of eighty-six corpses of gassed men and women were delivered to Hirt.

As it became increasingly apparent that Strasbourg was likely to soon be captured by the advancing Allies, Hirt wrote to Sievers asking what was to be done with his collection. The bodies had been embalmed for a year, completely untouched. In turn, Sievers wrote to Himmler, setting out the options—either preserve the collection in whole or in part, or simply destroy it. In the circumstances it was not a difficult decision for Himmler to make. By 16 October 1944 the Strasbourg collection had been "dismantled" and the corpses cremated.[963]

Skeletons were only part of Hirt's proposed collection. In late 1941 he had written to Sievers:

> We have a nearly complete collection of skulls of all races and peoples at our disposal. Of the Jewish race, however, only very few specimens of skulls are available, with the result that it is impossible to arrive at precise conclusions from examining them. The war in the east now presents us with the opportunity to overcome this deficiency. By procuring the skulls of the Jewish-Bolshevik commissars, who represent the prototype of the repulsive but characteristic subhuman, we have the chance now to compile good, scientific documents.
>
> The best practical method for obtaining and collecting this skull material could be handled by directing the *Wehrmacht* to turn over alive all captured Jewish-Bolshevik commissars to the Field Police. They, in turn, are to be given special directives to inform a certain office at regular intervals of the numbers and place of detention of these captured Jews and to give them close attention and care until a special delegate arrives. This special delegate, who will be in charge of securing the material (a junior physician of the *Wehrmacht* or even the Field Police or a student of medicine equipped with a motor car and driver), will be required to take a previously established series of photographs, make anthropological measurements and, in addition, determine as far as possible other personal data of the prisoners.
>
> Following the subsequently induced death of the Jew, whose head should not be damaged, the physician will separate the head from the body and will forward it to its proper point of destination in a hermetically sealed tin can, especially made for this purpose and filled with a conserving fluid. Having arrived at the laboratory, the comparison tests and anatomical research on the skull, as well as determination of the race membership and of pathological features of the skull form, the form and size of the brain, etc., can proceed by means of photos, measurements and other data supplied on the head, and the skull itself.[964]

In part Hirt was probably able to boast of his "nearly complete collection of skulls of all races and peoples", because in late 1940, Himmler had arranged for Franz Joseph Gall's eighteenth-century collection of skulls to be sent from Paris to the racial-

963 Mitscherlich and Mielke, *Doctors of Infamy*, pp. 81–89.

964 http://www.nizkor.org/hweb/imt/tgmwc/tgmwc-20/tgmwc-20-198-04.shtml (Accessed 15 January 2009).

biological institute at Tübingen University.[965] Himmler's obsession with such pseudoscience was apparently boundless.

Among other of the typically weird interests of Himmler's were the "racial characteristics" of Gypsies. In a document he issued in 1937 on the so-called "Fight against the Gypsy Nuisance", Himmler declared that "the proper method of attacking the Gypsy problem seems to be to treat it as a matter of race...It has therefore become necessary to distinguish between pure and part-Gypsies [*Zigeunermischlinge*] in the final solution of the Gypsy question."[966] Initially, Nazi anti-Gypsy measures, which included compulsory sterilisation under the 1933 "Law for the Prevention of Genetically Diseased Offspring", and the prohibition of marriage between Gypsies and "Aryans" through the 1935 Nuremberg "Law for the Protection of German Blood and German Honour", as well as the "Law for the Protection of the Hereditary Health of the German Nation" (the "Marriage Health Law") of the same year, had been justified on alleged criminal and asocial grounds, but with the introduction of the racist element, persecution entered a new and alarming phase.[967] In December 1938 Himmler issued another decree entitled "Combating the Gypsy Plague", in which he made it clear that the Gypsy problem was to be resolved by reference to the "inner characteristics of that race."[968]

For the Nazis, identifying Jews on the basis of the religious affiliation of parents and grandparents appeared to be relatively easy, although in practice it proved an intractable problem. Gypsies presented significantly greater difficulties, for they were Christian, and there had been a much higher level of intermarriage than with the Jews; moreover, many Gypsies had given up the traditional ways and were living sedentary lives. To resolve this issue, in spring 1936 a division of the *Reich* Health Office was established in Berlin. Under the title of the "Racial Hygiene and Demographic Biology Research Unit"("*Rassenhygienische und bevölkerungsbiologische Forschungsstelle*"), and headed by the University of Tübingen neurologist and psychiatrist, Dr Robert Ritter, its task was to collect data about Germany's "travellers", particularly "pure" Gypsies and *Zigeunermischlinge* (in polite terms, Gypsies of mixed ancestry, but as the Nazis interpreted the demeaning label, Gipsy mongrels).[969]

Despite the eventual extermination of tens of thousands of those defined by the Nazis as Gypsies, Himmler wished to preserve a few clans of pure Sinti, Roma, and Bohemian Lalleri under the auspices of *Ahnenerbe*.[970] This interest arose because, in accordance with the prevalent inane racial theory, the Sinti and Roma had originated from India, and had therefore once been "Aryan". According to Ritter's research, only 10 percent could claim to be pure Gypsies. It was this 10 percent (or at least part of this 10

---

965 Saul Friedländer, *The Years of Extermination: Nazi Germany and the Jews, 1939–1945* (New York: HarperCollins, 2007), p. 166.

966 Michael Burleigh and Wolfgang Wipperman, *The Racial State: Germany 1933–1945* (Cambridge: Cambridge University Press, 1991), p. 121.

967 Ulrich Herbert (ed), *National Socialist Extermination Policies—Contemporary German Perspectives and Controversies* (New York: Berghahn Books, 2000), p. 191. Although the Nuremberg laws had been directed primarily against Jews, the Minister of the Interior, Wilhelm Frick, had pointed out that there were others who contaminated the purity of the race—"Gypsies, Negroes, and their bastards." (Lewy, *Himmler and the 'Racially Pure Gypsies'*, p. 201.)

968 Lewy, *Himmler and the 'Racially Pure Gypsies'*, p. 201.

969 Ibid., p. 202.

970 Burleigh and Wipperman, *The Racial State*, p. 125.

percent) that Himmler was interested in preserving as a kind of ethnic oddity.[971] Ritter had proposed that this itinerant minority be allowed limited and police supervised freedom of movement, and in October 1942, Arthur Nebe head of the *Kripo*, who had been instructed by Himmler to co-operate closely with *Ahnenerbe*, issued a regulation, which read in part:

> The *Reichsführer*-SS intends that in the future racially pure Gypsies be allowed a certain freedom of movement, so that they can itinerate in a fixed area, live according to their customs and mores, and follow an appropriate traditional occupation...*Zigeunermischlinge*, who from the point of view of the Gypsies are good *Mischlinge*, shall be returned to specific racially pure Sinti Gypsy clans. If they apply for membership in a racially pure clan and the latter has no objections, they shall be assigned the same status as racially pure Gypsies.

Lists of racially pure Gypsies were to be sent to the *Kripo* head office in Berlin. Nebe allegedly said that while he had no option but to announce Himmler's plans, nothing would come of them. He was right. Ritter's proposals might have been feasible in peacetime, but were impossible in wartime. Few of the Gypsies owned caravans, even fewer possessed horses; how were they to become self-sufficient? With the rationing of food and other essentials, how could their activities be regulated? Whether Himmler came to realise the impractical nature of his proposal is not known, but the fate of German Gypsies deemed racially "impure" was sealed when on 16 December 1942 he issued an order for their deportation to Auschwitz-Birkenau.[972] The deportations began in March 1943 and continued until the summer of 1944.[973]

On 3 December 1942, Martin Bormann, who had got wind of Himmler's plans, wrote to him:

> Special treatment for the racially pure Gypsies would represent a fundamental departure from presently applied measures for fighting the Gypsy plague and would not be understood by the population and the lower ranks of the party leadership. The *Führer*, too, would not approve of it if a segment of the Gypsies is given back their old freedom.[974]

In fact, it is proposed, Hitler had little real interest in Gypsies, racially pure or otherwise. For him, there was only one racial problem in Germany of any real significance, and that concerned the Jews.[975] There is a contrasting opinion that Hitler was directly involved in the issuing of Himmler's deportation order of 16 December; the timing of the correspondence and consultations (Himmler met with Hitler on the afternoon of 6 December 1942 and Bormann the same evening)[976] lends credence to this theory.[977] In any event, Himmler's decree of October 1942 remained in place, and racially acceptable Gypsies were exempted from deportation. Some sources suggest that as many as 50 percent or perhaps more of German Gypsies escaped deportation as a result of Himm-

---

971 Ibid., pp. 113–125.

972 Michael Berenbaum and Abraham J Peck (eds.), *The Holocaust and History: The Known, the Unknown, the Disputed, and the Reexamined* (Bloomington: Indiana University Press, 2002), p. 321.

973 Herbert, *National Socialist Extermination Policies*, p. 201

974 Lewy, *Himmler and the 'Racially Pure Gypsies'*, p. 209.

975 Ibid., p. 202.

976 Ibid., p. 209.

977 Robert Gellately, and Nathan Stoltzfus (eds.), *Social Outsiders in Nazi Germany* (Princeton: Princeton University Press, 2001), p. 226.

ler's ruling.[978] Others suggest that a much smaller number survived.[979] While this debate about the lives of thousands of innocent people went on, *Ahnenerbe* continued its dreadful work.

The second of Himmler's eugenic institutions was *Lebensborn e.V.* (Fount of Life [registered society]), an entity devoted to the production of supposedly superior children from the mating of racially suitable parents. In September 1936 Himmler issued an edict setting out the philosophy behind the establishment of this new organisation:

> ...We have fought in vain if political victory was not to be followed by births of good blood. The question of multiplicity of children is not a private affair of the individual, but his duty towards his ancestors and our people...The existence of sound marriage is futile if it does not result in the creation of numerous descendants. I expect that here, too, the SS, and especially the SS leadership corps, will serve as guiding examples.
>
> The minimum number of children for a good sound marriage is four. Should unfortunate circumstances deny a married couple their own children, then every SS leader should adopt racially and hereditarily valuable children, educate them in the spirit of National Socialism, and let them have education corresponding to their abilities...The organisation *Lebensborn e.V.*serves the SS leaders in the selection and adoption of qualified children."[980]

The agency was funded in part through the NSV (*Nationalsozialistische Volkswohlfahrt*, the National Socialist People's Welfare Organization, commonly interpreted within this context as "The *Reich* Adoption Service") as well as through the compulsory membership of all SS members, who in many cases contributed in ways other than financial. Provided they passed a test proving that they were racially suitable, both married and unmarried expectant mothers were admitted to *Lebensborn* maternity homes to give birth.[981] The newly born illegitimate were then subjected to the same kind of racial testing as their mother; those failing either became part of the children's "euthanasia" statistics, or in some cases were returned to their natural mother for upbringing. Those deemed acceptably "Aryan" were either given to foster parents, or remained in the home until such time as suitable foster parents were found. In 1941, Hitler stated: "I do not doubt for a moment, despite certain people's scepticism, that within a hundred years or so from now all the German élite will be a product of the SS—for only the SS practises racial selection."[982]

Whilst facilities for expectant mothers were exceptionally good in the *Lebensborn* homes, once in the care of the nurses who were to be responsible for their everyday needs, the abandoned children were brought up in a particularly cold and unfeeling atmosphere. They lacked the love and attention normally lavished on their offspring by doting parents. Instead they were raised in a regimented and disciplined manner, treat-

---

978 Lewy, *Himmler and the 'Racially Pure Gypsies'*, p. 210.

979 Berenbaum and Peck, *The Holocaust and History*, p. 321.

980 Roderick Stackelberg and Sally Anne Winkle, *The Nazi Germany Sourcebook: An Anthology of Texts* (London: Routledge, 2002), p. 200.

981 The wives of SS-men were of course admitted without recourse to such testing. They had already been pre-approved by Himmler's consent to their marriage.

982 Burleigh and Wipperman, *The Racial State*, p. 274.

ed as simply another product removed from a seemingly endless human conveyor belt. The long term psychological effect this was to have on the children is incalculable.

Originally administered by the Families [or Clans] Department (*Sippenamt*) of the SS Race and Settlement Office (*Rasse- und Siedlungshauptamt-SS –RuSHA*), in 1938 Himmler took personal control of *Lebensborn* and installed it in his own office under the supervision of Karl Wolff. The compulsory subscriptions for SS members, deducted from their salaries, were based upon a convoluted arbitrary amalgam of age, marital status, and procreativity, such that only a mind like Himmler's could fashion. So far as Himmler was concerned, *Lebensborn* formed an intrinsic part of the Nazi's cherished genetic goals, aimed at producing a master race of *Herrenmenschen*. Others viewed the operation with rather more cynicism. Rumours abounded that *Lebensborn's* maternity homes were little more than SS stud farms. That certainly was, and is, the commonly held (if incorrect) perception of them.[983] In fact the *Lebensborn* homes were an essential component of Himmler's unsuccessful drive to eradicate abortion, since they provided outstanding facilities for pregnant women who might otherwise be tempted to terminate their pregnancy.[984] If termination did occur, and the unborn child would have been of racially valuable stock, from Himmler's perspective every such abortion represented a disastrous loss of precious material to the nation. *Lebensborn* can thus be seen as a practical example of positive, if typically inhuman, Nazi eugenics.

An illustration of the impact Himmler's inane ideas could have on the most trivial matters was his belief that the English upper classes possessed good figures because they ate porridge for breakfast. Therefore, he commanded, members of the SS and pregnant women in *Lebensborn* homes would perforce begin each day with porridge.[985] However, the war took its toll; in January 1944 Himmler decreed that in future the *Lebensborn* porridge was to be made with water, not milk, as had previously been the case.[986] Other pronouncements of the *Reichsführer* were less likely to be greeted with mirth (or indigestion). In October 1939, shortly after the outbreak of war, he called on the women of the *Reich* to do their duty just as their men folk were doing, albeit in a somewhat different fashion:

> But beyond the boundaries of otherwise perhaps bourgeois laws and customs, even outside marriage, it can be a lofty task for German women and maidens of good blood, not lightly, but in deepest moral seriousness, to become mothers of children of soldiers called to war...Also the men and women, who serve the state in the homeland, have at this precise time the sacred obligation to become again mothers and fathers of children

This pronouncement was viewed by some as a licence for wholesale fornication and infidelity, for it seemed to imply that in time of war, old-fashioned mores not only could, but should be disregarded. But how would the soldier returning from the front feel about the wife who had taken heed of Himmler's words, and produced one or more children for the Fatherland through dalliance with a new (racially valuable) partner? Was it now to be the duty of men working in wartime industries to seek out and seduce

---

983 Padfield, *Himmler*, p. 166.
984 Jill Stephenson, *Women in Nazi Germany* (Harlow: Pearson Education Limited, 2001), p. 39.
985 Padfield, *Himmler*, p. 202.
986 Ibid., p. 477.

the wives and girlfriends of fighting men? Of course the idea was absurd, and can be viewed as yet another example of Himmler's obsession with all things sexual.[987] In the wake of German losses on the eastern front, Himmler's ideas became even more grandiose. He envisaged the creation of a *Lebensborn* central office in Munich which would function as a kind of bureau to find partners for "the approximately 400,000 [women] there are already who, because of the men who have fallen in the war, cannot acquire husbands," a proposal which also came to nothing.[988]

Lebensborn did not operate solely in Germany. Homes were established in Austria, Belgium, the Netherlands, France, Luxembourg, Denmark, and Norway,[989] where in 1942, Josef Terboven, *Reich* Commissar for Norway, granted the organisation the right to care for the illegitimate result of union between German soldiers and Norwegian women. There were eventually to be eight *Lebensborn* homes established in Norway. Himmler secretly ordered the agency to "Germanise" racially suitable Norwegian children by removing them from their mothers and transferring them to Germany for upbringing. This, so Himmler believed, would improve the racial stock of the *Reich* through the introduction of fresh 'Nordic blood.' It has been estimated that during or immediately following the Second World War, 10,000–12,000 children were born in Norway of native mothers and German fathers.[990] In Denmark a figure of more than 5,500 such births has been quoted.[991]

There was an ongoing conflict of interests between the NSV and *Lebensborn* over the question of the illegitimate children born in other countries as a result of the fraternisation of German soldiers and local women. In the Netherlands, *Reich* Commissar Arthur Seyss-Inquart decided that the NSV was to be the responsible authority in such affairs, as a result of which Dutch NSV maternity homes had registered their one thousandth birth by July 1943. After the war the Dutch authorities went to great lengths to conceal the German paternity of many children, so much so that even today there are still individuals living in the Netherlands who are unaware of the true identity of their father.[992]

In eastern Europe, children up to the age of six were abducted on a vastly greater scale. At least 350 Polish children were found by *Lebensborn* to be ethnically suitable for adoption by German foster parents as a result of this "fishing for blood." In May 1940 Himmler had prepared a memorandum that touched on this subject from a different angle, and which apparently met with the *Führer's* approval. Parents wishing to educate their children above the most primitive level in the soon to be occupied eastern nations, or those already occupied, would make an application to the SS. Thereafter:

> ...The decision will be made on whether the child is racially immaculate and suitable for our conditions. If we recognise such a child as of our blood, the parents will be notified that the child should attend school in Germany and remain in Germany permanently...The parents of this child of good blood will then be faced with the choice of either surrendering the child...or the parents commit themselves to go to Germa-

987 Ibid., pp. 279–280.

988 Ibid., p. 368.

989 http://www.jewishvirtuallibrary.org/jsource/Holocaust/Lebensborn.html (Accessed 27 September 2009).

990 Kjersti Ericsson and Eva Simonsen, *On the Border: The contested children of the Second World War* (Childhood, Vol. 15, No.3, 2008), p. 399.

991 Ibid., p. 402.

992 Ibid., p. 402.

> ny and become loyal citizens...because we believe that they are really our own blood, which has found its way into a foreign nationality through the mistakes of German history, and we are convinced that our *Weltanschauung* and our ideals will re-echo in the racially equal souls of these children.[993]

This sounds a good deal more humane in theory than it would prove in practice, for in reality the NSV were infinitely more ruthless than Himmler's ramblings suggested. By 1942 there were 600 NSV kindergartens in the *Warthegau* alone, from where, after being subjected to the usual round of testing for racial suitability, children were sent to "Germanisation" centres in the *Reich.* There they were given new identities and forbidden to use their native tongue, before being shipped out to foster parents. Among the children selected for "Germanisation" were thirteen from the Czech village of Lidice, destroyed in an act of vengeance following the assassination of Reinhard Heydrich.[994] It has been reliably estimated that as many as 200,000 eastern European children were carried off in this manner; only 15–20 percent of them were reunited with their legitimate parents after the war.

The case of Folker Heinecke may be considered typical. Born Alexander Litau in Alnowa in the Crimea, he was kidnapped from his parents in 1942 when he was two years of age. He was first taken to Lodz in Poland, where he was examined to find out if he was "worthy" of "Aryanisation". Post-war investigation of the files maintained on him by the Nazis indicate that he was meticulously examined in order to determine whether or not he had "Jewish Aspects". Satisfied that he was capable of being "Germanised", he was shipped to his new parents in Germany. It was only with the opening of the National Socialist archive at Bad Arolsen that Herr Heinecke/Litau was finally able to discover who he really was.

The NSV were involved in another example of Nazi eugenics in action, when in 1942, "Eastern Confinement Homes" or "Child Collection Points" were established for the purpose of selecting the newly-born children of Polish and Soviet female forced labourers. Racially "especially valuable" children were taken for adoption by *Lebensborn*; those considered merely "valuable" by the NSV. The "worthless" were starved to death in "Care Centres for Foreign Children."[995] On this subject, Erich Hilgenfeldt, a former Berlin councillor who rose to become head of the NSV, wrote to Himmler in August 1943:

---

993 Padfield, *Himmler*, pp. 302–303.

994 Ibid., p. 387.

995 An interesting sidelight on the ignorance of the British Foreign Office regarding conditions in occupied Europe are the comments by Victor Frederick William Cavendish-Bentinck, chairman of the Joint Intelligence Committee, appended in August 1943 to a Polish report on Nazi barbarism:

> In my opinion it is incorrect to describe Polish Information regarding German atrocities as 'trustworthy'. The Poles, and to a far greater extent the Jews, tend to exaggerate German atrocities in order to stoke us up. They seem to have succeeded...I do not believe that there is any evidence which would be accepted in a Law Court that Polish children have been killed on the spot by Germans when their parents were being deported to work in Germany, nor that Polish children have been sold to German settlers. As regards putting Poles to death in gas chambers, I do not believe that there is any evidence that this has been done...

[Robert Jan van Pelt, *The Case for Auschwitz—Evidence from the Irving Trial* (Bloomington: Indiana University Press, 2002),p. 127].

In my opinion, the present handling of the question is impossible. It is a case of either/or. Either one decides that the children are not to live—in which case one should not allow them to slowly starve, thereby denying many litres of milk for general consumption, for there are ways in which this can be done without pain and suffering. Or, on the contrary, one's object is to raise the children in order to use them later as a source of labour. In which case, it is necessary to feed them so that one day their labour will be of value.[996]

It is only in recent times that the full impact this cruel and heartless scheme had on the product of Himmler's racial fantasies has begun to be appreciated. Many *Lebensborn* children were ostracised or worse in childhood. On reaching middle-age a number of them have begun to question their parentage. Brought up by their natural mother and perhaps a foster-father or in the case of the kidnapped children, foster-parents, they have a burning desire to discover who their progenitors were—and what they did. The answer is often far from pleasant, especially insofar as their biological father is concerned. Perhaps even worse is the predicament of the children stolen from their now long dead natural parents in distant countries. For them, discovering their true identity and surviving blood relations is an almost hopeless task.

Long after it had ceased to be a major element of domestic Nazi racial hygienic policy, Himmler's involvement in the search for methods of mass sterilisation was based more upon short-term economic considerations than eugenics, although there can be no doubting Himmler's enthusiasm for the latter. Unquestionably there were longer term eugenic implications for such ideas so far as any remaining Jews and the captive Slav population were concerned,[997] but in 1942 an acute shortage of labour brought to the fore the concept of exploiting the existing slave-labour force whilst at the same time preventing their ability to multiply.

The idea of utilising X-rays for such a purpose had been advanced by a number of individuals for some time. In his post-war Nuremberg testimony, Viktor Brack stated that in 1941 it was an "open secret in higher party circles that those in power intended to exterminate the entire Jewish population in Germany and [the] occupied territories." Because murder on such a scale was "unworthy of party leaders and humanity more generally", being the good Samaritans that they were, Brack and Hefelmann instead advocated deporting the Jews to Madagascar.[998] However, when that idea foundered:

---

996 Burleigh and Wipperman, *The Racial State*, pp. 65–73.

997 Franciszek Piper and Teresa Swiebocka, *Auschwitz Nazi Death Camp* (Oswiecim: The Auschwitz-Birkenau State Museum in Oswiecim, 1996), p. 89.

998 Brack's claimed involvement in the so-called "Madagascar Plan" was at best an exaggeration, if not a downright lie. The idea of using Madagascar as a place to dump European Jews had been mooted by anti-Semites as early as 1885. A Polish commission visited the island in 1937 to investigate the possibility of forcibly settling Jews there. In Nazi circles, the suggestion had first been put forward as a supposedly serious option in June 1940 by the "Jewish expert" in the Foreign Office, Franz Rademacher. [Christopher R Browning, *The Path to Genocide: Essays on Launching the Final Solution* (Cambridge: Cambridge University Press, 1997) p. 18]. Eichmann at the RSHA was contemporaneously putting together an SS version of the scheme which Heydrich forwarded to the Foreign Office in August 1940. The idea had been circulating among the Party hierarchy long before this totally impractical proposal surfaced in detailed form. In a diary note made two years earlier, Goebbels mentioned a discussion of this very subject with Hitler. Other leading Nazis like Alfred Rosenberg and Julius Streicher were also

> ...we came up with the idea that sterilisation might provide the solution to the Jewish question. Given that sterilisation is a rather complicated business, we hit upon the idea of sterilisation by X-rays. In 1941 I suggested to Bouhler the sterilisation of Jews by X-rays; this idea was also rejected, however. Bouhler said that sterilisation by X-rays was not an option, because Hitler was against it. I worked on this programme further and finally came up with a new plan.[999]

Himmler met with Brack in January 1941 to discuss the progress of the "euthanasia" programme. It was at this meeting that Himmler suggested to Brack that he institute research into the possibility of sterilising large numbers of people through the use of X-rays.[1000] In March 1941, Brack sent a lengthy report to Himmler, which read in part:

> The experiments in this field are concluded...If any persons are to be sterilised permanently this result can only be attained by applying X-rays in a dosage high enough to produce castration with all its consequences, since high X-ray dosages destroy the internal secretion of the ovary, or of the testicles, respectively. Lower dosages would only temporarily paralyze the procreative capacity...The actual dosage can be given in various ways, and the irradiation can take place quite imperceptibly...If the X-ray intensity is too high, those parts of the skin which the X-rays have reached will exhibit symptoms of burns—varying in severity in individual cases—in the course of the following days or weeks.
>
> One practical way of proceeding would be, for instance, to let the person treated approach a counter, where they could be asked to answer some questions or to fill in forms, which would take them 2 or 3 minutes. The official sitting behind the counter could operate the installation in such a way as to turn a switch which would activate the two valves simultaneously (since the irradiation has to operate from both sides). With a two-valve installation about 150–200 persons could then be sterilised per day, and therefore, with 20 such installations as many as 3,000–4,000 persons per day.
>
> ...Mass sterilisation by means of X-rays can be carried out without difficulty. However, it seems to be impossible to do this in such a way that the persons concerned do not sooner or later realise with certainty that they have been sterilised or castrated by X-rays."[1001]

It would appear that Himmler was unenthusiastic about pursuing the idea at the time, but Brack persisted with it. On 23 June 1942 he wrote to Himmler:

> ...Among 10 millions of Jews in Europe there are, I believe, at least 2–3 millions of men and women who are fit enough to work. Considering the extraordinary difficulties the labour problem presents us with, I hold the view that those 2–3 millions should be specially selected and preserved. This can, however, only be done if at the same time they are rendered incapable of propagation. About a year ago I reported to you that agents of mine had completed the experiments necessary for this purpose. I would like to recall these facts once more. Sterilisation, as normally performed on persons with hereditary diseases, is here out of the question, because it takes too long and is too expensive. Castration by X-ray, however, is not only relatively cheap, but can also be performed on many thousands in the shortest time...

---

considering the Madagascar option in 1938. [Michael Wildt, *An Uncompromising Generation—The Nazi Leadership of the Reich Security Main Office* (Madison: University of Wisconsin Press, 2009), pp. 248–249]. What is certain is that the "Madagascar Plan" was a subject well-known not only to high ranking Nazis, but even to the potential deportees. In a diary note of 7 July 1940, Victor Klemperer recorded the comment of a friend as a rumour spread that the British government had fallen: "Now they'll make peace and we'll be packed off to Madagascar." [Victor Klemperer, *I Shall Bear Witness : The Diaries of Victor Klemperer 1933–41* (London: Weidenfield & Nicolson, 1998), p. 332].

999 Proctor, *Racial Hygiene*, p. 206

1000 Ulf Schmidt, *Karl Brandt: The Nazi Doctor—Medicine and Power in the Third Reich* (London: Hambledon Continuum, 2007), p. 153.

1001 Dick de Mildt, Dick, *In the Name of the People: Perpetrators of Genocide in the Reflection of Their Post-War Prosecution in West Germany. The 'Euthanasia' and 'Aktion Reinhard' Trial Cases* (The Hague: Martinus Nijhoff Publishers, 1996), pp. 384–385, note 40.

...Should you, *Reichsführer*, decide to choose this way in the interest of preservation of labour, then *Reichsleiter* Bouhler would be prepared to place all physicians and other personnel needed for this work at your disposal. Likewise he requested me to inform you that I would then have to order the apparatus so urgently needed with the greatest speed."[1002]

Brack did not specify the fate of 7–8 million Jews who were not to be sterilised, but his support for their extermination was implied in his comment earlier in the letter of the necessity of completing "the whole Jewish action".[1003] His claimed humanitarianism is thus exposed as merely another example of the planned economic exploitation of the victims. In fact, at the very time Brack wrote this report, the extermination camps of *Aktion Reinhard* were murdering Jews in vast numbers.

Between receipt of Brack's two missives, Himmler's interest in mass sterilisation had been aroused by an idea of Dr Adolf Pokorny, a Viennese born dermatologist and military physician, who had contacted Himmler in October 1941 suggesting the potential use of a herb, *Caladium seguinum*, to produce sterility in humans.[1004] Pokorny drew Himmler's attention to research by Dr Gerhard Madaus, whose homeopathic remedies held a natural attraction for Himmler. Madaus claimed to have discovered that the herb was efficacious in causing sterility in small mammals—males permanently, females temporarily.[1005] Pokorny was even more ambitious than Brack:

If we are able to discover as quickly as possible the means that could imperceptibly cause sterilisation in a relatively short period of time, then we will have acquired a new, very effective weapon. The broad possibilities are alone suggested by the thought that three million Bolsheviks currently held as German prisoners could be sterilised, and as a result, they would stand at our disposal as workers, but without the ability to multiply.[1006]

Whether experiments were conducted using the herb for the purpose of sterilisation is uncertain, but in any event the idea was completely impractical in wartime conditions.[1007] Oswald Pohl, head of the WVHA, informed Himmler in June 1942 that since the plant was native to North America, it was impossible to import it in sufficient quantity in the then current circumstances. Attempts to grow it in bulk had not been successful, and would require much larger hothouses than had been utilised to date, with no guarantee of success. Although Himmler urged that experiments be conducted with whatever supplies of the drug were available, Pokorny's suggestion does not appear to have been further pursued with any great enthusiasm. However, the concept did stimu-

1002 Ibid., pp. 238–239.

1003 Lifton, *The Nazi Doctors*, p. 280.

1004 Ibid., p. 275. Pokorny (also somewhat confusingly described as a Sudeten *Volksdeutsch*), whose former wife was Jewish (they divorced in 1935), was among those indicted at the Medical Trial in Nuremberg. He was acquitted of participation in the sterilisation programme. His half-Jewish children arrived in Great Britain on a June 1939 *Kindertransport*. (Weindling, *Nazi Medicine and the Nuremberg Trials*, p. 166).

1005 Mitscherlich and Mielke, *Doctors of Infamy*, pp. 133–134.

1006 Piper and Swiebocka, *Auschwitz Nazi Death Camp*, p. 89.

1007 Rudolf Brandt, Himmler's personal adjutant, testified that experiments with *Caladium seguinum* were made on concentration camp inmates. A report of 24 August 1942 from Karl Gund, Deputy *Gauleiter* of Niederdonau (Lower Austria) refers to the possibility of the "necessary investigations and human experiments" being made on the inmates of the Gypsy camp at Lackenbach in Niederdonau, but no details of any such experiments have emerged to date. (Mitscherlich and Mielke, *Doctors of Infamy, Medical Crimes*, pp. 133–135).

late Himmler into serious consideration of a mass sterilisation programme. To this end, two gruesome medical experiments were organised on Auschwitz inmates, both mainly conducted in block 10 of the main camp, although there were also stations set up in other blocks and in Birkenau. One experiment, conducted by Carl Clauberg, was concerned with female sterilisation through the injection of a caustic substance into the cervix, the intention being to obstruct the fallopian tubes. The second had a direct connection with T4, since it involved both male and female sterilisation by the use of X-rays, as had been suggested by Brack, and was conducted by the former chief physician at Grafeneck and Sonnenstein, Horst Schumann.[1008]

Both experiments were doomed to failure, but not before indescribable suffering had been endured by the mostly Jewish prisoner "guinea pigs", followed by the death of many of them.[1009] Although Clauberg was a professor of some distinction and chief physician at the Women's Hospital of Kiel University, as well as head of a clinic for gynaecological complaints at Königshütte (Chorzow), he cut a ridiculous figure at Auschwitz when he arrived there in late 1942. He was described by witnesses as "short, bald, and unlikable", and "...a small, ugly, funny-looking, more or less deformed person. He wanted to imitate Prussian officers but he looked like a salesman in a general's hat...He was absurd."

Absurd or not, Clauberg was feared by the women he mutilated, for he literally held the power of life and death over them. When he appeared on the ward women were "overcome with anxiety and terror...They considered what Dr Clauberg was doing as the actions of a murderer."[1010] It was not an inaccurate description. In many cases women who survived his experiments were killed directly afterwards by means of a phenol injection or were sent to the gas chamber, with some corpses being dissected in order to evaluate the results of this "research".[1011]

After his initial consultation with Himmler, Clauberg had believed that female prisoners would be sent from Ravensbrück to his clinic at Königshütte for him to experiment upon. Himmler, on the other hand, was under the impression that Clauberg would be working at Ravensbrück itself. Ernst-Robert von Grawitz, Chief Doctor of the SS, quickly disabused Himmler of the notion:

> In view of the unprecedented significance which such a technique would have in the sense of a negative population policy and the consequent importance of promoting with all means a perfect solution to the problem, allow me, *Reichsführer*, to propose that Professor Clauberg establish a suitable research institute in or near Königshütte and incorporate in it a women's concentration camp for about ten persons.

Auschwitz provided the ideal compromise; closer than Ravensbrück to Königshütte, and with a virtually limitless quantity of experimental subjects.[1012] Despite boasting to Himmler that the success of "the method I have conceived for sterilising women without an operation is already nearing completion" and raising the prospect that "...in a

1008 Yisrael Gutman and Michael Berenbaum (eds.), *Anatomy of the Auschwitz Death Camp* (Bloomington: Indiana University Press, 1998), pp. 304–307.

1009 A female prisoner-doctor who spent a year in block 10 described it as a place that was somewhere between hell and a mental institution. (Lifton, *The Nazi Doctors*, p. 270).

1010 Ibid., p. 273.

1011 Piper and Swiebocka, *Auschwitz Nazi Death Camp*, p. 90.

1012 Padfield, *Himmler*, p. 333.

short period of time...one trained physician located in a reasonably equipped outpost, assisted by ten staff members, will most likely be able to sterilise in the course of a single day several hundred or even a thousand women," Clauberg had made only slow progress, in large part because of his arrogance and unpleasant personality. As the Russians approached Auschwitz, Clauberg moved his operations to Ravensbrück, where he continued to experiment on female Jewish prisoners transferred from Auschwitz, and on Gypsy women.[1013]

Clauberg was captured by the Russians in June 1945. Three years later he was put on trial in the Soviet Union, and having been convicted of "mass extermination of Soviet citizens", was sentenced to 25 years' imprisonment.[1014] However, the death of Stalin in 1953 and a series of diplomatic agreements resulted in the release of Clauberg, along with other German prisoners of the Soviets. Clauberg arrived back in Germany in October 1955 to resume practicing medicine, to the horror of many. After two months he was arrested. Pressure from many sources, including the outraged protest of a group of former Auschwitz prisoner doctors that "such medical practitioners who...put themselves at the service of National Socialism to destroy human lives...are today in a position to practice once more the profession which they have profaned in such a scandalous manner," forced the German Chamber of Medicine to eventually revoke Clauberg's medical licence. But before he could be brought to trial, Clauberg died in prison in Kiel on 9 August 1957, probably of a heart attack, although there is speculation that he may have been killed by former medical colleagues to prevent his "naming names" in court.[1015] As we shall see, such conspiracy theories were not unusual when men accused of the most heinous crimes died whilst in post-war custody—and before they could publicly testify about their wartime colleagues.

Horst Schumann's activities before and after his participation in mass sterilisation experiments are detailed elsewhere in this volume. He visited Auschwitz, probably for the first time in July 1941, when he selected 573 sick prisoners for transportation to Sonnenstein and gassing under the umbrella of 14f13.[1016] Clauberg and Schumann were pursuing the same objective by differing and competing means, and the physical contrast between the two of them was as great as the gulf that existed between their choice of method; Schumann was tall, well-built, and considered by some to be handsome. Others thought him distant and unfeeling.[1017] He arrived back in Auschwitz for a rather longer stay in November 1942, by now holding the rank of an *Oberleutnant* in the *Luftwaffe*, although in reality his sponsors were Brack and T4.[1018]

1013 Piper and Swiebocka, *Auschwitz Nazi Death Camp*, p. 91

1014 Israel Gutman (ed), *Encyclopedia of the Holocaust* (New York: Macmillan Publishing Company, 1990), p. 302.

1015 Lifton, *The Nazi Doctors*, p. 277–278.

1016 There were no functioning gas chambers in Auschwitz at that time. The first gassing experiments with *Zyklon B* were conducted in the basement of block 11 of the *Stammlager* in late summer 1941. Shortly thereafter the former mortuary attached to the camp crematorium was converted to form the first camp gas chamber proper [Crematorium I]. (Piper and Swiebocka, *Auschwitz Nazi Death Camp*, pp. 157–159).

1017 Lifton, *The Nazi Doctors*, p. 280.

1018 Henry Friedlander, *The Origins of Nazi Genocide: From Euthanasia to the Final Solution* (Chapel Hill: University of North Carolina Press, 1995), p. 198.

Schumann's attempts at sterilisation using X-rays were accommodated at first in barrack 30 of the women's camp (Bla) at Birkenau (Clauberg had begun his "research" there as well), although surgical procedures were conducted at blocks 10, 21 and 28 of the *Stammlager* (Auschwitz I, the main camp). Rudolf Höss, the former commandant of Auschwitz, stated that he received orders from Himmler to supply Schumann with whatever he required, both in terms of male and female prisoner "guinea-pigs", as well as anything else necessary. Groups made up of several dozen Jewish male and female prisoners were sent to barrack 30, where they were exposed to varying doses of radiation.[1019] Despite often suffering severe burns as a result, the prisoners were still forced to work as usual. Many died as a consequence of Schumann's experiments; others, incapacitated by exposure to the X-rays, fell victim to the gas chamber following "selections" of the unfit within the camp. A former prisoner described how

> ...during my stay at the hospital I met a formerly well-known Jew from Warsaw...His entire family had been murdered at the camp, and he himself had become a victim of some criminal experiment based on irradiating the testicles with X-rays. He was severely crippled and suffering terribly. He explained to me that such experiments were being carried out on an entire group of Jews, 50 men and the same number of women. After a short interval, they were all gassed.

A Greek Jewess deported from Thessalonica to Birkenau, described her ordeal at the hands of Schumacher (who had no surgical experience) and his prisoner-doctor assistants:

> ...In summer 1943, during the morning roll-call, the female *Blockältester* summoned a number of girls, including me...Under guard, we were taken to Auschwitz, to block 10. Other Greek women and girls were already there. One day Dr Schumann appeared...and ordered all the girls to step forward. He pointed to several girls, including me, and took down our numbers. The next day, we were taken...back to Birkenau and irradiated...Two plates [were] attached to my body, one to my stomach and the other to my back...On the same day we were taken back to Auschwitz. On the way, all of us vomited. We did not know what had happened to us. After several days, festering wounds appeared in the areas where [the plates] had been attached...After around two months we were sent back to work at Birkenau, although our wounds had not yet healed...One day, during the morning roll-call, my number was called, along with a number of other girls who had been irradiated. Once again we were taken to Auschwitz, to block 10...There we were examined by Dr Samuel [a Jewish prisoner-doctor who was later murdered in the camp]...Sometime in November 1943, I was taken, along with nine other girls, to the neighbouring block 21, and we were operated on there...After the operation I was taken back to block 10. We all lay in the same room and screamed in pain...After one or two months...Dr Schumann appeared at the block, observed our wounds, and ordered us back to work at Birkenau, no matter how difficult it still was for us to move..."[1020]

For women, Schumann's technique involved the removal of the irradiated ovaries, for men the excision of one or both testes. It is estimated that 1,000 male and female prisoners underwent these X-ray procedures, of which about 200 were subjected to removal of their sex organs.[1021] How many died is not known, but the number of those expiring as a result of Schumann's experiments must have amounted to several hundred. The total of Clauberg's victims is even less certain, but a figure of 700 is sometimes quoted—as is a figure of "several thousands."[1022] It is not possible to estimate

1019 The machinery was operated by Schumann personally. (Padfield, *Himmler*, p. 439).

1020 Piper and Swiebocka, *Auschwitz Nazi Death Camp*, pp. 91–94.

1021 Gutman and Berenbaum, *Anatomy of the Auschwitz Death Camp*, p. 308.

1022 Lifton, *The Nazi Doctors*, p. 277.

how many of those who survived these experiments subsequently fell victim to the gas chambers.

Schumann called a halt to his Auschwitz activities in early 1944, writing to Himmler in April of that year that he considered surgical castration to be preferable to X-rays as a method of sterilisation. Notwithstanding these comments, Schumann did subsequently conduct further X-ray sterilisation procedures on Gypsy girls at Ravensbrück.[1023] In December 1944, a few weeks before the evacuation of Auschwitz, Schumann's X-ray device was removed. The Czech prisoner who had maintained and serviced it, Stanislaw Slezak, a man who knew too much, was transferred to Mauthausen in January 1945 and shot.[1024]

1023 Gutman and Berenbaum, *Anatomy of the Auschwitz Death Camp*, p. 308

1024 Piper and Swiebocka, *Auschwitz Nazi Death Camp*, p. 94.

CHAPTER 11: RETRIBUTION

The defendants...are not all perverts. They are not ignorant men. Most of them are trained physicians and some of them are distinguished scientists. Yet these defendants, all of whom were fully able to comprehend the nature of their acts, and most of whom were exceptionally qualified to form a moral judgement in this respect, are responsible for wholesale murder and unspeakable cruel torture. - Telford Taylor[1025]

But Fascist ideology and Hitlerism have transformed the consciences of the young, and those under thirty-five are willing to commit any crime for any purpose ordered by their leader. - Eugène Tisserant, June 1940[1026]

We see these cultured and learned men hesitating at times, but none the less making steady progress, step by step, along the path to the final solution. They did not go the whole way. Those who stopped closed their eyes, or rather they blinded themselves to the truth. They knew nothing, just as they believed Hitler knew nothing. These learned men wanted to know nothing, and so there came into being a remarkable community of self-blinded internal exiles coexisting with the annihilators, those who did go all the way to the final solution. - Benno Müller-Hill[1027]

To find the way back to the law one must first find the way to justice - Eugène Aroneanu[1028]

Although it is apparent that the majority of those involved in Nazi eugenics and the perpetration of associated criminal acts were men, particularly at a senior level, those implicated insofar as "euthanasia" was concerned did include a significant number of women. More specifically, these medical crimes could not have been undertaken without the participation of literally hundreds of nurses, most of whom were female. How could they, members of the most selfless and dedicated profession, rationalise killing helpless patients?[1029] It is true that to a material extent they were misled by the very people they most relied upon for guidance—doctors and other senior medical personnel, an estimated 400 of whom had directly participated in the programme. Eugenics was not a passing fancy, these nurses were told, but a "science" whose importance had finally been recognized. If in the past nothing had been done to "protect the eugenically valuable" members of the nation, thanks to Adolf Hitler's "insight and infectious energy", a way had been shown "to avoid the certain decline of our Volk." If the prevention of infectious disease was preferable to its cure, ran the argument, was it not sensible to apply the same logic to genetic disorders?[1030] This is not, of course, to excuse either the nurses' actions or their lack of compassion. They were not automatons, devoid of human sensibilities; many later acknowledged their feelings of guilt at being accomplices, or worse, to the murders. One specific case was particularly enlightening in this respect.

1025 Ulf Schmidt, *Karl Brandt: The Nazi Doctor—Medicine and Power in the Third Reich* (London: Hambledon Continuum, 2007), p. 245.

1026 Saul Friedländer, *The Years of Extermination: Nazi Germany and the Jews, 1939–1945* (New York: HarperCollins, 2007), p. 74.

1027 Benno Müller-Hill, *Murderous Science: Elimination by Scientific Selection of Jews, Gypsies, and Others in Germany 1933–1945* (Woodbury: Cold Spring Harbor Laboratory Press, 1998), p. 24.

1028 Eugène Aroneanu, *Inside the Concentration Camps: Eyewitness Accounts of Life in Hitler's Death Camps* (Westport: Praeger, 1996) p.xvii.

1029 It has been estimated that nurses intentionally killed more than 10,000 individuals during the Nazi epoch. [Susan Benedict and Jochen Kuhla, *Nurses' Participation in the Euthanasia Programs of Nazi Germany* (Western Journal of Nursing Research, Vol. 21(2), 1999), p. 246].

1030 Bronwyn Rebekah McFarland-Icke, *Nurses in Nazi Germany* (Chichester: Princeton University Press, 1999), p. 147–148

In 1965, a trial was held in Munich of fourteen female nurses who had served at the Meseritz-Obrawalde mental hospital. They were accused of "...cooperation with the 'euthanasia' programme by killing thousands of mentally disabled patients through the administering of overdoses of Veronal or Luminal, or by injections with Morphia-Scopolamine or plain air, respectively." Their testimony provides an insight into the ease with which, given suitable social and political circumstances, it is possible to convince healers to become killers. And if those dedicated to the curative could be so effortlessly converted to the practice of murder, how much easier was it to convince the non-medical personnel of *Aktion Reinhard* to kill on a vastly greater scale?

Luise Erdmann. [the main defendant in the Munich trial, accused of participation in the killing of 210 patients]:

> Through the behaviour of Dr. Wernicke I realized that incurable patients were to be released by giving them Veronal or another medicine. I also declare that was I was not informed by Dr. Wernicke or any other person at the home about euthanasia. I wasn't sworn to secrecy in this respect... I was of the opinion it was taken for granted or believed that I would approve of euthanasia. My attitude to euthanasia was, should I become incurably ill—I don't make any distinction between mental or physical illness—I would consider it as a release if a physician or, on the direction of a physician, another person, would give me a dose releasing me from everything. Despite my attitude to euthanasia, I have—when confronted with the problem—fought out serious inner conflicts. Euthanasia, in the form I experienced it at that time, was after all, a killing of people and I asked myself if a legislator had the right at all to order or permit the killing of people. Never, however, did I hear about a corresponding law on the use of euthanasia but, on the other hand, Dr. Mootz explained to me once that there was no need for reservation as, should the situation arise, he would cover up for me. From this statement I concluded that there had to be some legality for euthanasia.

Erna Elfriede E. [accused of participation in the killing of 200 patients]:

> They didn't make me swear about a secret matter of the Reich and I wasn't sworn to silence.... I considered the killings as injustice. Something like that was not supposed to happen, because nobody was allowed to order it. I was brought up as a Christian. I already learned as a child what one may and may not do. I learned that one mustn't steal and mustn't kill.

When asked why she didn't refuse to participate in the killings, she replied:

> Because I was ordered to do it. When I am asked again, why I didn't refuse, although I realized that it was an injustice, I can't give an answer to this question. I do and did in the past have a strong feeling of guilt but it is impossible for me to give a reason for the fact that I didn't refuse. It simply was ordered and I had to execute the orders.

Martha Elisabeth G. [accused of killing 28 patients]:

> I felt guilty about it at that time and, although I didn't do any killings by myself, I did help and I had a certain feeling of guilt. I'm only an ordinary nurse...and never realised that, legally speaking, I had become implicated in the killings. When I had to assist in the killings, I acted under duress and never with the intention to kill a person. At that time, nobody would have helped us at Obrawalde if we had refused to do the work, and there wasn't anybody to pour one's heart out to, and who could we trust? As a sort of slave, we were completely at the mercy of the rulers and their political line.

Gertrude F. [accused of killing 5 patients]:

> When I did it by preparing the medicine, I did it without any knowledge of legal consequences...I wasn't able to see a direct connection between my work and the killings...Still today, I haven't completely become aware of my wrongdoing.[1031]

All fourteen women accused in the *Munchner Schwesternprozess* (Nurses' Trial in Munich), were found not guilty of acting as an accessory to murder. The court ruled that the defendants lacked the intellectual and moral capacity to understand that killing patients was illegal,[1032] which may be considered both a rather peculiar interpretation of nursing responsibilities as well as an extremely beneficial judgement so far as the defendants were concerned. It is illuminating to contrast the sentences handed down on these nurses with those passed in the immediate aftermath of war on Helene Wieczorek and Amanda Ratajczak, who had served in the same institution and had committed identical crimes.

It has been argued that it was no great surprise that so many of these supposed "carers" were prepared to become killers. They were "bitter, frustrated, disillusioned, tired, underpaid and undervalued"[1033]—in other words, easy prey for a regime prepared to exploit their vulnerability. Precisely when they became aware of this exploitation was a question that only the individual could answer, for many had been uncomfortable with what was going on around them from the earliest days of "euthanasia", even if they then had no direct involvement in or complete knowledge of what was happening to their patients. But there are some things it is difficult to admit, even to oneself. Of course, whatever they may have known or not known in 1939–41, so far as the nurses of Meseritz-Obrawalde were concerned, there was no doubt about their knowledge of and complicity in the killings from 1943 onward.[1034]

Some nurses could find personal reasons for participating in murder. As described above, Anna Katschenka assisted in the killing of handicapped children at the Am Spiegelgrund hospital in Vienna. She had been successfully treated for depression by the hospital's psychiatrist, Erwin Jekelius, to whom she became attached and on whom she became dependant. When in 1941 she found out that nurses were required at Am Spiegelgrund, she applied to be transferred there in order to be with Jekelius, who, on her arrival, told her of children's "euthanasia", and asked her to administer lethal medication to young patients. This she agreed to do, accepting at face value Jekelius' assurance that only the most severely handicapped children would be selected for extermination. When Jekelius was called-up by the *Wehrmacht*, Katschenka transferred her dependence to his successor, Ernst Illing, and continued to kill children.[1035] A post-war Austrian court found Katschenka intelligent but impressionable, and sentenced her to eight

---

1031 Benedict and Kuhla, *Nurses Participation in the Euthanasia Programs of Nazi Germany*, pp. 255–258.

1032 Michael S Bryant, *Confronting the "Good Death": Nazi Euthanasia on Trial 1945–1953* (Boulder: University Press of Colorado, 2005), p. 213. The judgement stated that, in the opinion of the court, the defendants "on the basis of their education and activity had developed limitless trust in the doctors and also, on account of their mental rigidity, were equipped with only a below-average critical capacity." (McFarland-Icke, *Nurses*, p. 11).

1033 Michael Burleigh, *Death and Deliverance : 'Euthanasia' in Germany c. 1900–1945* (London: Pan Books, 2002), p. 238.

1034 McFarland-Icke, *Nurses in Nazi Germany*, pp. 222–223.

1035 Sylvia Anne Hoskins, *Nurses and National Socialism—a Moral Dilemma: One Historical Example of a Route to Euthanasia* (Nursing Ethics, 12 (1), 2005), p. 87.

year's imprisonment; her co-defendant, Illing, was sentenced to death and executed in 1946. Jekelius died in Soviet captivity in 1952.[1036]

Doctors had flocked to join the Nazis. In 1929, the National Socialist Physicians' League had been formed with 44 members. Immediately prior to the *Machtergreifung*, membership had risen to 2,786; by October 1933, it was 11,000, and a year later the waiting list to join was so great that potential members were asked to withhold their application until the backlog had been cleared. By 1942, records indicate that membership of the league was almost 40,000; a year later the figure had grown to 46,000. Nor was it possible to rationalise membership as being on purely medical grounds. It has been estimated that 45 percent of all doctors eventually joined the Nazi party itself, with 26 percent of all male doctors joining the SA and 7 percent the SS.[1037] As Leonardo Conti commented, "Doctors, among all the professions, were the earliest and most active participants in the National Socialist movement."[1038] And as Karl Brandt attested: "Every individual doctor was responsible for what he did in the course of those measures which led to euthanasia. Each doctor was absolutely responsible for his judgement."[1039]

The extent to which the medical profession as a whole had rationalized their homicidal behaviour is perhaps best illustrated by the testimony of some defendants at their post-war trials. Valentin Faltlhauser insisted that, for him, "the decisive motive was compassion." Paediatrician Ernst Wentzler stated, "I had the feeling that my activity was something positive, and that I had made a small contribution to human progress."[1040] Karl Brandt affirmed the following:

> Do you think it was a pleasure for me to receive the order to permit euthanasia? For fifteen years I had toiled at the sickbed and every patient was to me like a brother. I worried about every sick child as if it had been my own...I fully realize the problem; it is as old as mankind, but it is not a crime against man nor humanity. It is pity for the incurable, literally. Here I cannot believe like a clergyman or think as a jurist. I am a doctor, and I see the law of nature as being the law of reason. In my heart there is a love of mankind, and so it is in my conscience. That is why I am a doctor!...Death can mean deliverance. Death is life—just as much as birth. It was never meant to be murder.[1041]

In the immediate aftermath of war, the courts were not fooled by such apparent display of contrition. Telford Taylor, chief of counsel for the prosecution at the Nuremberg Medical Trial, was scathing in his description of the prominent physicians who appeared before the court.[1042] However, despite the fine words and admirable sentiments

---

1036 Ernst Klee, *Das Personenlexikon zum Dritten Reich–Wer war was vor und nach 1945* (Frankfurt am Main: S.Fischer, 2005), p. 286.

1037 Robert N Proctor, *Racial Hygiene: Medicine Under the Nazis* (Cambridge Massachusetts: Harvard University Press, 1989), pp. 65–66.

1038 Ibid., p. 94.

1039 Paul Julian Weindling, *Nazi Medicine and the Nuremberg Trials—From Medical War Crimes to Informed Consent* (Basingstoke: Palgrave Macmillan, 2006), p. 252.

1040 Burleigh, *Death and Deliverance*, p. 103.

1041 http://www.nightingalealliance.org/pdf/Lessons_from_History.pdf (Accessed 4 January 2007).

1042 Although, in the main, attention then and thereafter was focused on evidence at the Medical Trial concerning medical experiments on involuntary human patients, Taylor had no doubt that all medical atrocities were linked as part of an overall genocidal strategy, with both sterilisation and "euthanasia" providing essential components of the policy. (Weindling, *Nazi Medicine and the Nuremberg Trials*, pp. 5–6).

of the victorious Allies, the scale of retribution for the great majority of the perpetrators of Nazi crimes was neither commensurate with the offences under consideration, nor swift in its application. In particular, justice for the victims of "euthanasia" was often simply ignored. To an appreciable extent this failure of due process arose initially as a result a misinterpretation of the nature of the programme, its application, and its illegality. Subsequently it came about as a consequence of the compelling impetus of "cold war" politics and the establishment of the new West German state. So far as the latter nation was concerned, the "euthanasia" perpetrators shared the good fortune of thousands of war criminals, who either evaded punishment altogether, or at worst served brief terms of imprisonment, sometimes with the minor added inconvenience of a temporary loss of civil rights. As has been observed, these later judgements were the kind of sentences second rate cheque forgers might have been expected to receive.[1043]

Perhaps Richard Toellner, medical historian at the University of Münster, best expressed the guilt of the supposed healers: "A medical profession that accepts mass murder of sick people as normal, and to a large degree approves of it as a necessary, justified act for the sake of the community, has failed and betrayed its mission."[1044] And as Professor Thomas Szasz has commented:

> "Schizophrenia" is a strategic label as "Jew" was in Nazi Germany. If you want to exclude people from the social order, you must justify this to others, but especially to yourself. So you invent a justificatory rhetoric. That's what the really nasty psychiatric words are all about: they are justificatory rhetoric, labelling a package "garbage", it means "take it away! Get it out of my sight!" etc. That's what the word "Jew" meant in Nazi Germany; it did not mean a person with a certain kind of religious belief. It meant "vermin!", "gas him!" I am afraid that "schizophrenic" and "sociopathic personality" and many other psychiatric diagnostic terms mean exactly the same thing; they mean "human garbage," "take him away!", "get him out of my sight."[1045]

With the end of the war and the collapse of the Nazi regime, Germany was left without any effective or acceptable form of government. In order to establish and maintain order, on 5 June 1945 the Allies issued a declaration by which the duties and responsibilities of governing the defeated country were vested in them:

> The Governments of the United States of America, the Union of Soviet Socialist Republics and the United Kingdom, and the Provisional Government of the French Republic, hereby assume supreme authority with respect to Germany, including all the powers possessed by the German Government, the High Command and any state, municipal, or local government or authority.

Although each of the four victor nations had authority within their own zone of occupation, matters that affected Germany "as a whole" were to be decided jointly by all four military Commanders-in-Chief, who for this purpose formed a single governing body called the Control Council.

1043 Donald Bloxham, *The Final Solution: A Genocide* (Oxford: Oxford University Press, 2009), p. 201.

1044 Götz Aly, Peter Chroust, Christian Pross, *Cleansing the Fatherland: Nazi Medicine and Racial Hygiene* (Baltimore: The Johns Hopkins University Press, Baltimore, 1994), p. xiii

1045 Guido Bolaffi, Raffaele Bracalenti, Peter Braham, Sandro Gindro, (eds.) *Dictionary of Race, Ethnicity & Culture*, (London: Sage Publications, 2003), p.xviii.

The Allies derived their authority to conduct post-war trials from Control Council Law number 10. The Moscow Declaration of 30 October 1943, "Concerning Responsibility of Hitlerites for Committed Atrocities", and the London Agreement of 8 August 1945, "Concerning Prosecution and Punishment of Major War Criminals of the European Axis", had laid down the basis on which the Allies would seek to prosecute Nazi crimes. Apart from defining the categories of crime committed, law 10 laid down the degrees of punishment to be meted out to the guilty, the legal authority of zonal and International Military Tribunals (IMTs), and the intention that those accused of a crime alleged to have been committed in a country other than Germany should be tried by the country in which the alleged crime had taken place.[1046] In the increasingly hostile and confrontational atmosphere of the cold war, this latter intention was not extensively pursued in practice and by 1948 had been effectively abandoned by the British and Americans.[1047]

As indictments were raised by the Allies, there was the anticipated accusation of "victor's justice" from those who had been members of and loyally served a criminal regime and their politically motivated friends and supporters (then and now), as if in the prevailing circumstances any other kind of justice was feasible. There were, however, some thorny legal issues. At the forefront of these was the concept of *nulla crimen sine lege* (no crime without a law), and the associated unacceptable notion of *ex post facto* justice (a law created after the event cannot make something a crime.) Both of these abstractions failed so far as "euthanasia" crimes were concerned because (a), as has already been illustrated, the Hitler authorization was not a law, and therefore throughout the Nazi regime, "euthanasia" was illegal and a punishable offence under the then existing legal code; and (b), even if some perpetrators had believed that the authorization did have the status of a law, "euthanasia" remained a crime, with or without a civil law authorizing it, by virtue of the violation of "natural law." In September 1945, the German lawyer, Gustav Radbruch, wrote:

> There are basic legal principles that are stronger than every existing law...These basic principles are called the law of nature or reason. If laws consciously violate the ends of justice...then these laws lack validity. Then the people owe them no obedience, then jurists must summon the courage to deny them the character of law.[1048]

To the defendants, there was another concept that was even more difficult to grasp. The fundamental Anglo-Saxon legal principle of presumption of innocence (an accused remains innocent until proven guilty) was unknown in Germany, where examining magistrates were responsible for establishing probable guilt in a pre-trial process held before a case came to actual trial.[1049]

Despite these caveats, the horrific nature of the crimes under consideration demanded some form of judicial process, whatever its real or imagined imperfections may have been, and the courtrooms of the Allies provided the only viable option. German

---

1046 http://www.ess.uwe.ac.uk/documents/cntrl10.htm (Accessed 6 February 2009). The ruling regarding jurisdiction only applied where alleged crimes were restricted to a specific geographical region (Schmidt, *Karl Brandt*, p. 339).

1047 Weindling, *Nazi Medicine and the Nuremberg Trials*, p. 308.

1048 Bryant, *Confronting the "Good Death"*, p. 113.

1049 Weindling, *Nazi Medicine and the Nuremberg Trials*, p. 161.

judges who had held office under the Nazis could hardly be relied upon to truly serve the cause of justice, although in time such reliance was to become evident. Many former Nazi judges of both the civil and criminal courts were to serve out their full period of office in the new democracy of West Germany.[1050] In part, this explains the attitude of courts of the Federal Republic towards alleged perpetrators when the newly constituted nation inherited the responsibility of trying the accused from the Allies.[1051] But in 1945, the alternative to trials conducted by the occupiers would have been to simply let the guilty walk away from their crimes unscathed. Such a course of action was unthinkable. If the problems associated with arranging and conducting war crimes trials were unprecedented for the Allies, the scale of criminality involved dictated that there had to be accountability for the horrors that had occurred.

Yet how was this to be achieved? In 1945, international law concerning the conduct of warfare was derived from the Hague Conventions of 1899 and 1907. These set out what were to be considered war crimes, but it was apparent that whilst there was no difficuly in bringing charges against an allegedly guilty nation and its citizens for crimes committed against that nation's foreign enemies and enemies' citizens, the crimes as defined in the conventions could not be applied to an errant country for actions taken by that country agaist its own citizens. Such a concept was unthinkable in 1907, even if it had become a reality by 1945. Only by directly linking national illegal acts to international war crimes was a prosecution feasible. If the Allies could not find grounds for bringing charges against those guilty of the appalling crimes committed against German nationals by the German government, German law, demeaned and perverted by the Nazis, certainly could not do so.

The United States was the driving force behind the intended prosecutions, utilising a combination of any or all of the four headings eventually adopted by the Allies—(1) the common plan or conspiracy; (2) crimes against peace; (3) war crimes; and (4) crimes against humanity. The latter was an all-encompassing general category intended to include crimes committed against civilians, such as murder, torture, the annihilation of specific ethnic groups, and any other acts that could not be included under any of the other three classifications.[1052] It has been persuasively argued that when considering

---

1050 Peter E Quint, *The Imperfect Union: Constitutional Structures of German Unification* (Princeton: Princeton University Press, 1997), p. 185. Despite the thousands of death sentences handed down by them as servants of a criminal regime, few if any of these Nazi jurists were convicted of wrongdoing by the courts of the Federal Republic.

1051 Donald Bloxham, *Genocide on Trial: War Crimes Trials and the Formation of Holocaust History and Memory* (Oxford: Oxford University Press, 2003), p. 201 The eminent German jurist Ernst Müller-Meiningen commented in 1958:
Would it have been possible to avoid this entire dung-hill cleaning? Only if one had fundamentally reorganized the courts at war's end—by replacing personnel—and had handed the administration of justice over to a new generation. Because this was not possible at the time, the undeniable and inextricable film of past injustice stills clings to us. As before, this places the German rule of law in question. [Philipp Gassert and Alan E Steinweis (eds.), *Coping With the Nazi Past: West German Debates on Nazism and Generational Conflict 1955–1975* (New York: Berghahn Books, 2007), p. 53].

1052 Weindling, *Nazi Medicine and the Nuremberg Trials*, p. 33. The IMT Charter defined crimes against humanity as "...murder, extermination, enslavement, deportation, and other inhumane acts committed against any civilian population, before or during the war, or persecutions on political, racial, or religious grounds in execution of or in connection with any crime within the jurisdiction of the Tribunal,

those implicated in "euthanasia", the distinction between (3) and (4) became blurred, almost to the point of invisibility. "Euthanasia" was perceived by the Americans as being essentially a war crime—if hospitalised civilians were murdered, it was postulated that this was as a result of a planned war of expansion, with the wards vacated by the victims being required to house the casualties, specifically the military casualties, such a war would inevitably produce.[1053] This had to be the case if there was to be successful prosecution of the accused under the provisions of the Hague Conventions. There had to be a demonstrable link between "euthanasia" and Nazi aggression. Whilst this factor (the freeing-up of hospital facilities) did provide an essential element of the programme, and, as has been noted, was put forward as their understanding of its necessity by a number of participants, it was by no means the sole motivation for T4. Rather, it served as a smoke screen, disguising the reality of the long pre-history of "euthanasia" as an essential component of Nazi ideology on racial as well as economic grounds.

The truth was that in general terms there was no desire among the Allies for the prosecution of Germans for the crime of murdering or otherwise maltreating their fellow Germans, Jewish or otherwise. Indeed there was considerable doubt about whether such acts could be described as criminal where relevant Nazi legislation had been enacted (for example, that regarding compulsory sterilisation, and, it was thought at the time, "euthanasia"). This was a subject that extensively exercised the Allies' finest legal minds. In June 1944 the Lord Chancellor, amongst the most senior government functionaries, prepared a memorandum for the British cabinet concerning the United Nations War Crimes Commission. It outlined the position to be adopted by the government in the proposed was crimes trials that were to follow the winning of the war and the cessation of hostilities. Could the Commission do anything to "investigate the massacre or other horrible ill-treatment of German Jews in Germany"? Since this was clearly not a war crime, "for a war crime essentially involves maltreatment by the enemy of those who are not fellow-subjects of the enemy," it presented a difficult problem. After all, "...to seek to try by any sort of Allied or Inter-Allied court German officials for acts done in Germany to German subjects (albeit Jews) or to persons ordinarily resident in Germany would raise great legal difficulties. One could not appeal to the laws and usages of war or international law. It would be useless to appeal to German law, which is the law under which the officials were living and to which they were subject at the time."

He went on to suggest that as a consequence of this assessment, he "would be inclined to oppose any suggestions that the present Commission should collect evidence of maltreatment of Jews in Germany itself, i.e., in circumstances which would not constitute an offence against the laws and usages of war. We run the risk of getting into considerable confusion if the Commission is allowed to concern itself with incidents which are not war crimes at all. If it is felt desirable to collect evidence of what has been done in Germany, the arrangements for doing this would, I think, be best made otherwise

---

whether or not in violation of domestic law of the country where perpetrated." (Bryant, *Confronting the "Good Death"*, p. 71).

1053 Bryant, *Confronting the "Good Death"*, p. 104.

than through the Commission....The fate of Hitler and Co. is a political question and must be decided by the principal Governments themselves, and not by a court of judges, whether national or international."[1054]

How then was justice for the victims be served? Clearly not by the action, or inaction, of the Lord Chancellor. The answer to this apparently intractable problem was the addition of an all-encompassing charge to the indictments laid before the defendants by the IMT at Nuremberg, one designed to overcome the legal difficulties presented by the conduct of a regime the like of which the world had never before seen. It was the singularity of Nazism which resulted in post-war accountability for what came to be known as crimes against humanity, an accountability, however imperfect, which remains with us to this day.

Debate surrounding the implementation of legislation concerning crimes against humanity can be traced back to Robespierre and the French Revolution , as well as to Belgian atrocities in the Congo a century later, but it was in connection with the Turkish annihilation of their Armenian population in 1915 that it made its first appearance in the twentieth century. Concerns over issues of Allied sovereignty ensured that, despite the accusations of Turkish guilt, no meaningful prosecution of those responsible for the Armenian genocide was attempted, and it remains a contentious and much disputed issue, although the evidence that such atrocities occurred is overwhelming.

In 1943, in the certain knowledge of what was happening under Axis rule, the United Nations War Crimes Commission began to consider the possibility of a new category of international crime. What eventually resulted was the brainchild of a man named Hersch Lauterpacht, a professor of international law born in a small town in Poland, who was to serve as a member of the British prosecting team at Nuremberg and rose to become a Judge at the International Court of Justice, before being knighted for his services to the legal profession. The definition of this newly defined crime was quite straightforward:

> Murder, extermination, enslavement, deportation, and other inhuman acts committed against any civilian population, before or during the war, or persecutions on political, racial or religious grounds in execution of or in connection with any crime within the jurisdiction of the Tribunal, whether or not in violation of the domestic law of the court where perpetrated.[1055]

Despite the debate surrounding the production of this definition and its eventual adoption (continuing concerns of the British and American governments regarding issues of sovereignty notwithstanding), the prime concern in the immediate aftermath of the war, as it had been during it, was to track down and prosecute those responsible for the mistreatment and killing of Allied military personnel.[1056] It was only as evidence of the sheer magnitude of Nazi atrocities emerged that attention became more clearly focused on the barbarity inflicted upon civilians. Even then, there was little reference made to citizens of the former *Reich*. Thus, at the Hadamar "euthanasia" trial held by

---

1054 War Cabinet, United Nations War Crimes Commission, *Memorandum of Lord Chancellor to the Cabinet, (W.P. (44) 294, 1944)*.

1055 Dinah L. Shelton (ed.), *Encyclopedia of Genocide and Crimes Against Humanity Vol.1* (New York: Macmillan Reference, 2004).pp. 209–211.

1056 Weindling, *Nazi Medicine and the Nuremberg Trials*, p. 47.

the United States in October 1945, the charges brought against the defendants related solely to the killing of Polish and Russian workers. The charge consequently read:

> ...In that Alfons Klein, Adolf Wahlmann, Heinrich Ruoff, Karl Willig, Adolf Merkle, Irmgard Huber and Philipp Blum, acting jointly and in pursuance of a common intent and acting for and on behalf of the then German *Reich*, did, from on or about 1 July, 1944, to on or about 1 April, 1945, at Hadamar, Germany, wilfully, deliberately and wrongfully, aid, abet, and participate in the killing of human beings of Polish and Russian nationality, their exact names and number being unknown but aggregating in excess of 400, and who were then and there confined by the then German *Reich* as an exercise of belligerent control.[1057]

When defence counsel asked a witness to estimate the number of German mental patients killed between the re-commencement of "euthanasia" operations in August 1942 and March 1945, in words mirroring those of the Lord Chancellor, the *prosecutor* objected, saying:

> The charge doesn't cover Germans at all. We could be here a month trying that. Counsel's clients are not on trial for any German nationals they may have killed. This is confined entirely to the Russian and the Polish.[1058]

In reviewing the trial, the Judge Advocate confirmed that it had been shown by the evidence that "as many as 10,000 Germans alleged to be mentally ill were admitted to Hadamar and there put to death". However, since the accused were not tried for the deaths of these people, "it is not deemed necessary to do more than than state the above facts as a prelude to the important elements of the present case."[1059] The wording was unfortunate, for it seemed to imply that the killing of 10,000 German civilians by their own government was not only legal, it was unimportant.

It was the selfsame questions of interpretation that faced the British at the so-called "Belsen Trial" in September 1945. Although the trial itself was as concerned with atrocities committed at many other camps, particularly Auschwitz-Birkenau, revealing to the public for the first time the true horror of extermination *en masse*, both prosecution and defence found themselves in agreement regarding the applicable law: "We are not, of course, concerned in this trial with atrocities by Germans against Germans." (Prosecution)[1060] "The victims must be Allied nationals. It was no part of the business of the Court to punish crimes committed by one German against another, or to punish Germans for crimes committed against their allies." (Defence)[1061] " ...it was not the task of the Court to judge the policy of the extermination or persecution of the Jews..." (Defence)[1062] "According to German law, Poland as a sovereign state had ceased to exist and that previous Polish nationals from that part of Poland annexed by Germany were, as a result, German nationals...A German could not commit a war crime against another

---

1057 http://www.ess.uwe.ac.uk/WCC/hadamar.htm (Accessed 5 January 2008).

1058 Bryant, *Confronting the "Good Death"*, p. 81.

1059 http://nuremberg.law.harvard.edu/documents/160-trial-proceedings-extractsfrom?q=issue:%22Euthanasia+program+%28Action+14+f+13%29%22# p. 3 (Accessed 26 December 2016).

1060 Law Reports of Trials of War Criminals, Volume ii, *The Belsen Trial*, (London: The United Nations War Crimes Commission, 1947). p. 8

1061 Ibid. p. 74.

1062 Ibid. p. 81.

German." (Defence).[1063] "The court was not, of course, concerned with what Germans did to Germans during the war, but it was concerned with the protection of Allied subjects from German ill-treatment during the war." (Prosecution)[1064]

More than two years later this attitude still prevailed. At the so-called RuSHA trial held at Nuremberg between 20 October 1947 and 10 March 1948, one of the defendants, Richard Hildebrandt, stood accused of " special responsibility for and participation in the extermination of thousands of German nationals pursuant to the so-called 'euthanasia' programme' of the Third Reich." The evidence submitted by the prosecution was that a unit commanded by Hildebrandt had killed thousands of mentally sick Germans in the Danzig region. Incredibly, the Tribunal dismissed this charge on the grounds that the administration of death under Nazi legislation against persons who were solely citizens of the Third Reich did not constitute a crime against humanity, a verdict worthy of the Nazis themselves.[1065]

There was a flawed understanding of the true meaning of crimes against humanity, as a percipient observer noted in 1946:

> Traditionally, we have been inclined to think of crimes against humanity as crimes against common law. It would follow, then, that the world would have no reason to be alarmed since every nation has its penal code...[However]...crimes against humanity are only vaguely related to crimes against common law. As long as society is subordinate to common law, the victim always retains the possibility of seeking help from public authority. In the cases of crimes against humanity, those concerned stand there completely powerless. No one interferes, not the police, not the mayor, not anybody. Public authorities are no longer able to provide protection to those who innocently come into conflict with the penal code.[1066]

As has been suggested, the overriding reason for this abrogation of the fundamental concept of justice was the matter of sovereignty—at first the threat to that of the Allies, particularly the United States, and subsequently the possible impairment that might be caused to that of West Germany after the new state inherited the responsibility of prosecuting the alleged criminals.[1067] In the first instance, this surfaced most visibly in the American legal interpretation of the "euthanasia" dispensation. Hadamar was described by U.S. Army lawyers in pre-trial documentation as being part of the "mercy killing programme" established "in accordance with the existing laws of Germany, which permitted mercy killings and which became effective in 1939." Under international law, the U.S. was "only concerned with the murder of ... foreign forced labourers." The court was not convinced of the legality of this "euthanasia" law that had been so readily accepted by Army counsel, stating:

> ...As a matter of fact the existence of this alleged German law or decree was not established by the Defence. At the most it was proved, through the testimony of a former Chief Prosecutor of Wiesbaden, that there was an "administrative order" from Hitler's office which permitted it.

Unfortunately, the judgement went on to qualify its otherwise accurate dictum:

1063 Ibid. p. 92.

1064 Ibid. p. 106

1065 http://www.ess.uwe.ac.uk/WCC/greifelt3.htm (Accessed 22 December 2009).

1066 Eugène Aroneanu, *Inside the Concentration Camps: Eyewitness Accounts of Life in Hitler's Death Camps* (Westport: Praeger, 1996) p.xii.

1067 Bryant, *Confronting the "Good Death"*, pp. 63–106.

There was the additional consideration that even if such a decree existed and if it made the killing of incurable persons legal, such provisions could not legalise the killing of other than German nationals, because under general rules of interpretation a rule of this kind would have to be interpreted strictly. Moreover, the accused could not prove that their victims had actually been incurable persons.[1068]

In essence, this assumption of the legality of "euthanasia" presented an intractable problem. The inevitable conclusion was that only defendants who had killed non-German mentally healthy patients could be brought to trial; prosecuting those who had murdered German patients, whether mentally healthy or disabled, would be a *de facto* breach of German sovereignty. To appear to interfere in the domestic affairs of a foreign country would lay the United States and Great Britain open to the possibility of being subject to such a process themselves in certain circumstances. Both nations has reason to be nervous about their recent records, America for its treatment of its ethnic minorities and sterilisation laws, Britain for its colonial history. Could sitting in judgement in respect of crimes committed by Germans on Germans create a precedent, leaving the Allies perhaps vulnerable to similar charges themselves in the future? Thus, it seemed, if an act was perceived to be legal under German law when committed, then prosecution of the perpetrators by the Allies was deemed to be out of the question.[1069] This dichotomy became even more apparent when considering the issue of enforced sterilisation. Unlike "euthanasia", a valid law permitting sterilisation was in existence, albeit one passed under a Nazi government. Sterilising individuals without receiving their prior consent may have been morally indefensible, but it was not illegal in the *Reich* between 1933 and 1945.[1070] The physicians who had conducted such procedures were therefore able to resist calls for their prosecution.[1071] The only recourse for the younger members of an indignant post-war population was the "name and shame" technique such as that practised by his students against Karl Heinrich Bauer, described below.

The Hadamar and Belsen cases were thus paradigmatic in establishing U.S. and British attitudes towards "euthanasia" crimes, and brought to the forefront the ever present conflict between law and justice. For whilst it was possible for lawyers to argue about the legality or otherwise of certain acts, there could be no argument concerning the morality, or lack of it, of those selfsame deeds . If justice was to be served at all, it was surely to be in the name of commonly accepted standards of decency and compassion. Regrettably that was not always to be the case.

The reliance placed by the Allies on the conspiracy doctrine was completely misplaced. It was a concept literally quite foreign to continental law, and one which was to

1068 http://www.ess.uwe.ac.uk/WCC/hadamar.htm (Accessed 5 January 2008).

1069 Bryant, *Confronting the "Good Death"*, pp. 78–80.

1070 It was not until 1998 that the International Criminal Court defined enforced sterilisation performed as part of a widespread or systematic attack directed against any civilian population as a "crime against humanity". (Anne-Marie De Brouwer, *Supranational Criminal Prosecution of Sexual Violence: the ICC and the practice of the ICTY and the ICTR* (Mortsel: Intersentia 2005), p. 85).

1071 Not the least of the problems facing the Allies in considering this question was the fact that similar sterilisation laws had recently been upheld in the United States. (Proctor, *Racial Hygiene*, p. 117). This point was not lost on the defendants at the Nuremberg Medical Trial, who made much of alleged Allied hypocrisy. (Weindling, *Nazi Medicine and the Nuremberg Trials*, pp. 201–203, p. 241).

prove singularly unsuccessful when argued before the International Military Tribunal in Nuremberg. There, of the 22 defendants indicted under the conspiracy charge, no fewer than 14 were found not guilty. By way of contrast, in the 52 other charges laid against the defendants at the only proceedings conducted jointly by all four occupying powers, 44 guilty verdicts were returned.[1072] The logic behind the conspiracy charge was simple; proving guilt of a common plan to conduct aggressive expansionist war would of itself confirm guilt on all the other charges—crimes against peace, war crimes, and crimes against humanity would flow naturally as a consequence of the alleged conspiracy. So far as "euthanasia" was concerned, it had to be seen as part of that alleged conspiracy if trials and judgements were to have validity. U.S. law at the time stated: "Whoever directly commits any act constituting an offence defined in any law, or aids, abets, counsels, commands, induces, or procures its commission is a principal."[1073] Applying this definition would invalidate any distinction between degrees of criminality. It presumed that any person participating in a crime, whether as principal or accessory, was equally guilty. Thus, linking "euthanasia" to the common plan or conspiracy became an essential plank in any prosecution. That this was conceptually quite inaccurate only became evident much later.

The conspiracy accusation appeared again as part of the indictment at the Nuremberg Medical Trial. In place of the IMT's "Crimes against Peace" was the charge of "Membership in Criminal Organizations"; the two indictments were otherwise identical. But the idea of a conspiracy of physicians was even more difficult to maintain than one pertaining to the hierarchy of the Third Reich. Thus, on a motion of the defence (and no doubt following the failure of the conspiracy charge at the IMT), as the Medical Trial progressed the accusation of a common design was dropped.[1074] In fact, experience shows that whilst accusations of conspiracy abound and are easily assumed, they are notoriously difficult to prove without what is, in effect, a "smoking gun".

So far as Britain and the United States were concerned, following the cessation of hostilities and a brief interlude of determined pursuit and prosecution of those responsible for criminal acts, an overwhelming desire ruled to abandon the trial of perpetrators to the Germans themselves. Confrontation with a fresh enemy in the shape of their recent ally, the Soviet Union, necessitated the support of West Germany as a "bulwark against communism," a phrase reminiscent of the Third *Reich*. The post-war trials conducted by the Allies had not been welcomed in Germany,[1075] and further trials were now of little interest in Britain and America, if not the subject of actual opposition.[1076]

1072 Ann Tusa and John Tusa, *The Nuremberg Trial* (London: Macmillan Publishers Limited, 1984), p. 504.

1073 Bryant, *Confronting the "Good Death"*, p. 85. A significant number of statutes contained in title 18 U.S. Code are concerned with various aspects of conspiracy as an indictable offence. See also the Racketeer Influenced and Corrupt Organizations Act (commonly referred to as RICO Act or RICO), a United States federal law that provides for extended penalties for criminal acts performed as part of an ongoing criminal organisation. [John Madinger, *Money Laundering: A Guide for Criminal Investigators* (Boca Raton: CRC Press, 2006), p. 92.]

1074 Alexander Mitscherlich and Fred Mielke, *Doctors of Infamy: The Story of the Nazi Medical Crimes* (New York: Henry Schuman, 1949), p. 3; Tusa and Tusa, *The Nuremberg Trial*, p. 417; Weindling, *Nazi Medicine and the Nuremberg Trials*, p. 201.

1075 In 1946, 70 percent of the German population approved of war crimes trials. By 1950 an equal percentage opposed them. (Bryant, *Confronting the "Good Death"* p. 109.)

1076 Bloxham, *Genocide on Trial* pp. 153–161

The historian John Wheeler-Bennett, one of those responsible for the prosecution of the German generals and diplomats, considered that by 1948 "the British people were bored to death with war crimes trials."[1077] In October of that year Britain announced that no further such trials would be conducted in the zone of Germany under its control.[1078] The following year trials in the American zone ended.[1079] Notwithstanding the ambitious wartime declarations concerning the retribution to be meted out to those guilty of war crimes, post-war *realpolitik* dictated a different response. Following an amnesty in 1950/51 for many of those who had been convicted of war crimes in trials conducted by the United States, Telford Taylor bitterly wrote that this represented "the embodiment of political expediency...[dealing] a blow to the principles of international law and concepts of humanity for which we fought the war."[1080]

The courts of the former West Germany were particularly lenient in their sentencing of individuals who had been responsible in part or in whole for the murder of thousands of individuals. To a significant extent this was due to the nature of the legal principles developed by the judicial system of the Federal Republic.[1081] An accused could not be convicted solely on the basis of membership of a unit or organisation that had taken part in a crime. To be convicted, actual participation in a criminal act had to be proved. In cases that did result in a conviction, but in which the accused had not acted on their own initiative but had simply followed the orders of their superiors, the courts as a rule concluded that the accused had been an accessory to the crime and not an actual perpetrator. The requirement was to prove the "base motive" of a killer in order that his action be declared outright murder, even though he may have personally fired the fatal bullet, or dropped the canisters of *Zyklon* B into the gas chamber.[1082] In simplistic terms this meant that if, for example, a defendant succeeded in claiming that when gassing 100,000 victims he was merely following orders and had not acted out of "malice, cruelty, and base motives", the court could find him an accomplice, and sentence him accordingly. If, on the other hand, he had killed a single person and the evidence proved that he had acted out of malice, cruelty, or a base motive, the court could rule him a perpetrator and impose a much more severe punishment. So far as actual sentencing was concerned, the distinction made between perpetrator and accomplice in cases of murder was of overwhelming importance. A guilty verdict for a perpetrator could mean a sentence of life imprisonment. In contrast, to be found guilty of being an accomplice would result in a sentence of not less than three years and a maximum of

---

1077 David S Wyman (ed), *The World Reacts to the Holocaust* (Baltimore: The John Hopkins University Press, 1996), p. 612.

1078 Ibid., p. 613.

1079 Ibid., p. 717.

1080 Bloxham, *Genocide on Trial*, p. 162.

1081 The suspicion remains that the West German legal code was deliberately manipulated and where necessary amended to benefit those who were likely to be accused of criminal offences committed during the Nazi era. (Michael Wildt, *An Uncompromising Generation—The Nazi Leadership of the Reich Security Main Office* (Madison: University of Wisconsin Press, 2009), p. 417). However, following the trial of John Demjanjuk in 2011 the definition of "criminal guilt" and "perpetrator" was changed. German courts are no longer required to prove that an accused was responsible for or had been directly involved in a Nazi-era related crime. Mere presence at the time and place a crime was committed is now sufficient to make the accused an accessory to that crime and therefore culpable of a criminal offence

1082 Bloxham, *Genocide on Trial*, p. 201

five years.[1083] Almost inevitably, after time served whilst detained in custody awaiting trial was taken into account, the guilty party left the court a free man immediately following the reading of the verdict.

So far as questions of culpability are concerned, whether justice and the supposed rights of the victims were served by this interpretation is open to question, but there is no doubt that this concept worked to the considerable benefit of the defendants. As was demonstrated:

> An ever smaller number of perpetrators, mostly from the subordinate ranks of the Nazi regime's hierarchy, were convicted. Mid- and high-ranking officials were considered accessories and thus could count on either lesser punishments or stays of the proceedings brought against them. As a result, the image of the individual, abnormal perpetrator solidified alongside the corresponding view of top-level Nazi bureaucrats as helpless victims unjustly delivered for judgement before the courts.[1084]

Until it was reviewed and effectively revoked in 1975, this differentiation between degrees of guilt produced a scenario in which only those at the very apex of Nazi government—Hitler, Himmler, and Heydrich—could be considered perpetrators.[1085] Everybody below them in the chain of command could claim to have been an accomplice—they were "only obeying orders". This was in fact predominately the position taken by the courts of the Federal Republic throughout the 1950s and 1960s, a stance much, it must be said, in accordance with the views of the general population.[1086] Had they survived long enough to be tried in West Germany at some time during those decades, it is interesting to speculate whether Himmler and Heydrich might themselves have mounted such a manifestly absurd defence. After all, they too might have argued, they were "only obeying orders". Which would conveniently have left Adolf Hitler as the sole person legally accountable, in anything other than derisory terms, for the murder of millions.

In order to better understand the German judicial process as applied to those accused in post-war trials of Nazi crimes, it is necessary to have some comprehension of the manner in which justice was (and is) administered. These defendants in the courts of the Federal Republic appeared before a *Schwurgericht,* a tribunal consisting of a combination of laymen and professional judges who sit together in a single panel that deliberates and decides on all issues of verdict and sentence. Trial by jury, as exemplified by Anglo-Saxon law, was abolished in Germany in 1924.

The *Schwurgericht* is made up of three professional judges and two lay-judges. This setting is determined by law and is the same throughout Germany. In all questions relating to the criminal case under consideration, the two lay-judges are the equal of the professionals, in that they take part in the entire decision making process. The *Schwurgericht* is chaired by a professional judge and decisions are made by a simple majority.

1083 Wildt, *An Uncompromising Generation*, p. 415.

1084 Philipp Gassert and Alan E Steinweis (eds.), *Coping With the Nazi Past: West German Debates on Nazism and Generational Conflict 1955–1975* (New York: Berghahn Books, 2007), p. 59.

1085 Wildt, *An Uncompromising Generation*, p. 537, note 32.

1086 Ibid., p. 412.

During the main proceedings, all issues relevant to the case are investigated and judges can (and often do) question witnesses. Decisions (that is verdicts) are made behind closed doors and are always presented in the judgment as unanimous. That is to say the court speaks with one voice, and it is impossible to determine who had the most influence on which decision. As already mentioned, deliberations take place in the seclusion and privacy of the judges' chambers, as is common to most, if not all legal systems, and are not made public.

The *Bundesgerichtshof* (Federal Court of Appeal) is made up of five professional judges, and so there is no lay-element involved. Any appeal to the *Bundesgerichtshof* from the *Schwurgericht* is limited to questions of law only. The *Bundesgerichtshof* deals only with these questions and does not "repeat", so to speak, the case itself. That is, the facts of the case are considered proven by the court of the first instance; only the correct interpretation of the law is at stake in the *Bundesgerichtshof's* deliberations. The *Bundesgerichtshof* is in fact a so-called *cassation* court, a term common to the court of appeal in a number of countries. In Great Britain and the United States it is the equivalent of an appellate court.[1087]

In Austria, whose record regarding the prosecution of war criminals has been immeasurably worse than that of Germany, the *Schwurgericht* is made up of three professional judges and eight jurors. It is the latter who decide upon questions of guilt or innocence. The professional judges rule on matters of law and procedure; they are responsible for pronouncing sentence, and can also unanimously challenge the decision of the jurors to the Supreme Court. Having carefully nurtured the idea that Austria had been the first victim of German aggression and therefore bore no responsibility for the crimes of the Nazi era, successive post-war Austrian governments were notably reluctant to acknowledge the criminality of certain of their country's nationals. It was not until 1991 that the Austrian federal chancellor, Franz Vranitzky, assumed responsibility for "the harm which Austrian citizens had done to other human beings and peoples". He went on:

> We must also not forget that there were not a few Austrians who in the name of [the Third *Reich*] brought great suffering to others, who took part in persecutions and crimes of this *Reich*... [some of whom] were in prominent positions. Our citizens cannot distance themselves even today from a moral responsibility for these deeds.[1088]

In considering the nature of the verdicts, particularly so far as the sentencing of those found guilty was concerned, it may be that certain extraneous factors came into play. Apart from international political considerations and the questionable personal histories mentioned earlier of both the professional and lay-judges, it is conceivable there was a kind of knee-jerk legal reaction to the excesses of National Socialist "justice". After Roland Freisler and the horrors of the *Volksgerichtshof* (People's Court), it was necessary to show that the law could be compassionate and merciful, rather than brutal and merciless. It can be argued that the application of these humanitarian principles, admirable though it may have been, was often taken too far. It may well have

---

1087 I am indebted to Dick de Mildt for providing this explanation of the *Schwurgericht* and *Bundesgerichtshof.*

1088 Wyman, *The World Reacts to the Holocaust*, p. 505.

been impossible to make the punishment fit the crime, but passing derisory sentences in proven (and sometimes admitted) cases of mass murder was in nobody's long term interest, other than the criminal's.

It is impossible to arrive at an acceptable figure for the number of citizens of the *Reich* who were involved, to a greater or lesser degree, in the criminal deeds of the National Socialist government.[1089] What can be stated with certainty is that between 1945 and 1992, 103,823 individuals were investigated in connection with such matters in the Federal Republic.[1090] Impressive as it sounds, put into perspective that figure represents about 0.12 percent of the total 1939 population of the Greater *Reich* of 87.1 million.[1091] To that total must of course be added the number of similar investigations conducted in the former Democratic Republic and Austria,[1092] but even assuming a quadrupling of the Federal Republic's volume of enquiries, the total of potential criminals examined would be less than one half of one percent of the total 1939 Greater *Reich* population. Clearly, given the scale of criminal activities involved, that would have represented no more than a fraction of those responsible. Some who were never called to account for their actions were dead, killed in battle, in air raids, or by their own hand, some had fled to other more congenial climes, some had been prosecuted by the Allies. Other arrests were prevented by the failure of the West German police to take the fingerprints of men accused of thousands of murders.[1093] But what of those who were investigated in the Federal Republic, where some reliable statistics are available?

6,487 of the investigations resulted in a trial and conviction. Of those convictions, only 974 related to Nazi capital crimes committed during the Second World War. Thus, of the more than 100,000 investigations commenced, less than one percent resulted in a conviction for wartime killings. On closer examination, the figures are even less admi-

---

1089 One estimate puts the number of individuals convicted of crimes associated with the Nazi regime at 50,000. (Proctor, *Racial Hygiene*, p. 401, note 4). No source is quoted for this figure, which presumably refers to the global total of convictions. Caution should be exercised when considering the reliability of any such statistics. For example, in the Soviet occupied zone of Germany alone, nearly 18,000 individuals were sentenced in secret post-war proceedings, some of which, it is believed, were motivated by political rather than judicial considerations. These were, of course, in addition to the well publicized public trials of those accused of Nazi crimes. (de Mildt, *In the Name of the People*, p. 19).

1090 de Mildt, *In the Name of the People*, p. 20. Although the Federal Republic of Germany was not formally established until 1949, for the sake of convenience, the area that comprised the American, British, and French zones of occupation between 1945 and 1949, and the post 1989 unified West and East German States, are referred to here and elsewhere as "The Federal Republic."

1091 http://www.tacitus.nu/historical-atlas/population/germany.htm (Accessed 18 October 2008).

1092 Statistics for Austria suggest a total of some 80,000 preliminary investigations resulting in 23,495 convictions, which, on the face of things and given the relative size of populations, is rather more impressive than the Federal Republic's efforts. However, it should be born in mind that Austria was under Allied occupation from 1945 to 1955. Post 1955 only 20 defendants were convicted of Nazi crimes by Austrian courts, compared with around 600 by West German and 122 by East German courts. (Raul Hilberg, *The Destruction of the European Jews* (New Haven: Yale University Press, 2003), p. 1170); http://www.nachkriegsjustiz.at/service/archiv/en_garscha_bologna2002.php (Accessed 18 October 2008).

1093 Michael Burleigh and Wolfgang Wipperman, *The Racial State: Germany 1933–1945* (Cambridge: Cambridge University Press, 1991), p. 166.

rable. Of the 974 convictions, just 472 related to the murder of Jews.[1094] It is true that there were prosecutions both of German nationals and of their own citizens in other countries that had suffered under Nazi occupation, but it is apparent that there had been little accountability for the death of millions.

The figures regarding "euthanasia" are even more revealing. Between 1945 and 1988, there were 31 trials in the Federal Republic relating to the "euthanasia" *Aktion.* More than one defendant was involved in each of these trials.[1095] A total of 106 individuals were tried for the murder of mentally sick patients. 45 of the defendants were found guilty; with two exceptions (Hilde Wernicke and Helene Wieczorek, who were sentenced to death and executed in 1946),[1096] and in two cases where no punishment was imposed despite a guilty verdict, they received prison sentences varying in length from one year to life imprisonment. In other pre-1949 cases where a death sentence was initially imposed, it was subsequently reduced to a term of imprisonment. The remaining 61 defendants were acquitted.[1097]

It was not solely members of the medical profession at the killing centres or in positions of leadership in the administration of the "euthanasia" programme who bore the guilt. For the system to function it was necessary for physicians and others at the grassroots level to participate—which many did more than willingly.[1098] Karl Schnurer was a medical officer in the Wolfsberg district of Austria. He dispatched the "feeble-minded" and others deemed unworthy of life from the region of which he was *Kreisärzteführer*, or district leading doctor, to Klagenfurt, the regional capital. There they were either killed with lethal injections, or sent on to Hartheim or another T4 establishment to be gassed. In 1945 Schnurer was arrested by the British occupation authorities, accused—in his own words—of "murdering twelve women". He was briefly imprisoned before being released. No formal charges appear to have ever been brought against him. The British had withdrawn his medical licence, but in post-war Austria Schnurer practised as a dentist, a prominent and respected citizen and unrepentant National Socialist.[1099]

Perhaps even more disturbing than the leniency of post-war courts when sentencing the guilty, was the complete absence of any kind of meaningful investigation into the activities of the many physicians who had been enthusiastic members of the Nazi Party, and/or the SA and SS, and who, on the cessation of hostilities, blithely resumed their positions as leaders of the German medical profession. Men such as Wilhelm Heim, in 1983 president of the West Berlin Chamber of Physicians, a man who as an SA member

---

1094 de Mildt, *In the Name of the People*, pp. 20–21

1095 Ibid., p. 404.

1096 Bryant, *Confronting the "Good Death"*, pp. 118–120.

1097 de Mildt, *In the Name of the People*, p. 83.

1098 By 1937, nearly 44 percent of German doctors were members of the Nazi Party, a far greater proportion relative to their overall numbers than other professions such as teachers or lawyers. In the same year, 1.3 percent of all SS-men were doctors—an overrepresentation by a factor of seven compared to their percentage of the total population. On the same basis, only lawyers achieved a higher ratio of overrepresentation in the SS (twenty-five to one) than the medical profession. (Bryant, *Confronting the "Good Death"*, p. 4–5.)

1099 Berndt Rieger, *Creator of Nazi Death Camps—The Life of Odilo Globocnik* (London: Vallentine Mitchell, 2007), pp. 11–18. Wolfsberg was a hotbed of Nazism, a town frequented at one time or another by a disproportionately large number of the men who would go on to become among the principal perpetrators of the Holocaust.

in 1933 had been involved in the persecution of Jewish physicians in that city, or Herman Voss, sometime professor of anatomy at Posen University, who recorded in his diary how looking into the furnace at the remains of four members of the Polish resistance executed by the *Gestapo* made him "feel very calm and comfortable." Voss, who conducted experiments on the content of the blood in the spleen, recorded "How nice it would be to chase the whole [Polish] population through such furnaces!" After the war Voss held the prestigious chair of anatomy at the University of Jena and thereafter was appointed professor emeritus at the Greifswald anatomical institute.[1100] Another eminent surgeon, Karl Heinrich Bauer, in 1946 appointed Rector of the newly reopened Heidelberg University, and subsequently made director of the National Cancer Institute at that seat of learning, in 1968 was confronted by his students with his alleged involvement in compulsory sterilisation. The extent of Bauer's direct participation in such matters is disputed, but there is no dissent concerning his support for the Sterilisation Law.[1101] Bauer had not only been a vocal supporter of sterilisation, he had co-authored a guide to the surgical procedures required to conform with the new Sterilisation Law, and had allegedly performed such operations at the direction of the Breslau Hereditary Health Court. Moreover, it was claimed that Bauer had also been well aware of the "euthanasia" programme, although knowledge alone was not, of course, in itself a crime.[1102]

The real issue with these men and hundreds like them was the absence of any of the aforementioned accountability for what they had said, written, or done. It can be offered in their defence that in essence their position was no different from members of the judiciary or the civil service. Somebody had to be found to fill the void left on the collapse of the Nazi regime, and innocents within the professions were rare indeed in the former *Reich*. Nonetheless, the uncomfortable feeling remains that rather than being held answerable for their actions, some seem almost to have been rewarded by post-war society for the services they had rendered to an odious and criminal regime. In particular there seemed to be a fundamental misconception regarding the nature of the comprehensive involvement of the medical profession in the crimes of National Socialism. As has been observed:

> Whereas it is known that some German physicians and scientists tortured and killed innocent persons, many histories of the Nazi era concentrate on the direct involvement in killing of a relatively small number of biomedical scientists and thus ignore or downplay the broad complicity of German medicine and science with Nazism. Ironically, even the post-war legal proceedings against war criminals tended to lend weight to the view that relatively few doctors or scientists killed, tortured or murdered in the name of the Nazi state.[1103]

1100 George J Annas and Michael A Grodin, *The Nazi Doctors and the Nuremberg Code: Human Rights in Human Experimentation* (New York: Oxford University Press, 1992), pp. 34–41; Aly, Chroust, Pross, *Cleansing the Fatherland*, p. 99.

1101 Robert N Proctor, *The Nazi War on Cancer* (Princeton: Princeton University Press, 2000), pp. 59–60.

1102 Steven P Remy, *The Heidelberg Myth: the Nazification and Denazification of a German University* (Harvard: Harvard University Press, 2002), p. 89.

1103 Arthur L Caplan, (ed), *When Medicine Went Mad: Bioethics and the Holocaust* (Totowa: Humana Press, 1992).

If a balanced and nuanced view of post-war German judicial proceedings is taken, ultimately it is difficult to disagree with the observations of Dr Dick de Mildt:

> I wouldn't say that, in regard to the "euthanasia" and *Aktion Reinhard* functionaries, justice wasn't administered. In my view, it was, but not necessarily in accordance with our interpretation of the concept. Now, of course, this is not the time or the place to plunge into an in-depth discussion on the fascinating question of what would have constituted Justice (with a capital J) in connection to the NS-trials. Suffice it to say here that my ponderings over this issue during the past years have led me to question whether the German courts could have done a substantially better job, under the given circumstances (e.g. within the confines of existing German criminal law). I'm not so sure about an unequivocal answer to that question. Our indignation about the quality of justice as administered by the German courts (and I'm referring here solely to the courts' judgments, not to overall prosecution policies, which were essentially beyond the grasp of the courts) is in itself understandable enough, given the crimes under consideration, but this doesn't necessarily imply that these feelings allow for a fair and sound judgement of these courts. There are, I feel, many factors which have to be taken into account here. One of them concerns the question whether crimes such as those of the *Aktion Reinhard* state-organized type of mega-murders can be addressed in an appropriate way by regular criminal courts (such as was the case with the German ones). Indeed, given the experiences with international criminal courts such as the International Criminal Tribunal for the former Yugoslavia, one may well ask whether such crimes can be "suitably" addressed by any type of court.
>
> As in any western constitutional state, German courts and judges were independent and their judgments were the result of the application of given German criminal law to the extraordinarily complex issue of trying (often decades later) individual mass murderers hired by a previous German government. These complicated circumstances in themselves were, I feel, quite enough to produce results that could hardly "do justice" to the enormous crimes under consideration.[1104]

This very point was addressed by Hannah Arendt in her report on the trial of Adolf Eichmann. She concluded that those proceedings had demonstrated "the inadequacy of the prevailing legal system and of current juridical concepts to deal with the facts of administrative massacres organized by the state apparatus."[1105] Arendt's observations were made more than fifty years ago; they could equally be applied to the attempts to find an acceptable judicial response to the crimes of Nazism both before and since Eichmann's trial. More controversially, Arendt also considered that while "certain political responsibilities among nations might some day be adjudicated in an international court [it is] inconceivable that such a court would be a criminal tribunal which pronounces on the guilt or innocence of individuals."[1106] Which naturally once more begs the question, where then is justice for the victims to be found?

1104 Dick de Mildt in private correspondence with the author, 30 May 2008.

1105 Hannah Arendt, *Eichmann in Jerusalem: A Report on the Banality of Evil* (New York: Penguin Books, 1977), p. 294.

1106 Ibid., p. 298.

CHAPTER 12: CONCLUSION

The thing died at last, and the stench of it stank to the sky. It might be thought that so terrible a savour would never altogether leave the memories of men, but men's memories are unstable things. It may be that gradually these dazed dupes will gather again together, and attempt to believe their dreams and disbelieve their eyes.
– G K Chesterton[1107]

When you think of the long and gloomy history of man, you will find far more hideous crimes have been committed in the name of obedience than have been committed in the name of rebellion. – C P Snow[1108]

Circumstances rule men; men do not rule circumstances. – Herodotus[1109]

Cursed is he who says 'Vengeance!'
Vengeance for the blood of a small child
Satan has not yet created.
– Bialik, Al Ha'shehitah [On the Slaughter][1110]

The first phase of "euthanasia" provided the paradigm for *Aktion Reinhard*. From the creation of gas chambers to their camouflaging as shower rooms, from the creation of the *Die Schleuse*, the "sluice" or "the tube", to the use of carbon monoxide as the killing agent of choice, from the plundering of the victims' remains to the commonality of personnel, the parallels are evident. It can even be argued that the use of the "euthanasia" transit institutions such as Eichberg and Wittenau was replicated, although largely for different reasons, by the *Aktion Reinhard* transit ghettos like Piaski and Izbica. In both cases potential victims were "stored" until the appropriate killing site was ready to accommodate them. But what are often overlooked are the vast economic benefits common to both programmes. The need to release hospital beds, as well as medical staff and supplies, is regarded by some scholars as an excuse rather than a reason for the "euthanasia" programme. They believe that the overwhelming motivation for the killing of the mentally and physically disabled was a eugenically inspired racism:

> ...The object was to create a utopian society organized in accordance with the principles of race. A key concern in this endeavour was the "purification of the body of the nation" from "alien", "hereditarily ill", or "asocial" elements. Racial "purification" was an integral part of wider "social" policies designed to create a "healthy", performance-oriented, "Aryan" national community.[1111]

Yet the same source defined the criteria for being a member of the "national community" as including not just "racial purity" and biological health, but also "socio-economic" performance.[1112] In other words, this was not simply a matter of eugenics. There were real economic benefits to the Nazi war machine in killing "useless eaters". That the authorization of "euthanasia" coincided with the outbreak of hostilities was

1107 Michael W Perry, (ed), *G K Chesterton—Eugenics and Other Evils: An Argument Against the Scientifically Organised State* (Seattle: Inkling Books, 2001), p. 183.

1108 Philip Zimbardo, *The Lucifer Effect: How Good People Turn Evil* (London: Rider Books, 2007), p. 273.

1109 Richard Alan Krieger, *Civilization's Quotations: Life's Ideal* (New York: Algora Publishing, 2002), p. 322.

1110 Shai Lavi, *"The Jews are Coming": Vengeance and Revenge in post-Nazi Europe* (Law, Culture and the Humanities, No.1, 2005), p. 282

1111 Michael Burleigh, and Wolfgang Wipperman, *The Racial State: Germany 1933–1945* (Cambridge: Cambridge University Press, 1991), p. 3.

1112 Ibid., p. 306.

not an accident. Whilst it is true that the fog of war was a major factor in masking the activities of the killing centres, the need to make available the facilities utilised by those considered "valueless", facilities which were now required for tending to the inevitable casualties of the new war, was of even greater importance:

> The euthanasia programme for adults was very much bound up with clearing the decks in order to wage war...Mentally and physically disabled people were killed to save money and resources, or to create physical space for ethnic German repatriates and/or civilian and military casualties...In other words, these policies reflected a non-medical agenda...[1113]

The assumption that the sterilisation campaign was predominately attributable to the application of a policy of racial hygiene rather than economic issues seems justified. Yet recent research at the University of Heidelberg has produced compelling evidence in support of the proposition that the overwhelming motivation for the "euthanasia" campaign was not eugenics in any of its various manifestations, but rather the consistent application of a ruthless economic stratagem.[1114] It was only after reunification that the medical records held by the East German *Staatssicherheit* (*Stasi*, or State Security Service) of approximately 30,000 "euthanasia" victims were discovered. To date in excess of 3,000 of these files have been examined by the Heidelberg researchers, to produce a revealing picture. Although there were numerous criteria applied in deciding whether a patient should or should not be gassed, far and away the most important consideration was the individual's ability to work. The risk of being murdered was up to eight times greater for those considered incapable of productive labour than for those who were so able. Even more striking was the discovery that heredity apparently played no statistically significant part in arriving at the decision of whether or not to kill the patient being evaluated. As has been observed: "Patients in mental institutions were killed because it was cheaper than caring for them. The most important criterion for allowing them to live was their ability to work."[1115] The evidence suggests that the second, "decentralized" phase of "euthanasia" appears to have motivated by exactly the same considerations as the first stage, namely the ability, or lack of it, to be a productive member of society, either at the time or at any point in the foreseeable future. These findings gainsay the frequently expressed assumption that the activities of T4 in particular were undertaken "to protect the health of the Aryan race", rather than for crass commercial considerations.[1116] So far as children were concerned, the determining factor for or against selection also appears to have been fundamentally an economic one, in this case the "educability" of the child. That is to say, if allowed to live, was it probable that the child would be self-sufficient, and on reaching adulthood be of value to the state? If the answer was in the negative, the child was more likely to be murdered irrespective of any hereditary considerations.

1113 Michael Burleigh, *Death and Deliverance : 'Euthanasia' in Germany c. 1900–1945*, p. 4.

1114 I am grateful to Dr Herwig Czech for his permission to access and quote his privately obtained unpublished paper "Medical Crimes, Eugenics, and the Limits of the 'Racial State' Paradigm in the Third Reich" as the source of much of the material in the first part of this chapter.

1115 Götz Aly, Peter Chroust, Christian Pross, *Cleansing the Fatherland: Nazi Medicine and Racial Hygiene* (Baltimore: The Johns Hopkins University Press, Baltimore, 1994), p. 23.

1116 Michael Berenbaum and Abraham J Peck (eds.), *The Holocaust and History: The Known, the Unknown, the Disputed, and the Reexamined* (Bloomington: Indiana University Press, 2002), p. 59.

In fact, for all the theorising and supposed medical vindication, similar post-war research into the surviving T4 files in the German Federal Archives has confirmed that there was no significant link between factors such as the "hereditary nature of the illness" or "socially conspicuous behaviour before admission to the asylum" and selection for killing. The study revealed that the most important criterion was a negative assessment of an individual's capacity to work within the hospital itself. Other decisive criteria included a similar negative appraisal of a patient's behaviour while institutionalised, and cases where the patient had been in care for more than four years. From this it would appear that the primary justification for murder rapidly became economic, if it had not been so from inception. Or as Bouhler and Brandt agreed following their meeting at Berchtesgaden on 10 March 1941, the programme would undertake "[The] elimination of all those who are unable to perform productive tasks even within the asylum, and not only those who are mentally dead."[1117]

All of this directly contradicts earlier conclusions drawn by many eminent historians, such as:

> Nor were these patients murdered to free hospital space or to save money; the killers were motivated by an ideological obsession to create a homogeneous and robust nation based on race, They wanted to purge the handicapped from the national gene pool...Heredity determined the selection of the victims.[1118]

Or:

> For Hitler, typically...the economic argument used by the eugenics lobby in the medical profession and others weighed less heavily than questions of "racial hygiene" and the "future maintenance of our ethnic strength, indeed of our ethnic nationhood altogether."[1119]

It would appear that, so far as "euthanasia" was concerned, there was a vast gulf between eugenic theory and practice, since the Nazis themselves seem to have been somewhat confused regarding the relative importance of eugenic as opposed to economic principles when applied to questions of hereditary and racial health care. In "Guidelines on the Evaluation of Hereditary Health", issued in 1940 by the Ministry of the Interior, it was stated: "In the selection according to eugenic criteria, the evaluation of personal productivity must be of decisive importance." Moreover, "A gifted and productive family will have to be considered valuable for the *Volksgemeinschaft*, even if isolated cases of hereditary defects etc. have occurred."

It follows that, whilst it is indisputable that the killing of the Jews was primarily motivated by and inseparable from a venomous racism, the contribution made to their annihilation by economic considerations also cannot be ignored. Arthur Greiser, *Gauleiter* of the *Warthegau*, had ordered that the Jews be placed in ghettos "until what they have amassed is given back in exchange for food and then they will be expelled over the border."[1120] If at the time of making this remark expulsion was a feasible option or repre-

---

1117 http://www.memorialgugging.at/pdf/Czech_MedizinverbrechenGugging_engl.pdf (Accessed 19 September 2009.)

1118 Friedlander, Henry, *The Origins of Nazi Genocide: From Euthanasia to the Final Solution* (Chapel Hill: University of North Carolina Press, 1995), pp XI-XII.

1119 Ian Kershaw, *Hitler 1936–45: Nemesis* (London: Allen Lane, The Penguin Press, 2000), p. 255.

1120 David Cesarani (ed), *The Holocaust: Critical Concepts in Historical Studies, Volume 2* (London: Routledge, 2004), p. 215.

sented a potential future solution, *Aktion Reinhard* and Greiser himself were to give a deadlier meaning to the concept of expulsion "over the border". This was the essential economic thinking behind the ghettos—press the Jews into a restricted area, steal from them what can be stolen, forbid them to practise their professions or engage in gainful employment, maintain rations at starvation level, and before they eventually die, they will be forced to exchange whatever remains of their wealth for food. Which, to a great extent, and until it was determined to accelerate the final, killing phase, is exactly what happened. At one level, the Holocaust can be viewed as nothing more than a gigantic exercise in exploitation and theft, culminating in an orgy of mass murder when there was nothing left to steal or exploit.

Nazi theology dictated that Jews, like the disabled victims of "euthanasia", were incapable of anything except the most rudimentary labour, if even that. The Jews, cooped up in their ghettos, too, were "useless eaters", and were therefore to be condemned *a priori*. On 16 July 1941, Rolf-Heinz Höppner, the SS officer responsible for the expulsion of Jews and Poles from the Warthegau to the *Generalgouvernement*, wrote to Adolf Eichmann from Posen setting out a proposed solution to the "Jewish problem". Although Höppner was primarily concerned with resolving matters in the Warthegau, and not all of the suggestions contained in his letter reached fruition, one paragraph in particular provides a clear indication of the prevalent thinking—and not simply regarding those falling under Höppner's jurisdiction:

> This winter there is a danger that it will not be possible to feed all the Jews. It should therefore seriously be considered whether the most humane solution would not be to eliminate those Jews unfit for work by some fast-working method. That would in any case be more agreeable than leaving them to die of starvation.[1121]

This was genocide masquerading as humanitarianism, just as "euthanasia" was depicted as somehow advancing the cause of medicine in general, and psychiatry in particular, rather than the unvarnished murder of thousands on economic grounds it had largely been in reality. The apologists of annihilation deluded nobody—including themselves.

It is an oversimplification to assume a deliberately planned concatenation in Nazi eugenic policy, a chain that led with frightening logic from sterilisation to "euthanasia" and thus to the application of continent-wide genocide. The National Socialist state was neither so organised nor so methodical. To those responsible for administering governmental policy, these were not systematic incremental increases in the intensity of a pre-ordained scheme. Each phase merged almost imperceptibly with the next. To the sibylline devotees of racial hygiene, all were logical steps on the utopian path, the ultimate aim of which was the purification of the race. So far as these enthusiasts were concerned, if this policy simultaneously provided certain economic benefits, so much the better.

1121 Eugen Kogon, Hermann Langbein, Adalbert Rückerl, (eds.), *Nazi Mass Murder: A Documentary History of the Use of Poison Gas* (New Haven: Yale University Press, 1993), p. 74. Emphasis added.

Whilst each stage of this progression towards mass murder may have been an exemplar of negative eugenics, the consequences of a century's theorizing in a dozen different nations, the development of individual phases of the policy was not wholly interdependent. "Euthanasia" was not an inevitable corollary of involuntary sterilisation; in the 1930s several countries has compulsory sterilisation policies in place similar to that of Germany. In none of these other countries was "euthanasia" ever seriously contemplated. In the same way, it is unlikely that the gathering at Brandenburg for the first experimental "euthanasia" gassing in winter 1939/40 did so in anticipation of using a modified version of this technique to murder millions of Jews. Tempting though it may be to detect some kind of structured progression in these measures, they were simply representations of different facets of the same racially and economically driven ideology.

On the other hand, it is evident that in Wilhemine and post First World War Germany in particular, enthusiasm for "euthanasia" ran side by side with the desirability of compulsory sterilisation of the "inferior", on both racial and economic grounds. Haeckel, Ploetz, Schallmayer and their many disciples were of a mind. Long before the advent of National Socialism the connection between sterilisation and the permanent elimination of the "worthless" existed as a concept in the published works of many academics and scientists, both in Germany and in other countries. Whilst there were some who coupled this to anti-Semitism of varying degrees of malevolence, it took a Hitler to provide the required extreme degree of racism to ultimately morph from these eugenic theories to full-blown genocide.

That such a hatred of the Jews would result in their exclusion from German society was inevitable under Hitler, who from his earliest days had made his intentions clear. There is reason to believe that initially this exclusion was viewed in terms of emigration, but when, in the face of an indifferent world and the outbreak of war this "voluntary" emigration became impossible, enforced deportation to one of a number of possible destinations was considered. It is apparent that this expulsion was intended to subsequently result in more than simple removal; Jews were not expected to thrive in tropical Madagascar or frozen Siberia under what was unlikely to have been a benign SS administration. What was intended was genocide "by other means". Hitler was perfectly aware of the nature of the Turkish annihilation of Armenians in 1915–1918, an act of "ethnic cleansing" that would have provided a suitable precedent. As he was reputed to have said to a gathering of SS personnel in August 1939, on instructing them to kill without mercy in the forthcoming invasion of Poland, "Who remembers now the massacres of the Armenians?"[1122] Quite when proposed expulsion to some distant location and more gradual liquidation instead became a matter of immediate extirpation within European boundaries is a subject of continuing debate. By mid-1940 Heydrich for one was talking ambiguously in terms of "a territorial Final Solution to the Jewish Question", which, like Hitler's oft repeated threats of annihilation, could be interpreted as referring to either of those options.[1123]

1122 Ervin Staub, *The Roots of Evil: The Origins of Genocide and Other Group Violence* (Cambridge: Cambridge University Press, 1989), p. 187; p. 309, note 64.

1123 Berenbaum and Peck, *The Holocaust and History*, p. 171.

Even before exile to remote regions in the Indian Ocean or the Soviet Union was under serious consideration, there briefly existed the so-called "Nisko Plan", a scheme to deport the *Reich's* Jews and Gypsies to an inhospitable region of eastern Poland near Lublin on an interim basis, prior to their expulsion further eastwards. By the end of September 1939 Himmler had obtained Hitler's approval for a programme of demographic engineering which was eventually to develop into *Generalplan Ost*. This was eugenics on a grand scale, intended in time to involve the deracination of millions of people. By mid-October 1939 the first transports were rolling, only for the "Nisko Plan" to be abruptly abandoned on Himmler's orders. He had now also been given the responsibility for the repatriation of *Volksdeutsche* to the *Heimat*, a matter that took priority over the ejection of Jews and Gypsies from German soil, at least for the time being.[1124] Their turn would come later.

The first phase of "euthanasia", viewed as the logical manifestation of National Socialist eugenics, should therefore be seen as an indispensable contributor to the development of a pan-European genocidal policy, paving the way in terms of both method and personnel. In summer 1941, when the hitherto nebulous concept of the "Final Solution of the Jewish Problem" began to take concrete form, Himmler consulted with the Chief Physician of the SS, *Gruppenführer* Dr Ernst Grawitz concerning the best method of implementing mass murder.[1125] Grawitz advised the use of gas chambers, doubtless based upon the experience gained by T4.[1126] Bottled carbon monoxide would be replaced by petroleum engine exhaust fumes and hydrocyanic acid, Polish killing centres were to succeed their German and Austrian counterparts, victims were to be delivered from every part of occupied Europe instead of merely from the *Reich*, but it was the T4 *Aktion* that provided the template for the industrialization of murder. However, although the scale of killing was to increase dramatically through their use, it is worth bearing in mind that gas chambers and gas vans probably accounted for no more than 50 percent of the victims of the *Shoah*.[1127] Nonetheless, whilst the *Einsatzgruppen* and their auxiliaries systematically killed an estimated 1.4 million Jews, the overwhelming majority by shooting,[1128] gassing provided an appealing alternative. It was cost effective, efficient, and required minimal staffing (there were never more than a handful of German personnel present in any of the extermination camps at any one time).[1129] Ded-

1124 Christopher R Browning, *Nazi Policy, Jewish Workers, German Killers* (Cambridge: Cambridge University Press, 2000), pp. 4–7; Israel Gutman (ed), *Encyclopedia of the Holocaust* (New York: Macmillan Publishing Company, 1990), pp. 1062–1065.

1125 Grawitz had been appointed President of the German Red Cross in January 1937, an appointment comparable in cynicism with that of Reinhard Heydrich as head of Interpol in 1940. (Ulf,Schmidt, *Karl Brandt: The Nazi Doctor—Medicine and Power in the Third Reich* (London: Hambledon Continuum, 2007), p. 81)

1126 Raul Hilberg, *The Destruction of the European Jews* (New Haven: Yale University Press, 2003), p. 932.

1127 The most recent conservative estimates of the number of Jews killed in the major extermination camps are: Belzec 430,000, Treblinka 700,000, Sobibor 170,000, Chelmno 150,000, Auschwitz-Birkenau 950,000, a total of approximately 2.4 million, the overwhelming majority of whom were gassed. To this number must be added Jews killed in camps such as Majdanek, Janowska , Maly Trostinec, and many others . Although there were gas chambers or gas vans operating in some of these camps, gassing was usually not the principle killing method employed.

1128 Hilberg, *The Destruction of European Jews*, p. 1320.

1129 450 individuals were assigned to *Aktion Reinhard*. At any one time 20–35 SS-men were stationed in each of the death camps. [Yitzhak Arad, *Belzec, Sobibor, Treblinka: The Operation Reinhard Death*

icated killing sites were also easier to disguise, situated as they were behind barbed wire and fencing rather than in the open-air like the killing pits of the *Einsatzgruppen*, and were somewhat less stressful for the perpetrators in terms of the physical act of murder. The *Vernichtungslager* also provided one other vital component; instead of having to transport the executioners to the victims, now the camps brought the victims to the executioners. This represented a sea change in the practice of genocide.

Still, once the decision to murder millions had been taken, method became largely a secondary consideration. The Jews were doomed, whatever the *modus operandi*. For example, of the 60,000 Jews estimated to have been resident in the Kolomja district of Galicia when the German invaders arrived, approximately one-third were deported to Belzec for extermination. The remaining 40,000 were shot in a series of '*Aktionen*' conducted mainly in the Scheparow Forest, but also at various other killing sites within the Kolomja district itself. During the course of a period of little more than the twelve months following October 1941, this once thriving Jewish community had been completely destroyed. By 1945, just 200 Jews from Kolomja had survived the slaughter.[1130]

On a national scale, some 80 percent of Lithuanian Jewry, which in spring 1941 had numbered some 250,000 souls, were killed in the six months following the German invasion of the country, the vast majority of them shot in and around the towns and villages in which they had lived.[1131] Once it had begun, the compulsion to destroy the Jews overrode all other concerns—economic, military, and political, so that when in November 1943 the gas chambers of *Aktion Reinhard* had all been dismantled, 42,000 Jews could be shot in the Lublin district alone during the two days of *Aktion Erntefest*.[1132] More than two years earlier, months before the extermination camps had even begun to function, 33,700 Jews had been killed at Babi Yar outside of Kiev in the course of another two day operation.[1133] In order to make room for Jews newly deported from the *Reich*, a three day shooting spree in November and December 1941 accounted for some 27,000 Jewish victims in Riga.[1134] At around this time, and for the same reason, tens of thousands of Jews, both native and foreign, were murdered in cities such as Kovno and Minsk. The daily count of those shot in these mass executions was anything between 5,000–13,000.[1135] Clearly arranging for such slaughter at short notice (as was usually the case) presented few problems for the murderers.[1136]

Camps (Bloomington: Indiana University Press, 1987), pp. 17–19]. Many men served in more than one camp, some in all three. There were also 90–130 guards of mainly Ukrainian nationality at each camp. (Ibid, p. 22). Compare this to the 3,000 members of the *Einsatzgruppen*, who were aided in their killing *Aktionen* by several thousand foreign acolytes [*Hilfswillige* (Auxiliaries) or "Hiwis"], as well as by the many thousands of men of the Police Battalions.

1130 http://www.jewishgen.org/Yizkor/galicia/gal005.html (Accessed 22 August 2009).

1131 Joshua Rubenstein, Ilya Altman, *The Unknown Black Book: The Holocaust in the German-Occupied Soviet Territories* (Bloomington: Indiana University Press, 2008), pp. 277–278.

1132 Tomasz Kranz, *Extermination of Jews at the Majdanek Concentration Camp* (Lublin: Panstwowe Muzeum na Majdanku, 2007), p. 70.

1133 Saul Friedländer, *The Years of Extermination: Nazi Germany and the Jews, 1939–1945* (New York: HarperCollins, 2007), p. 197.

1134 Ibid., p. 262.

1135 Ibid., p. 267.

1136 Jeckeln, the HSSPF *Ostland*, was instructed by Himmler on 12 November to kill the Jews of the Riga ghetto. The killings began on 30 November. (Friedländer, *The Years of Extermination*, p. 261).

Nor was shooting always necessary for the implementation of genocide. It has been estimated that at least 35–40 per cent of Jewish deaths in occupied Poland were a result of the living conditions imposed by the Germans.[1137] In the Warsaw ghetto, 43,239 died during 1941, overwhelmingly from a combination of malnutrition, disease, exposure, and exhaustion brought on by the appalling working and living conditions, a mortality rate of more than 10 percent. By the time the first transports departed from Warsaw for Treblinka on 22 July 1942, the death rate in the ghetto was running at an annualised rate of more than 14 percent. This represented another example of a gradual and inevitable process of extermination "by other means".[1138] It was a process repeated in countless eastern European ghettos. Although German realisation of the necessity to maintain a productive workforce eventually forced a marginal improvement in the lot of the ghettoised Jews, there can be no doubt that, even allowing for this amelioration, mortality rates within the ghettos would have continued to have been extraordinarily high. As Donald Bloxham has observed: "Contrary to popular conception, extermination camps complemented rather than replaced the more intimate method of murder."[1139]

Would there then have been fewer victims of the *Shoah* had the gas chambers of Belzec, Sobibor, and Treblinka not existed? Perhaps, but such was the exterminatory imperative that it is just as likely that the Nazis would have resorted to other, more conventional methods of mass murder to achieve their goal. The killing would have been less efficient, but just as inevitable. What is certain is that the extermination camps of *Aktion Reinhard* could not have functioned as they did without the know-how gained from the first phase of "euthanasia". Apart from demonstrating a functional process for the rapid disposal of a large number of "undesirables", not the least significant aspect of the "euthanasia" programme was the indication it provided to Hitler, as well as to government and Party leaders, of just how much could be achieved behind the mask of "official secrecy", providing as that did the smokescreen necessary to furnish time enough for the desired objectives be attained. Of course the concealment could not last; eventually the truth would emerge. But it could last long enough.

Even more importantly, T4 supplied the kernel of professional killers who were to undertake *Aktion Reinhard* in Poland. These men were long term exponents of the physical process of murder and the disposal of corpses. Some had been recruited because they were considered ideologically reliable, or were recommended by a relative or friend. Others began their careers in the concentration camps or, particularly, in the police. Four of the six men who were to be the principal commandants of the *Aktion Reinhard* camps had been career policemen.[1140] All of those involved at the Polish killing centres quickly became desensitized killers (if they were not already such), rapidly descending into a criminal nightmare of sadistic brutality and death.

1137 Glass, James M, "*Life Unworthy of Life*"—*Racial Phobia and Mass Murder in Hitler's Germany*, (New York: Basic Books, 1997), p.xvi.

1138 Yisrael Gutman, *The Jews of Warsaw 1939–1943* (Bloomington: Indiana University Press, 1989), pp. 63–65.

1139 Donald Bloxham, *The Final Solution: A Genocide* (Oxford: Oxford University Press, 2009), p. 251.

1140 Christian Wirth, Gottlieb Hering, Franz Stangle, and Franz Reichleitner.

As already described, following the official cessation of the euthanasia programme a substantial number of T4 personnel were sent to the eastern front as part of the enigmatic *Organisation Todt* mission.[1141] According to the testimony of Pauline Kneissler, their duties included providing fatal injections to severely wounded members of the *Wehrmacht*. In his post-war report, Leo Alexander went even further, stating:

> Certain classes of patients with mental diseases who were capable of performing labour, particularly members of the armed forces suffering from psychopathy or neurosis, were sent to concentration camps to be worked to death, or to be reassigned to punishment battalions and to be exterminated in the process of removal of mine fields.[1142]

More life unworthy of life.[1143] Alexander also made the somewhat startling claim that doctors serving on U-boats were ordered to use lethal injections to execute any crew members considered "troublemakers".[1144] Given the presence on board German submarines of ex-T4 physicians like Ernst Baumhard and Günther Hennecke, this may not have been as unlikely as it sounds.

Initially 92 out of the estimated four hundred T4 operatives were made available by T4 to Odilo Globocnik in Lublin to operate the gas chambers of Belzec, Sobibor and Treblinka.[1145] That number was swollen in the summer of 1942 as other T4 members were ordered eastwards.[1146] An analysis of 131 men known to have served at the principal *Aktion Reinhard* camps, indicates that just four of them had not graduated from T4.[1147] Of the 46 known SS-men who worked in Sobibor, 39 had a T4 past.[1148] The actual number of ex-T4 operatives participating in *Aktion Reinhard* was probably greater—not all of those assigned to the death camps have been identified, any more than the names all of those employed in some capacity by T4 are known. For all the attempted camouflage, given the size and scope of the organisation's activities it is reasonable to assume that there were many individuals who were either involved in or otherwise aware of T4's activities , but whose identities remain a mystery. The degree of knowledge would obviously have varied from person to person, but that all embroiled at any level knew *something* seems undeniable. Letters and reports had to be typed, deliveries made, everyday routine tasks performed. Moreover, the circle of those who knew must surely have been significantly wider than just those directly involved. For all of the threats of

---

1141 Allers mentions a total of four hundred T4 personnel being sent to Russia, a figure which would have included the majority of those employed at the killing centres. (Gitta Sereny, *Into That Darkness—From Mercy Killing to Mass Murder* (London: Pimlico, 1995), p. 84).

1142 Leo Alexander, *Medical Science Under Dictatorship* (The New England Journal of Medicine 241, 1949), p. 41.

1143 There is another source suggesting that "euthanasia" was also practised on mentally disturbed members of the *Wehrmacht* within the *Reich* itself. In February 1943, Siegfried Handloser, head of the military medical service, ordered that "war hysterics who cannot be cured of their symptoms through treatment are to be committed to the hospital sections of mental institutions." (Aly, Chroust, Pross, *Cleansing the Fatherland*, pp. 89). It is not difficult to imagine what the result of such confinement would have been.

1144 Alexander, *Medical Science Under Dictatorship*, p. 42.

1145 Christopher R Browning, *Ordinary Men: Reserve Police Battalion 101 and the Final Solution in Poland* (New York: HarperCollins, 1993), p. 163.

1146 Robin O'Neil, *Belzec: Stepping Stone to Genocide* (New York: JewishGen Inc, 2008), pp. 92–93.

1147 Ibid., pp. 326–329.

1148 http://stichtingsobibor.nl/index.php/339/opstand-in-sobibor/ (Accessed 18 November 2008).

concentration camps and death penalties, human patterns of behaviour dictate that people will talk about their work to each other and to their families and friends. Knowledge was of course no offence in itself. However, post-war German and Austrian claims of ignorance concerning Nazi criminality have often lacked plausibility as a consequence of this avowed lack of awareness.

Whatever the precise figures regarding staffing may have been, it is certain that when, where, and how to implement the Jewish genocide had been decided, the KdF was able to draw upon the experience and know-how of T4 personnel to fulfil the demand for an experienced bunch of killers.[1149] Fritz Bleich (another to join the *Organisation Todt* mission to the eastern front), who was registrar at the KdF, and was thus in a better position than most to provide relevant testimony, stated that ex-T4 personnel often returned on leave from the *Aktion Reinhard* camps to Berlin. They would tell him of the killing of Jews and the methods used, showing off photographs of themselves beating, torturing, and murdering their victims. Friedrich Haus put together a collection of these photographs in an album for the delectation of a selected few.[1150]

The connection between T4 and *Aktion Reinhard* is undeniable. Why specific T4 functionaries were chosen for *Aktion Reinhard* is not known, although it is probable that this was a case of some men being more adept and enthusiastic killers than others, or simply because those chosen were familiar to their superiors. As Heinrich Barbl,[1151] who worked first at Hartheim, and then Belzec and Sobibor commented: "The method employed in the camps [of *Aktion Reinhard*] was the same as the one utilized in the castle at Hartheim, except that those killed there [in the camps] were all Jews."[1152]

But as a number of eminent historians have pointed out, the symbiosis between "euthanasia" and the "Final Solution" went beyond these obvious connections. It is not simply a question of drawing parallels between the murder of large numbers of people in specialised killing centres by means of poison gas common to both the "euthanasia" programme, and the *Aktion Reinhard* and other camps. In establishing the logistics of extermination, T4 was the paradigm in terms of organisation of labour, deception, and putative secrecy.[1153] Others go further than this, suggesting that the absence of an unambiguous written order by Hitler for the "Final Solution", a red herring so beloved of Holocaust deniers, may be attributable, at least in part, to the lessons learned from the population's eventual reaction to the "euthanasia" programme. It rapidly became apparent that however strenuous the efforts, killing on this scale simply could not be kept secret. Rather, it caused disquiet among the populace, which might have been exacerbated if the written *Führer* "euthanasia" authorization had become common

1149 Himmler allegedly told the *Aktion Reinhard* personnel that what he expected of them was "*superhuman-inhuman*". (Hilberg, *Destruction of the European Jews*, p. 961).

1150 Schmidt, *Karl Brandt*, p. 171. This sounds plausible. Other testimony by Bleich should be treated with caution.

1151 Barbl was interrogated several times by the Austrian police, but apparently never put on trial (Dick de Mildt, *In the Name of the People: Perpetrators of Genocide in the Reflection of Their Post-War Prosecution in West Germany. The 'Euthanasia' and 'Aktion Reinhard' Trial Cases* (The Hague: Martinus Nijhoff Publishers, 1996), p. 394, note39).

1152 David Cesarani (ed), *The Final Solution: Origins and Implementation* (London: Routledge, 1996) p. 55.

1153 Peter Longerich, *The Unwritten Order: Hitler's Role in the Final Solution* (Stroud: Tempus Publishing Limited, 2003), p. 81.

knowledge.[1154] What might be the reaction of the citizenry if a document bearing Hitler's signature ordering the destruction of the Jews existed, and somehow became widely known? Better by far to maintain Hitler's God-like façade of plausible deniability and have nothing in writing, thereby ensuring that others could purportedly take responsibility for the excesses of the regime.

The question arises as to whether Nazi "euthanasia" can be regarded as genocidal in and of itself, whether indeed "euthanasia" was not simply a prologue but the first chapter of Nazi genocide."[1155] It has been posed with increasing frequency since the term "genocide", a combination of the Greek word *genes* (race) and the Latin word *cide* (killing), was created by Raphael Lemkin in 1944 in his book *Axis Rule in Occupied Europe: Laws of Occupation, Analysis of Government, Proposals for Redress.*[1156] Lemkin fought to include genocide as one of the charges to be included in the indictments at the International Military Tribunal at Nuremberg, but the term was to make only peripheral appearance at the trial, and was not to acquire a legal definition until the adoption of the *Convention on the Prevention and Punishment of the Crime of Genocide* by the United Nations in 1948, the first truly meaningful human rights treaty. That was a triumph for Lemkin and his determined campaigning for his beliefs, even if the ratification of the treaty by many nations was a slow and painful process.[1157]

The Convention defined genocide in the following terms:

> Any of the following acts committed with intent to destroy, in whole or in part, a national, ethnical, racial or religious group, as such: (a) killing members of the group; (b) causing serious bodily or mental harm to members of the group; (c) deliberately inflicting on the group conditions of life, calculated to bring about its physical destruction in whole or in part; (d) imposing measures intended to prevent births within the group;(e) forcibly transferring children of the group to another group.

It is interesting to contrast this definition with that of crimes against humanity set out above. In essence Lemkin and Hersch Lauterpacht approached the same problem from different directions. Both were concerned with the persecution and annihilation of entire communities, but whilst for Lemkin this was about the group in its entirety, for Lauterpacht the focus was on the individual.The difference was one of intent. Was the intention to destroy the group? If so, it was genocide. If not, it was a crime against hu-

---

1154 Ibid., p. 82.

1155 Friedlander, *The Origins of Nazi Genocide*, p.xii.

1156 Dan Stone (ed), *The Historiography of Genocide* (Basingstoke: Palgrave Macmillan, 2008), p. 9. Lemkin refined the term further, defining killing and abortion as "ktonotechnics" and sterilisation and castration as "sterotechnics", both derived from the Greek—*'ktonos'* = murder, *'steirosis'* = infertility . Neither term has entered common usage, any more than "ktonology" as a definition of medicalised killing, although the expression "thanatology" to describe a delight in death has. [Paul Julian Weindling, *Nazi Medicine and the Nuremberg Trials—From Medical War Crimes to Informed Consent* (Basingstoke: Palgrave Macmillan, 2006), pp. 284–285]. These definitions were all coined as the nature and extent of Nazi crimes were revealed; contemporary language was deemed inadequate to describe such atrocities.

1157 Philippe Sands, *East West Street: On the Origins of Genocide and Crimes Against Humanity* (London: Weidenfeld & Nicolson, 2016), p. 377

manity.[1158] One may well ask whether this terminology is important. It certainly mattered little to the victim

Despite the wording of the Convention, no entirely acceptable definition of the term "genocide" has yet been found. Perhaps the simplest and most appropriate characterization of genocide is "group destruction without regard to the means of destruction or the type of group destroyed."[1159] If that definition is applied to the Nazi "euthanasia" programme, there can be little doubt it was indeed genocidal. However, as with many other subjects, there is a clear distinction to be made between the legal and the moral interpretations of genocide, since, even if the existence of natural law is acknowledged, it is individual national communities that in practice determine the boundaries of legality, whilst questions of morality are, of necessity, much more amorphous. Thus German doctors and others could claim that because they believed that "euthanasia" had the force of law, they were justified in murdering their patients. Not only were their actions legal, they would proudly declare, but since they were also motivated by humanitarianism, rather than being judged genocidal they should be considered morally impeccable too. In early post-war judgements, the courts refused to accept that defence, even if such an enabling law had indeed existed. Regrettably, as time passed, and in the interests of political expediency, attitudes changed. Morality, on the other hand, has a different perspective. Questions of legality may safely be left to lawyers. What is at issue here is human behaviour as a trope—or put another way, the immutable application of that natural law, defined as "unchanging moral principles common to all people by virtue of their nature as human beings". Whatever "justice" may decide, man cannot escape the verdict of history.

The terminology of genocide may be relatively new, but the actions certainly are not. In 1648, a campaign to exterminate the Jews of eastern Europe was waged by a band of Ukrainian Cossacks (subsequently joined by Crimean Tatars and Russians) under the leadership of Bohdan Khmelnytsky. Over the next eight years, tens of thousands of Jews were murdered in Poland and the western Russian Empire. Massacres occurred in all of the major cities of the region—Polotsk, Vitebsk, Minsk, Vilna, and Lvov, amongst others.[1160] Not surprisingly there is no unanimity so far as the number of Khmelnytsky's victims is concerned; estimates ranging from between 6,000 and 14,000 at the lower end of the scale to a clearly impossible 500,000 at the upper. However, given the overall Jewish population of the region at the time of about 350,000, it is conceivable that Jewish fatalities were proportionately greater as a result of this genocide than in Europe as a whole during the Holocaust.[1161] Ironically, whilst Khmelnytsky (whose motives were as much economic as religious or racial—the Jews were perceived as allies of the much hated Poles) is regarded in Jewish history as being second only to Hitler as a persecutor, he is generally considered by Ukrainians to be a hero—"the father of the nation." His portrait even appears on a Ukrainian banknote.

1158 Ibid., p.xxii–xxix

1159 Stone, *The Historiography of Genocide*, p. 89.

1160 Josephine Bacon, *The Illustrated Atlas of Jewish Civilization: 4,000 Years of Jewish History* (Royston: Eagle Editions, 1998), p. 136.

1161 Richard S Levy (ed), *Antisemitism: A Historical Encyclopedia of Prejudice and Persecution* (Santa Barbara: ABC-CLIO Ltd, 2005), pp. 116–117.

If not named as such at the time it was inflicted, genocide has been a common historical event, with its perpetrators often hailed as heroic for their actions.[1162] This is certainly how Odilo Globocnik saw himself in killing not less than 1.3 million Jews as part of *Aktion Reinhard*. Kurt Gerstein reported Globocnik's pride in his achievement:

> Gentlemen, if there was ever, after us, a generation so cowardly and so soft that they could not understand our work which is so good, so necessary, then, gentlemen, all of National Socialism will have been in vain. We ought, on the contrary, to bury bronze tablets [with the corpses] stating that it was we who had the courage to carry out this gigantic task![1163]

There is no doubt that Globocnik's sentiments were shared by the bureaucrats, medical personnel, and other operatives of T4, as well as by all those other eugenic theoreticians and fellow-travellers who had preceded them. They believed their ideas and/or their work was not only justified but critical for the improvement of the treatment of sickness, both mental and physical, and the creation of the brave new world eugenics promised. The genuineness of that belief by many should not be underestimated. When the "stop order" was issued in August 1941, many physicians regarded the period of relative calm as an opportunity to "raise the reputation of psychiatry". As far as they were concerned, the space created by the murder of "hopeless" patients and the associated economic benefits and freeing of personnel was to be utilized to develop methods of therapy and metamorphose mental hospitals into clinics.[1164] In his post-war evidence, Hans Hefelmannn stated:

> Right at the beginning of the asylum operation during the Second World War, a general consensus was reached that the institutional manpower, medications, and therapeutic opportunities recovered through euthanasia should benefit the 80 percent of institutional inmates that would probably remain. To this end, a psychiatric expert, Professor [Carl] Schneider of Heidelberg, consented to be appointed to the Reich Association to further expand therapy and research. This circumstance underlined the moral justification for carrying out the euthanasia measures.[1165]

The Central Clearing Office for Mental institutions was intended to be more than the commercial arm of T4; through its financial input it was hoped to completely reform the German institutional system, standardizing legal and fiscal regulations. In due course, rather than the existing hotchpotch of state and charitable institutions and the complexities of legislation that varied from region to region, all of the components of mental health care would be gathered together in a single all-powerful entity, which Hans-Joachim Becker christened "The Reich Office for Mental Hospitals."[1166]

It remains difficult to comprehend that individuals like Paul Nitsche or Carl Schneider could countenance "euthanasia" as part of a lifelong struggle to improve conditions in mental institutions. It is easy to assume that all Nazi eugenic policies were driven by

---

1162 Stone, *The Historiography of Genocide*, p. 73.

1163 Yitzhak Arad, *Belzec, Sobibor, Treblinka: The Operation Reinhard Death Camps* (Bloomington: Indiana University Press, 1987), p. 101. At the time these remarks were made (August 1942), victims were still being buried at the three principal *Aktion Reinhard* camps. Exhumation and cremation of corpses and the immediate cremation of new victims only began in Sobibor in autumn 1942, and later still at Belzec and Treblinka. (Ibid., pp. 170–178).

1164 Aly, Chroust, Pross, *Cleansing the Fatherland*, p. 46.

1165 Ibid. Emphasis added

1166 Ibid., p. 179–180.

some kind of blood lust. Whilst for a minority that may have been true in the shape of the eugenically inspired racism that resulted in the extermination of Jews, Gypsies, and others, there is also the suggestion that for men like Nitsche and many of his colleagues, "euthanasia" simply formed part of a long term strategy for the reformation of German healthcare. There is also an alternative view that this was merely an *ex post facto* rationalization of a literally murderous administration by men among the most guilty for the realization of its objectives. These were not modernizing idealists, it is proposed, but cold-blooded killers.[1167] Whatever interpretation is placed upon it, this attempted justification of murder is perhaps the most horrifying aspect of "euthanasia", for to suggest that killing patients somehow advanced the cause of medical knowledge is a concept that should always have been dismissed out of hand. This was a principle that the courts of the Federal Republic in particular frequently chose to misunderstand, for few of the perpetrators were brought to justice, and even fewer expressed regret for their crimes. The great majority continued to practice their professions after the war with an apparently clear conscience and little or no fear of prosecution.[1168]

This then also raises the issue of the possible connection between modernity and Nazism, a subject which has given birth to almost as many differing opinions as an acceptable definition of genocide.[1169] Modernity has been defined as

> ... a shorthand term for modern society or industrial civilization. Portrayed in more detail, it is associated with (1) a certain set of attitudes towards the world, the idea of the world as open to transformation by human intervention; (2) a complex of economic institutions, especially industrial production and a market economy; (3) a certain range of political institutions, including the nation-state and mass democracy. Largely as a result of these characteristics, modernity is vastly more dynamic than any previous type of social order. It is a society—more technically, a complex of institutions—which unlike any preceding culture lives in the future rather than the past.[1170]

Some consider the Third Reich to have been a prime example of modernity based upon one interpretation of Nazi social policy—that this was revolutionary in both conception and practice. This view predicates that Hitler's introduction of social benefits such as child allowances, public housing, better educational opportunities for all and the like, are representations of modernity in action. It is opposed by others who suggest that, far from being a modern political and sociological movement, National Socialism represented a reversion to barbarism, that in fact there was little, if anything "modern" about it, and that the benefits introduced by the Nazi regime were motivated solely by racial rather than altruistic considerations.[1171] And finally, there are those that argue that Nazism in practice was simultaneously an example of both modernity and anti-modernity.[1172] Given the "blood and soil" (*Blut und Boden*) views of so many leading Nazi theorists and their influential predecessors, such as Richard Wagner, it seems curious that the proposition has been subject to such debate. The National Socialists and

---

1167 Michael Burleigh, *Ethics and extermination: Reflections on Nazi genocide*, (Cambridge: Cambridge University Press, 1997), pp. 128–129.

1168 Aly, Chroust, Pross, *Cleansing the Fatherland*, pp. 229–230.

1169 Stone, *The Historiography of Genocide*, pp. 156–193.

1170 Anthony Giddens, and Christopher Pierson, *Conversations with Anthony Giddens: Making Sense of Modernity*, (Stanford: Stanford University Press, 1998).p. 94.

1171 Burleigh and Wipperman, *The Racial State*, pp. 39.

1172 Ibid., pp. 1–22.

those who inspired them regarded modernity as a Jewish trait, an influence which was poisonous to Germanism. The idea of modernity in economics, the arts, the professions or any other aspect of society was anathema to them. The virulently anti-Semitic French Catholic, Georges Bernanos, had precisely described this attitude when in 1931 he had written: "In this engineers' paradise, naked and smooth like a laboratory, the Jewish imagination is the only one able to produce these monstrous flowers."[1173] The reasoning was straightforward; conceptually modernity was foreign, and the prime example of the foreigner was the Jew. Ergo, any example of modernity was, by definition, Jewish in origin.[1174]

There is a danger that in attempting to define the indefinable, not only does our comprehension of the true horror of this vilest of regimes disappear, but we run the risk of failing to appreciate the singularity of the manner in which National Socialism functioned. All dictatorships are not the same—nor are all genocides. In fact, each is different, although they may all have some features in common. The proponents of the theory of modernity suggest that Nazism lurks beneath the surface of every society in equal measure, but there is reason to believe that only the unique circumstances of Germany in 1933 could empower an Adolf Hitler and permit his hideous ideas to flourish.[1175]

There is also a persuasive argument that Orthodox Judaism provided many examples of practical eugenics, from dietary prohibitions and personal hygiene to fecundity and purity of the bloodline. Indeed, there was considerable support for eugenics amongst pre-war Zionists on the grounds that, by its very nature, the Diaspora was dysgenic. Intermarriage and other factors contravening *Halacha*, Jewish religious law, would inevitably occur in a non-Jewish state, resulting in the decline of the Jewish race (if Judaism can be so described).[1176] In 1921, Zewi Parnass, an enthusiastic Jewish sup-

---

1173 Friedländer, Saul, *The Years of Persecution: Nazi Germany and the Jews 1933–39* (London: Phoenix, 2007), p. 212.

1174 Marius Turda and Paul J Weindling, (eds.), *"Blood and Homeland". Eugenics and Racial Nationalism in Central and Southeast Europe, 1900–1940*, (Budapest: Central European University Press, 2007), p. 355.

1175 Burleigh and Wipperman, *The Racial State*, pp. 304–307.

1176 An attempt to determine whether the long standing question, "who is a Jew?" is a matter of race or religion (if either) surfaced in, of all places, an English courtroom in 2008. The case was concerned with deciding if the policy of a Jewish school contravened the 1976 Race Relations Act in prioritising the admission of children only recognized as being Jewish in accordance with a ruling of the Orthodox "Office of the Chief Rabbi". Essentially, the matter to be determined was whether the child of a woman who had converted to Judaism in a manner not approved by the aforementioned "Office of the Chief Rabbi" was *Halachically*, that is matrilineally, Jewish. If such a child was refused admission to the school on religious grounds, that was perfectly legal; if, on the other hand, refusal had been on the grounds of ethnicity, the Act had been contravened. Judgement was reached that this was a question of religious practice rather than ethnic or racial origin, a decision that was surprisingly overturned on appeal. In fact, the Appeal Court concluded: "the requirement that if a pupil is to qualify for admission his mother must be Jewish, whether by descent or by conversion, is a test of ethnicity which contravenes the Race Relations Act 1976," a decision which undermined one of the fundamental principles of Judaism. [http://www.bailii.org/ew/cases/EWHC/Admin/2008/1535.html; and http://www.bailii.org/ew/cases/EWCA/Civ/2009/626.html (Accessed 9 July 2009)]. By a narrow margin the Supreme Court confirmed the Appeal Court's decision. This judgement may well result in either an amendment to the Race Relations Act permitting the school's long-standing policy to be reinstituted, or alternatively in a complete reassessment of the principle of faith schools. If nothing else, this case has provided evidence of the pitfalls inherent in framing anti-discriminatory legislation. [http://www.supremecourt.gov.uk/docs/uksc_2009_0136_ps.pdf (Accessed 16 December 2009)].

porter of eugenics, wrote: "Our religious regulations indicate that hygiene, and particularly race hygiene, is what we were aiming for in social life."[1177] Taking this reasoning to its logical conclusion, it has been suggested that Moses can be seen as the "father of eugenics", thousands of years before Dalton and his followers.[1178] If this view prevails, then in real terms eugenics and racial hygiene are about as far removed from modernity as it is possible to be.[1179]

The Second World War sounded the death knell for eugenics as it had been understood only a generation earlier. The public recoiled in horror as the liberation of Nazi extermination and concentration camps and other killing centres, as well as the concomitant war crimes trials, revealed the consequences of racial hygiene in practice. In a history of the eugenics movement published to coincide with his retirement as Secretary of the British Eugenics Society in 1952, Carlos Paton Blacker acknowledged that Galton's eugenic vision had, to all intents and purposes, proved no more than an illusion. The reason for this was not hard to find. Nazism had perverted Galton's message to the extent that eugenics in its entirety had been thoroughly discredited.[1180] In a post-war environment, other ways were sought to tackle the pressing social issues of the

1177 Turda and Weindling, "*Blood and Homeland*", p. 288.

1178 Ibid., pp. 283–297. It might be more accurate to ascribe the dubious honour of fatherhood in this matter to Abraham: "Abraham's servant, Eliezer, so the Midrash states, desired to offer his own daughter to Isaac, but his master sternly rebuked him, saying: `Thou art cursed, and my son is blessed, and it does not behoove the cursed to mate with the blessed, and thus deteriorate the quality of the race.'" This quotation is taken from an essay published in 1916 by Rabbi Max Reichler, in which he enthusiastically endorsed eugenic practice. The ethicist, Dr. Stanton Coit, is approvingly cited: "The Jews, ancient and modern have always understood the science of eugenics, and have governed themselves in accordance with it; hence the preservation of the Jewish race." Furthermore, questions of heredity could be viewed in both religious and eugenic terms

> Rabbi Akiba...claims that "a father bequeaths to his child beauty, health, wealth, wisdom and longevity." Similarly, ugliness, sickness, poverty, stupidity and the tendency to premature death, are transmitted from father to offspring. Hence we are told that when Moses desired to know why some of the righteous suffer in health and material prosperity, while others prosper and reap success; and again, why some of the wicked suffer, while others enjoy success and material well-being; God explained that the righteous and wicked who thrive and flourish, are usually the descendants of righteous parents, while those who suffer and fail materially are the descendants of wicked parents. Rabbi Reichler's championing of eugenics was in keeping with the *Zeitgeist*.

[Rabbi Max Reichler, Rabbi Joel Blau, Rev. Dr. D.de Sola Pool, *Jewish Eugenics and Other Essays: Three Papers Read Before The New York Board of Jewish Ministers 1915* (New York: Bloch Publishing Company, 1916) pp. 7–23].

1179 In his Nuremberg cell, for rather obvious reasons Alfred Rosenberg attempted to rationalise the murder of millions by pointing out to G.M.Gilbert, the prison psychologist, the ostensible similarities between Jewish and Nazi eugenic practice: "National Socialism wasn't based on racial prejudice", Rosenberg brazenly declared. "We just wanted to maintain our own racial and national solidarity...The Jews wanted to maintain their identity as a people, and I say more power to them, but so do we...It is only natural for the members of a group to feel a common bond and protect themselves and their identity." [G. M. Gilbert, *Nuremberg Diary* (New York: Da Capo Press, 1995, p. 120].

1180 Richard A Soloway, *Demography and Degeneration: Eugenics and the Declining Birthrate in Twentieth-Century Britain* (Chapel Hill: University of North Carolina Press, 1995), p. 350. Founded in 1908 as the Eugenics Education Society, so disreputable had the very word "eugenics" become, that in 1989 the Society's name was changed to the Galton Institute.

day. No longer would eugenics dominate social welfare and policy debates as it had done in the previous half-century. Politics changed the face of eugenics as geneticists attempted to distance themselves from the class- and race-based events of recent history. "Faced with the evidence of...barbarous policies, many who had once been tolerant of, if not impressed by, eugenic arguments rushed to condemn them."[1181] Perceptions were upended. Where, to the eugenicists, the rights of the individual were to be sacrificed on the altar of racial betterment, now attention was entirely concentrated on the person. If previously eugenics, whether "positive" or "negative", had been concerned with social issues on a national scale, now the focus was on the individual, and on personal choice. Which is not to say that eugenic ideas and principles ceased to be preached and practised after 1945. But like pupae, they had metamorphosed. If less overtly racist and discriminatory than before, most eugenicists still proffered the phantasmagoria of a perfect, man-made society, one in which the scientist would not simply replicate God, he would replace Him.[1182]

Viewed objectively, the real concern with much of eugenics as originally conceived was not its ends, elements of which—the prevention and elimination of sickness and disease rather than its cure, a massive improvement in the quality of life for an entire underclass– were commendable, but its means. Until recently, science was simply not equipped to resolve the problems it recognized. Knowledge of heredity and human behaviour was inadequate, technology non-existent. The problems were real; the solutions proffered were not. And the connection between eugenics and racism, inevitable in the political climate of the first half of the twentieth century, ensured that the practice of eugenics at that time could only lead to the charnel house. The nagging question remains, however, of how it was possible for so many of the best and the brightest of their generation to blindly follow Hitler's nihilistic path. Some differing factors are worthy of consideration.

In an otherwise excellent recently published biography of Karl Brandt, it is suggested:

> The killing of thousands of handicapped people had, it seemed, completely desensitized these men [that is medical professors and physicians] over the years. Unable to distinguish between right and wrong, their moral conscience had corroded to such an extent that they were willing to commit wholesale murder as long as the operation was loosely sanctioned by the regime.[1183]

The inability to distinguish between right and wrong is the principal test applied in law by the M'Naghten Rules when determining pleas of insanity, viz:

> The jurors ought to be told in all cases that every man is presumed to be sane, and to possess a sufficient degree of reason to be responsible for his crimes, until the contrary be proved to their satisfaction; and that to establish a defence on the ground of insanity, it must be clearly proved that, at the time of the committing of the act, the party accused was labouring under such a defect of reason, from disease of the

1181 Rachel Iredale, *Eugenics and its Relevance to Contemporary Health Care*, (Nursing Ethics, 7 (3), 2000), p. 209.

1182 Dennis Sewell, *The Political Gene—How Darwin's Ideas Changed Politics* (London: Picador, 2009), *passim*, especially pp. 213–234.

1183 Schmidt, *Karl Brandt*, p. 245. Emphasis added.

mind, as not to know the nature and quality of the act he was doing; or, if he did know it, that he did not know he was doing what was wrong.[1184]

The proposition that much of the medical profession of Germany was suffering from a collective bout of insanity, that the physicians who staffed the camps, murdered the mentally and physically handicapped and advanced theories of racial hygiene were simply lunatics, is surely unsustainable. Rather, the extermination of the incurably ill, according to T4 psychiatrist Friedrich Panse, put all perpetrators in a state of being "drunk with elation". In 1973, The University Clinic of Psychiatry, Düsseldorf, published an obituary of Panse. Their eulogy culminated with the sentence. "A life in the service of the suffering people...is completed."[1185] No suggestion of being unable to distinguish right from wrong in that statement.

Hannah Arendt also touched upon this supposed lack of a moral compass amongst the Nazi killers when she wrote:

Those few who were still able to tell right from wrong went really by their own judgements, and they did so freely; there were no rules to be abided by, under which the particular cases with which they were confronted could be subsumed. They had to decide each instance as it arose, because no rules existed for the unprecedented.[1186]

This too was a less than satisfactory analysis, for there were rules, illogical, barbaric, and by no means always observed, but rules nonetheless. What was occurring was not an inability to distinguish between good and evil, right and wrong, but a complete inversion of the these terms. It is circumstances that dictate how these abstract concepts are to be interpreted and the manner in which they are to be applied. As conditions change, so does the meaning we ascribe to these words. A simple example of the practical application of this is our attitude to war.

Modern society has evolved on the basis that killing your neighbour is an offence that should (a) be prohibited and (b) punished if it occurs. This is not so much a question of religious precept as of common sense. If we did not accept certain rational guidelines we would live in a state of perpetual anarchy. So, by and large, we do not murder each other. However, when war is declared, society reverses the norms. Far from prohibiting such actions, individuals are not merely permitted to kill whosoever the state declares the enemy to be (and who is often yesterday's friend) but are positively and vehemently encouraged to do so. The moral and ethical code is turned completely upside down. What was previously considered essential for all—the preservation of life—now becomes undesirable for many. Arendt considered this dichotomy in terms of a distinction between the actions of the state and those of the individual. Since the state is responsible for maintaining the very existence of a nation, it is not subject to the law in the same manner as the ordinary citizen. Commonly acknowledged crimes committed in the name of the state may be considered measures necessary for that state's continued existence. She continued:

1184 Michael Molan, *Sourcebook on Criminal Law*, (London: Cavendish Publishing Limited, 2001), p. 245.

1185 http://www.irren-offensive.de/kleespeech.htm (Accessed 23 April 2008).

1186 Hannah Arendt, *Eichmann in Jerusalem: A Report on the Banality of Evil* (New York: Penguin Books, 1977), p. 295.

> In a normal political and legal system, such crimes occur as an exception to the rule and are not subject to legal penalty...because the existence of the state itself is at stake, and no outside political entity has the right to deny a state its existence or prescribe how it is to preserve it. However...in a state founded upon criminal principles the situation is reversed.[1187]

As has been proven in endless measure, no state in history was more criminal than Nazi Germany. Moreover, Hitler took this assumption of *raison d'état* one step further and applied it not only to "outside political entities", but to the state's own citizens. Arendt consequently went on to question whether in fact one could apply the same principles of legality to a government "in which crime and violence are exceptions and borderline cases to a political order in which crime is legal and the rule."[1188] Clearly one could not, and because one could not, the defence of an exigency to obey superior orders for acts of state-sanctioned murder was unsustainable. Or should have been. As has been seen, post-war German courts did not always fully apply this precept.

Taken to its logical extreme, if a regime was criminal then all of that regime's commands and decrees would by definition be unlawful, and obeying them was in itself an illegal act. Apart from the likely personal consequences, it would obviously be completely impractical for citizens to refuse to obey every statute introduced by such a government. Some laws have to be obeyed if a state is to function at all. But as has been demonstrated, it was possible to draw a line in the sand and say "so far and no further." Even among the *Einsatzgruppen* there is no evidence to support the assertion that individuals were forced to kill innocent civilians out of the belief of subsequent imprisonment or death if they refused to do so.[1189] Nobody was compelled to kill, other than those who did so out of personal conviction. It came down to a question of choice. What is it then that provided the stimulus needed to permit the killers to behave as they did?

There has been a good deal of research into the question of individual obedience to authority. Probably the best known examples of this are Stanley Milgram's experiments conducted at Yale University in 1960–1963, in which a group of volunteers were paid a modest sum to participate in what was described as an investigation into the effects of punishment on learning. Each volunteer acted as a "teacher", instructed to administer an electric shock of an increasing degree of intensity via an elaborate piece of equipment to a "pupil" whenever the pupil answered a question incorrectly. The pupil, out of sight of the teacher in an adjoining room, would emit screams of anguish as the severity of the electric shock increased with every incorrect answer. If the teacher refused to continue, the experimenter in charge of the supposed investigation would insist with mounting firmness that he must do so. In fact, only the teacher was genuine; there were no electric shocks, and the pupil was trained to deliberately give incorrect answers and act as if in increasing agony with every concomitant increase in the dosage of electricity, until a point was reached at which he made no sound at all, feigning unconscious-

---

1187 Ibid., p. 291.

1188 Ibid., pp. 291–292.

1189 Ernst Klee, Willi Dressen, Volker Riess (eds.), *The Good Old Days—The Holocaust as Seen by Its Perpetrators and Bystanders* (New York: Konecky & Konecky, 1991), pp. 75–86.

ness—or worse. The issue to be determined was just how far the teacher was prepared to go in administering punishment at the command of the experimenter, that is the authority figure. The answers were disturbing. In the initial experiment, no less than 65 percent of the teachers obeyed the authority figure's order that they steadily increase the supposed shocks until they reached a potentially life-threatening maximum.

Milgram went on to conduct many further studies, varying the parameters slightly with each. Whilst the number of "teachers" who were prepared to administer the maximum shock differed, in every experiment there were at least some who were willing to do so. Moreover, in similar experiments involving scores of individuals of varying age and gender, the results have all been comparable. The conclusion reached was that obedience to authority is deeply embedded within the psyche of at least some of us.[1190] Yet Milgram himself was quick to point out the danger in reading too much into his research. "Is the obedience observed in the laboratory in any way comparable to that seen in Nazi Germany? (Is a match flame comparable to the Chicago fire of 1898?)," he questioned rhetorically.[1191] In fact, there were several areas in which the conditions relating to Milgram's experiments differed significantly from those prevailing in genocidal circumstances. Nonetheless, this research has been of great importance in at least one respect; it indicates just how easy it is to convince some people, purely on the basis of "superior orders", to inflict excruciating pain on others for whom they bear no personal animosity. As Milgram commented:

> After witnessing hundreds of people submit to the authority in our experiments, I must conclude that Arendt's conception of the banality of evil comes closer to the truth than one might imagine...That is, perhaps, the most fundamental lesson of our study: ordinary people, simply doing their jobs, and without any particular hostility on their part, can become agents in a terrible destructive process...While there are enormous differences of circumstance and scope, a common psychological process is centrally involved in both [my laboratory experiments and events in Nazi Germany].[1192]

Another experiment demonstrated the ease with which it is possible to inculcate prejudice. In a small American town, teacher Jane Elliott, divided her class of 8–9 year-olds between the blue-eyed (deemed the "superior") and the brown-eyed (to be considered "inferior"), then produced fictitious "evidence" to support these categorizations. The blue-eyed children were given special privileges, whilst the brown-eyed had to obey rules that accentuated their second-class status, including the wearing of a ribbon to enable identification from a distance. Once the experiment was underway, the previously friendly "blue-eyes" refused to play with the "brown eyes", and suggested to the school authorities that the latter might be thieves. Fights broke out between the two groups. Within a single day the "blue eyes" became arrogant and domineering, the "brown-eyes" depressed, sullen, and angry. Their "brown-eyes" schoolwork suffered as their self-esteem was dissipated. A few days later the roles were reversed. Elliott had been mistaken, she told her class—it was the brown-eyed who were superior and the blue-eyed less valuable. And, extraordinarily, the patterns of behaviour were reversed in parallel fashion, despite the "brown-eyes" having themselves already experienced

1190 James Waller, *Becoming Evil. How Ordinary People Commit Genocide and Mass Killing* (Oxford: Oxford University Press, 2002), pp. 102–106; Zimbardo, *The Lucifer Effect*, pp. 266–276.

1191 Waller, *Becoming Evil, p. 107.*

1192 Ibid., *p. 108.*

the effects of discrimination. Overnight the *Übermensch* became the *Untermensch*, and vice-versa. The exercise, subsequently repeated many times with groups of different ages, indicates how simple it is to implant irrational prejudice and bigotry in the individual.[1193]

A further example of the ease with which it is possible to determine behaviour, as well as the infinite and immediate corruptibility of power, was the famous 1971 "Stanford Prison Experiment", in which twenty-four healthy and intelligent male undergraduates were arbitrarily divided into groups of prisoners and guards. No specific instructions were given to either group about how they were to play out their roles, and there was no evident distinction between the individuals making up the two bodies, yet within an extraordinarily short period of time these otherwise normal students had divided on the one hand into guards who seemingly derived pleasure from insulting, threatening, humiliating, and dehumanizing their "prisoners", and on the other into detainees who were passive, dependant, helpless, and depressed. So concerned became the organizers of the experiment at the evident sadism of the guards that what had been intended to last two weeks was abandoned after only six days.[1194]

One other piece of research is particularly germane to the subject under review. In a real hospital, each of twenty-two nurses received a call from a staff doctor they had never met. Completely against regulations, which required that the physician sign the order for any drug usage beforehand, the nurses were instructed to administer twice the maximum advised dosage of a particular drug to patients, solely on the basis of the physician's telephoned instructions. No less than twenty-one of the nurses had started to pour the possibly lethal prescribed dosage (which was in fact a harmless placebo), before the person conducting the research stopped them. In a survey of a large sample of registered nurses, 46 percent stated that they had at sometime carried out the orders of a doctor that the nurse felt might have had harmful consequences for the patient.[1195] Again, it is possible to read too much into such experiments when making comparisons with conditions in the "euthanasia" killing centres of Nazi Germany, but there is no gainsaying the powerful influence physicians wield over the nurses trained to obey a medical superior's orders.

In his groundbreaking work, *Ordinary Men*, Christopher Browning examined the activities of Reserve Police Battalion 101 in Poland. Between July 1942 and November 1943, these ordinary men shot a minimum of 38,000 Jews and dispatched at least a further 45,200 to Treblinka and death.[1196] What was these policemen's rationale, their motivation? Browning presents an absorbing, and highly disturbing account, for with it he raises the spectre of our own potential for perpetrating evil. He concludes his book with the words:

> Everywhere society conditions people to respect and defer to authority, and indeed could scarcely function otherwise. Everywhere people seek career advancement. In every modern society, the complexity of life and the resulting bureaucratization and specialization attenuate the sense of personal responsibility of those implementing official policy. Within virtually every social collective, the peer group exerts tre-

---

1193 William Peters, *A Class Divided, Then and Now* (New Haven: Yale University Press, 1987), *passim*.
1194 Zimbardo, *The Lucifer Effect*, pp. 23–257; http://www.prisonexp.org/ (Accessed 12 April 2009).
1195 Zimbardo, *The Lucifer Effect*, p. 277.
1196 Browning, *Ordinary Men*, pp. 191–192.

mendous pressures on behaviour and sets moral norms. If the men of Reserve Police Battalion 101 could become killers under such circumstances, what group of men cannot?[1197]

Attempting to answer that question has produced both an impressive quantity and quality of scholarly debate. Certainly, everything in the above quotation can be applied equally to all of the perpetrators of "euthanasia" and the *Shoah.* In an earlier analysis of perpetrator motivation, Leo Alexander went some way towards endorsing Browning's view by describing the driving force in the case of members of the SS as being

> ...a process of reinforcement of group cohesion...called *Blukitt* (blood-cement)...This motivation, with which one is familiar in ordinary crimes, applies also to war crimes and to ideologically conditioned crimes against humanity—namely, that fear and cowardice, especially fear of punishment or of ostracism by the group, are often more important motives than simple ferocity or aggressiveness.[1198]

Putting the psychopaths among them to one side, however, one is bound to ask whether the actions of so many can be attributed solely or even mainly to fear of punishment or peer pressure? Was it to seek the approval of their contemporaries, or out of concern for their own well being, that Hermann Pfanmüller exhibited the emaciated body of a child he was starving to death at Eglfing-Haar "like a dead rabbit"[1199], or that Mina Wörle was prepared to continue killing children at Kaufbeuren-Irsee for weeks after the war had ended, when even she must have realized that the regime she had previously served was no more, and that any oath she might have sworn was no longer binding? There are no satisfactory answers to such questions. In truth, the more we learn of human behaviour, the less we understand. For some, an "education" in "euthanasia" merely served as a precursor to a greatly expanded killing programme. As Franz Suchomel explained:

> What is true, of course, is that the people who were involved in the actual killing process in the institutes, those who worked in the crematoria—we called them *'die Brenner '*—became calloused, inured to feeling. And they were the ones who were afterwards the first to be sent to Poland.[1200]

But that ambition, greed, envy, and sometimes unreasoning hatred also played a part is apparent from the biographical data contained in this manuscript. Self-justification too; as Hannah Arendt commented:

> ...the problem was how to overcome not so much their conscience as the animal pity by which all men are afflicted in the presence of human suffering...the trick...was very simple and very effective; it consisted in turning these instincts around, as it were, directing them toward the self. So that instead of saying: "What horrible things I did to people!" the murderers would be able to say: "What horrible things I had to watch in pursuance of my duties, how heavily the task weighed upon my shoulders!"[1201]

Himmler used this technique to considerable effect in his infamous Posen speeches, made firstly to senior SS personnel and then to the assembled *Reichsleiters* and *Gauleiters* in October 1943, in which he openly described the ongoing policy of genocide.

1197 Ibid., p. 189.

1198 Alexander, *Medical Science Under Dictatorship*, p. 44.

1199 Robert Jay Lifton, *The Nazi Doctors: Medical Killing and the Psychology of Genocide* (London: Papermac, 1990), p. 62.

1200 Sereny, *Into That Darkness*, p. 83.

1201 Arendt, *Eichmann in Jerusalem*, p. 106.

These speeches had been foreshadowed by another address Himmler made to SS personnel on 7 September 1940 in Metz. There he was comparing the deportation of French citizens (predominately Jewish) from Alsace and Lorraine to unoccupied France with the actions of the *Einsatzgruppen* following the invasion of Poland:

> Exactly the same thing happened on Poland in weather 40 degrees below zero, where we had to had to haul away thousands, ten thousands, hundred thousands; where we had to have the toughness—you should hear this but also forget it immediately—to shoot thousands of leading Poles, where we had to have the toughness, otherwise they would have taken revenge on us later...[1202]

In the case of the non-medically qualified killers, those physically perpetrating murder in the institutions and camps, their lives and personalities were depressingly mundane, as has been illustrated. To commit such atrocities on a daily basis it was doubtless necessary to use such "ordinary men", although at least some of them were perhaps better described as "ordinary psychopaths" (in Michael Burleigh's turn of phrase[1203]), if not suffering from an even more malevolent personality disorder.[1204] Primo Levi, perhaps the best known and most gifted Holocaust survivor, said this of his tormentors:

> ...They were made of our same cloth, they were average human beings, averagely intelligent, averagely wicked: save for exceptions, they were not monsters, they had our faces, but they had been reared badly. They were, for the greater part, diligent followers and functionaries: some fanatically convinced of the Nazi doctrine, many indifferent, or fearful of punishment, or desirous of a good career, or too obedient...Some, very few in truth, had changes of heart, requested transfers to the front lines, gave cautious help to prisoners or chose suicide. Let it be clear that to a greater or lesser degree all were responsible...[1205]

Others have made similar observations; "Nothing would be more mistaken than to see the SS as a sadistic horde driven to abuse and torture thousands of human beings by instinct, passion, or some thirst for pleasure. Those who acted in this way were a small minority," recalled a former Auschwitz prisoner. An inmate of the medical experiment section at Ravensbrück commented that the staff were "not a very inspiring lot but, on the whole, not too different from what might be found in any hospital."[1206]

Ausrottung—"extermination", a word which frequently recurs in Hitler's writings and speeches, is important to any appreciation of his *Weltanschauung*. A man who saw everything in terms of absolutes, the idea of compromise was alien to him. Thus war,

---

1202 Michael Wildt, *An Uncompromising Generation—The Nazi Leadership of the Reich Security Main Office* (Madison: University of Wisconsin Press, 2009), pp. 262–263.

1203 Burleigh, *Death and Deliverance*, p. 223.

1204 "A distinction, however, should be made between the ordinary psychopath (who acts antisocially, is frequently impulsive, and whose crimes are characterized by an affective reaction, consciously related to actively pursuing material gains) and the malignant type of psychopath (a clearly predatory, violent individual whose goal is the gratification of vengeful or sexual sadistic fantasies)." [Richard N Kocsis (ed), *Serial Murder and the Psychology of Violent Crimes: An International Perspective* (Totowa: Humana Press, 2007), p. 86]. Both types were present in the "euthanasia" killing centres and the extermination camps.

1205 Primo Levi, *The Drowned and the Saved* (London: Abacus Books, 2002), pp. 169–170.

1206 Tzvetan Todorov, *Facing the Extreme: Moral Life in the Concentration Camps* (London: Phoenix, 1996), p. 122.

like every other facet of life's struggle, was seen in terms of total victory or total defeat. In this Hitler, although born an Austrian, was merely the inheritor of a traditional Prussian militarism that could be traced back to a time prior to the creation of the German nation state, and which was displayed in all of its implacable brutality in a number of colonial expeditions (particularly in Africa), as well as in the *Kaiserreich's* conduct of the First World War.[1207] If then the campaign against the perceived enemies of Nazism is viewed as a kind of social warfare, (and it is worth bearing in mind that Lucy Dawidowicz titled her book on the *Shoah*, *The War Against the Jews*), it becomes possible to view the actions of the perpetrators of Nazi crimes as being perfectly moral, that is "good" from a National Socialist perspective. At the same time, permitting the genetic and racial enemies of the *Reich* to continue to exist and propagate could be viewed as "evil". Both of these aspects were elements necessary for the successful waging of a total "war", the ultimate aim of which was the complete annihilation, the *Ausrottung* of the enemy.

If experience teaches anything, it is that some individuals are capable of rationalizing any kind of conduct, however extreme. Given suitable circumstances, it would therefore appear that all that is required to convert at least some members of society from the side of the angels is a regime dedicated to producing a sufficiently intense level of quasi-religious indoctrination. When the perceived "war" was over, and Nazism defeated, it was possible for these individuals to immediately revert to more commonly accepted standards of civilized behaviour. Those responsible for compulsory sterilisation and mass murder who escaped retribution did not continue to practice those acts in the post-war world. In general, they immediately abandoned the dogma they had so loudly championed and executed, in many cases going on to become highly regarded members of their respective professions, or sinking back into the obscurity from which they had emerged. It is noteworthy that few of them expressed any remorse for their previous espousal of Nazi ideals.

Of course, individual circumstances dictated that there were other factors contributing to the motivation of a great number of the men and women under consideration. Some retained a burning conviction regarding the verisimilitude of Nazi ideology to the end of their life. But in general, and unfortunately it is only possible to arrive at any conclusions in the most general terms, it seems more appropriate to consider the actions of at least some perpetrators as a reflection of this inversion of moral and ethical standards rather than, for example, peer pressure alone, although it is reasonable to suggest that in some instances this may also have been one of a number of contributory factors. Be that as it may, what was really at issue was not an inability to distinguish between right and wrong, the "natural law" that remains a constant, and of which the perpetrators remained fully aware, but rather the conception of what these terms had come to mean in a totalitarian state. Of even greater concern is that this behavioural rationalization is not something limited to Nazism, as is apparent from the evidence provided by pre- and post Second World War atrocities committed in places like Armenia, Rwanda, Bosnia, and elsewhere. However, the National Socialists were unique in one

1207 Isabel V Hull,., *Absolute Destruction—Military Culture and the Practices of War in Imperial Germany* (Ithaca: Cornell University Press, 2006), *passim.*

important respect. Genocide generally targets very specific national, religious or social elements within the Genocidaires' own community. What distinguished the Nazis was the sheer range of their perceived enemies—Jews, Slavs, the disabled, both mental and physical, the non-white, homosexuals, Jehovah's Witnesses, asocials, communists, social democrats, members of the clergy—the list was endless. Moreover, the potential victims were not restricted to those of German nationality. The Nazis killed with complete disregard to questions of citizenship. In regions under the dominion of the Third *Reich*, any group, any person could become a target.

In his study of the leadership cadre of the RSHA, Michael Wildt found that, not surprisingly, a significant majority (more than 75 percent) of this intellectual (in the broadest sense) elite of Nazism shared much in common. Most had been born after 1900, too late to see service in the Great War, but old enough to have lived through the turmoil and instability of the Weimar years.[1208] The products of universities and colleges which became hotbeds of nationalism and extremism in the 1920s and 1930s, they were profoundly mistrustful of parliamentary democracy, believing not in the virtues of debate and rational argument, but rather in action, ruthlessly applied, as the function and purpose of government. They regarded themselves not as part of the despised *bourgeoisie*, but rather as a leadership in waiting, men who would break the mould of what they regarded as a corrupt and degenerate society. Their byword was "action"; swift, decisive, merciless. In their philosophy the end would always justify the means, the deed vindicate the idea. They would achieve authority and power not through debate and reason but via the application of a fanatical determination to achieve their objective, no matter what the cost might be. As Himmler expressed their philosophy: "The word impossible must not exist and will never exist for us."[1209] Leni Riefenstahl had summarised their *Weltanschauung* in the title of her film of the 1934 Nuremberg Party Congress—*Triumph des Willens* (*Triumph of the Will*). Academically brilliant, but morally bankrupt, such would-be leaders required only one thing—a *Führer*, one who shared their world view and who would empower this generation of mini-*Führers* to inherit what they regarded as their rightful place in society.[1210] It was to the misfortune of mankind that they found such a *Führer*, one with ideas even more extreme than their own.

The similarities between the RSHA leadership and the senior technical and medical staff of T4 are apparent. Of the individuals comprising the latter categories considered in the appendix to this volume, almost 60 percent had been born post-1900. Some 72 percent had received the benefit of a higher education, during the course of which those born after 1900 would undoubtedly have been exposed to the political radicalism prevalent in Germany's centres of learning in the decades following their birth. Evidently there was a commonality of background, influence, and belief among many of those who were to form the nucleus of the future Nazi administration, whether as policemen, bureaucrats, or physicians. It would be a mistake to assume that for these men (and a handful of women) there was anything other than a burning conviction that what they did was not only morally and ethically justified, but sociologically and politically essen-

1208 Wildt, *An Uncompromising Generation*, p. 10.
1209 Ibid., p. 440.
1210 Ibid., pp. 76–80.

tial. They became criminals out of certitude rather than convenience, and certainly not, as many of them concluded of others, as a result of some hereditary deficiency. Few of them indeed ever admitted to the error of their ways.

So far as the administrative *Schreibtischtäters*, the desk murderers, were concerned, there is no reason to believe that, in the great majority of cases, they were enthusiastic supporters of a eugenically inspired policy to kill the sick and disabled, any more than there is to assume that they were driven by a militant anti-Semitism in implementing the planned annihilation of Jewry. That is not to say that eugenic principle on the one hand and racist propaganda on the other did not have a part to play; however, the driving force behind the activities of these individuals was not a burning hatred of their victims, but the application of a longstanding professional routine added to complete indifference as to the consequences of their actions.[1211] Their guilt lay in precisely the post-war defence they put forward—they were "only obeying orders."

It may be concluded from these observations that it is unlikely that there is ever any single rationale to account for the joint actions of a group of individuals. The motivation to support a particular cause is as personal as tastes in food. What delights one, disgusts another. That there were people in Germany who found Nazism and its policies abhorrent, who were opposed to "euthanasia" and the persecution of racial minorities, is beyond doubt. But in a dictatorship as ruthless as Hitler's, it required an extraordinary degree of courage to display that opposition. It would be wrong to condemn anybody for failing to exhibit such valour. Equally it is right and proper to censure those who endorsed the regime and carried out its murderous policies. It is worth repeating that none were compelled to do so, except out of personal conviction. But as Donald Bloxham has persuasively argued: "In every case, different agents worked towards the new governing norm, whether out of shared values, fear, licence, careerism, greed, sadism, weakness, or, more realistically, a combination of more than one of these."[1212]

Between the two extremes of perpetration and dissent lay the majority of German citizens, who were neither perpetrators nor dissenters. They were bystanders, indifferent and apathetic onlookers, happy to go along with the perceived benefits that Nazism brought, and simply turn their back on its excesses. If they claimed not to know what was being done in their name, it was because they did not want to know. Assuming that there was anything to the outmoded concept of collective guilt, this related to sins of omission rather than commission. Regrettably, that was, and is, hardly an uncommon position for a preponderance of the population to adopt in any modern society. Ignorance may not be bliss, but it can be a good deal more comfortable than knowledge. Whatever may have been the case in the past, given the wealth of evidence now available none can any longer take refuge in a claimed lack of awareness of Nazi eugenics and its consequences.

What of Chesterton's vision? After seeing the practical consequences of more than a decade of eugenic theory, did men indeed "attempt to believe their dreams and disbe-

1211 Enzo Traverso, *The Origins of Nazi Violence* (New York: The New Press, 2003), p. 43.
1212 Bloxham, *The Final Solution*, p. 297.

lieve their eyes"? Sad to relate, in the years immediately following the Second World War, that seems to have been the case. A British Member of Parliament, Evan Durbin, described as "the Labour Party's most interesting thinker of the 1940s and arguably of the twentieth century",[1213] concluded that "people were far more wicked, i.e. mentally ill, than was commonly supposed...as a whole we were all very sick and very stupid...selective breeding was probably the answer."[1214] Durbin arrived at this conclusion four months after the defeat of Nazi Germany, the existence and eugenic policies of which state might have given him pause for thought about exactly who was going to do the selecting, the criteria to be applied, and the probable repercussions of such a policy. In September 1945 he certainly didn't have to look far to find an unhappy precedent for the approach he proposed. It seems even more extraordinary to relate that the launching of the British National Health Service in July 1948, hailed by one eminent historian as "one of the finest institutions ever built by anybody anywhere",[1215] produced an analysis of the new creation by the *Manchester Guardian* which Wilhelm Schallmayer himself might have penned. The newspaper dyspeptically pronounced that taken together with the simultaneously introduced Social Security provisions, the measures aimed

> ... at evening up the weak with the strong, whether the weakness is inherent or accidental, [and] at eliminating selective elimination. This policy risks an increase in the proportion of the less gifted. Its logical counterpart is to take every possible step, especially in the field of child health, to prevent all accidental deficiencies and to minimise all inherent ones.[1216]

It may safely be assumed that the *Manchester Guardian* did not consider its readers to be among "the less gifted". Not surprisingly, the newspaper refrained from suggesting exactly how those "accidental" and "inherent" deficiencies were to be prevented or minimised.

Nor did the countless number of Nazism's victims arouse much sympathy in certain circles. In a typical display of insensitivity, the British right-wing philosopher, social critic, admirer of Hitler, and anti-Semite, Anthony Mario Ludovici, resurrected a familiar phrase in 1961, when he wrote:

> Men must learn again to feel in their hearts contempt and repugnance for biological depravity; and when this lesson has been learnt and the taste displayed in mating correspondingly chastened, there will be no need to argue over the pros and cons of a lethal chamber for human rubbish, for morbidity and defect will insensibly and inevitably diminish to the extent of ceasing to be a social problem.[1217]

Fifty years after its first appearance, George Bernard Shaw's "lethal chamber" was back in vogue. Having enjoyed a measure of repute with his earlier works, passages such as the above thankfully hastened Ludovici's rapid decline into post-war obscurity.

In many of the countries that had been occupied by Germany during the Second World War, questions of the "quality" of children resulting from the union of German

---

1213 David Kynaston, *Austerity Britain 1945–51* (London: Bloomsbury, 2008), p. 38.

1214 Ibid., p. 130.

1215 Peter Hennessy, *Never Again: Britain 1945–51* (London: Jonathan Cape, 1992), p. 144.

1216 http://archive.guardian.co.uk/Repository/ml.asp?Locale=english-skin- (Accessed 24 August 2009).

1217 Dan Stone, *Breeding Superman: Nietzsche, Race and Eugenics in Edwardian and Interwar Britain* (Liverpool: Liverpool University Press, 2002), p. 133.

fathers and native mothers were raised, based upon dubious concepts such as "blood", "nation" and "race". The same kind of eugenic and racially hygienic thinking common to the Nazi regime became fashionable. In Norway, the discredited concept of "feeble-mindedness" again appeared, with the suggestion that the children of such couplings were in some way mentally and morally deficient. A report written by a respected Norwegian psychiatrist as late as 1999 presumed that among the mothers of these children of mixed nationalities would be found

> ... a disproportionate number of feeble minded persons, in addition to some asocial psychopaths—partly insane. These persons have defective minds that for a great part must be considered hereditary, and there is some risk that these hereditary dispositions will become manifest in their offspring.

Supposed "feeble-mindedness" was considered a major social problem in Norway both before and after the Second World War, with the result that, just as in Nazi Germany, eugenic methods such as sterilisation and the internment of so-called "inferior" individuals were adopted as remedies for the perceived malady. It was reasoned that as a consequence of the mother's lax morals and intellectual deficiency, the child carried a fundamental hereditary flaw, which might manifest itself sometime in the future by the child adopting a Nazi mentality. This professional assessment of the children was clearly based upon eugenic ideas of both racial and mental hygiene. Being labelled "feeble-minded" became an immense burden for these children. Some were ostracised by their teachers, who seemed to assume that trying to teach them was a waste of time. No attempt at rational diagnosis was made; children simply received the impression that they were somehow intellectually inferior, and that this in one way or another related to their parentage. As a result of this application of eugenic theory, many of these children have spent the rest of their lives struggling against a deep-rooted sense of inferiority.[1218]

In Germany itself, the post-war treatment of so-called "occupation" children was another example of the continued application of the principles of racial hygiene in a non-National Socialist environment. About 5,000 children were born to German mothers and black American GI fathers. These so-called *Mischlingskinder* (how familiar that term sounds) presented a whole series of problems. What was their nationality? Were they really German, American, perhaps even African? Scientific expertise was needed in order to assess the biological and mental state of the children, to predict how they were likely to adapt to society in the future and how they were to be educated. An extensive testing programme was carried out, much of which simply replicated pre-war Nazi psychological and anthropological eugenic assumptions. Paradoxically, some children achieved extraordinarily high test results, leaving those conducting the tests with the task of explaining the theoretically inexplicable.

Two conflicting ideas of the manner in which these children were to be treated emerged. One suggestion was that in order to protect them from prejudice, the children be sent for adoption in some suitable foreign country—a variation on the Nazi practice of deportation of the undesirable. The other, more humane proposal, was that it was

---

1218 Kjersti Ericsson and Eva Simonsen, *On the Border: The contested children of the Second World War* (Childhood, Vol. 15, No.3, 2008), pp. 407–409.

time for the German nation to come to terms with its racist past. In what was surely a supreme example of irony, just as the Nazis had used the media so extensively to promote their theories of "Aryan" supremacy, a wide-ranging public campaign was launched which sought to educate the German public about the virtues of handling the "problem" of *Mischlingskinder* with compassion. Unfortunately, the films, books, and pamphlets that were produced for the campaign do not appear to have served their intended purpose; instead, they may have promoted existing racial prejudice and stereotyping.[1219]

The year 2009 saw the commemoration of both the 200th anniversary of the birth of Charles Darwin and the 150th anniversary of the publication of *On the Origins of Species*, events celebrated by a seemingly endless series of newspaper and magazine articles, books, films, and radio and television programmes. The Royal Mail produced a special series of colourful postage stamps illustrating Darwin's contributions to zoology, ornithology, geology, botany, and anthropology. The Royal Mint struck a commemorative two pound coin bearing the image of Darwin staring down a chimpanzee. Considered by all except the most blinkered creationists as one of the most important and influential of scientific minds, Darwin's legacy continues to impinge upon all of our lives every day, and in many different ways.

In 1909, the centenary of Darwin's birth involved commemoration in a commensurate, if naturally technologically somewhat different manner. The great and the good gathered at Cambridge University on 22–24 June of that year to honour Darwin's memory and his immense contribution to scientific knowledge. Among them was the ex-Prime Minister, Arthur James Balfour, who in a speech delivered at a celebratory banquet, said, *inter alia*:

> ...The problem of life, is the one which it is impossible for us to evade, which it may be impossible for us ultimately to solve, but in dealing with it in its larger manifestations Charles Darwin made greater strides than any man in the history of the world had made before him.

Or since, many would suggest. Notwithstanding the fulsomeness of his praise, Balfour did not overlook the consequences of Darwin's hypothesis:

> From the very nature of the case his great generalisation, from the very fact of its magnitude, produced, as was inevitable, violent controversy, and human nature in 1859 and 1860 was not different from human nature in 1909, and violent controversy then, as now, was prolific, and must be prolific, in misrepresentation.

It is not hindsight alone that leads to the conclusion that so far as both human nature and misrepresentation are concerned, nothing much has changed in the more than one hundred years since Balfour spoke.

At the same centennial banquet, Darwin's son, William Erasmus, provided a glimpse of his father's character:

1219 Ericsson and Simonsen, *On the Border*, pp. 410–411.

> I have been thinking over the characteristics of my father which are quite apart from the qualities on which his influence and his success as a man of science depended, and I think the quality which stands out in my mind most pre-eminently is his abhorrence of anything approaching to oppression or cruelty, and especially of slavery; combined with this he had an enthusiasm for liberty of the individual and for liberal principles.[1220]

What then would Darwin make of the abuses of the 20th century carried out, in large part, in his name? If his ideas had provided the anlage from which much that followed was derived, it is safe to say he would have been appalled at the manner in which the research to which he had devoted most of his life and the science which bore his name came to be exploited by so many men who were his intellectual and moral inferiors. The mutation of Darwinism into its so-called "social" application, thence to racial hygiene, eugenics, and eventually genocide, would doubtless have been anathema to him . But once out, there was no forcing the genie of Darwinism back into the bottle. Like other groundbreaking scientific studies, it was exploited by ruthless and callous men for the greater good of a powerful few, rather than the benefit of the powerless many. The inevitable conclusion must be that if he was personally largely blameless for the subsequent excesses of those who so devalued his conclusions in the interests of racial and political dominance, and whose excrescence reached its apotheosis with the policies and actions of Adolf Hitler and Nazism, Darwin's findings laid the foundation for many of the evils of the century that followed the publication of his dissertations. For even he was not immune to making what would today be considered, at the very least, highly contentious observations of a kind that were to prove a justification for the stratagems of despots such as Hitler. "We civilized men", Darwin wrote in *The Descent of Man*,

> ...build asylums for the imbecile, the maimed, and the sick; we institute poor-laws; and our medical men exert their utmost skill to save the life of every one to the last moment. Thus the weak members of civilised societies propagate their kind. No one who has attended to the breeding of domestic animals will doubt that this must be highly injurious to the race of man.[1221]

He did not suggest the alternative to such humanitarianism. Those who came after him were not so reticent. Nor, despite his abhorrence of slavery, did Darwin refrain from endorsing the quintessential Victorian view of the inferiority of non-European races. "The breath of civilization" was "poisonous to savages," he wrote in a note. In another passage in *The Descent of Man*, he described the death of natives in British colonies as the inevitable concomitant to the advance of that civilization, a consequence which endorsed his theory of natural selection. It was an attitude that did nothing to curb the growth of social Darwinism with all that pernicious myth was to entail.[1222] In short, in common with the rest of humanity, Darwin was a mass of contradictions, a man typical in many ways of his times.

1220 http://darwin-online.org.uk/content/frameset?itemID=A279&viewtype=text&pageseq=1 (Accessed 18 April 2009).

1221 Jonathan Peter Spiro, *Defending the Master Race: Conservation, Eugenics, and the Legacy of Madison Grant*, (Burlington: University of Vermont Press, 2008), p. 122

1222 Traverso, *The Origins of Nazi Violence*, p. 60.

Nazi eugenics sought to create a world based upon a nihilistic vision of inequality, persecution and thanatology. It made much of Europe a necropolis, and had ambition to extend that empire of death to the rest of the globe. That it ultimately failed in its objectives is a cause for rejoicing. But our celebration should be muted. For, whilst it would be comforting to end on a note of optimism, to at least hope that something of lasting value has been absorbed from the abominations of the past, that racism, anti-Semitism, discrimination and all of the other iniquities implicit in Nazi eugenics were banished with its demise, a glance at any newspaper or television bulletin shows that this is sadly not the case. We have to acknowledge that despite the nobility of the sentiment, all men (and women) are not created equal nor, despite Rousseau's famous pronouncement, are they all born free. We are not all equally intelligent, equally physically able, or equally artistically talented, whilst freedom remains a tantalising mirage for millions. It is unrealistic to think otherwise. Nazism sought to exploit many of these individual disparities in the name of eugenics. Whilst continuing to recognize the essential veracity of Darwin's findings, we should by now have learned to acknowledge and respect such differences rather than abusing them. But it is all too evident that we have not.

It goes without saying that the one undeniable equality we should all share is the equal right to live our lives in peace, to seek happiness and fulfilment as we individually characterize those otherwise indefinable objectives, free to pursue whatever talents we may possess without fear of prejudice or persecution on the grounds of race, religion, disability or any other perceived "inferiority", still less out of any deemed "economic necessity". If this in its own way is as Panglossian a vision as that of eugenicists like Jenő Vámos, it is surely one that we should always keep before us.

APPENDIX—
PERPETRATORS AND ACCOMPLICES

What does a perfect group of followers do? It doesn't think, and it doesn't feel any more—it follows. - Victor Klemperer[1223]

All "*Führers*" were incurable phantoms. They robbed us of our freedom, first inwardly, then outwardly. But their existence was possible because so many people no longer wanted to be free, to be responsible for themselves. - Karl Jaspers[1224]

But the actual murderers and those who sent them out had accomplices. German scholarship provided the ideas and techniques which led to and justified this unparalleled slaughter. - Max Weinrich[1225]

The number of victims of robbers, highwaymen, rapists, gangsters and other criminals at any period of history is negligible compared to the massive numbers of those cheerfully slain in the name of the true religion, just policy, or correct ideology... the crimes of violence committed for selfish, personal motives are historically insignificant compared to those committed *ad majorem gloriam Dei*, out of a self-sacrificing devotion to a flag, a leader, a religious faith or a political conviction." - Arthur Koestler[1226]

Hannah Arendt sub-titled her study of the trial of Adolf Eichmann "A Report on the Banality of Evil." There was no doubting the banal nature of Eichmann's appearance in the Jerusalem courtroom. Balding, suited, in horn-rimmed spectacles, he resembled little more than a stereotype of the travelling salesman for the Vacuum Oil Company he had been prior to seeking an alternative career in the Nazi Party.[1227] On the surface, polite, respectful, and obsequious, his appearance and attitude, however, were parts of a carefully contrived defence strategy. The public had expected a monster; instead they saw everyman.[1228]

"The banality of evil," and what this phrase has come to imply, is intrinsic to any study of Nazi eugenics and its consequences, indeed of Nazism in its entirety. The phrase is often used to inadequately describe evil behaviour that is in all other respects incomprehensible, but it is pertinent to question the meaning Arendt actually ascribed to it. As a figure, Eichmann may well have appeared banal. But his acts were not. In truth, Eichmann was an efficient, rather dull bureaucrat, who, bereft of his SS uniform and consequent limitless powers over life and death, would not even rate a footnote in Holocaust history. Circumstances, rather than talent or ability made him, although there is no doubting his dedication to murder on a gigantic scale when fate placed him in a position to play God. It is at least open to question as to whether he could have ever have been anything other than an enthusiastic administrator, a classic example of the

1223 Victor Klemperer, *The Language of the Third Reich: LTI—Lingua Tertii Imperii—A Philologist's Notebook* (London, Continuum, 2006), p. 228

1224 David S Wyman (ed), *The World Reacts to the Holocaust* (Baltimore: The John Hopkins University Press, 1996), p. 413.

1225 Robert N Proctor, *Racial Hygiene: Medicine Under the Nazis* (Cambridge Massachusetts: Harvard University Press, 1989) p. 222.

1226 http://www.worldofquotes.com/author/Arthur+Koestler/1/index.html (Accessed 29 December 2016).

1227 David Cesarani, *Eichmann: His Life and Crimes* (London: William Heinemann, 2004) p. 23.

1228 Ibid., p. 257.

Schreibtischtäter—the desk murderer, corrupted by limitless power. Accused of killing a boy in the garden of his villa in Budapest in 1944, Eichmann vehemently denied having ever personally killed anybody—indeed, this became a cornerstone of his defence.[1229] Such an accusation was an offence to his "honour". Although it was apparent that this in no way lessened his monstrous crimes, a fact surely evident even to Eichmann himself, it does raise the issue of the designation Arendt chose. What has generated a seemingly endless debate in the more than half-century since she coined the phrase is not her description of Eichmann's acts, but the subtitle of her book, which is not so much "A Report on the Banality of Evil", but rather "A Report on the Banality of the *Evildoer*." Such banality becomes increasingly apparent as the history of the perpetrators of Nazi crimes is studied. In fact, this was quite evident to Arendt herself, for she concluded "that it would have been very comforting indeed to believe that Eichmann was a monster...[but]...the trouble...was precisely that so many were like him, and that the many were neither perverted nor sadistic, that they were, and still are, terribly and terrifyingly normal."[1230] . Of course, much depends on what is understood by "normal" in this context, but what was true of Eichmann was also true of the great majority of those responsible for both the theory of National Socialist racial hygiene and its practice.

If most of these perpetrators of the crimes under consideration in this text were not sadistic monsters, neither were the majority of them true *Schreibtischtäter* like Eichmann, long-distance executioners who were content to issue orders for the killing of millions to other operatives in the field without soiling their own hands. Their guilt lay somewhere in between; they were often directly and personally involved in killing the helpless, and if deriving little or no personal gratification from it, or acting out of what German law describes as "base motives", still finding ample justification for their actions.[1231] In an almost classic display of hubris, Paul Nitsche, psychiatric head of mass murder, and a leading light in the production of the "euthanasia" films already detailed, once commented: "Isn't it wonderful to get rid of all the ballast collecting in the asylums? Now we can perform some real therapy."[1232] What this meant in reality was spelt out at the very commencement of T4 in 1939. The number of psychiatric beds was targeted to be reduced by fifty percent; the average length of a patient's hospitalisation was to be severely reduced, treatment was to become more radical, and long term patients were to be exterminated.[1233] This was to be literally a case of kill or cure.

What follows does not pretend to be a comprehensive lexicon of those accused of perpetrating the crimes previously described. Such a task is beyond the scope of this, or it may be presumed, any other work. Rather it should be viewed as an attempt to pro-

---

1229 Ibid., p. 280.

1230 Hannah Arendt, *Eichmann in Jerusalem: A Report on the Banality of Evil* (New York: Penguin Books, 1977). p. 276.

1231 As Henry Friedlander comments, the term *Schreibtischtäter* is usually taken to mean those bureaucrats who never saw any of the victims their actions affected, but this was certainly not true of the managers of T4. [Henry Friedlander, *The Origins of Nazi Genocide: From Euthanasia to the Final Solution* (Chapel Hill: University of North Carolina Press, 1995), p. 194].

1232 http://www.irren-offensive.de/kleespeech.htm (Accessed 10 January 2007).

1233 Götz Aly, Peter Chroust, Christian Pross, *Cleansing the Fatherland: Nazi Medicine and Racial Hygiene* (Baltimore: The Johns Hopkins University Press, Baltimore, 1994), p. 39.

vide brief biographical details of a representative cross-section of those accused of participating or impacting upon, directly or indirectly, a gigantic criminal enterprise, and, where known, their eventual fate. Their number includes those who designed and implemented the programme and those who executed it in killing centres and extermination camps. Here there professors and doctors, administrators and policemen, technicians and nurses, and others who were employed in the most mundane of occupations prior to their involvement with T4. Some were simply murderers, a few of whom were undoubtedly motivated by the pleasure they derived from killing, but the majority participated for reasons it is still difficult, if not impossible, to truly comprehend. Examining the education, political and otherwise, careers, ambitions, influences, and ultimate destiny of these individuals provides not only a portrait of who they were, what they did, and how they were, or were not, brought to account for their actions, but also presents a vignette of their interlocking relationships and mutual devotion to eugenic goals. Hopefully, it also imparts some indication of the size and scope of this iniquitous undertaking. Complete information about every individual included is by no means easily available or accessible. In some cases very little is known. But what is known paints a frightening picture.

The categories that follow, of (a) theorists, (b) technocrats, (c) members of the medical profession, and (d) operatives, are not intended to be definitive, but rather illustrative. It is difficult in some cases to precisely allocate an individual to one group rather than another, since there was frequently an overlapping of duties and responsibilities. However, these designations attempt to generally distinguish between (a) the pioneers of German racial hygiene and eugenics; (b) the higher ranking and administrative staff of T4; (c) the doctors and nurses who manned senior positions in T4, in most cases for at least some of the time at one or more killing centres; and (d) the rank-and-file operatives, those at the "sharp end" of "euthanasia" and *Aktion Reinhard,* who were responsible for administering the instructions that filtered down to them from above.

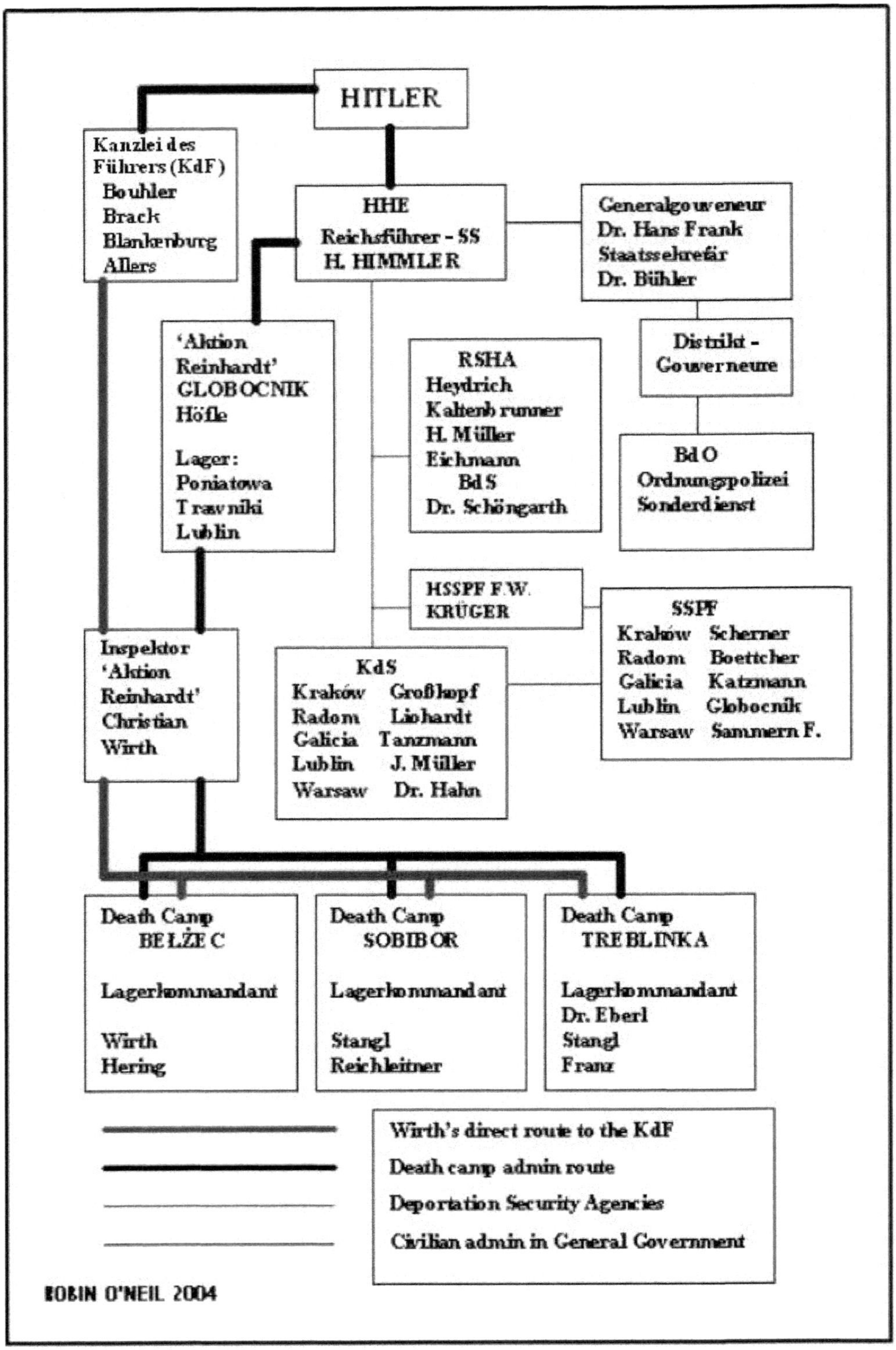
HITLER
Kanzlei des Führers (KdF)
Bouhler
Brack
Blankenburg
Allers
HHE
Reichsführer - SS
H. HIMMLER
Generalgouveneur
Dr. Hans Frank
Staatssekretär
Dr. Bühler
'Aktion Reinhardt'
GLOBOCNIK
Höfle
Lager:
Poniatowa
Trawniki
Lublin
RSHA
Heydrich
Kaltenbrunner
H. Müller
Eichmann
BdS
Dr. Schöngarth
Distrikt - Gouverneure
BdO
Ordnungspolizei
Sonderdienst
HSSPF F.W. KRÜGER
SSPF
Kraków Scherner
Radom Boettcher
Galicia Katzmann
Lublin Globocnik
Warsaw Sammern F.
Inspektor 'Aktion Reinhardt' Christian Wirth
KdS
Kraków Großkopf
Radom Liphardt
Galicia Tanzmann
Lublin J. Müller
Warsaw Dr. Hahn
Death Camp BEŁŻEC
Lagerkommandant
Wirth
Hering
Death Camp SOBIBOR
Lagerkommandant
Stangl
Reichleitner
Death Camp TREBLINKA
Lagerkommandant
Dr. Eberl
Stangl
Franz
Wirth's direct route to the KdF
Death camp admin route
Deportation Security Agencies
Civilian admin in General Government
ROBIN O'NEIL 2004

Theorists

Eugen Fischer

Source: Wikimedia Commons / Bundesarchiv, Bild 183-1998-0817-502 / CC-BY-SA 3.0 https://creativecommons.org/licenses/by-sa/3.0/de/deed.en

Eugen Fischer (1874–1967)[1234] was born in Karlsruhe. Together with Fritz Lenz (q.v.) and Wilhelm Schallmayer (q.v.), he studied under the influential zoologist August Weismann in Freiburg.[1235] In 1900 he attained the status of *Privatdozent*, the approximate equivalent of an associate professorship, and in 1912 was appointed ao professor, before in 1918 taking up the chair in anatomy at Freiburg University. His subsequent academic career has been described above.

Fischer was a conservative nationalist, and although he did not join the Nazi party until 1940, he was a spiritual National Socialist long before then, endorsing Nazism for realizing (in his opinion) that it was the qualities of race that were responsible for a nation's culture.[1236] He served as a judge in the Berlin Higher Hereditary Health Court, and was considered an expert witness concerning the issue of paternity in civil lawsuits.[1237] He was also one of those involved in assessing the racial characteristics of the so-called *Rheinlandbastarde*.[1238] He made his views clear in a speech of 20 June 1939:

> When a people wants, somehow or other, to preserve its own nature, it must reject alien racial elements, and when these have already insinuated themselves, it must suppress them and eliminate them. The Jew is such an alien and, therefore, when he wants to insinuate himself, he must be warded off. This is self-defence. In saying this, I do not characterize every Jew as inferior, as Negroes are, and I do not underestimate the greatest enemy with whom we have to fight. But I reject Jewry with every means in my power, and without reserve, in order to preserve the hereditary endowment of my people.[1239]

Fischer retired from the Kaiser Wilhelm Institute in 1942; two years later that establishment's name was changed to the "Eugen Fischer Institute" in his honour.[1240] In

1234 Ernst Klee, *Das Personenlexikon zum Dritten Reich–Wer war was vor und nach 1945* (Frankfurt am Main: S. Fischer, 2005) p. 151.

1235 Proctor, *Racial Hygiene*, p. 33.

1236 Fischer had also been a great admirer of Mussolini, considering him, until Hitler came along, the only politician who was likely to "really carry out eugenic measures". [Hans-Walter Schmuhl, *The Kaiser Wilhelm Institute for Anthropology, Human Heredity, and Eugenics, 1927–1945: Crossing Boundaries* (New York: Springer-Verlag, 2008), p. 116].

1237 Proctor, *Racial Hygiene*, p. 40–41.

1238 Ibid., p. 113.

1239 Benno Müller-Hill, *Murderous Science: Elimination by Scientific Selection of Jews, Gypsies, and Others in Germany 1933–1945* (Woodbury: Cold Spring Harbor Laboratory Press, 1998), p. 39.

1240 Richard Weikart, *From Darwin to Hitler. Evolutionary Ethics, Eugenics, and Racism in Germany* (New York: Palgrave MacMillan, 2004) p. 226. The Kaiser Wilhelm Society, founded in 1911, was the umbrella body for the scientific institutes and other organizations operating under its authority. After the end

the same year Fischer was awarded the *Adlerschild des Deutschen Reiches*[1241] in recognition of his eminence as "the founder of human genetics".[1242] Otmar Freiherr von Verschuer (q.v.) recognized Fischer's contribution as "groundbreaking for the scientific basis of the hereditary and racial cultivation of the National Socialist state."[1243] Fischer's beliefs can be summarised in his statement: "All human traits—normal or pathological, physical or mental—are shaped by hereditary factors."[1244] After the war Fischer continued to lecture on anthropology and other subjects, and to contribute to academic publications.[1245] Despite his Nazi associations, in 1952 Fischer was made an honorary member of the German Anthropological Society.[1246]

Hans Friedrich Karl Günther (1891–1968) was born in Freiburg, the son of a musician. He studied comparative linguistics at Freiburg University, obtaining his doctorate in 1914. At the outbreak of the First World War he enlisted in the army, but due to sickness was quickly hospitalized and then discharged. Thereafter he served with the Red Cross.

After the war, Günther went to Dresden, where he became a teacher, and in 1919 produced his first book, a work entitled "*Ritter, Tod und Teufel*" ("*The Knight, Death and the Devil: The Heroic Idea*"), a title taken from an engraving by Albrecht Dürer.[1247] In 1920 the Nazi publisher Julius Friedrich Lehmann[1248] asked Günther to write a book about the German people from a racial-biological point of view. The result, *Rassenkunde des deutschen Volkes* ("*Racial Biology of the German People*"), appeared in 1922. By 1935 it had reached its 16th edition, and 91,000 copies had been printed.

In an article entitled "Hatred" published in 1921, Günther dismissed the idea of an all-embracing love of mankind. Rather one should extend love to those who were members of the racial community, but hatred towards all those who were not.[1249] This was a message that Nazism readily embraced. Günther claimed that the Nordic was the preeminent race, establishing valuable cultures everywhere. The antithesis of the Nordic was the Jew, "a thing of ferment and disturbance, a wedge driven by Asia into the European structure." Günther proposed that the Nordic peoples should unite to secure glob-

of the Second World War the organization was renamed The Max Planck Society. The Institute of Anthropology, Human Heredity, and Eugenics ceased to exist in 1945.

1241 This medal, a leftover from Weimar days, was awarded to outstanding contributors to the arts, culture, the humanities and natural sciences, and economics. Only 38 were awarded by the Nazi regime.

1242 Proctor, *Racial Hygiene*, p. 345, note 60.

1243 Michael Wildt, *An Uncompromising Generation—The Nazi Leadership of the Reich Security Main Office* (Madison: University of Wisconsin Press, 2009) p. 394.

1244 Proctor, *Racial Hygiene*, p. 286.

1245 Ibid., p. 300.

1246 Roderick Stackelberg, *The Routledge Companion to Nazi Germany* (London: Routledge, 2007)p. 194.

1247 It is also the title of a fictionalised biography of Hermann Göring by Ella Leffland.

1248 As Germany's leading publisher of medical works, Swiss born Lehmann, who joined the NSDAP in 1920, was hugely influential in the burgeoning racial hygiene/eugenic movement, so much so that in 1934 he became the first member of the Nazi party to receive the Golden Medal of Honour, aka the "Golden Party Badge" *[Goldene Ehrenzeichen]*. (Proctor, *Racial Hygiene*, pp. 26–27). In the same year he was awarded the Nazi's highest scientific award, the *Adlerschild des Deutschen Reiches*. Lehmann died in 1935. (Klee, *Das Personenlexikon zum Dritten Reich*, p. 362).

1249 Weikart, *From Darwin to Hitler*, p. 222.

al dominance, another theory that was hardly likely to find disfavour with Hitler and his followers.

In 1923 Günther moved to Scandinavia, where he was recognized as being among the most prominent theorists in matters of racial biology. By 1930, however, he had returned to Germany to take up the chair in Social Anthropology at Jena University, a position that had been specifically created for him. Hitler, Göring, Frick, and other prominent Nazis attended his inaugural lecture; that same evening Nazi students held a torchlight procession in his honour.[1250] Gunther joined the Nazi party in 1932 and went on to receive a number of honours from the regime, including the Goethe Medal for Art and Science. In 1935 he was appointed Director of the Institute for Racial Studies, People Biology, and Rural Sociology in Berlin Dahlem (*Anstalt für Rassenkunde, Völkerbiologie und ländliche Soziologie*).

By now nicknamed *Rassen*-Günther (Racial-Gunther), he remained in Jena until taking up a professorship at Freiburg University in 1939, where he continued to teach until 1944. At the end of the war Günther was removed from his professorship and was imprisoned by the French occupying authorities for three years. On his release he continued to write and issued new editions of his books. He never relinquished his National Socialist worldview, and continued to promote sterilisation as a solution to mankind's ills. His last work, an autobiography entitled "My Impression of Adolf Hitler" ("*Mein Eindruck von Adolf Hitler*"), described as "a good example of repression and selective perception", was published posthumously in 1969.[1251]

Ernst Haeckel
Source: Wikimedia Commons / Public Domain

Ernst Heinrich Philipp August Haeckel (1834–1919) was born in Potsdam. He studied medicine in Würzburg, Berlin, and Vienna, obtaining his doctorate in 1857 and his licence to practice shortly thereafter. After abandoning a career as a physician, he obtained a degree in zoology and eventually became professor of comparative anatomy and director of the Zoological Institute at the University of Jena, a position he occupied from1862 to 1909. In 1865 he was appointed to the university's zoology chair. This was a position which had been specially established for him, in the same way that Günther's professorship in Social Anthropology was to be at the same university 65 years later.[1252]

Haeckel was a prolific researcher, illustrator and author, producing no less than 42 books and innumerable other publications by the time of his sixtieth birthday. His contribution towards nineteenth-century perceptions of Darwinism have already been considered above. Although today regarded as highly dubious, Haeckel's conclusions regarding eu-

1250 Gerwin Strobl, *The Bard of Eugenics: Shakespeare and Racial Activism in the Third Reich* (Journal of Contemporary History, Vol. 34, no.3, 1999), p. 326.
1251 http://www.vethist.idehist.uu.se/Newsletter_pdf/NewsL_38.pdf (Accessed 29 September 2008).
1252 http://www.gennet.org/facts/haeckel.html (Accessed 4 October 2008).

genic matters were enormously influential in their day, particularly with men who were themselves to become important contributors to National Socialist racial hygienic theory. Ironically, his own views were considered sufficiently unacceptable to the Nazis for them to have his books banned on the grounds that they incorporated an unacceptable element of "democratic-liberalistic spirit."[1253]

Fritz Lenz (1949)
Source: Archiv der Max-Planck-Gesellschaft, Berlin-Dahlem

Fritz Lenz (1887–1976)[1254] enjoyed a long and successful career, as already described. His fundamental philosophy can be summarized in a quotation from his doctoral thesis of 1912: "The only way to eliminate genetic illness is through the negative selection of the afflicted families."[1255] It is not difficult to understand why such pronouncements made him the leading light of what came to be called "scientific racism", and the Nazis' favourite eugenic theoretician, even though he only became a member of the NSDAP in 1937, and the Nazi Physicians' League in 1940. To the end of his life he continued to believe that the eugenic hypothesis of racial differences had been scientifically proven, and that the Holocaust was merely another example of social Darwinism in practice. Hailed on the occasion of his eighty-fifth birthday by the neo-Nazi journal *Neue Anthropologie* as the grandfather of German racial hygiene,[1256] in 1951 he had written: "

> I also have sympathy for the chimpanzees and gorillas, and it distresses me that they are expected to become extinct like so many other animal species and native peoples. To me the fate of millions of Jews is also very painful; but all this does not permit us to regard questions of biology differently or to look at them other than purely objectively.[1257]

1253 Proctor, *Racial Hygiene*, p. 340, note 19.

1254 Klee, *Das Personenlexikon zum Dritten Reich*, p. 366.

1255 Proctor, *Racial Hygiene*, p. 49.

1256 Ibid., p. 48. As opposed to August Forel, who had been hailed as the grandfather of German eugenics. Well, we all had two grandfathers.

1257 Klee, *Das Personenlexikon zum Dritten Reich*, p. 367.

Alfred Ploetz

Source: http://www.kumc.edu/dc/pc/ploetz.jpg
Courtesy of the Clendening History of Medicine Library, University of Kansas Medical Center

Alfred Ploetz (1860–1940)[1258] was born in Swinemünde, Pomerania (now Swinoujscie, Poland), but grew up and attended school in Breslau (now Wroclaw, Poland) before enrolling at the University of Zurich. Initially a student of political economics, after a six month sojourn in the United States he returned to Zurich and medicine, qualifying as a doctor in 1890. He then went back to America for a short period, opening a medical practice in Springfield, Massachusetts, but found life there a disappointment and quickly returned to Europe.[1259] His subsequent career has been outlined above. He joined the Nazi party in 1937, and on his death three years later was eulogized by Otmar Freiherr von Verschuer (q.v.), who praised his "inner sympathy and enthusiasm [for] the National Socialist Movement".

Rudolf Ramm

Source: Wikimedia Commons / Public Domain

Rudolf Ramm (1887–1945)[1260] was born in Dortmund. He studied medicine and pharmacy in Strasbourg, Munich, and Cologne, in due course obtaining his licence to practice as a physician and pharmacist. In 1921 he began working as a general practitioner and in 1929 was appointed *Gau* Chairman of the National Socialist Doctor's League of the Rheinpfalz. In the same year he became a member of the town council of Pirmasens, going on to officiate as mayor of the town between 1934 and 1937. In 1930 he joined the NSDAP as well as the SS, and in 1932 was elected to the *Reichstag* as the party's representative. Between 1932 and 1936 he was Nazi *Kreisleiter* in Pirmasens, and from 1933 became a member of both Leonard Conti's (q.v.) staff and the medical faculty of the University of Berlin.

Ramm became an influential member of the Nazi medical elite, holding a number of important positions before eventually taking on responsibility for maintaining the standards of German medical education.[1261] His idea of medical ethics can be gathered from some of his published comments; enforced sterilisation of the "valueless" was beneficial, "euthanasia" merciful for the incurably sick and insane, and moreover "an obligation to the *Volk*."[1262] A true physician "must not only be a Party member on the

---

1258 Ibid., p. 466.

1259 Mark B Adams (ed), *The Wellborn Science: Eugenics in Germany, France, Brazil, and Russia* (New York: Oxford University Press, 1990), pp. 14–16.

1260 Klee, *Das Personenlexikon zum Dritten Reich*, p. 478.

1261 Ulf Schmidt, *Karl Brandt: The Nazi Doctor—Medicine and Power in the Third Reich* (London: Hambledon Continuum, 2007), pp. 176–177.

1262 Robert Jay Lifton, *The Nazi Doctors: Medical Killing and the Psychology of Genocide* (London: Papermac, 1990) , p. 30.

outside, but rather must be convinced in his heart of hearts of the biological laws that form the centre of his life." In addition to accepting the inviolability of these laws, Ramm wrote, the doctor's duty was to continually preach them to unbelievers.[1263] Jews had spread "spiritually poisonous ideas" and had destroyed the "genetic life" of the German nation.[1264] In 1942 he informed his students that: "The faster and more thoroughly [the Jewish question] is solved, the more rapid (and better) will be the reformation of the European continent upon a racial basis, and the happier will be its future."[1265]

Ramm died on 9 August 1945, thereby escaping probable arrest and trial for a career largely devoted to endorsing Nazi eugenic and racial policies.

Ernst Rüdin (1874–1952)[1266] psychiatrist, geneticist, eugenicist, and brother of Alfred Ploetz's (q.v.) first wife, was born in St Gallen, Switzerland. Rüdin began his career in psychiatry at Emil Kraepelin's clinic in Munich.[1267] In1931 he was appointed managing director of the Kaiser Wilhelm Institute for Psychiatric Research, which Kraepelin had founded in that city as "The German Research Institute of Psychiatry in Munich" in 1917.[1268] At the time of the foundation of the Institute, Rüdin had been made head of the department dealing with psychiatric research on heredity.

After 1933 the National Socialist government and party endorsed Rüdin's work by supplying financial and manpower support. Considered the spiritual father of the "Law for the Prevention of Genetically Diseased Offspring,"[1269] in 1934 he wrote the official commentary on that legislation, as well as subsequently serving in the Hereditary Health Courts.[1270] Rüdin was unstinting in his praise of Nazi policies. In January 1943 he not only commended the sterilisation law and the Nuremberg racial laws, but acclaimed the Nazi's "combat against parasitic foreign-blooded races, like the Jews and Gypsies."[1271] He was another to be awarded the *Adlerschild des Deutschen Reiches*, in his case in 1944 for his efforts as a "path breaker in the field of human hereditary care."[1272] He dabbled in providing special research facilities at his institute for the *Ahnenerbe* department of the SS, was a supporter of aviation medical experiments, and was not averse to utilizing "euthanasia" victims for the purpose of brain research.[1273]

1263 Ibid.p. 33.

1264 Proctor, *Racial Hygiene*, p. 90.

1265 Ibid.p. 175.

1266 Klee, *Das Personenlexikon zum Dritten Reich*, p. 513.

1267 In 1908, Kraepelin had deprecated modern social welfare initiatives that were keeping alive weak and sick individuals who in earlier times would have long since perished. (Weikart, *From Darwin to Hitler*, p. 85).

1268 The institution was incorporated into the Kaiser Wilhelm Institute in 1924. In 1954 it became known as the Max Planck Institute of Psychiatry. [[http://www.mpipsykl.mpg.de/en/institute/history/inde x.html (Accessed 3 November 2008)].

1269 Dick de Mildt, *In the Name of the People: Perpetrators of Genocide in the Reflection of Their Post-War Prosecution in West Germany. The 'Euthanasia' and 'Aktion Reinhard' Trial Cases* (The Hague: Martinus Nijhoff Publishers, 1996), p. 358, note 43.

1270 Weikart, *From Darwin to Hitler*, p. 225.

1271 Ibid.p. 226.

1272 Proctor, *Racial Hygiene*, p. 345, note 60.

1273 Paul Julian Weindling, *Nazi Medicine and the Nuremberg Trials—From Medical War Crimes to Informed Consent* (Basingstoke: Palgrave Macmillan, 2006), p. 237.

Following a denunciation by the psychiatrist Theo Lang, who had defected to Switzerland in around 1941, Rüdin was arrested in Munich by the Americans in December 1945, but released eleven months later. He was considered a candidate for prosecution at the Nuremberg Medical Trial, but in the event his contribution to Nazi racial hygiene policies and practice was not considered worthy of legal proceedings. Apart from his age and alleged health problems, the issue of the sterilisation of German citizens, in which Rüdin had played such a prominent role, was considered too sensitive to form part of the main thrust of the trial.[1274]

Even today, Rüdin's research on the genetics of schizophrenia, which established a theoretical basis for his eugenics work, continues to be cited in matters of psychiatric genetics without reference to his involvement in eugenics.

Falk Ruttke (1894–1955) was a lawyer born in Halle. A former *Freikorps* recruit, he joined the Nazi party in 1932 and the SS in 1933.[1275] As a member of the Committee for Population and Race Policies at the Ministry of the Interior he co-authored a commentary on the Sterilisation Law together with Arthur Gütt (q.v) and Ernst Rüdin (q.v.), a matter which brought him into conflict with Gerhard Wagner (q.v.). Wagner was dissatisfied with the commentary on the grounds that, in essence, it did not adequately reflect Nazi racism, a dispute that resulted in some amendments to the Sterilisation Law in 1938, but ultimately was resolved by the introduction of "euthanasia".[1276]

Throughout the 1930s Ruttke was a prominent advocate of National Socialist ideology, appearing at a variety of international conferences and seminars to espouse the virtues not only of the Sterilisation Law, but other legislation as disparate as The Law to Reduce Unemployment, The Decree for the Granting of Marriage Loans, the Law for the New Formation of the German Farmerstock, and their ilk. Germany's progress, indeed its continued existence, Ruttke proclaimed, was based upon such new legislation continuing to promote a wholly racist vision of social hygiene. At a 1936 international gathering of eugenicists, he declared:

> Hereditary traits are not only given to us, but carry a moral obligation to pursue the highest biological development possible. This not only calls for work on behalf of the *Volk*, into which the individual is born and with which he is connected through blood ties, but also on behalf of all humankind. This is thus extremely important work toward the maintenance of peace.[1277]

Ruttke continued his involvement in matters of racial law throughout the Nazi era.

1274 Ibid.p. 41.

1275 Klee, *Das Personenlexikon zum Dritten Reich*, p. 516.

1276 Proctor, *Racial Hygiene*, pp. 115–117.

1277 Stefan Kühl, *The Nazi Connection: Eugenics, American Racism, and German National Socialism* (New York: Oxford University Press, 2002), pp. 28–31.

Wilhelm Schallmayer
Source: Wikimedia Commons / Public Domain

Wilhelm Schallmayer (1857–1919) was born in Mindelheim, a small town in Bavaria. After a year's voluntary army service, he attended the universities of Würzburg, Munich, and Leipzig, finally settling on a career in medicine after having flirted with a variety of other subjects. In 1883, Schallmayer became a licensed medical practitioner, eventually commencing to work in that capacity at Kaufbeuren in 1887. By then he was a prototypical eugenicist, for whom the all important question was "whether the physical development of the human race, upon which the continuation of all cultural progress depends, is presently advancing or declining." The answer was obvious to him:

> Even if medical technology grew to such an extent that malfunctioning human organs could be safely replaced by healthy human, animal, or laboratory-produced ones, the following generations would not be more efficient, rather just the opposite: the more advanced therapeutic medicine becomes, the more succeeding generations will have need of it. Therapeutic medicine affects the improvement of national health in about the same way as poor-relief contributes to the improvement of national welfare. Both encourage an increase in the dependent [population].... Medicine, insofar as it aims at treatment rather than prevention, contributes nothing to the gradual advance of human productivity and human happiness. It aids the individual but at the expense of the human race.

Since he considered his work as general practitioner useless, if not actually detrimental to the improvement of the *Volk*, he decided to leave Kaufbeuren and specialize in urology and gynaecology instead. After year's training in Vienna, Leipzig, and Dresden, he began a lucrative practice in Düsseldorf, where apart from a two year break during which he worked as a ship's doctor, he remained until 1897. In that year he settled in Munich, an independent scholar intent upon preaching the virtues of eugenics to his fellow citizens. But it was winning the Krupp Prize competition in 1903 which brought him to national prominence. Much of the rest of Schallmayer's life was devoted to defending his thesis, for as has been demonstrated, there was little unanimity of opinion among eugenicists. Of Schallmayer's importance as the intellectual patriarch of German racial hygiene, and thus his influence on the course of twentieth century history, however, there can be no doubt.[1278]

Reinhold Ralph Ernst Otmar Freiherr von Verschuer (1896–1969)[1279] was born in Richelsdorfer Hütte, a small town near Fulda in central Germany. In August 1914 he volunteered for the army, and served with distinction in the Great War, being thrice wounded and winning the Iron Cross first and second class, among other medals. In 1919 he began to study medicine in Marburg, and in 1920 became involved in the Kapp *Putsch*, a failed right-wing attempt to overthrow the Weimar government of Friedrich Ebert. Taken together with his membership of the Thule Society, a forerunner of the NSDAP, he thus evinced his political inclinations at an early age. He continued his medi-

---

1278 Sheila Faith Weiss, *Race Hygiene and National Efficiency: The Eugenics of Wilhelm Schallmayer* (Berkeley: University of California Press, 1987), p. 38ff.

1279 Klee, *Das Personenlexikon zum Dritten Reich*, p. 639.

cal studies in Hamburg, Munich, and Freiburg and in 1923 was awarded his doctorate. From 1923 to 1927 he worked as an assistant physician at the medical clinic of the University of Tübingen, where he began his research into the hereditary and environmental characteristics of twins.[1280] This was a subject which was to obsess him for the rest of his days. In Verschuer's opinion, twin research was the "sovereign method for genetic research in humans."[1281]

In October 1927 Verschuer became a lecturer at the University of Berlin and was also appointed a department head at the Kaiser Wilhelm Institute of Anthropology, Human Heredity, and Eugenics (*Kaiser-Wilhelm-Institut für Anthropologie, menschliche Erblehre, und Eugenik*), becoming director of the Institute in 1942. In July 1932 he was involved in discussions with the Prussian authorities concerning the introduction of a sterilisation law, a matter soon superseded by the advent of Hitler's government. Verschuer remained a staunch supporter of racial hygiene, and approved of the Nuremberg Laws, although he did not join the NSDAP until 1940, and only became a member of the Nazi Medical Association in 1942. In 1934 he became the first editor of *'Der Erbarzt'* ('The Genetic Doctor'), a journal devoted to the discussion of sterilisation and related issues. "Every doctor must be a genetic doctor", Verschuer declared.[1282] In 1935 he was appointed director of the Institute for Genetic Biology and Racial Hygiene (*Institut für Erbbiologie und Rassenhygiene*) in Frankfurt, and held a number of other positions in organisations and institutions dedicated to the pursuit of Nazi eugenics.

Verschuer's most infamous pupil was Josef Mengele, who was awarded his PhD under Verschuer's tutelage. Mengele shared his mentor's enthusiasm for research into twins, as was evident from Verschuer's report on his protégé: "My assistant, Dr [Josef] Mengele (M.D., PhD) has joined me in this branch of research. He is presently employed as *Hauptsturmführer* and camp physician in the concentration camp at Auschwitz." Mengele supplied "material" from murdered Auschwitz inmates to Verschuer at the Kaiser Wilhelm Institute.[1283]

Although described by Allied investigators as "a man who remained in the background, but in spite of all, [was] one of those principally responsible for the theory of racial hygiene and executions by gas," in 1946 a denazification tribunal deemed Verschuer a "*Mitläufer*" ("collaborator" or "fellow traveller") rather than a criminal, and fined him 600 *Reichsmarks*. Apart from a brief spell of house arrest, this represented the complete extent of his encounter with the judicial system insofar as his eugenic activities were concerned.[1284] He subsequently tried to have his branch of the Kaiser Wilhelm Institute re-established in Frankfurt, without success. Instead some recognised him for what he was, the head of the commission responsible for such matters writing:

> Verschuer should not be considered a collaborator, but one of the most dangerous activists of the Third *Reich*. An objective judgment of the investigative committee must recognize this, and thereby take action to guarantee that this man does not come into contact with German youth as a university teacher, or with the broader population as a scientist in the fields of genetics and anthropology.

1280 http://www.bautz.de/bbkl/v/verschuer_o.shtml (Accessed 04 November 2008).
1281 Proctor, *Racial Hygiene*, p. 43.
1282 Ibid.pp. 104–105.
1283 Ibid.p. 44.
1284 Weindling, *Nazi Medicine and the Nuremberg Trials*, p. 239.

Notwithstanding these observations, in 1951 Verschuer was appointed professor of human genetics at the University of Münster, and went on to hold several other important academic positions, his past apparently both forgiven and forgotten. After a serious car accident in September 1968, Verschuer fell into a coma. He died a year later without regaining consciousness. His obituaries made no mention of his extensive and influential participation in Nazi eugenic affairs.[1285]

Technocrats

Dieter Allers

Source: Ghetto Fighters House Museum, Israel, Photo Archive

August Dietrich ("Dieter") Allers (1910–1975)[1286] was born in Kiel and studied law at the universities of Jena and Berlin. As a student he joined the Nazi party in 1932 and the SA in 1934,[1287] subsequently becoming a civil servant in the Prussian government in 1937. He was conscripted on the outbreak of war, and as a non-commissioned officer, was sent to Poland to train recruits. As previously described, following a chance encounter between his mother and Werner Blankenburg (q.v.), Allers was recruited to T4.

He rapidly became the organization's chief legal expert, heading its financial bureau the *Zentralverrechnungsstelle Heil- und Pflegeanstalten* (Central Clearing Office for Mental Institutions) and succeeding T4's first legal advisor, Dr Gerhard Bohne (q.v.), who had resigned his position in June 1940 in a dispute over the corruptibility of his colleagues. In spring 1941, Allers was promoted to the position of general manager of T4, effectively becoming principal assistant to Viktor Brack (q.v.), his deputy, Werner Blankenburg (q.v.), and T4's chief medical advisor, Paul Nitsche (q.v.). Aller's office coordinated efforts to disguise the killings, which involved deliberately misinforming both the relatives of the victims as well as the agencies involved in committing patients and paying for their care.[1288] Allers also succeeded Bohne as head of the *Gemeinnützige Stiftung für Anstaltspflege* (Charitable Foundation for Institutional Care), usually referred as the *Stiftung* (Foundation), which was the T4 front organization responsible for all personnel affairs and financial matters. Although Allers supervised activities, acting in effect as managing director of T4, he was not directly responsible for the day to day administration, leaving this to Friedrich Tillmann (q.v.).[1289] As part of his duties, Allers visited all six of the main "euthanasia" killing centres on numerous occasions, ensuring the smooth running of the operation. Following the reorganisation of the killing in August 1941, he was a member of the T4 *Organisation Todt* Russian expedition the following winter.

1285 Proctor, *Racial Hygiene*, pp. 307–308.

1286 Klee, *Das Personenlexikon zum Dritten Reich*, p. 12.

1287 Friedlander, *The Origins of Nazi Genocide*, p. 70.

1288 Ibid., p. 71.

1289 Ibid.

That Allers was immersed in the business of murder is evident from a letter he wrote in 1943 to Rudolf Lonauer (q.v.), the physician-in-charge at Hartheim, in which he stated that Herbert Linden (q.v.) wanted to transfer a "mentally ill" Russian named Boris Mirkolo to Hartheim and that Lonauer "should have no difficulty understanding the purpose." It seems unlikely that Mirkolo, who spoke German, actually suffered from any mental illness. At Hartheim he was shot and then, since he was not yet dead, dragged into the gas chamber and gassed.[1290] Linden's reason for wanting Mirkolo murdered is unknown.

It is certain that Allers visited the *Aktion Reinhard* extermination camps, although the extent of his involvement in that operation has not been clarified. What is known is that he dispatched Adolf Kaufmann (q.v.) to Belzec to collect sealed crates, probably containing gold, jewellery or banknotes, and take them directly to Berlin.[1291] He remained manager of the "euthanasia" project until May 1944, when together with other T4 members he was posted to Trieste [replacing the shot Christian Wirth[1292] (q.v.)], allegedly to conduct anti-partisan activities. In fact, it has been established that his unit was concerned with the rounding up of Italian Jews for extermination.[1293]

Thirty years after these events, Allers remained an unrepentant Nazi. On "euthanasia":

> Well, as far as I myself was concerned, the idea of euthanasia was not new to me; I had read quite a bit about it. Good heavens, it's been discussed, and on the cards, for centuries. What was intended at the time has been completely distorted since[1294]...People have completely misunderstood: now it is constantly being misinterpreted. Just look at the world now: don't you think something very much like this will have to happen?[1295]

On murdering Jews: "Originally what they wanted to do was put into practice an old Polish plan...one third to be killed; one third resettled somewhere; and one-third to be allowed to assimilate."[1296] On that reasonable basis "only" one million Polish Jews would have been murdered, rather than the three million eventual victims.

Allers was arrested by the British military in August 1945, and remained in various camps until his release in February 1947, at which time he became the manager of a mining company. He was re-arrested by the Americans in April 1948, and handed over to the German authorities. It would appear his new captors either had no knowledge of Aller's activities, or alternatively possessed no desire to delve into them too deeply, for he was released again in September 1949. The investigation into his participation in the "euthanasia" programme was discontinued in May 1950. Allers now entered into the law practice of his friend Johannes Plöger (former legal counsellor to Victor Brack), and in November 1951 became legal advisor to the German Shipyard in Hamburg. He continued to be involved in right-wing politics and maintained contact with his T4 *Alte*

1290 Ibid., p. 165.

1291 http://www.jewishgen.org/Yizkor/belzec1/bel120.html#9r (Accessed 13 April 2008).

1292 Gitta Sereny, *Into That Darkness—From Mercy Killing to Mass Murder* (London: Pimlico, 1995), p. 88.

1293 In 1975, a trial against Allers was commenced by the Italian authorities in Trieste. Allers died in March of that year before a verdict could be reached (de Mildt, *In the Name of the People*, p. 344, note 13).

1294 Sereny, *Into That Darkness*, pp. 79–80.

1295 Ibid., p. 57.

1296 Ibid., p. 89.

Kameraden, as well as running "*Die Stille Hilfe*" ("Silent Help"), the secret organization set up to aid ex-SS members.[1297] In 1954 he wrote to Reinhold Vorberg (q.v.) in confident mood:

> Take a look around and think back to the time around 1947. At that time, all of us envisioned a miserable personal future for ourselves. Today most of us have again become something, and it is my opinion we should not demand too much. Certainly quite a few people who held totally different positions (before) have landed well on their feet.[1298]

The confidence was misplaced. In August 1962 Allers was arrested once again on suspicion of his involvement in the "euthanasia" programme, but was released on bail in May 1963 to await trial. He lost his job with the German shipyard, but continued with his Hamburg law practice. In October 1963 it was decided to link his case with that of Vorberg, but it was to be April 1967 before the trial of the two opened in Frankfurt am Main, where it was to last twenty months.

The court had no doubt that both men had participated in the killing of tens of thousands of mentally ill patients (Allers' *Aktion Reinhard* activities were unfortunately not investigated in any depth), but were less convinced about their involvement in *Sonderbehandlung* 14f13. In December 1968, Allers was sentenced to eight years' imprisonment, confirmed in October 1972 by the Federal High Court. However, since the term of his pre-trial confinement was deducted from the sentence, Allers was released after the verdict had been read out in court.[1299]

Allers may be regarded as representative of the majority of those who became embroiled first in "euthanasia", and then in *Aktion Reinhard*. Wartime life was undoubtedly more comfortable (and lucrative) in a Berlin office than in a barrack in Poland, and career prospects infinitely rosier in the KdF than in the army. Allers was no idealist committed to the practical application of eugenic theory, nor was he a virulent anti-Semite. Instead, he was a *petit bourgeois* opportunist, who saw the main chance, and seized it. As he himself put it regarding his colleagues, "it wasn't a matter of careful or scientific selection of these people."[1300] In only slightly different circumstances he would have remained a nonentity, just another small-town lawyer. But once he had crossed the genocidal bridge and was able to accept murder as simply a further task to be efficiently accomplished, there were no limits to his ambition, or to his capacity for evil.

Georg Andreae (1888–?)[1301] was born in Göttingen. He held a doctorate in law rather than medicine, but nonetheless controlled the asylums at Wunstorf, Hildesheim, Göttingen, Lüneburg, and Osnabrück from his office in Hannover.[1302] A dedicated Christian, Andreae had been a senior civil servant since 1926. He joined the NSDAP in 1933, and in July of that year became a sponsoring member of the SS. He had done so, he later

1297 Weindling, *Nazi Medicine and the Nuremberg Trials*, p. 306.

1298 Michael S Bryant, *Confronting the "Good Death": Nazi Euthanasia on Trial 1945–1953* (Boulder: University Press of Colorado, 2005), p. 217.

1299 de Mildt, *In the Name of the People*, pp. 86–89.

1300 Ibid., p. 226.

1301 Klee, *Das Personenlexikon zum Dritten Reich*, p. 16.

1302 Michael Burleigh, *Death and Deliverance: "Euthanasia" in Germany c. 1900–1945* (London: Pan Books, 2002), p. 133.

claimed, inspired, "by the hope that the National Socialist movement would finally evolve from its stormy early phase into a more peaceful, less violent development." When he discovered that such aspirations were a pipe dream, he desisted from further participation in party affairs.[1303] However, by that time and the ending of the first phase of "euthanasia", an estimated 1,500 of the 7,000 patients under his jurisdiction had been deported from the Hannover institutions to Hadamar, and death.[1304] Andreae was well aware of their fate.

Andreae appeared in court alongside Ludwig Gessner (q.v.) in Hannover in July 1950, accused of aiding and abetting murder and crimes against humanity.[1305] With little documentary evidence concerning "euthanasia" in the province having survived the war, much reliance was placed upon oral testimony. Andreae declared that he had always been opposed to "euthanasia", which he described as a "monstrous crime". Witnesses described him as an "exemplary civil servant, whose heart beats warmly for his patients whom, as a leading administrative official, he had cared for over several decades."[1306] He had even confronted Werner Heyde (q.v.) in an (unsuccessful) attempt to prevent the implementation of "euthanasia" in Hannoverian institutions.[1307] How could such a decent man be convicted? He wasn't. All charges against him were dismissed.

Hans-Joachim Becker (1909–?)[1308] [nicknamed *'Millionen'* ('Millions')], was born in Kassel, and entered his father's dairy equipment business on leaving school. As economic conditions worsened he had to leave his job and take up employment as an unpaid factory worker. In 1930 he found work in the civil service in Kassel, and became a member of the Nazi party in 1937. He studied higher public and financial administration, passing his examinations in 1940. In June of that year he was called up, but was discharged as medically unfit after just a few days. Shortly thereafter he was transferred to a local government office in Danzig where he was not happy. He applied for relocation to the Ministry of the Interior in Posen or Strasbourg, but instead found himself transferred to the *Stiftung* in Berlin.[1309] He had no idea what that organization was, but in January 1941 found himself employed by T4 as assistant to Allers, whose financial expert he was to become.

It was Becker who was primarily responsible for organizing T4's finances into the extremely profitable enterprise they grew to become, earning him the nick-name of *'Millionen-Becker'*. He was appointed director of the Central Clearing Office for Mental Institutions, which, as Becker himself described it, was essentially "the financial intermediary between the cost carriers and the so-called intermediate institutions and euthanasia facilities, the names and existence of which were to be kept secret."[1310]

1303 de Mildt, *In the Name of the People*, p. 168.

1304 Bryant, *Confronting the "Good Death"*, p. 184.

1305 Ibid., p. 180.

1306 de Mildt, *In the Name of the People*, p. 167–169.

1307 Burleigh, *Death and Deliverance*, p. 134.

1308 Klee, *Das Personenlexikon zum Dritten Reich*, p. 34.

1309 There is a suggestion that Becker's enlistment in T4 came about as a result of the intercession of his cousin, who was married to Herbert Linden. (de Mildt, *In the Name of the People*, p. 345, note 23). See also Aly, Chroust, Pross, *Cleansing the Fatherland*, p. 166 and Friedlander, *The Origins of Nazi Genocide*, p. 74.

1310 Aly, Chroust, Pross, *Cleansing the Fatherland*, pp. 176–177.

Apart from standardizing various regional rates with cost carriers such as local authorities, insurance companies, welfare agencies and pension funds, then paying those costs to the T4 institutions involved,[1311] he also enriched T4 operations to a significant extent by creating a sophisticated system of fraud, whereby the nursing-fees of murdered patients continued to be collected long after they were dead. He performed many other chores in order to generate funds for T4, not the least of which was the plundering of gold dentures from the mouths of victims. These were regularly sent from the killing centres to Becker in Berlin for smelting. It was due in large part to Becker's input that T4 not only became self-financing, but was also able to donate funds to other sympathizing agencies.[1312] He claimed to have collected 6–8 million *Reichsmarks* per annum on average, in one year actually raking in 10 million *Reichsmarks*.[1313]

Becker worked as a personal assistant to Paul Nitsche (q.v.) , and in 1943 was given the responsibility of supervising Hartheim, where in the following year he was in charge of the gassing of 3,228 inmates from Mauthausen and Gusen. Among his final tasks at Hartheim was the dismantling of the gassing apparatus and the reconversion of the institution to a "normal" mental hospital.

At the war's end, Becker avoided arrest, and even found employment with the Americans as a translator in Thuringia. When the American turned that region over to the Soviets in June 1945 and left, Becker left with them, returning to Kassel. In January 1947, the police in Thuringia issued a warrant for his arrest, but the Kassel police chief found "no grounds whatsoever for a police arrest." Nonetheless, shortly thereafter an investigation into Becker's activities was commenced by the Chief Prosecutor's Office in Kassel, but abandoned in May 1950 because of an inability to find " a causal relationship between his activities and the resultant killings."

And there matters rested for the next sixteen years while Becker continued to work for the Americans in Bad Kreuznach. In June 1966 his arrest in connection with his involvement in "euthanasia" was ordered by the Frankfurt court. He was released on bail, but the Americans had decided to dispense with his services, and he remained unemployed for most of the time until his trial commenced on 20 August 1969. Although Becker admitted to much of his T4 career, he sought to minimize his participation in the collection of gold dentures and the killings at Hartheim. However, the court had little difficulty in dismissing his defence, which included a claim that Werner Heyde (q.v.) had shown him the *Führer* authorization, an act that Becker contended had legitimized his actions. Instead, the court ruled that "no human being with normal feelings can consider mass killings as right and as a just cause."

In the opinion of the court, rather than being a fanatical National Socialist, Becker was merely a "weak character". It was considered that without Nazism he would not have become a criminal; on the other hand, without people like him, Nazism could never have realized its horrific crimes. Consequently, he was sentenced to the minimum term of imprisonment the court could impose—10 years for being an accessory to the murder of 24,540 mentally ill patients and 3,228 concentration camp inmates.[1314] He

1311 Ibid., p. 179.
1312 Ibid., p. 177.
1313 Friedlander, *The Origins of Nazi Genocide*, p. 74.
1314 Klee, *Das Personenlexikon zum Dritten Reich*, p. 34.

served three years in prison before entering hospital suffering from "psychological strain." In September 1974, four years after sentencing, Becker's remaining prison term was suspended.[1315]

Otto Friedrich (Fritz) Bernotat (1890–1951)[1316] Born in Mittel-Jodupp, East Prussia (now Czarnowo Srednie, Poland), *Landesrat* Fritz Bernotat was the representative for mental institutions in the state of Hesse. Together with his superior, *Landeshauptmann* Wilhelm Traupel (1891–1946) he had indicated his determination to eliminate "ballast existences" as early as 1936/1937. Having begun his local government career in 1922 he joined the NSDAP in 1928, a career move that saw him rapidly promoted to the post of head of the Hessian department of political affairs, welfare services agency, and the central administration of institutions, among other positions.[1317] In 1931 he became a member of the SS.

Responsible for all Hessian asylums, Bernotat was one of the most brutal and fanatical of Nazi functionaries, even going so far as to arrange for his Jewish mistress to be murdered in Eichberg by a lethal injection.[1318] He would repeatedly tell medical staff: "Why don't you beat patients to death, then you are rid of all of them?" He and Traupel supervised a devastating lowering of the standard of care in Hessian hospitals through the reduction of nursing fees and the appointment of young Nazi doctors who, so one hospital director was told, "were better with the needle."[1319] On their engagement, all recruits to T4 were asked if they belonged to a church. If they did, Bernotat demanded of those he interviewed that they give up their membership, for "they could not serve two masters."[1320] He terrorized his workforce, once stating at an Eichberg staff meeting that if he heard even a hint that a nurse had spoken about conditions at the institution, or said anything at all in public, he would make that person accountable and "stop at nothing."[1321]

In December 1940, Bernotat employed his brother-in-law, Fritz Scherwing, a master mechanic, to assist in the installation of the pipe-work and ventilation system in the gas chamber at Hadamar. Scherwing, clearly an innocent abroad in a cruel and heartless world, denied at his post-war trial of having any knowledge about the true purpose of the work he had undertaken, despite receiving repeated warnings from Bernotat concerning the strict secrecy surrounding his labours, and the presence of a crematorium adjoining the "disinfection room." The court chose to believe Scherwing, and acquitted him for lack of evidence.[1322]

A month prior to the commencement of Schwering's employment (that is while Hadamar was still operating as a "normal" mental hospital) Bernotat had gathered the

---

1315 de Mildt, *In the Name of the People*, pp. 89–95; Aly, Chroust, Pross, *Cleansing the Fatherland*, pp. 176–184.

1316 Klee, *Das Personenlexikon zum Dritten Reich*, p. 43.

1317 http://www.chgs.umn.edu/histories/documentary/hadamar/policy.html (Accessed 12 April 2008).

1318 Comparison has been made between the character of Bernotat and Christian Wirth. (Friedlander, *The Origins of Nazi Genocide*, pp. 203–204).

1319 de Mildt, *In the Name of the People*, p. 51.

1320 Ibid., p. 198.

1321 Bronwyn Rebekah McFarland-Icke, *Nurses in Nazi Germany* (Chichester: Princeton University Press, 1999), p. 229.

1322 de Mildt, *In the Name of the People*, p. 194. Burleigh, *Death and Deliverance*, pp. 144–145.

entire staff together for the purpose of swearing them to absolute secrecy about forthcoming events at the institution. He threatened that any transgressions would be severely punished by detention in a concentration camp, or even execution.[1323] There is some doubt concerning whether he broached the subject of "euthanasia" at that time, but the evacuation of patients, Schwering's activities, and the arrival of T4 staff must have provided a reasonably clear indication of the State's intentions.[1324]

When the war ended Bernotat assumed a false name and lived at Neuhof, near Fulda, where he died a natural death in 1951. He was never brought before a court to account for his participation in the murder of thousands of individuals.[1325]

Werner Blankenburg (1905–1957)[1326] [T4 pseudonym "Brenner", post T4 pseudonym "Bieleke"][1327] was born in Caputh, Brandenburg. A Nazi party and SA member since 1929, he worked as a sales representative before joining the KdF.[1328] Despite becoming a member of the NSDAP at a relatively early date, he had held no previous party post, although he had achieved the rank of SA *Oberführer.* Together with Reinhold Vorberg (q.v.), Blankenburg had been recruited from private business in order to become a senior member of the T4 staff,[1329] in Blankenburg's case to be head of Amt IIa of the KdF, one of the sub-divisions responsible for the implementation of the "euthanasia" operation. His immediate superior, and overall head of Amt II was Viktor Brack (q.v.).[1330] Blankenburg was Brack's permanent deputy and in the latter's absence conducted the affairs of Amt II. He was thus completely immersed in the criminal activities of the KdF in general, and of T4 in particular.

After Brack left the KdF for the front as a *Sturmbannführer* in the SS-Division Prinz Eugen in 1942, Blankenburg succeeded him as day-today manager of T4.[1331] Brack had been corresponding with Himmler concerning the possibility of the use of X-rays for the mass sterilisation of Jews (see chapter 10). When Himmler evinced interest in the proposal, directing that the method be tried in at least one concentration camp, Blankenburg replied to him in August 1942: "As permanent deputy of *Oberführer* Brack, I shall immediately take the necessary measures and get in touch with the Chiefs of the Main Offices for the Concentration Camps." Some 20 months later, Blankenburg again wrote to Himmler explaining that the method had not been successful, and that "operative castration requires not more than 6–7 minutes, and can therefore be performed more reliably and quicker than the castration by X-rays." Blankenburg concluded by assuring

1323 Dire warnings of this kind were commonplace in T4. So far as is known, nobody was executed for violating the secrecy oath. The harshest punishment administered was incarceration in a concentration camp for a few months, admittedly no holiday, but not capital punishment either. (de Mildt, *In the Name of the People,* p. 338, note 69).

1324 de Mildt, *In the Name of the People* p. 205.

1325 http://motlc.wiesenthal.com/site/pp.asp?c=gvKVLcMVIuG&b=395193 (Accessed 12 April 2008).

1326 Klee, *Das Personenlexikon zum Dritten Reich,* p. 52.

1327 de Mildt, *In the Name of the People,* p. 395, note 63.

1328 Friedlander, *The Origins of Nazi Genocide,* p. 69.

1329 Ibid., pp. 192–193.

1330 Ibid., p. 40.

1331 Ibid., p. 69.

Himmler that he (Blankenburg) would soon be able to submit a report on the continuation of this work.[1332]

Blankenburg was as deeply implicated in the murders and theft of *Aktion Reinhard* as all of the other senior T4 personnel. When he visited the *Reinhard* offices in Lublin to temporarily deposit some "important items for the front," the duty supervisor contacted Odilo Globocnik (in charge of the *Aktion*) to obtain the necessary permission. Globocnik abruptly told the supervisor "to mind his own business". A few days later, Blankenburg returned, removed the items and, in loading them onto a motor vehicle, one of the crates was damaged, dropping gold items onto the ground. The supervisor concluded that these items were the result of "Jewish measures".[1333] It is not difficult to surmise where Blankenburg had obtained the contents of the crates. Franz Suchomel (q.v.), in charge of the "gold Jews" at Treblinka, stated that a messenger arrived at the camp one day bearing an order from Blankenburg to hand over one million *Reichsmarks*. "We filled his suitcase with a million marks, and he returned with it to Berlin," Suchomel testified.[1334] Blankenburg's citation for a medal in 1942 stated: "Without his decisive contribution, it would not have been possible to complete this special assignment, so important to the war effort, on such a broad basis and with the necessary speed".[1335] The "special assignment" in question was self-evident.

In April 1945, Blankenburg was evacuated from Berlin to Bavaria together with other members of the KdF, including Viktor Brack. When the war ended he adopted the identity of "Werner Bieleke" (his wife's maiden name), and lived in the Wangen district of Stuttgart. He worked as a bank clerk in Ludwigsburg, and later as representative for a textile company in Freudenstadt. Despite being on the "most wanted" list of war criminals from 1945 until the time his death, he managed to escape arrest, even keeping in contact with his parents in an old people's home in Ulm, as well as with former T4 colleagues.

At the request of his wife, Blankenburg had been officially declared dead by a magistrates' court in Berlin-Tempelhof in 1956. The date of his death was given as 31 December 1945. In 1950, the Stuttgart *Kriminalpolizei* began to take an interest in the wartime activities of "Herr Bieleke" and he quickly disappeared into the Swabian Alps, south of the city. There he sought refuge with a former comrade from T4 in the market town of Münsingen. He returned to Stuttgart about a year later to resume his previous occupation. Blankenburg died in Stuttgart-Wangen on 28 November 1957[1336] and is buried there under the name of Werner Bieleke. At the funeral service numerous former members of *Aktion* T4 were present, among them Dietrich Allers (q.v.) and Erwin Lambert (q.v.).

1332 http://nuremberg.law.harvard.edu/php/search.php?DI=1&FieldFlag=1&PAuthors=118 (Accessed 19 May 2008).

1333 http://www.jewishgen.org/Yizkor/belzec1/bel120.html#8#8 (Accessed 19 May 2008).

1334 Yitzhak Arad, *Belzec, Sobibor, Treblinka: The Operation Reinhard Death Camps* (Bloomington: Indiana University Press, 1987), p. 161.

1335 Aly, Chroust, Pross, *Cleansing the Fatherland*, p. 39.

1336 Klee, *Das Personenlexikon zum Dritten Reich*, p. 52.

Gerhard Bohne (1902–1981) [post T4 pseudonym "Kurt Adolf Rüdinger"][1337] was born in Braunschweig and studied law at the University of Cologne, qualifying as a lawyer in 1924 and obtaining his doctorate in 1928. After working in the civil service for several years, he commenced practicing law in Berlin in 1930. He returned to the civil service in 1935, serving as an administrative judge in a government economic agency. Bohne joined the NSDAP and SA in 1930, leaving the latter in 1935 to become a member of the SS.[1338]

In September 1939 Bohne was recruited by T4; together with Herbert Linden (q.v.) he represented Department IV of the Ministry of the Interior. T4's first legal expert, and thus responsible for creating the concern's administrative organization, he resigned in June 1940 in a dispute over the corruption prevalent amongst his colleagues,[1339] and was replaced in 1941 by the rather brighter and apparently more corruptible Dieter Allers (q.v.).[1340] Bohne's accusations were deemed to be somewhat exaggerated and led to his dismissal from the SS in December 1943.

In January 1943 Bohne was conscripted into the *Wehrmacht*, finishing the war in American captivity in Italy, from where he was released in late 1946. He immigrated to Argentina in 1949, then returned to Germany in 1955, settling first in Cologne and then in Düsseldorf, where in 1956 he began to practice law. Arrested in 1959 and charged with the murder of at least 15,000 people, Bohne was released from pre-trial confinement on medical grounds on 15 March 1963. Bohne's symptoms of "illness" were in fact caused by a fellow doctor overdosing him with quinine and caffeine. Despite Bohne having surrendered his passport, four months later his lawyer informed the court that Bohne was now seeking to recuperate in Buenos Aires,[1341] where he proceeded to live under the name of Kurt Adolf Rüdinger.[1342]

Arrested by the Argentinean police in February 1964 and eventually extradited to Germany in November 1966, Bohne stood trial together with Allers, Vorberg (q.v.), and Adolf Kaufmann (q.v.). However, before the case ended he was deemed "unfit to stand trial" on medical grounds. In October 1968 the proceedings against him were temporarily suspended, and nine months later the suspension became final.[1343] Miraculously, Bohne managed to survive until 1981.

1337 Ibid., p. 62.
1338 Friedlander, *The Origins of Nazi Genocide*, p. 70.
1339 de Mildt, *In the Name of the People*, p. 86.
1340 Aly, Chroust, Pross, *Cleansing the Fatherland*, p. 37.
1341 de Mildt, *In the Name of the People*, p. 82.
1342 Burleigh, *Death and Deliverance*, pp. 274–275.
1343 de Mildt, *In the Name of the People*, p. 342–343, note 15.

Phillip Bouhler
Source: Wikimedia Commons / Bundesarchiv, Bild 146-1983-094-01 / CC-BY-SA 3.0 https://creativecommons.org/licenses/by-sa/3.0/de/deed.en

Phillip Bouhler (1899–1945)[1344] was born in Munich, the son of a retired colonel. He served in the German army during the Great War, eventually achieving the rank of *Leutnant* before being wounded and invalided out of the army in 1917. For a short time he became a philosophy student, but by 1921 he had joined the Nazi party, member number 12, and was a contributor to the *Völkischer Beobachter*. In 1922 he was appointed second secretary of the NSDAP.

Bouhler participated in the failed 1923 *Beerhall Putsch* and was consequently awarded the *Blutordensträger* (Blood Order). During the subsequent period, during which the party was banned, he acted as business manager of a Nazi cover organisation in south Germany, "The Greater German Racial Community". Following the relaunching of the party in 1925, he became *Reich* Business Manager, working with Max Amann, the publisher of *Mein Kampf.*[1345] After the *Machtergreifung*, Bouhler was made a *Reichsleiter*, the highest position attainable in any Nazi Organisation, second only to the office of *Führer.*[1346] He was also appointed a Member of the Reichstag for Westphalia and became Commissioner for the Protection of National Socialist Literature,[1347] an office Goebbels, Rosenberg, and others undoubtedly considered their own. Despite the splendid titles he was clearly a man in search of an assignment of greater importance. After serving a brief term as Police Chairman of Munich, one such was found for him with the creation of the KdF, whose head he became in October 1934. Two years later he was appointed an SS-*Obergruppenführer*.

Pale and bespectacled, Bouhler remained a shadowy figure, always polite, always self-effacing,[1348] the archetypal obsequious civil servant—at least to the casual observer. Initially he shared an office at Schellingstrasse 50[1349] with Rudolf Hess, his ostensibly immediate superior, before directing the KdF from the Brown House. He wrote several books on Nazism, and another bearing the title *Napoleon – Kometenpfad eines Genies* ("Napoleon—the Comet-Path of a Genius"), which was apparently a favourite of Hitler's.[1350] Despite his undoubted Nazi convictions, Bouhler lacked aggression. He was

---

1344 Klee, *Das Personenlexikon zum Dritten Reich*, p. 68.

1345 William Sheridan Allen (ed), *The Infancy of Nazism: The Memoirs of Ex-Gauleiter Albert Krebs 1923–1933* (New York: New Viewpoints, 1976), pp. 246–248.

1346 Goebbels for one was far from impressed by Bouhler: "Bouhler is stirring up trouble for me with the *Führer*...But I defend myself vigorously. These little nonentities think that one *Reichsleiter* is equal to another. But it is not the title that counts, it is the quality." [Fred Taylor (ed), *The Goebbels Diaries 1939–1941* (London: Sphere Books Limited, 1983), p. 300, entry for 6 April 1941].

1347 Stephen H Roberts, *The House That Hitler Built* (New York: Harper & Brothers,1938), p. 79.

1348 Allen, *The Infancy of Nazism*, pp. 246–248.

1349 This was the address of the NSDAP headquarters in Munich prior to the acquisition of Barlow Palace, the so-called "Brown House" in 1930.

1350 Gerald Reitlinger, *The Final Solution: The Attempt to Exterminate the Jews of Europe 1939–1945* (Northvale: Jason Aronson Inc, 1987), p. 127.

essentially an executor of orders, allegedly sharing "the view held by Hess and many other sincere idealists within the party that its 'great goals' justified every suppression of your own ideas, even of your own conscience."[1351] In that he unquestionably succeeded. If Brack (q.v.) is to be believed, in March 1941 when the so-called "Madagascar Plan" was supposedly still considered as a possible "Final Solution to the Jewish Problem", it was Bouhler whose name was suggested to Hitler as governor of the island.[1352] Given Bouhler's activities at the KdF before and after that date, it requires little imagination to envision the fate of the Jews intended to be deported to that tropical wilderness under his stewardship .

Even after "euthanasia" had officially been stopped in August 1941, Bouhler retained the hope of regaining the services of the T4 functionaries who had been loaned to Globocnik for the purpose of *Aktion Reinhard*.[1353] It was intended that the programme recommence at the successful conclusion of the war, but Bouhler was irate that Globocnik had used his well-trained men in connection with an assignment which would make them unfit, in Bouhler's opinion, to return to their former service.[1354]

As the war drew to a close Bouhler sought Göring's protection, following the *Reichsmarschall* to Austria. The sometime nominated successor to the *Führer* and the former joint head of the "euthanasia" programme both only succeeded in postponing the inevitable. Together with his wife, Bouhler committed suicide on 19 May 1945 whilst in American custody in transit to the Dachau concentration camp. Göring followed suit on 15 October 1946 following his conviction by the IMT at Nuremberg.

Viktor Brack
Source: Wikimedia Commons / Public Domain

Viktor Brack (1904–1948)[1355] [T4 pseudonym "Jennerwein", post T4 pseudonym "Hermann Ober"] was born in Haaren, near Aachen, the son of a physician. He studied agriculture for a time in 1923, before switching to economic science at the Technical High School in Munich, receiving his diploma in the latter subject in 1928.[1356] Thereafter, for a time he administered the estate attached to his father's sanatorium and was also a test driver for BMW. Active from an early age in right-wing politics, he joined the SA in 1923, and the Nazi Party and SS in 1929.[1357] In 1930 and 1931 he was often Himmler's chauffeur (he knew the Himmler family well, since Brack's father

1351 Allen, *The Infancy of Nazism*, pp. 246–248.
1352 Reitlinger, *The Final Solution*, p. 79.
1353 Whilst killing on a gigantic scale in Poland these men remained on the payroll of T4.
1354 Gerald Reitlinger, *The SS, Alibi of a Nation: 1922–1945* (New York: Viking Press, 1957)., p. 282.
1355 Klee, *Das Personenlexikon zum Dritten Reich*, p. 68.
1356 Bryant, *Confronting the "Good Death"*, p. 99.
1357 It may be considered curious that one could be a member of the SA (or indeed the SS) *without* joining the NSDAP. This has been explained as follows: "If at the time of National Socialism one was politically incriminated or suspect one could, without difficulty, become an SA member, but under no circumstances a Party member, because in regard to Party membership, and even ordinary Party membership, much higher political qualifications were required than in the case of the SA. There were certainly many SA members who joined this organization only to escape to some extent the persecution they had to expect because of their incriminating political record." [http://www.tinyurls.co.uk/U13892

had delivered one of Himmler's children),[1358] before Bouhler (q.v.), then *Reich* Business Manager of the NSDAP, employed Brack on a full time basis at the "Brown House" in Munich in the summer of 1932. During 1933, Brack acted as Bouhler's adjutant; when Bouhler moved to Berlin as head of the KdF in 1934, Brack moved with him.

In 1936 Brack was placed in charge of Amt II of the KdF with the title of *Reichsamtsleiter*. This department examined complaints arriving from all sections of the population. By now Brack was also officially Bouhler's deputy, acting as liaison between the KdF and the SS, where he received steady promotion, achieving the rank of *Oberführer* in 1940.[1359] Despite his lack of any medical qualification, Brack was also given the responsibility of liaising directly with the Ministry of Health on behalf of the KdF.[1360] He was thus an integral and vital cog in the machinations of the Chancellery, the individual primarily responsible for the overall organization and day-to-day activities of T4. It was an occupation that placed him at the heart of both the "euthanasia" programme and the "Final Solution". Amongst countless other tasks, together with Werner Blankenburg (q.v.), he was responsible for interviewing all potential T4 personnel. Moreover, as early as the summer of 1939 he had been discussing with his cohorts the use of carbon monoxide as the killing agent of choice for use in the forthcoming "mercy killings".[1361]

Having largely been given the responsibility of organizing "euthanasia", Brack claimed to have read a good deal about the subject, and decided to visit the asylums at Buch and Brandenburg-Görden to judge the condition of his potential victims for himself. Not surprisingly, he found the sight of inmates distressing and fully meriting of the release he was about to offer them. After all, as he explained at his trial:

> The life of the insane person has, for himself and his relatives, lost all purpose, and consists only of pain and misery. Just as the soul belongs in the helping hands of the priest, so the body belongs in the helping hands of the physician. Only so can the sick person really be assisted...It is [the doctor's] duty to free the person from his unworthy condition, so—I might even say—from his prison.[1362]

The court was not impressed with Brack the humanitarian, particularly after he freely admitted that he had been a participant in the original discussions concerning implementation of the "euthanasia" programme, as well as being present at the initial and many subsequent meetings of medical experts, administrators, and other operatives. He had frequently acted as Bouhler's representative, taking vital decisions based upon his own personal assessments and individual authority, in effect assuming the role of chief executive officer to Bouhler's chairman. Brack even went so far as to accept that his involvement in "euthanasia" was such that it might be presumed he was the key man at T4—an accurate, if unwise, display of vanity.[1363]

(Accessed 24 December 2008)]. The same criterion applied to the SS: "...among 118 SS officers, 89 were members of the NSDAP and 29 were not. It is interesting that even in the SS one third of the officer class (all volunteers) were *not* members of the Nazi Party." [http://cgi.stanford.edu/group/wais/cgi-bin/?p=5471 (Accessed 6 March 2008)].

1358 Friedlander, *The Origins of Nazi Genocide*, p. 68.

1359 http://140.247.226.138/NurTranscript/TranscriptPages/177_138.html (Accessed 26 June 2008).

1360 Bryant, *Confronting the "Good Death"*, p. 99.

1361 de Mildt, *In the Name of the People*, p. 228.

1362 Burleigh, *Death and Deliverance*, pp. 262–263.

1363 http://individual.utoronto.ca/jarekg/Ravensbruck/VictorBrack.html (Accessed 27 June 2008).

Another example of Brack's immersion in practical Nazism was his espousal of the economic advantages of sterilisation. On the principle that by such methods, valuable labour could be utilized whilst ensuring the ultimate disappearance of Nazism's arch enemy, Brack had been eager to suggest methods of mass sterilisation for Jews as an alternative to extermination, a subject which is examined in greater detail in chapter 10.

In evidence given in 1962, Josef Oberhauser (q.v.) described a visit by Brack to Lublin in May 1942. The Belzec extermination camp had been inoperative for the previous six weeks, but now Brack met with Globocnik to discuss the resumption of killing. When Globocnik complained of a lack of staff, according to Oberhauser, Brack replied:

> The euthanasia programme had now stopped and that the people from T4 would from now on be detailed to him [meaning Globocnik] on a regular basis so that the decisions taken at the Wannsee conference could be implemented. The decision had been taken to set up two further extermination camps, which would be ready by 1 August 1942, namely Treblinka and Sobibor The large-scale extermination programme was due to start on 1 August 1942.[1364]

Presumably because there was little killing left for him to organise, Brack left T4 in 1942, joining the *Waffen*-SS as a *Sturmbannführer*, and apparently remained on active duty until the war's end.[1365] When arrested in May 1945, he claimed his name was Hermann Ober (his wife's maiden name), but his true identity was quickly discovered.[1366] He was arraigned before an American military tribunal at Nuremberg in 1946, one of 23 defendants in the Medical Trial. It says much for Brack's qualities that he chose SS-*Obergruppenführer und General der Waffen-SS* Karl Wolff as a character witness. Wolff, for several years Himmler's Chief of Personal Staff and the *Waffen-SS* liaison officer at Hitler's military headquarters, had this to say about his former colleague: "I can only call Brack an extremely decent, extremely obliging man, to whom any thought of committing or taking part in an inhumane act would be alien."[1367]

Evidently the court did not find Wolff's testimony credible. Sentenced to death in 1947 for his manifest crimes, Brack was executed in 1948.[1368]

Ludwig Gessner (1886–1958)[1369] was born in Crumstadt, a village near Darmstadt, and held a doctorate in chemistry. He joined the NSDAP in 1930, and in 1933 was appointed to the position of *Landeshauptmann* (effectively, Governor) of the province of Hannover. In this capacity he was the immediate superior of Georg Andreae (q.v.).[1370]

1364 Ernst Klee, Willi Dressen, Volker Riess (eds.), *The Good Old Days—The Holocaust as Seen by Its Perpetrators and Bystanders* (New York: Konecky & Konecky, 1991), pp. 229–230. Oberhauser's memory regarding dates, like most other matters, was unreliable, probably deliberately so. Sobibor commenced killing operations at the beginning of May 1942, while the first transport to Treblinka arrived on 23 July 1942. However, Brack's alleged promise of a further supply of T4 staff seems quite plausible, since that is exactly what transpired during summer 1942.

1365 http://individual.utoronto.ca/jarekg/Ravensbruck/VictorBrack.html (Accessed 27 June 2008).

1366 Weindling, Nazi Medicine and the Nuremberg Trials, p. 140 (Accessed 8 September 2010).

1367 http://nuremberg.law.harvard.edu/php/pflip.php?caseid=HLSL_NMT01&docnum=315&numpages=3&startpage=

1368 Yisrael Gutman (ed), *Encyclopedia of the Holocaust* (New York: Macmillan Publishing Company, 1990), p. 238

1369 Klee, *Das Personenlexikon zum Dritten Reich*, p. 182.

1370 de Mildt, *In the Name of the People*, p. 167.

Gessner began to hear rumours of a comprehensive planned "euthanasia" programme in early summer 1940, although even before then he was aware that patients were being killed in Pomerania. He visited Hans Hefelmannn (q.v.) and Herbert Linden (q.v.) in Berlin, who confirmed the intention of implementing the policy in Hannover.[1371] Gessner claimed to have been far from convinced about the legality of this, and wrote to the Minister of the Interior, Wilhelm Frick, setting out his objections. Unfortunately no copy of this document survived the war. Together with Andreae, Gessner thereafter continued to make strenuous efforts to prevent, or at least minimize the impact of "euthanasia" on the institutions under their control—or so they both testified.

Despite a career that obviously had relied much on his longstanding enthusiasm for Nazism, at his trial in July 1950 the court considered Gessner to be of exemplary character. He could not have resigned his position since, "Berlin would not only have accepted [his] withdrawal, but would have encouraged it, because it would have vacated his position for a more ideologically suitable replacement." So, like Andreae, and despite the demise of so many patients in their joint care, Gessner was acquitted of all charges.[1372]

Friedrich Haus (1909–1945). According to the testimony of his sister, Haus joined the NSDAP and SA in 1930 and was a friend and colleague of Viktor Brack (q.v.). Haus was one of the so-called "old fighters" (*Alten Kämpfer*), and wore the golden party badge (*Goldene Ehrenzeichen*). This was awarded to the first one hundred thousand members of the party, but was also conferred on other individuals at Hitler's discretion. In about 1934 Haus became a full-time employee of the SS central office in Munich.[1373]

Haus became head of T4's Personnel Department in 1940. This office was responsible for hiring all T4 staff and administering the oath of secrecy to them.[1374] In 1943 Haus joined either the *Wehrmacht* or the *Waffen*-SS, and was succeeded at T4 by Arnold Oels (q.v.). Haus died on 8 August 1945 as the result of a traffic accident. In a list of names kept by Friedrich Lorent (q.v.) in a bank vault, Haus' death is noted as "suicide".[1375]

Hans Hefelmann (1906–1986)[1376] [T4 joint pseudonym with Richard von Hegener (q.v.), "Dr Klein"][1377] was born in Dresden, the son of a textile manufacturer. A member of the NSDAP since 1931, he received his doctorate in agricultural engineering in 1932. After spending some time working for his father and in private business, he became an employee of the Nazi party's economics department headed by Rudolf Hess, moving in January 1936 to the KdF. There he was appointed head of Amt IIb in 1937, a section responsible for *Reich* ministerial matters (other than those relating to the *Wehrmacht* and the police), as well as clemency petitions. Hefelmann's brief was to settle such matters

---

1371 Burleigh, *Death and Deliverance*, p. 133–134.

1372 Bryant, *Confronting the "Good Death"*, pp. 179–187.

1373 Ernst Klee, *Was sie taten–Was sie wurden: Ärzte, Juristen und andere Beteiligte am Kranken- oder Judenmord* (Frankfurt am Main: Fischer Taschenbuch Verlag, 1986), p. 294, note 103.

1374 Friedlander, *The Origins of Nazi Genocide*, p. 72

1375 Klee, *Was sie taten–Was sie wurden*, p. 81.

1376 Klee, *Das Personenlexikon zum Dritten Reich*, p. 236.

1377 Bryant, *Confronting the "Good Death"*, p. 32.

in as informal a manner as possible.[1378] However, Viktor Brack (q.v.) decided to also allocate the organization of children's "euthanasia" to this office, with the programme to be directed by Hefelmann and his deputy, Richard von Hegener (q.v.).[1379]

In the T4 pecking order, Hefelmann was senior to both Werner Heyde (q.v.) and Paul Nitsche (q.v.);[1380] together with Brack and Herbert Linden (q.v.), he had been involved in the planning of children's "euthanasia" since early 1939.[1381] Later, his participation would extend to adult "euthanasia".[1382] For his services to the mass-murder industry, in 1942 Hefelmann was awarded a medal—the War Service Cross Second Class. Philip Bouhler's (q.v.) citation stated:

> In addition to his especially important contributions relating to matters of public health handled by *Hauptamt* II, party comrade Dr Hefelmann provided the intellectual basis for the implementation of a special task important to the war effort and assigned by the Führer. He directs a separate department with independent responsibility for this special task.[1383]

This "special task" was, of course, the killing of children. According to Werner Heyde, in September or October 1939, discussions ensued between interested parties concerning the manner in which the simultaneous death of so many victims of "euthanasia" could be disguised. Hefelmann came up with the idea of a series of fictitious fatal rail and road accidents, at which point Heyde claimed to have refused to participate if such a ludicrous suggestion was adopted.

Like other leading lights of T4, as the nature of "euthanasia" changed, and *Aktion Reinhard* drew to a close, in 1943 Hefelmann was conscripted to serve in the *Wehrmacht*. He was discharged in March 1944 suffering from malaria and jaundice, to return to T4. In January 1945, together with von Hegener and others, he left Berlin. Hefelmann became director of a refugee camp established at the former asylum in Stadtroda. The medical director of the hospital, Gerhard Kloos (q.v.), having himself been involved in children's "euthanasia", knew Hefelmann well. But Hefelmann's stay was brief. He moved first to Munich, then to Innsbruck, before in June 1948, with the assistance of the Catholic charitable organization, Caritas International, he immigrated to Argentina. There he worked in a variety of jobs before finally becoming manager of a German-language bookshop. He returned to the Federal Republic in 1955, and a year later was managing director of a textile business in Waging am See in Bavaria.[1384]

After Heyde's arrest in 1959, reasoning with justification that his name was bound to crop up in any forthcoming proceedings (he had made no attempt to disguise his identity), Hefelmann fled again, this time to Spain. But he was unhappy there, and so returned to Germany in August 1960, where he surrendered himself to the public prosecutor's office in Munich.[1385] Accused of participation in the murder of more than 70,000 adults and at least 5,000 children, Hefelmann's trial opened in February 1964. With the

---

1378 Burleigh, *Death and Deliverance*, p. 97.
1379 Friedlander, *The Origins of Nazi Genocide*, p. 41.
1380 Aly, Chroust, Pross, *Cleansing the Fatherland*, p. 36.
1381 Friedlander, *The Origins of Nazi Genocide*, p. 42.
1382 Ibid., p. 69.
1383 Ibid. pp. 41–42.
1384 Klee, *Das Personenlexikon zum Dritten Reich*, p. 236.
1385 Burleigh, *Death and Deliverance*, p. 269.

aid of a helpful opinion supplied by his good friend Dr Kloos, by then medical director of the Göttingen hospital, six months later Hefelmann was diagnosed as suffering from a progressive "metabolic disorder unknown in detail", which had mysteriously developed from the moment of his arrest. Medical experts announced that the unknown illness had resulted in a serious deterioration in Hefelmann's mental and physical capacity, and would result in his demise within two years. Moreover, in their opinion, Hefelmann was no longer able to follow the proceedings with the requisite attention, and was therefore incapable of adequately defending himself. Other medical opinion was more sceptical.

It took a further eight years of legal wrangling before Hefelmann was finally declared unfit to stand trial; the case against him was abandoned in October 1972. Hefelmann was once again a free man. Amazingly, despite the pessimistic 1964 medical prognosis, Hefelmann survived until 1986. He did not, however, go unpunished; because of his illness, his driver's licence was cancelled.[1386]

Richard von Hegener (1905–1981)[1387] [T4 joint pseudonym with Hans Hefelmann (q.v.), "Dr Klein"] born in Sensburg, East Prussia, was the son of an army officer. He left school in 1923 to work first for the Dresdner Bank, then for a trucking company, before becoming a statistician for the Association of German Iron and Steel Producers. He joined the NSDAP in 1931, becoming active in local party affairs.[1388] He was unemployed prior to being recruited by the KdF in 1937 through the auspices of Hans Reiter, president of the *Reich* Health Office, who was his brother-in-law.[1389] Initially working in an office dealing with marriage applications from "*Mischlinge*" (individuals of mixed Jewish and "Aryan" descent—literally, "mongrels"), in due course he was promoted to the position of Hefelmann's deputy both in Amt IIb and in the children's euthanasia programme.[1390]

Together with Hefelmann, von Hegener was responsible for the establishment of the *Kinderfachabteilungen*, or "special children's wards", specifically created for the purpose of killing children within existing mental facilities. The pair were also responsible for sifting the forms received under the aegis of the decree of 18 August 1939 requiring doctors and midwives to report various categories of handicapped children. The fact that neither Hefelmann nor von Hegener had any medical training whatsoever was no impediment to their deciding which cases should be passed to the panel of three experts who finally decided on the fate of the child.[1391] In similar terms, and for identical reasons, in 1942 he and Hefelmann were both recommended by Philip Bouhler (q.v.) for the same decoration—the War Service Cross Second Class.[1392] However, the activities of both men had not been limited to solely murdering children. Both were also par-

1386 de Mildt, *In the Name of the People*, p. 33.
1387 Klee, *Das Personenlexikon zum Dritten Reich*, p. 237
1388 Friedlander, *The Origins of Nazi Genocide*, p. 192.
1389 Ibid., p. 193.
1390 Ibid., p. 42.
1391 de Mildt, *In the Name of the People*, p. 67.
1392 Friedlander, *The Origins of Nazi Genocide*, p. 42.

ticipants in adult "euthanasia", attending meetings and visiting facilities. In practice, there was little administrative distinction between the killing of children and adults.[1393]

After the war von Hegener was first employed as a farm worker and then as a carpenter. Dropping the "von", he subsequently worked at the Ministry of Trade and Supply in Mecklenburg, where he was quickly promoted. He was arrested in 1951 and charged with crimes against humanity. In 1952, the court in Magdeburg sentenced him to life imprisonment, but he was released in 1956, and with the assistance of his former superior Dieter Allers (q.v.), found employment at the German Shipyard in Hamburg. He testified at the trials of Hefelmann and Allers, but was never subject to further prosecution himself.

Adolf Gustav Kaufmann (1902–1974)[1394] the son of a railway inspector, was born in Przemysl, Poland, then part of the Habsburg Empire. He attended school at Ried im Innkreis in Upper Austria before serving in the Austrian navy during the Great War. At the war's conclusion he trained to become a mechanic. Between 1921 and 1934 he was employed in the commercial department of the Austrian Federal Railways. In 1923 he had joined the SA and the *Deutsche Nationalsozialistische Arbeiterpartei* (DNSAP), the predecessor of the Nazi party in Austria. He became a member of the NSDAP in Linz in 1926. After being arrested for possession of weapons and explosives, in 1934 he lost his job with Austrian Railways and spent four months in custody before moving to Germany.

From 1935, Kaufmann worked full-time for the SA; in 1937 he was appointed *Gauinspektor* for Pomerania. Called-up for naval service in 1939, he was granted leave of absence in January 1940 when his personal friend of twenty years, Viktor Brack (q.v.), enlisted him in T4. Kaufman was appointed head of the inspection department,[1395] responsible, among other things, for selecting the institutions that were to become killing centres, refurbishing five out of six of them as necessary for their new task,[1396] and looking after the welfare of those employed at these facilities. He personally supervised the transformation of Brandenburg into a killing centre before turning the new resource over to Irmfried Eberl (q.v.).[1397] Kaufmann visited the killing centres from time to time to ensure all was in order, although despite his regular inspection trips, he claimed never to have seen an actual gassing.[1398] In addition to all of this, he found time to establish a holiday home for T4 personnel at Weissenbach am Attersee in Austria.[1399] He also managed to ensure that his brothers Rudolf and Reinhold were employed by either the KdF or T4.

With the completion of the first phase of "euthanasia", on 31 January 1942 Kaufmann returned to Pomerania as *Gauamtsleiter*. From October 1942 he was a deputy district head of the Nazi Party in the General District of Taurien in the office of the *Reich* Commissioner Ukraine. As the tide of war turned, a year later he returned to Pomera-

1393 Ibid., p. 69.
1394 Klee, *Das Personenlexikon zum Dritten Reich*, p. 301.
1395 Klee, *Was sie taten–Was sie wurden*, p. 70.
1396 Friedlander, *The Origins of Nazi Genocide*, p. 194.
1397 Ibid., p. 89.
1398 Ibid., p. 194.
1399 Ibid., p. 72.

nia. In July 1944 he was appointed NSDAP-*Oberbereichsleiter* in Budweis, then in Czechoslovakia.

Having enjoyed a risk-free wartime career at a variety of jobs within the Nazi party, after the war Kaufmann worked for BMW and as a representative for a manufacturer of laboratory equipment. He was arrested in July 1965, and together with Dieter Allers (q.v.), Reinhold Vorberg (q.v.), and Gerhard Bohne (q.v.), was arraigned before the court in Frankfurt in April 1967. Having been unsuccessful in persuading the court to abandon proceedings against him because of his wife's alleged bedridden medical condition (his plea was not helped when a letter from her was discovered in which she detailed how she had painted the entire house in his absence)[1400] two months later the case against Kaufmann was suspended after he had suffered a heart attack. In September 1969 he was declared permanently unfit to stand trial.

Friedrich Lorent

Source: Ghetto Fighters House Museum, Israel, Photo Archive

Friedrich Wilhelm Siegmund Robert Lorent (1905–1988)[1401] the son of a coffee merchant, was born in Bremen. Following an accident he left secondary school and became an agricultural apprentice, but in view of the then unfavourable economic prospects for agriculture, he gave up his apprenticeship and instead joined his father's firm. In December 1930 he became a member of the Nazi party and in October 1932 joined the SA, by whom he was subsequently employed. By January 1934 he had risen to the position of leading manager of the Hannover SA under Viktor Lutze. Following the so-called "Night of the Long Knives" and the murder of Ernst Röhm in June 1934, Lutze was appointed head of the SA and moved to Berlin, with Lorent accompanying him to become treasurer of the SA head office. Here he met and befriended Viktor Brack (q.v.), an association which led to a falling-out between Lutze and Lorent, with the latter demoted in 1936 to the post of accountant with the National Socialist Relief Organisation for War Casualties (*NSKOV*), a position he held until the outbreak of war.

Lorent served in the military from November 1939 until May 1940, at which point he was transferred by *NSKOV* to Warsaw and Cracow, where he was employed as accountant to a "trust company" (*Treuhandverwertungsgesellschaft*), one of the organisations set up by the Nazis to handle the theft of Polish and (primarily) Jewish property. In autumn 1941 Lorent was appointed manager of a formerly Polish-owned company in Warsaw dealing in building materials. Shortly afterwards he met Brack in Berlin. When he complained to Brack about working conditions in Poland, Brack offered him the position as manager of the Economic Department of T4; despite a substantial resulting reduction in income, Lorent was happy to accept the job.

He commenced employment with T4 in February 1942, when the first phase of the "euthanasia" programme had already been terminated. Answerable only to Brack and Werner Blankenburg (q.v.), Lorent and the staff he controlled were responsible for the

1400 Burleigh, *Death and Deliverance*, p. 275.

1401 Klee, *Das Personenlexikon zum Dritten Reich*, p. 379.

entire financial and accounting functions of T4, including the ordering of the necessary gas and lethal medications, payment of the salaries of personnel in Germany and in the *Aktion Reinhard* camps, and overseeing the handling of the valuables of the Jewish victims of the death camps, among other tasks. Lorent was in a unique position to obtain virtually complete knowledge of the activities of T4 and its personnel, visiting most of the principle German "euthanasia" centres, as well as the Polish extermination camps.[1402] On at least one occasion he personally delivered the possessions of murdered Jews from Poland to Berlin.

Lorent was a conscientious and efficient administrator, priding himself on his incorruptibility.[1403] Near the end of the war, he was ordered to take the remaining assets of T4, some 1.2 million *Reichsmarks*, from Berlin to Bavaria. Due to advancing Allied troops he was unable to comply with this order, and in April 1945 was still carrying around a number of cheques and 87,000 *Reichsmarks* in cash (the fate of the rest of the cash is uncertain), money he used to pay the salaries of the T4 employees remaining in Berlin. In summer 1945 he passed the balance of the cash, by now reduced to little more than 16,000 *Reichsmarks* to the *Landrat* (District Administrator) of the Wesermarsch county, "for the care of political prisoners in need."

Lorent was interrogated by the American Counter Intelligence Corps (CIC), who in September 1946 pronounced him "administrative head of *'Stiftung'* and for all institutions as well as the camps in Poland and Italy...Well informed about institutions and camps." Despite his known involvement in such criminal affairs, for reasons unknown Lorent was released, and for the next twenty years remained a free man, living in the town of Nordenham under his own name whilst earning his living as a sales representative.

In September 1965, Lorent was arrested on suspicion of his involvement in "euthanasia". Together with Hans-Joachim Becker (q.v.), he was arraigned before the court in Frankfurt in 1969, accused (in Lorent's case) of complicity in the murder of 4,300 people. Despite his protestations of innocence the court considered Lorent's guilt proven beyond all reasonable doubt, yet still sentenced him to only 7 years' imprisonment. On the court's own admission this was a minimal sentence, which could "only approximately atone for the injustice committed." Notwithstanding these sentiments, having served two-thirds of his sentence, Lorent was released[1404]

Arnold Oels (1908[1405]–?) became a member of the SA in 1933 and subsequently of the NSDAP.[1406] At the beginning of the war he volunteered for the *Waffen*-SS. However, his

---

1402 Shortly after commencing employment with T4, Lorent was taken on a guided tour of Hadamar, Bernburg, Sonnenstein, and Hartheim, where he witnessed the gassing of concentration camp prisoners, part of *Sonderbehandlung 14f13*. Lorent claimed to have been deeply shocked by what he had witnessed. He professed to have been equally horrified by what he encountered in the *Aktion Reinhard* camps. Neither experience persuaded him to change his occupation, which, considering his relationship with Brack, was undoubtedly possible if he had indeed wished to do so. (de Mildt, *In the Name of the People*, pp. 93–94).

1403 de Mildt, *In the Name of the People*, p. 345, note27.

1404 Ibid., pp. 90–95.

1405 Klee, *Was sie taten–Was sie wurden*, p. 294, note 104.

1406 Friedlander, *The Origins of Nazi Genocide*, p. 191.

girlfriend, who worked at Hadamar, suggested he join T4,[1407] which he did in November 1940 as deputy to Friedrich Haus (q.v.) in the Personnel Department.[1408] When Haus left for the armed forces in 1943, Oels succeeded him as head of the department. Oels remained with T4 until March 1945, and after the war was employed as travelling salesman.[1409]

Fritz Schmiedel (?-?) was a graduate engineer. He was yet another of Viktor Brack's (q.v.) friends to be recruited to T4,[1410] where he succeeded Willy Schneider (q.v.) as head of the Finance Office in August 1941, before he was in turn succeeded by Friedrich Lorent (q.v.) in January 1942.[1411] In 1943 he joined the *Waffen*-SS. After 1945 he lived under a false name in France, before returning to the Federal Republic in 1951.[1412]

Willy Schneider (1900[1413]-?) the nephew of Alfred Ittner (q.v.), was the first head of T4's Finance Department, a position he relinquished in August 1941.

Gerhard Siebert (1905[1414]-?) was a cousin of Viktor Brack (q.v.) and Reinhold Vorberg (q.v.).[1415] He was born in Königsberg and trained as an engineer, joining the NSDAP in 1931. As an employee of T4 he was made Vorberg's deputy at *Gekrat*, succeeding the latter when Vorberg was dispatched to Russia as part of the *Organisation Todt* operation in late 1941.[1416]

Siebert visited Treblinka at least once, somewhat improbably claiming to have seen nothing of the extermination of the Jews occurring there. Rather he had merely met a Jewish tailor, who had repaired Siebert's clothes.[1417] Siebert maintained an equal degree of blamelessness concerning his activities at *Gekrat*, stating: "I am innocent of the killing of patients."[1418] His cousin Brack, he maintained was "a decent person" ("*ein anständiger Mensch*").[1419]

Like many other T4 personnel, Siebert spent some time in Trieste as part of *Aktion R*.[1420] After 1945 he worked for the Siemens-Schuckert electrical engineering company.

Friedrich Tillmann (1903–1964)[1421] was born in Cologne. On leaving school in 1921 he began a commercial apprenticeship. He had right-wing political sympathies from an early age and in 1923 joined the Nazi party. When the party was banned in November

---

1407 Ibid., p. 193.
1408 Klee, *Was sie taten–Was sie wurden*, p. 294, note 104.
1409 Ibid., p. 81.
1410 Friedlander, *The Origins of Nazi Genocide*, p. 193.
1411 Ibid., p. 71.
1412 Klee, *Was sie taten–Was sie wurden*, p. 81
1413 Friedlander, *The Origins of Nazi Genocide*, p. 191.
1414 Klee, *Das Personenlexikon zum Dritten Reich*, p. 581.
1415 Ernst Klee, "*Euthanasie*" *im NS-Staat: Die* "*Vernichtung lebensunwerten Lebens*" (Frankfurt am Main: Fischer Taschenbuch Verlag, 1991), p. 167.
1416 de Mildt, *In the Name of the People*, p. 85.
1417 Klee, *Was sie taten–Was sie wurden*, p. 295, note 111.
1418 Ibid., p. 81.
1419 Ibid., p. 82.
1420 Ibid., p. 295, note 111.
1421 Klee, *Das Personenlexikon zum Dritten Reich*, p. 626.

1923 after the abortive Munich "beer hall *putsch*", Tillmann continued to maintain connections with neo-Nazi organizations, and was readmitted as a party member in June 1925. He worked in Nazi youth associations, thereby ensuring that he was on good terms with many people who would be influential in furthering his career, one of whom was Viktor Brack (q.v.).

He met with little success in terms of his livelihood, so much so that in 1928 he was excluded from membership of the NSDAP because he could not afford to pay his subscriptions. He managed to join the party for a third and final time in 1933, and took up employment in the welfare department of the Cologne district. In late 1939, Tillmann met Herbert Linden (q.v.), who recruited him for T4. In February or March 1940, Tillmann succeeded Gerhard Bohne (q.v.) and was effectively responsible for the management of the Administrative Office at T4, although Dieter Allers (q.v.) was nominally head of that department.[1422] Tillmann took on this responsibility in addition to his job in Cologne.

Tillmann attended monthly meetings of the office managers at the individual killing centres, and on at least one occasion attended a gassing. When it was feared that notification of the deaths of too many people on the same day and in the same place might arouse suspicion, he was the person responsible for organizing the complex cover-up of the location and timing of the murder of the victims. For his dedication to the cause of T4, in 1942 Tillmann was awarded two medals.

After the war Tillmann was interned for a short time, and in 1949 unsuccessfully applied for re-employment with the city of Cologne. Instead he was engaged by a number of different children's homes, until in July 1960 he was arrested, accused of participation in the "killing of about 70,000 adult inmates of treatment and care institutions". He was then released to await trial,[1423] stating: "It is clear to me that my activity contributed to the smooth operation of euthanasia. However, I am not able to see that I thereby made myself punishable."[1424] He apparently had second thoughts, for on 12 February 1964, six days before proceedings against him were to commence, Tillmann jumped to his death from the eighth floor of a Cologne skyscraper.[1425] It is possible this was not the suicide it appeared; the next day, Werner Heyde (q.v.) was found dead in his cell in Butzbach, having apparently hanged himself.[1426] It is all too easy to detect a conspiracy where perhaps none existed, but there were many who had good cause to feel far from unhappy at the sudden demise of two of the leading members of T4.

Reinhold Vorberg (1904–1983)[1427] [T4 pseudonym "Hintertal",[1428] post T4 pseudonym "Heinz Vorberg"] was born in Kiel, and like Friedrich Tillmann (q.v.), on leaving school began a commercial apprenticeship, the completion of which three years lat-

1422 Friedlander, *The Origins of Nazi Genocide*, p. 71.
1423 Klee, *Was sie taten–Was sie wurden*, pp. 33–37.
1424 Klee, *Das Personenlexikon zum Dritten Reich*, p. 626.
1425 de Mildt, *In the Name of the People*, p. 82.
1426 Burleigh, *Death and Deliverance*, p. 273.
1427 Klee, *Das Personenlexikon zum Dritten Reich*, p. 645.
1428 The pseudonym was what passed for a joke in T4. "Vorberg" translates as "in front of the mountain." "Hintertal" means "behind the valley". (Schmidt, *Karl Brandt*, p. 136.)

er[1429] saw him join the so-called "*Schwarze Reichswehr*", an illegal paramilitary branch of the German army. Shortly afterwards he left that organization to work for the *Ostbank* in Königsberg, where he remained for about a year. 1927/28 saw Vorberg working as a sales representative in Spain. On his return to Germany the following year he assisted his mother for a time in managing family affairs in Königsberg, before in early 1930 he immigrated to a German colony in South-West Africa to pursue an unsuccessful career as a farmer. After that failure he returned to Germany a year later to set up another business with some friends in 1932, selling lighters and jewellery in Berlin. That enterprise was no more successful than Vorberg's other ventures into the world of commerce, it being declared bankrupt in 1935.[1430]

Vorberg had joined the Nazi party in 1929, and in about 1936/37 a party comrade offered him an unpaid job in the mailroom at the KdF. A little later he obtained paid employment at the KdF and shortly thereafter his cousin, Viktor Brack (q.v.), appointed him head of Amt IIc of the organisation, the department dealing with matters concerning the armed forces, the police, the SS, and churches.[1431] Vorberg was conscripted into the *Wehrmacht* on the outbreak of war, and served with an engineers' unit in Poland. However, on the conclusion of the Polish campaign his military service was suspended and he was recalled to Berlin, where Philip Bouhler (q.v.) informed him of the "euthanasia" programme, and offered him the position of head of T4's transport division, the *Gemeinnütziger Krankentransport GmbH*, or "*Gekrat*" (the "Charitable Society for the Transportation of the Sick Limited"). Vorberg accepted the offer with alacrity.

As general manager of *Gekrat*, and thus responsible for organizing all aspects of the transportation of victims, Vorberg was in constant communication with the leading officials of T4. He visited the killing centres, corresponded with other "feeder" institutions, organized the actual transports, and even arranged petrol rations for the bus drivers. He was of sufficient importance within the organisation to be Brack's third deputy, after Werner Blankenburg (q.v.) and Hans Hefelmann (q.v.).

After the suspension of "euthanasia" in August 1941, Vorberg was a member of the T4 "*Organisation Todt*" mission to Russia, but returned to Berlin a few months later to resume his position at the KdF, where he remained until almost the end of the war. In early 1945, together with other leading members of the Chancellery he was flown to Bavaria, where plans for a new office disintegrated as the war ended. Bearing false identity papers, Vorberg was captured by the Americans, but was released after a few weeks. He settled in the Munich region, and worked on a chicken farm, continuing to correspond with Brack so that when the latter was arrested the Americans found and arrested Vorberg too. The two of them were interned in the prisoner-of-war camp at Moosburg, where Vorburg contrived a false identity—"Heinz Vorberg", born 1 June 1906 in Juditten, East Prussia. Since this region was under Soviet occupation, it made it all but impossible to verify these bogus credentials.[1432]

Vorberg escaped from the camp and became a forester in Schleswig-Holstein until he was dismissed from this job in 1948. He was briefly detained by the British before

1429 de Mildt, *In the Name of the People*, p. 84.
1430 Ibid., p. 343, note 3.
1431 Friedlander, *The Origins of Nazi Genocide*, p. 42.
1432 de Mildt, *In the Name of the People*, p. 344, note 6.

settling in the town of Neuss and working in a fertilizer factory. After a while he moved to Freiburg, where together with Reinhold Siebert [brother of Gerhard (q.v.)], he opened an enamel factory. Needless to say this enterprise failed too, and it was not until 1952 that Vorberg found what he must have hoped was some kind of permanent employment at a tile factory in Bonn. But this too was not to last, for in 1961, with the probability of his arrest looming, Vorberg suddenly resigned from his job and fled to Spain. Here, under another Fascist regime he felt secure; however, as could be anticipated, he had misjudged the situation. In November 1962 he was extradited to Germany, and taken into pre-trial confinement.[1433]

When his trial began together with that of Dieter Allers (q.v.) in April 1967, Vorberg stood in the dock accused of participation in the murder of 70,273 persons. There was little doubt concerning the guilt of either defendant, although the court still found sufficient extenuating circumstances in Vorberg's case to restrict his sentence to ten years' imprisonment, rather than the life sentence it could have imposed. In October 1972, allowance having been made for the period of his pre-trial confinement, both in Germany and in Spain,[1434] Vorberg was released from custody.[1435]

There appears to have been little in Vorberg's miserable history as a businessman and would-be entrepreneur to recommend him to T4 for such a responsible position, other than his relatively early membership of the NSDAP and his familial ties to Brack and Siebert. If Vorberg can be seen as a classic example of the *Schreibtischtäter*, the "desk murderer", it is worth bearing in mind his presence at the experimental gassing at Brandenburg in winter 1939/40.[1436] He had seen for himself the consequences of the activities he organized so willingly at *Gekrat*.

Paul Werner (1900–1970)[1437] was born in Appenweier, Baden-Württemberg, the son of a railroad supervisor, and served in the armed forces during the last months of the Great War. At the war's conclusion he studied law at Heidelberg and Freiburg, and after qualification, was appointed an auxiliary public prosecutor in Baden, before going on to serve as a prosecutor in Offenburg and Pforzheim. In 1933 he became District Court Advisor in Lörrach, and joined the Nazi party in May of that year.

In early September 1933, Werner was appointed head of the state criminal police agency of Baden in Karlsruhe. His success in this position led to his 1937 promotion as deputy to Arthur Nebe, head of the newly formed *Reich* Criminal Office (*Reichskriminalpolizeiamtes—RKPA*) in Berlin. Werner enthusiastically endorsed the Criminal Police strategy of preventing crime on the basis of hereditary characteristics:

> If a criminal or asocial person has ancestors who also led a criminal or asocial life...the results of hereditary research have shown that the person's behaviour is hereditarily conditioned. Such a person must...be treated with differently than a person who...comes from a respectable family...The criminal is

1433 Ibid., pp. 84–86.
1434 Burleigh, *Death and Deliverance*, pp. 275–276.
1435 de Mildt, *In the Name of the People*, pp. 87–89.
1436 Friedlander, *The Origins of Nazi Genocide*, p. 87.
1437 Klee, *Das Personenlexikon zum Dritten Reich*, p. 670.

no longer regarded as an individual person, and his crimes are no longer regarded as individual crimes.[1438]

"Hereditary criminals" such as Gypsies were, in Werner's view, primarily a racial problem.[1439] In this he was later to find an enthusiastic accomplice in Robert Ritter (q.v).

On 27 September 1939 the *RKPA* became Department V of the *Reich* Security Main Office (*RSHA*). In 1963 Hans Hefelmann (q.v.) stated that Werner had been the *RKPA*'s contact with Reinhold Vorberg (q.v.), head of *Gekrat*. Together with Nebe, Werner had met in August 1939 with Viktor Brack (q.v.) at the KdF to discuss the proposed "euthanasia" programme. Nebe thereupon appointed Albert Widmann (q.v.) to advise T4 regarding suitable killing techniques. According to Hefelmann, Werner had been responsible for procuring killing agents such as morphine ampoules and Luminal tablets for T4.[1440] As mentioned earlier, it was Werner who had first interviewed Franz Stangl (q.v.) for T4, explaining to him the virtues of "euthanasia".

From spring 1942 to early 1943, Werner was employed by the German criminal police offices in Paris, Brussels and the Hague. In March 1943 he returned to the *RSHA* and remained there until the end of the war, following which his police career came to an end. Instead, following "denazification" (which determined he had been a "sympathizer"), he served in a senior capacity in the Baden-Württemberg Ministry of the Interior. In 1955 he was even proposed as head of the Federal Criminal Investigation Office (*Bundeskriminalamts*). Fortunately this promotion failed due to the opposition of officials of the Federal Ministry of the Interior.

Although a preliminary investigation into his wartime activities by the public prosecutor's office in Stuttgart was dropped in 1962, unlike many others Werner made no attempt to feign ignorance of the Holocaust. In a statement before the chief public prosecutor's office in Frankfurt in November 1962, Werner stated: "Naturally I knew of the *Einsatzkommandos* in the east and of Auschwitz."[1441]

Physicians, Nurses, and Scientists

Ernst Baumhard (1911–1943) [T4 pseudonym "Dr Jäger" at Grafeneck, "Dr Moos" at Hadamar][1442] was born in Ammendorf, near Halle, the son of a doctor. He joined the SA in 1934 and the Nazi party in 1937.[1443] After leaving school he studied medicine at the University of Halle, in 1938 becoming leader of the medical section of the university's Student Union. A totally committed Nazi, in May 1938 he was received by Adolf Hitler in person as a winner of *Reichsberufswettkampf*, a competition that was held from 1934 to

1438 Wildt, *An Uncompromising Generation*, p. 181.
1439 Omar Bartov (ed), *The Holocaust: Origins, Implementation, Aftermath* (London: Routledge, 2000) p. 69.
1440 Klee, *Das Personenlexikon zum Dritten Reich*, p. 670.
1441 Ibid.
1442 Ibid., p. 32.
1443 http://www.gedenkstaette-hadamar.de/webcom/show_article.php/_c-860/_lkm-643/i.html (Accessed 9 August 2008).

1939 for young Nazis who had displayed unswerving dedication to the cause of Nazism in the theory and practice of their vocation or profession.

Baumhard received his medical licence and doctorate on 1 September 1939,[1444] and worked for a short time at the *Barbarakrankenhaus* in Halle. He was recruited by T4 on 1 November 1939, and was among those present at the Brandenburg gassing demonstration in winter 1939–40.[1445] In early 1940 he was appointed Horst Schumann's (q.v.) assistant at Grafeneck, before succeeding the latter in April 1940. When Grafeneck closed in December 1940, Baumhard took up the position of physician-in-charge at Hadamar.[1446] Under Baumhard's direction, busloads of patients arrived daily at Hadamar for gassing from mid-January 1941.

Baumhard was party to a rather odd incident at Grafeneck involving the death of a nurse, Anne H. in mysterious circumstances. According to the somewhat unclear testimony of an eyewitness:

> In the summer of 1940, when I was in bed with a concussion, it was on 4 June 1940, a transport arrived in which there was someone who suffered from leprosy. According to the male nurse the leper's face had already been eaten away. The leper was shot immediately by Dr.Baumhard, to prevent infection. Dr. Baumhard ordered all of us to stay where we were. But head nurse H. from the Rhine country jumped to the other [meaning the dangerous] side and was hit fatally by the bullet. Dr. Baumhard, who told me about this himself, was in deep trouble because of the incident, and wanted to commit suicide.[1447]

After either falling out with Viktor Brack (q.v.) or simply seeking an escape from "euthanasia", Baumhard resigned from T4 in August 1941 to enlist in the navy as a U-boat doctor, and in this capacity was killed in action in June 1943.

August Becker (1900–1967)[1448] [nicknamed "*Rot*" ("Red")], was born in Staufenberg, Hesse. He was conscripted towards the end of the Great War, following which he studied chemistry and physics at the University of Giessen, obtaining a doctorate in the former subject in 1933. He joined the NSDAP in 1930 and the SS in 1931, and elected for a career within the Nazi movement, briefly working in 1934 for the *Gestapo* in Giessen before joining the SS regiment "Germania", with whom he continued to serve until 1938 when he was transferred to Berlin, to join the RSHA on the formation of that organisation as a chemist. In October 1939, Becker travelled to Posen, where in Fort VII he constructed a gas chamber. There he tested the relative efficacy of carbon monoxide and an agent similar to *Zyklon B* on patients from Polish mental hospitals, judging carbon monoxide the superior killing medium.[1449]

In December 1939, in his own words, Becker was loaned to T4 "...as expert for gassing during the destruction of mental patients in hospitals and nursing homes." As described above, in this capacity he attended the first experimental gassing at Brandenburg, and was responsible for the supply of carbon monoxide cylinders to the killing centres. Subsequently Becker returned to the RSHA, where he applied his expertise to

1444 Friedlander, *The Origins of Nazi Genocide*, p. 220.

1445 Ibid., p. 87.

1446 Ibid., p. 219.

1447 Klee, *"Euthanasie" im NS-Staat*, p. 195.

1448 Klee, *Das Personenlexikon zum Dritten Reich*, p. 34.

1449 Christopher R Browning, *Ordinary Men: Reserve Police Battalion 101 and the Final Solution in Poland* (New York: HarperCollins, 1993), p. 188.

improving the efficiency of the gas vans being used in the East to murder Jewish and Gypsy women and children. He visited all *Einsatzgruppen* operating in the Soviet Union to inspect and possibly maintain the gas vans in use with them.[1450]

After the war Becker did serve a three year sentence as a former member of the SS.[1451] In the 1960s it was intended that he be tried alongside Albert Widmann (q.v.), his fellow gassing expert, but proceedings against Becker were suspended and never re-commenced.[1452] In the knowledge that he was suffering from a terminal illness, Becker admitted: "Himmler wanted to use the people released from euthanasia who were experts in gassing, such as myself, in the great gassing programme getting underway in the east."[1453]

Herbert Becker (?-?) A former school and sports physician from Leipzig, and a close associate of Paul Nitsche (q.v.), Becker was head of the planning department of T4, which operated from spring 1941 until the winter of 1943–44. Becker reported directly to Herbert Linden (q.v.) in the latter's capacity as *Reich* Commissioner.[1454]

Ernst Beese (?-1945) A fanatical Nazi and eugenicist, Beese was appointed director of the Uchtspringe mental hospital on 1 April 1940. Under his control, between July 1940 and July 1941, Uchtspringe served as an intermediate institution for Brandenburg, before in August 1941 a children's killing ward was established at the hospital, supervised by Gerhard Wenzel (q.v.).

Although a physician, Beese was not considered a very good one. He had a bad reputation, both as a doctor and as a practitioner of "euthanasia"—or so Hildegard Wesse (q.v.) stated. He lacked all psychiatric training, and had no commitment to the welfare of his mentally ill patients, "who meant nothing to him at all." He had reputedly stated: "Euthanasia should be dealt with lock, stock, and barrel, without making too much fuss." According to Wesse, on his returning from a Berlin conference in late 1944, Beese informed her that adult "euthanasia" was now to commence at Uchtspringe. "In Berlin there was obviously no longer any order, which is why [Beese] said that we would conduct the selection ourselves."[1455] He would take charge of male patients; Wesse was to kill the women.[1456]

1450 Friedlander, *The Origins of Nazi Genocide*, pp. 210–211.

1451 Klee, *Was sie taten–Was sie wurden*, p. 152.

1452 de Mildt, *In the Name of the People*, p. 381, note 15.

1453 Christopher R Browning, *Fateful Months: Essays on the Emergence of the Final Solution* (New York: Holmes & Meier, 1991), p. 31.

1454 Aly, Chroust, Pross, *Cleansing the Fatherland*, p. 167. There appears to be little additional biographical information regarding this important member of the "euthanasia" programme. "I'm sorry to have to say to you that I can't help you with the particular Becker you're referring to. He doesn't surface in our digital databank on the German trials at all"—[http://www1.jur.uva.nl/junsv/index.htm]—"and, as you say, Klee hardly mentions him in his books. Since Klee is the undisputed expert on the post-war careers of types like Becker, the fact that he doesn't mention him at all in his *Was sie taten–Was sie wurden,* or in his *Personenlexikon zum dritten Reich* suggests to me that Becker, in one way or another, vanished from the face of the earth at the end of the war." (Dick de Mildt in private correspondence with the author, 20 May 2008).

1455 Schmidt, *Karl Brandt*,p. 237.

1456 de Mildt, *In the Name of the People*, pp. 135–136.

Beese died after suffering a stroke in June 1945, somewhat conveniently for the doctors who had killed at his direction, since he was thus unable to refute any of the charges levelled against him.[1457]

Friedrich Berner (1904–1945)[1458] [T4 pseudonym "Dr Barth"][1459] was born in Zwickau. He qualified as a doctor in 1931, joined the Nazi party in 1933 and the SS in 1934.[1460] In 1935 Berner was employed at the city hospital in Erfurt and a year later moved on to the city hospital in Mainz. In the same year he was recognized as a specialist in radiology, and from October 1938 he served as assistant physician in the radiology department of the University hospital at Frankfurt am Main. Two years later he had a post-doctoral lecturing qualification and lectured at the Medical Faculty in Frankfurt. Berner served in the *Luftwaffe* before appearing on a T4 list of personnel as one of the "doctors in the institutions," which in his case was Hadamar. His date of commencement with T4 is shown as 15 May 1941 and his date of cessation as 31 December 1941, during part of which time he was physician-in-chief at Hadamar. After leaving there he returned to Frankfurt and radiology, going on to command an *SS-Röntgensturmbann* (radiology battalion) in Posen. He was killed near Warthestadt (Wronka) on 2 March 1945.

Hans-Bodo Gorgass (q.v.) claimed that at Hadamar in mid-1941, Berner had told him that he (Berner) had a copy of the "euthanasia" law in his possession, but was unable to show it to Gorgass for reasons of security. Berner then made Gorgass swear an oath of secrecy, sealed with a handshake.[1461]

Hans Bertha (1901–1964)[1462] was born Bruck an der Mur, Austria. After graduating as a physician from the University of Graz in 1926, he was employed at several hospitals in Austria and Germany. He joined the then illegal Austrian Nazi party in 1933 and the SS in 1937. Following the *Anschluss*, in April 1938 Bertha was appointed provisional director of the psychiatric clinic at the University of Graz. He habilitated in 1939 in Graz, where the courses he held and papers he wrote on racial hygiene made him an obvious candidate for T4. He was appointed a *Gutachter* with effect from 30 September 1940.[1463] He held a number of important positions within the Austrian psychiatric *milieu*, and on 1 January 1944 became director of the Am Steinhof sanatorium in Vienna, and thus was directly responsible for the "mercy killings" committed from the time of his appointment until the end of the war.[1464] Between 1941 and 1945 more than 3,500 patients died of malnutrition at Steinhof. Bertha's enthusiasm for murder can be gleaned from the mortality statistic for that institution; in 1941 it was 13.9 percent; by 1945 it had risen to 42.7 percent.[1465] How many others died as a result of Bertha's less

1457 http://www.uchtspringe.de/synder.htm (Accessed 27 December 2008).
1458 Klee, *Das Personenlexikon zum Dritten Reich*, p. 42.
1459 Friedlander, *The Origins of Nazi Genocide*, p. 103.
1460 Ibid., p. 365, note 30.
1461 Bryant, *Confronting the "Good Death"*, p. 130.
1462 Klee, *Das Personenlexikon zum Dritten Reich*, p. 44.
1463 Klee. *"Euthanasie" im NS-Staat*, p. 228.
1464 http://www.doew.at/information/mitarbeiter/beitraege/rachyg.html (Accessed 23 November 2008).
1465 http://en.doew.braintrust.at/m22sm110.html (Accessed 21 November 2008).

direct participation in "euthanasia" is unknown, for together with Rudolf Lonauer (q.v.) he is considered one of the principle organizers of the programme in Austria.

Bertha received no punishment for his participation in Nazi mass murder. Even though incriminating evidence was available, no procedure was initiated against him. He went on to enjoy a distinguished academic career in post-war Austria.

Karl Bonhoeffer (1868–1948)[1466] of a quite a different character from the scoundrels and killers by whom he is surrounded here, was no National Socialist. However, he was, and still is, a controversial figure, whose prominence and support of some aspects of their eugenic policy lent an air of respectability to the Nazi's interpretation of racial hygiene.

Bonhoeffer was born in Neresheim and began his medical studies in 1887 in Tübingen, continuing these in Berlin and Munich. He became head of the Breslau Mental Hospital in 1904. In 1912 he moved to Berlin, where he was elected chair of the Department of Psychiatry and Neurology at the renowned Charité Hospital, a position he held until 1937 and his official retirement. He became chairman of the German Psychiatric Association,[1467] and was viewed as being among the most pre-eminent German psychiatrists of his generation. He was therefore in a position to use his considerable influence to oppose Nazi pseudo-scientific racial policy, but did not do so. To the contrary Bonhoeffer became a key player in the sterilisation of the "mentally inferior", delivering lectures on the subject to those responsible for implementing this policy[1468] and, like many others, did so voluntarily. Despite his retirement he continued to act as a consultant and as a judge in the Nazi Higher Hereditary Health Court.

In December 1941 he examined a Jewish *Mischling*, a "half breed" named Gottfried Hirschberg, who, 14 years before, had been admitted to a psychiatric unit. Even the *NS-Erbgesundheitsgericht* (the "Law Court for the Protection of German Blood and Honour") hesitated, since the individual showed no symptoms of disease and was capable of working normally, yet Bonhoeffer nevertheless advised sterilisation.[1469] As court consultant or judge, between 1934 and 1941 Bonhoeffer is known to have been involved in at least 126 sterilisation hearings, of which at least 57 resulted in the compulsory sterilisation of the individual concerned.[1470]

After 1945, Bonhoeffer participated in the reorganization of psychiatry in Berlin,[1471] and in 1957 the Wittenau mental hospital in Berlin was renamed the "Karl Bonhoeffer Clinic of Neurology" in his honour. Today, there are those who condemn Bonhoeffer for neither rejecting sterilisation, nor for working convincingly against its obligatory enforcement, but rather see him as being responsible for spurious "psychiatric scientifically" based diagnoses as a basis for enforced sterilisation. Whether, as has been

---

1466 Klee, *Das Personenlexikon zum Dritten Reich*, p. 64.

1467 Burleigh, *Death and Deliverance*, p. 11.

1468 Ibid., p. 12.

1469 http://www.irren-offensive.de/kleespeech.htm (Accessed 19 August 2008).

1470 Andreas Ströhle, M. D., Jana Wrase, Ph. D., Henry Malach, M.D., Christof Gestrich, Ph.D., and Andreas Heinz, M.D., *Images in Psychiatry: Karl Bonhoeffer [1868–1948]* (American Journal of Psychiatry, 165: May 2008), pp. 575–576.

1471 Ibid.

claimed, "Bonhoeffer can be seen as the 'missing link' to the Holocaust", is an even more debatable issue.[1472]

Margarete (or Margarethe or Maria) Hermine Borkowski (1885–?) was one of a number of nurses who appeared in court in Frankfurt at the second Hadamar trial held in 1947. She had begun working with youngsters at Hadamar in 1924 or 1927, and continued to do so until 1939, at which time she moved to the institution at Herborn, from where patients were transferred to Hadamar for killing.[1473] There is no evidence that Borkowski was implicated in "euthanasia" at this time, nor that she was ever a member of the Nazi party.

She was then employed successively by the mental hospitals at Kalmenhof and Weilmünster, before in January 1943 she returned to Hadamar. There, children were murdered under the supervision of Borkowski's superior, senior nurse Käthe Hakbarth. When the latter was absent, or in the course of her own night duty, Borkowski administered lethal medication to "about fifty" children. Although finding the task abhorrent, she accepted assurances that each case was carefully chosen, and, after all, *Befehl ist Befehl* ("an order is an order"). She was sentenced to imprisonment for a term of two years and six months.[1474]

Kurt Borm (1909–2005?)[1475] [T4 pseudonym "Dr Storm"] was born in Berlin. After leaving school he commenced his medical studies in Berlin in 1929, switching after a short time to Rostock, where he took his preliminary examination in medicine in 1932. He then returned to Berlin,[1476] where he eventually qualified as a physician in 1937, and was employed at the St Urban city hospital. He became a Nazi party member in December 1930, and joined the SS in 1933. Borm volunteered for the *Waffen*-SS in September 1939, resigning from his post at St Urban to serve as a medical examiner in the Sudetenland. Subsequently his duties took him to Prague and Munich, before in November 1940 he was assigned to the SS Medical Inspection Bureau in Berlin. One month later he was detached to the KdF for a "special assignment".

Werner Blankenburg (q.v.) met Borm at the KdF and told him that in order to relieve their suffering there was a Hitler-*Befehl* for the "euthanasia" of the incurably mentally ill. So as not to disturb the general population, the order must remain secret, as must Borm's participation in its implementation. With that Borm was dispatched to Sonnenstein, where Horst Schumann (q.v.) informed him of the nature of his duties. Schumann later testified that during his frequent absences from Sonnenstein, Borm acted as his deputy and supervised the entire institution, including gassing operations, a charge Borm vehemently denied, claiming to have "only" inspected the victims in order to come up with a mendacious but plausible cause of death and to have conducted other activities that were not in themselves directly murderous. He had never operated the gas valve, he said, and had no idea who was responsible for doing so, a wholly unbeliev-

---

1472 http://www.freedom-of-thought.de/may2/may2_2006.htm (Accessed 26 August 2008).

1473 Burleigh, *Death and Deliverance*, p. 240, de Mildt, *In the Name of the People*, p. 196.

1474 de Mildt, *In the Name of the People*, p. 371, note 96.

1475 Klee, *Das Personenlexikon zum Dritten Reich*, p. 65.

1476 He was a student of Karl Bonhoeffer's in Berlin. (Friedlander, *The Origins of Nazi Genocide*, p. 223).

able declaration. With the exception of occasional interludes at Bernburg as a substitute for Heinrich Bunke (q.v.), Borm remained at Sonnenstein until late August 1941.[1477] If he was dissatisfied with the demands killing helpless victims made on his professional capabilities, he apparently felt no legal or moral qualms about his involvement in their murder.

In winter 1941/42 Borm was one of the many T4 operatives sent to Russia as part of the *'Organisation Todt'* operation. He returned to the Berlin head office of T4 in March 1942 to work for Paul Nitsche (q.v.). On Blankenburg's recommendation, Borm was promoted to the rank of *Hauptsturmführer* (his previous rank had been *Obersturmführer*) in April 1943.[1478] He was active in so-called "wild euthanasia" and *Aktion* Brand, particularly in procuring and distributing large quantities of lethal drugs and medications.[1479] Borm remained an employee of T4 until the end of the war, at which time he left for Schleswig-Holstein and the Municipal hospital at Uetersen, where he eventually became a senior physician.

Borm was arrested in June 1962, but was released within a few weeks to return to Uetersen where he became a general practitioner. Initially Borm's case was linked with those of Bunke, Klaus Endruweit (q.v.) and Aquilin Ullrich (q.v.), but in the end was heard separately. The trial began in December 1971 and lasted a little less than six months. Even though it was proven that he had been an accessory to the murder of at least 6,652 psychiatric patients, Borm was acquitted on 6 June 1972 on the grounds that there was no irrefutable evidence that he was aware of the illegality of his actions.[1480] In the opinion of the court, Borm was "an obviously uncomplicated type, who is not inclined to engage in profound considerations regarding the orders or instructions given to him, or to straining his conscience to find out if these could possibly be illegal."[1481] When Borm's acquittal was confirmed by the Federal Court, it brought forth a storm of protest from artists and authors, including Heinrich Böll and Günter Grass.[1482]

Borm returned to Uetersen and general practice, but thereafter was largely shunned by the town's inhabitants. Not surprisingly, few were willing to become patients of a man so extensively implicated in mass murder.

1477 de Mildt, *In the Name of the People,* pp. 123–124.

1478 Borm claimed to have been asked by Blankenburg to serve in Lublin. Knowing of the extermination camps of the Lublin region, he refused—and suffered no consequences as a result of his refusal. (Friedlander, *The Origins of Nazi Genocide,* p. 244).

1479 de Mildt, *In the Name of the People,* p. 351, note 91.

1480 Burleigh, *Death and Deliverance,* p. 278.

1481 de Mildt, *In the Name of the People,* p. 125.

1482 Burleigh, *Death and Deliverance,* p. 278.

Karl Brandt
Source: Wikimedia Commons / Public Domain

Karl Brandt[1483] (1904–1948) was born in Mülhausen, Alsace (at that time part of Germany, now Mulhouse, France). His father was a policeman, although there were a number of medical connections on his mother's side of the family. Like George Renno (q.v.), Brandt and his family were expelled from Alsace in 1919. Brandt attended a number of different schools before taking up his medical studies in Jena in 1923 and continuing them in Freiburg (where he was a student of Alfred Hoche), Berlin, Weimar, and Munich, eventually passing his general medical examination in Freiburg in1928, and receiving his doctor's degree in 1929. Brandt had ambitions to be a surgeon, and so in 1928 took up a position at a hospital in Bochum specializing in bone fractures, where he was employed for the next several years.

By 1934 Brandt was working in the Surgical University Clinic in Berlin, and was appointed chief doctor there in 1936. He had joined the Nazi party in 1932, the SA in 1933, and the SS in 1934. In the late 1920's, Brandt met Anna Rehborn, his future wife. She was a swimming champion, and had known Hitler for some time. Hitler liked to be surrounded by attractive women, providing they knew their place and kept their collective mouths shut, and it was through Anna Rehborn that Brandt met the *Führer*. In summer 1934 Brandt became Hitler's "escort physician", accompanying him on his frequent travels.[1484] Brandt's dedication to the prevalent eugenic theory was never in doubt. In captivity in 1946, he wrote: "The more decisive eugenics, that is the prophylaxis and prevention of hereditary ill offspring, is being carried out in a comprehensive and consistent fashion, the less the whole problem of euthanasia needs to be debated."[1485]

As has already been described, Brandt was a major figure in "euthanasia" from the programme's earliest days. Through his relationship with Hitler, he became an important influence on many other aspects of Nazi policy. Like his good friend, Albert Speer, Brandt achieved a position of enormous power, although neither was initially perceived as being a political creature. In Hitler's *ad hoc* style of government that was not a necessary qualification. It was simply enough to ensure that one had the ear of the *Führer* and never ventured to disagree with his interminably expressed opinions. Brandt had much in common with Speer; they were both young (Speer was born in 1905), intelligent, talented, ambitious, cultured, industrious, loyal,—and convinced Nazis. It is only through the recent publication of the first complete biography of Brandt

1483 The following biographical material has largely been derived from Ulf Schmidt, *Karl Brandt: The Nazi Doctor—Medicine and Power in the Third Reich* (London: Hambledon Continuum, 2007).

1484 A number of different doctors treated Hitler for a variety of ailments over the years. Brandt's medical function as "escort physician" was limited to providing emergency treatment in the event of an accident. Unlike Theodor Morell, Ludwig Stumpfegger or Werner Haase, Brandt was never "Hitler's doctor" in the sense that he prescribed treatment for the *Führer's* numerous complaints, as he has many times been incorrectly described. (Weindling, *Nazi Medicine and the Nuremberg Trials*, p. 140).

1485 Schmidt, *Karl Brandt*, p. 78.

that the full extent of his importance in the cabal that surrounded Hitler has been revealed. Brandt became Hitler's envoy, a trouble-shooter sent to resolve difficult issues for which Hitler had neither the time, inclination, nor ability with which to deal.

After reaching the heights of "General Commissioner of the *Führer* for Health and Sanitation" in August 1942, a position whose jurisdiction was expanded in the ensuing years so as to make him the supreme medical authority of the Nazi state, Brandt's fall from grace followed a familiar course. In September 1944, his attempt to remove the charlatan, Theodor Morell, Hitler's personal physician, only resulted in his own loss of authority. On 2 April 1945, Brandt confronted Hitler with indisputable evidence that Germany's medical supplies either were, or shortly would be, exhausted. Bormann, Morell, and Goebbels, who were present when he made his report and were no friends of Brandt, accused him of defeatism the moment he left the room. Hitler ordered Brandt's arrest. After a "court martial", Brandt was condemned to death, but in the confusion surrounding the demise of the Third *Reich*, instead was escorted by a group of *Gestapo* men to northern Germany, where together with other members of the "Dönitz government" established after Hitler's death, he was arrested in Flensburg on 23 May 1945.

Brandt was the principal defendant at the Nuremberg Medical Trial, which opened on 9 December 1946. The trial concentrated on the medical experiments conducted on concentration camp inmates, in which Brandt had been so extensively implicated; the "euthanasia" programme was simply one in a long list of abominations. He was unrepentant concerning his role in T4, proclaiming to his interrogator:

> We German physicians look upon the state as an individual to whom we owe prime allegiance, and we therefore do not hesitate to destroy an aggregate of, for instance, a trillion cells in the form of a number of individual human beings if we believe they are harmful to the total organism—the state- or if we feel that the state will thrive without them.[1486]

His initial version of the "euthanasia" programme was interesting, since almost every word was a self-serving lie intended to minimize his involvement:

> (Brandt) stated that euthanasia cropped up in one way or another before the war and that various local Gauleiters were constantly advocating it. A legal formulation of this principle was rejected because of the difficulty of controlling it. In 1940 Hitler prohibited euthanasia. In 1944, however, (Brandt) received a notice from the Gestapo that there was a surprisingly high death rate in an insane asylum in Pomerania. (Brandt) requested information from the Gauleiter there and was told that he was misinformed. He passed this information on to the *Gestapo*. He has no further information.[1487]

Later, he came rather closer to a more accurate description of events, although his statement was still far removed from the truth:

> [The "euthanasia" programme] was [established] in 1939/40, after some kind of preliminary discussions in which I had not participated...The discussions did not come to any conclusion, so that in the autumn—at least as I recall after the Polish campaign—on the basis of newly submitted documents the Führer wanted a decree, and in this decree he gave the order to Mr Bouhler and me to authorize doctors to perform euthanasia for the incurably ill, after the most critical assessment of the illness.[1488]

1486 Ibid., p. 376.
1487 Ibid., p. 336.
1488 Ibid., p. 345.

In the courtroom Brandt attempted to claim the moral high ground with regard to the treatment of the thousands who had fallen victim to the policies he oversaw. "I see no justification because a person is sick or suffering, or because he can no longer work, to kill him, no matter what his nationality is or what his age is," he lied again.[1489] Brandt's guilt was never seriously in doubt. He was sentenced to death on 20 August 1947 and executed on 2 June 1948. How little he was prepared to acknowledge that guilt can be gleaned from words he uttered immediately before his execution: "It is no shame to stand on this scaffold. I served my fatherland as others before me."[1490]

Heinrich Bunke (1914–2001) [T4 pseudonym "Dr Rieper" at Brandenburg, "Dr Keller" at Bernburg] was born in Wohlde, near Celle.[1491] Instead of immediately entering university, in 1934 he first volunteered for Labour Service, then enlisted in the *Reichswehr*, the predecessor of the *Wehrmacht*. After his discharge from the army he joined the National Socialist Motor Corps (*NSKK*), and abandoning an earlier interest in theology, entered Kiel University in late 1935 as a medical student. He continued his studies at Freiburg University, joining the Nazi Party in 1937.

On the outbreak of war, Bunke was conscripted and served as an army physician, even though he had not completed his studies, and was therefore not licensed to practice medicine. However, in common with others who became part of the medical personnel of T4, he was granted the status of a "*Notapprobation* doctor", in effect an emergency licence to practice.

When asked if he could recommend any young doctors for participation in the programme, Bunke's friend Aquilin Ullrich (q.v.) had recommended Bunke to Werner Heyde (q.v.). In late June or early July 1940, Bunke received a written order from the KdF to contact Werner Blankenburg (q.v.), who summoned him to Berlin for an interview with Hans Hefelmann (q.v.), Paul Nitsche (q.v.), and Heyde. After a lengthy exposition of the virtues of "euthanasia" by this triumvirate, Bunke was asked whether he was prepared to participate in its implementation. Clearly overawed by the presence of such distinguished company, he protested that he was neither fully licensed nor had any knowledge of psychiatry. No matter; Ullrich's endorsement was sufficient, and in any event, others better qualified than he would make the important decisions. Bunke would merely be attached to a suitably experienced doctor. As Bunke procrastinated, Heyde made it perfectly clear that there was no compulsion to join T4. Eventually Bunke concluded that his prospects were rather brighter practising "euthanasia" than in the army, and agreed to join the programme. A few weeks later, his military service was suspended by order of the KdF.[1492]

From August 1940 Bunke worked at Brandenburg as an assistant to Irmfried Eberl (q.v.), before both were transferred to Bernburg in late October of that year. In May 1941 Bunke spent four to six weeks training at Berlin-Buch with Julius Hallervorden (q.v.) prior to commencing work on the removal of the brains of patients gassed at

---

1489 Weindling, *Nazi Medicine and the Nuremberg Trials*, p. 270.
1490 Ibid., p. 305.
1491 Klee, *Das Personenlexikon zum Dritten Reich*, p. 85.
1492 de Mildt, *In the Name of the People*, pp. 118–119.

Brandenburg.[1493] In October 1941 he was attached to *Organisation Todt* as a doctor, then returned to the military in December 1944.

In 1950 Bunke commenced practicing as a gynaecologist in Celle.[1494] Finally arrested in 1962, his pre-trial confinement was suspended and he returned to his practice while awaiting trial.[1495] The proceedings against Ullrich, Bunke, and Klaus Endruweit (q.v.) were conducted in concert. Notwithstanding the fact that Bunke had participated in the murder of "at least 4,950" individuals, he was acquitted, as were Ullrich and Endruweit. In the judgement of the court, all three defendants had acted "without culpability," being under the impression that the killing of mental patients was permissible and the victims were "without a natural will to live."[1496]

The German Supreme Court reversed the acquittal of all three defendants in 1970 and a new hearing was arranged.[1497] This time the trio were declared unfit to stand trial on medical grounds, although such incapacity did not appear to prevent them from retaining their individual medical licence. In 1987, after a protracted trial, Bunke and Ullrich were convicted of having been accessories to the murder of 11,000 and 4,500 people respectively. For this they were sentenced to terms of imprisonment of four years. In December 1988, the Federal court of appeal decided to reduce the number of victims to 9,200 in Bunke's case, and 2,340 in Ullrich's. The sentences were accordingly reduced to three years, thereby suggesting that a reduction in the number of victims somehow also reduced the seriousness of the crime.[1498] In the event, Bunke was released after serving 18 months in jail.[1499]

Werner Julius Eduard Catel (1894–1981)[1500] was a Mannheim born paediatrician. He served in the First World War before studying medicine in Freiburg and Halle, and in 1922 began working at the Leipzig University children's hospital, where he habilitated in 1926. When the head of the children's hospital, Georg Bessau, was appointed to the children's hospital of the Charité at the University of Berlin in 1932, Catel accompanied him. In October 1933 Catel was appointed professor of neurology and psychiatry at Leipzig University and director of the children's hospital there, positions he retained until 1946. Having previously been a member of other Nazi organizations, Catel joined the NSDAP itself in May 1937. As has already been described, Catel was one of the three *Obergutachter* in children's "euthanasia", responsible for an unquantifiable number of fatalities.

It seems almost beyond belief that denazification proceedings in 1947 described Catel as a "convinced anti-fascist and representative of humanity during the Third *Reich*."[1501] In the same year he became director of a paediatric hospital in Mammolshain. Two years later he was one of 13 defendants in proceedings brought in Hamburg

1493 Aly, Chroust, Pross, *Cleansing the Fatherland*, p. 225–226.
1494 Klee, *Das Personenlexikon zum Dritten Reich*, p. 85
1495 de Mildt, *In the Name of the People*,p. 118.
1496 Ibid.pp. 122–123.
1497 Bryant, *Confronting the "Good Death"*, p. 213.
1498 Burleigh, *Death and Deliverance*, pp. 278–280.
1499 Klee, *Das Personenlexikon zum Dritten Reich*, p. 85.
1500 Ibid. p. 91.
1501 de Mildt, *In the Name of the People*, p. 127.

in connection with the killing of children at the Rothenburgsort children's hospital. The trial was abandoned after the accused were deemed to have acted without "awareness of injustice."[1502] Thus one of the more notable participators in the murder of the young escaped retribution, to return to practising paediatrics. By 1954 he had been appointed professor of paediatric medicine at the University of Kiel. Incredibly, he never abandoned championing the value of the "mercy killing" of disabled children, so much so that public outrage forced his retirement in 1960.[1503]

Leonardo Conti
Source: Wikimedia Commons / Bundesarchiv, Bild 183-1989-0309-501 / CC-BY-SA 3.0 https://creativecommons.org/licenses/by-sa/3.0/de/deed.en

Leonardo Conti (1900–1945)[1504] was born in Lugano, Switzerland of a Swiss-Italian father and a German mother. He studied medicine in Germany, where in 1918 he was extensively involved in nationalistic and anti-Semitic organisations. He joined the *Freikorps* and participated in the Kapp *putsch* before becoming a member of the SA in 1923 and the SS in 1930. After qualifying as a physician in 1927, he became a general practitioner in Berlin, and joined the NSDAP in the same year, thereby strengthening his party connections. Gradually progressing via a variety of appointments of increasing importance, Conti was appointed *Reich* Health Leader, head of the *Reich* Physicians' Chamber, and following the death of Gerhard Wagner (q.v.) in 1939, State Secretary for Health in the Ministry of the Interior.[1505]

Conti played a vital role in the conception and initial realization of "euthanasia", but thereafter his importance to the programme and his overall political influence diminished as that of Karl Brandt (q.v.) grew. Although there is no doubt that Conti was kept aware of developments, and appended his signature to much of the relevant "euthanasia" paperwork, he left the daily chore of organizing cooperation with T4 to Herbert Linden (q.v.).[1506] In the never ending struggle for power in the Third *Reich*, Conti was to lose out to Brandt. Goebbels summarized the position in a diary entry of 1 September 1944: "Brandt has the ear of the Führer, while Conti has not been able to give a report to the Führer for years."[1507] By then, Martin Bormann, who regarded Conti as his protégé, was complaining in a letter to his wife:

> Today I tried my best to get Conti a hearing with the *Führer*...To my great distress, the *Führer*, who believes that every good thing he has been told of and about Brandt and every bad thing he has been told about Conti is the gospel truth, got very annoyed with my lack of understanding...I told the *Führer*...that it was an injustice to represent Conti as completely unqualified and inefficient or to exaggerate his weaknesses...The *Führer* thought differently; he spoke with high praise of Brandt...[1508]

1502 Ibid., p. 352, note 107.
1503 Burleigh, *Death and Deliverance*, pp. 273–274.
1504 Klee, *Das Personenlexikon zum Dritten Reich*, p. 96.
1505 Robert S Wistrich, *Who's Who in Nazi Germany* (London: Routledge, 1995), pp. 31–32.
1506 Friedlander, *The Origins of Nazi Genocide*, p. 190.
1507 Schmidt, *Karl Brandt*, p. 311.
1508 Ibid., p. 308.

After being arrested by the British on 19 May 1945 in Flensburg, Conti committed suicide at Nuremberg on 6 October 1945 by hanging himself with a towel tied to the bars of his cell.[1509] He left a note explaining somewhat bizarrely that he was taking his life because he had lied under oath whilst being interrogated about medical experiments.[1510] He would certainly have been among the most important defendants in the Medical Trial that commenced in Nuremberg on 9 December 1946. As was the case with other prominent Nazis, Conti's suicide whilst in custody inevitably raised suspicions, conspiracies being so much easier to assume than to prove.

Walter Creutz (1898–1964)[1511] and his actions can either be seen as an example of one way in which it was possible to resist Nazi government policy, at least to a limited extent, or alternatively as representing a cynical manipulation of the *status quo*. His conduct posed an ethical dilemma, the resolution of which remains divisive to this day.

Creutz was born in Osterfeld, a doctor's son. At the age of 18 he left the *Gymnasium* and was conscripted into the army. On his discharge in 1919 he studied medicine at the universities of Bonn and Münster, graduating in 1923. After gaining experience in different medical posts, he decided on a career as a psychiatrist and was employed at the mental hospitals at Bedburg-Hau and Duisburg-Hochfeld. Promoted to chief physician in 1930, he remained at the latter institution until 1935, when he was transferred to the Rhine provincial government's head office in Düsseldorf where, apart from his military service, he remained until the end of the war. In 1940 he was appointed a professor at the University Medical Academy in Düsseldorf; he had been attached there as a lecturer in psychiatry and neurology since 1934. He had joined the NSDAP and the SA in 1933, but was not active in either, allowing his SA membership to lapse in 1936.

Creutz was drafted into the *Wehrmacht* to serve as a medical officer at the beginning of the war, but in December 1940 his military duties were suspended and he returned to his previous Rhineland government position. He had already heard rumours concerning the "euthanasia" project, something to which he was strongly opposed, and knowing that such measures would inevitably be introduced in the Rhine region, he attempted to dissuade his superior, Heinrich Haake, from instituting them. Initially he was successful, but under pressure from a high powered committee headed by Werner Heyde (q.v.), Haake changed his position. To Creutz's disappointment, "euthanasia" would be initiated in the province.

Creutz claimed that he was faced with a difficult decision; either refuse to cooperate and attempt a return to the military, or stay and try to sabotage the programme from within. He chose the latter, and on 29 March 1941 called a meeting of the directors of all mental institutions under his jurisdiction to inform them of the forthcoming policy. Refusing to participate would be useless—the regime would simply find others who would be prepared to do so. Instead, they should remain at their posts and use every means at their disposal to save as many of their patients as they could. This strategy was successful—up to a point. According to Creutz's testimony, by September 1941, of 5,046 patients in his region selected for gassing, "only" 946 had been transported to the

1509 Ibid., p. 353.

1510 Raul Hilberg, *The Destruction of the European Jews* (New Haven: Yale University Press, 2003), p. 1156.

1511 Klee, *Das Personenlexikon zum Dritten Reich*, p. 97.

killing centres. He had also been instrumental in limiting the establishment of special killing wards for children in his province to just one, at Waldniel, but his tactics made it impossible to avoid some degree of collaboration with the murderers. In order to free hospital space, between 1942 and 1945 patients from the Rhineland were dispatched to a number of different killing institutions, mainly under the umbrella of *Aktion Brandt*. Whether Creutz was aware of their fate is at least debatable.

In November 1948, Creutz was acquitted by a Düsseldorf court of participation in the "euthanasia" *Aktion*. Although he had intentionally committed a criminal act, he had only done so in order to prevent a worse crime being enacted. The "duty to rescue" prevailed over "the duty to abstain from crime." However, the Supreme Court of the British Zone took a different view: "The attempt to oppose two groups of patients to one another, must fail...Human lives cannot be measured against each other."[1512] In January 1950 a retrial was ordered, resulting in Creutz's acquittal for a second time. Nearly 50 years later, to some Creutz did not appear the hero that the Düsseldorf court twice portrayed "Only this judiciary, itself well-schooled in murder, could extol to his patients a nine-hundred-fold murder accomplice as an idol of morality."[1513]

Max de Crinis

Source: https://neurologie.charite.de/fileadmin/user_upload/microsites/m_cc15/neurologie/Geschichte/Crinis.jpg / CC BY-SA 3.0 https://creativecommons.org/licenses/by-sa/3.0/

Maximinus Friedrich Alexander de Crinis (1889–1945),[1514] the product of a distinguished Austrian medical family, studied medicine at Graz and Innsbruck. In 1914 he became an assistant at the university of Graz psychiatric clinic, where he researched battlefield psychiatric casualties. Already extensively involved in Nationalist politics, and considered a fanatical anti-Semite, in 1931 he became a member of the then illegal Austrian Nazi party. He was arrested in 1934 in connection with Nazi terrorist activities, and was prohibited from teaching at Austrian universities.

He fled from Austria before the July 1934 *putsch* against Dollfuss (in which he may have been involved),[1515] spending some time in Yugoslavia treating those wounded in the attempted *coup d'état*. Later that year he was appointed professor of psychiatry at the University of Cologne, subsequently becoming *Führer* of the university teaching association. In 1936 he joined the SS, and in 1938 succeeded Karl Bonhoeffer (q.v.) both as professor of psychiatry and neurology at Berlin University, as well as director of the psychiatric clinic at the Charité hospital. In his inaugural lecture at the clinic, de Crinis was contemptuous of what he termed "those un-German, racially alien, so-called expert colleagues who have been purveying their immoralities in Vienna and other German cities until quite recently" (such as Sigmund Freud). De Crinis became a member of the steering committee of the Kaiser Wilhelm Institute for Neurological Research, a leading military psychiatric

1512 By way of contrast *vide* the reduction of sentences in the cases of Bunke and Ullrich.

1513 de Mildt, *In the Name of the People*,pp. 148–156, Bryant, *Confronting the "Good Death"*, pp. 151–164.

1514 Klee, *Das Personenlexikon zum Dritten Reich*, p. 97.

1515 Lifton, *The Nazi Doctors*, p. 121

advisor, and in early 1940[1516] head of the Office of Science at the Ministry of Education.[1517]

Although de Crinis' precise role in the "euthanasia" programme remains shadowy,[1518] there is no doubt that as a personal friend of Reinhard Heydrich he was heavily implicated in its implementation, playing a major role in all important decisions.[1519] Amongst other things, being among the most senior medical staff at T4,[1520] de Crinis may have been responsible for the final draft of the "euthanasia" sanction.[1521] However, he also went to great lengths to keep his involvement in the project a closely guarded secret, to the extent of even appearing to be an opponent of "euthanasia".[1522]

In November 1939, de Crinis participated in one of the more outré events of the Second World War, the so-called "Venlo Incident". De Crinis' best friend was Walter Schellenberg, a prominent member of the SD, and head of section IV E (counter-espionage) of Amt IV (the *Gestapo*) of the RSHA. Schellenberg was a man involved in all kinds of skulduggery. On the pretence that he was a member of a plot by high ranking *Wehrmacht* officers to get rid of Hitler and form a new German government, Schellenberg lured two British intelligence officers, Captain Sigismund Payne Best and Major Richard Stevens, to a series of meetings, some of which took place at the Café Backhus near Venlo in then neutral Holland. Heydrich suspected that there was in fact such a conspiracy within the *Wehrmacht*, and contact with British intelligence would perhaps prove this, as well as providing the opportunity of feeding false intelligence to the enemy.

Following the unsuccessful attempt on Hitler's life on 8 November 1939 (which Hitler was convinced had been engineered by the British secret service), Himmler ordered the cancellation of the cloak and dagger operation, and the arrest of Best and Stevens. German agents kidnapped the pair near the meeting place and hauled them back over the German border. It was quite a coup for Schellenberg, since Best had a list of British agents with him, and more were revealed by the subsequent interrogation of the British pair. De Crinis, invited to participate in the charade by Schellenberg, and using the pseudonym "Colonel Martini", played the role of the principal lieutenant of the leader of the plot. The whole episode had a Ruritanean quality about it, as Schellenberg's description of their clandestine meetings with Best and Stevens indicates:

> De Crinis and I agreed upon a system of signs whereby we could communicate with each other during the discussion with the Englishmen. If I removed my monocle with my left hand, that meant that he would immediately stop talking and leave me to pursue the conversation. If with the right hand, that I needed his support. The sign of an immediate breaking off of the conversation would be my having a migraine.[1523]

1516 Michael H Kater, *Doctors Under Hitler* (Chapel Hill: University of North Carolina Press, 2000), p. 128.

1517 Burleigh, *Death and Deliverance*, pp. 118–119.

1518 De Crinis has been described as the *éminence grise* of "euthanasia." [http://neurologie.med-network.de/geschichte/decrinis.html (Accessed 8 April 2008.)]

1519 Aly, Chroust, Pross, *Cleansing the Fatherland*, p. 37.

1520 Burleigh, *Death and Deliverance*, p. 119.

1521 Lifton, *The Nazi Doctors*, p. 63 (footnote).

1522 Ibid., p. 121.

1523 André Brissaud, *The Nazi Secret Service* (London: Corgi Books,1975), pp. 287–314.

De Crinis' other wartime activities were undoubtedly less romantic, and immeasurably deadlier. A true believer to the end, together with his wife he committed suicide by swallowing cyanide on 2 May 1945. In 1963, a historian of the Charité hospital commented that no other physician in the institution's history had been guilty of crimes comparable to de Crinis'.[1524]

Fritz Cropp (1887–1984)[1525] was born in Oldenburg. In 1920 he became a physician in Delmenhorst, and was a member of a number of different right-wing nationalist parties before joining the NSDAP and SA in 1931. Two years later Cropp was appointed head of district medicine in Oldenburg. In 1935 he received promotion to the National Health department of the *Reich* Ministry of the Interior in Berlin, subsequently succeeding Arthur Gütt (q.v.) as head of the department in 1939.[1526] In that capacity he was responsible for arranging the transfer of hospital patients to T4 through his subordinate, Herbert Linden (q.v.). From 1943 until the end of the war Cropp was general adviser on damage caused by Allied bombing.

After a short period of internment, Cropp returned to Delmenhorst in 1946 and worked as a general practitioner. In 1948, a denazification tribunal in Oldenburg barred him from any further political activity. The following year he became an adviser to the Central Committee of the Western Inner Mission in Bethel. When he retired in 1952 he received a pension based upon his former position as a ministerial official, yet another to escape any kind of meaningful retribution for his extensive involvement in "euthanasia".

Irmfried Eberl (1910–1948) [T4 pseudonym "Dr Schneider", possibly also "Dr Meyer"[1527]][1528] was born in Bregenz, Austria. He attended the University of Innsbruck, joining the then illegal Austrian Nazi party in 1931 and receiving his medical licence and doctorate in 1935. After completing his training at a number of different hospitals, he moved to Germany in 1936. By 1937 he had received accreditation as a physician in Germany, and subsequently worked for the emergency medical services in Berlin.

Eberl was an early recruit to T4. His dedication to Nazism and his undisguised enthusiasm for "euthanasia" made him an obvious choice. Ambition, too, certainly played its part. In 1938 he had married Ruth Rehm, an equally fanatical Nazi. She held a senior position in the women's section of the German Labour Front, and it is likely that she was influential in his burgeoning career.[1529] Eberl was present at the Brandenburg experimental gassing, where he was one of those responsible for opening a carbon monoxide cylinder.[1530] He went on to become the physician-in-charge at Brandenburg from the time "euthanasia" operations commenced there until September 1940, shortly after which he switched to occupying the same position at Bernburg.

1524 Lifton, *The Nazi Doctors*, p. 121.
1525 Klee, *Das Personenlexikon zum Dritten Reich*, p. 98.
1526 Friedlander, *The Origins of Nazi Genocide*, p. 64.
1527 Ibid., p. 103.
1528 Klee, *Das Personenlexikon zum Dritten Reich*, p. 123.
1529 Strous, Rael D, M.D., *Dr. Irmfried Eberl (1910–1948): Mass Murdering M.D.* (The Israel Medical Association Journal, Vol 11, April 2009) pp. 216–218.
1530 Friedlander, *The Origins of Nazi Genocide*, p. 210.

Eberl was another to join the *Organisation Todt* mission to the eastern front in winter 1941/42. During the late spring and early summer of 1942, Eberl supervised the construction of the Treblinka extermination camp, and was appointed the first commandant of the camp, a position he held from 23 July 1942, when the killings began, until the last week of August 1942. During that five week period, about 312,500 Jews were deported to Treblinka and murdered. It was a number far beyond the camp's (and Eberl's) capabilities. August Hingst, who served at the camp, stated: "Dr Eberl's ambition was to reach the highest possible numbers and exceed all other camps. So many transports arrived that the disembarkation and gassing of the people could no longer be handled." Eberl had completely lost control of the situation. When Odilo Globocnik and Christian Wirth (q.v.), the latter by then inspector of all of the *Aktion Reinhard* camps, arrived in Treblinka at the end of August, Globocnik was so furious at what he found that Eberl was immediately dismissed. Globocnik commented "that if Dr Eberl were not his fellow countryman" [Globocnik was also Austrian], "he would arrest him and bring him before as SS and police court." Wirth assumed control of Treblinka for a short time, before handing over to Franz Stangl (q.v.).[1531] There have been alternative theories advanced for Eberl's demotion, but the weight of evidence seems to suggest that incompetence was the prime, if not the sole factor.[1532]

Eberl returned to Bernburg and remained there until at least the time of the dismantling of the institution's crematorium in April 1943, doubtless engaged in gassing concentration camp prisoners as part of *Sonderbehandlung* 14f13. In January 1944 he was drafted into the *Wehrmacht*, although he continued to be paid by T4. His activities from that time until the end of the war are unclear. His wife Ruth having been killed in an air raid in 1944, Eberl remarried after the war and lived in Blaubeuren, near Ulm, where he operated a private medical practice until the time of his arrest in August 1947.[1533] In captivity Eberl initially attempted to deny his true identity, but after he was recognized and pointed out to the authorities by an ex-member of T4, he hanged himself in his cell at Ulm on 16 February 1948.[1534]

Sophie Ehrhardt (1902–1990)[1535] was born in Kasan or Kazan, today the capital of the Republic of Tatarstan, Russian Federation. A sociologist and anthropologist, she received a master's degree in zoology before obtaining her doctorate in anthropology in 1930. In 1935 she became an assistant to Hans Friedrich Karl Günther (q.v.), before in 1937 joining Robert Ritter's (q.v.) newly created "Eugenic and Population Biological Research Station of the *Reich* Health Office" at Tübingen University. This institution was concerned with conducting eugenic research into petty criminals, especially "those of alien race", notably Gypsies. In 1938 Ritter moved his organization to Berlin, taking Ehrhardt with him. She continued with her studies, which included both the Gypsies of

1531 Arad, *Belzec, Sobibor, Treblinka*, pp. 87ff.
1532 Friedlander, *The Origins of Nazi Genocide*, p. 299, Sereny, *Into That Darkness*, pp. 157–163.
1533 http://www.jewishgen.org/yizkor/belzec1/bel020.html (Accessed 11 November 2008).
1534 Burleigh, *Death and Deliverance*, p. 261.
1535 Klee, *Das Personenlexikon zum Dritten Reich*, p. 128.

East Prussia (concerning whom she published an article in the journal *Volk und Rasse* in 1942), and the Jews of the Lodz ghetto.[1536]

In 1950 Ehrhardt became an expert on paternity appraisal for the German Society for Anthropology, and in 1957 she was appointed Professor of Anthropology at Tübingen University. Two investigations in connection with her Nazi era activities were discontinued.

Klaus Endruweit (1913–1994)[1537] [T4 pseudonym "Dr Bader"][1538] was born in Tilsit (now Sovetsk, Russian Federation), the son of the head teacher at a school for deaf and dumb children. Whether this background had any effect on his subsequent choice of career is a matter of conjecture. After leaving the *Gymnasium* in 1933 he volunteered for labour service, and joined the SA. He never became a member of the Nazi party. In late 1933 he began to study medicine at Munich University, became a member of the Nazi student organization and befriended Aquilin Ullrich (q.v.). Together they served for a year in the *Reichswehr*, after which Endruweit continued his studies at the universities of Königsberg, Würzburg, and Berlin. He became a medical assistant with the Hitler Youth, and in 1938, together with Ulrich and other Würzburg students, set off on a "scientific" mission to Bessarabia, the purpose of which was to establish the racial characteristics of the *Volksdeutsch* inhabitants of that region.[1539] In August 1939 Endruweit entered the *Wehrmacht*, joining an army medical unit. Like Ullrich and Heinrich Bunke (q.v.) he had not yet completed his medical studies, so in 1939 instead was appointed a *Notapprobation* doctor.

Without obtaining his friend's consent, Ullrich had recommended Endruweit to Werner Heyde (q.v.). Thus in autumn 1940 Endruweit's military service was suspended, and he was ordered to report to T4 in Berlin, where he arrived in December of that year. Without further ado, and without any explanation as to the nature of his duties, either Werner Blankenburg (q.v.) or Hans Hefelmann (q.v.) sent him on to Sonnenstein to report to Horst Schumann (q.v.). Endruweit, who allegedly believed he was being assigned to the sick-bay of an armaments factory, was received not by Schumann, who was absent, but by Kurt Borm (q.v.), who informed him of the purpose Sonnenstein served. Endruweit was to replace Ewald Wortmann (q.v.), who had refused any further participation in "euthanasia". When Schumann returned he managed to convince a supposedly reluctant Endruweit that not only were the killings of immense humanitarian benefit, they were also perfectly legal. Nonetheless, so Endruweit claimed, he still harboured doubts, since he felt, not unreasonably, that "it [is] the duty of the doctor to help and not to kill." He claimed to have informed Schumann that he did not intend to stay long.

In fact he remained in Sonnenstein, with some interruptions, until October 1941, during which time thousands of patients were gassed there. In November or December 1941 Endruweit was transferred to *Organisation Todt*, and was initially posted to Bres-

1536 Friedlander, *The Origins of Nazi Genocide*, pp. 250–251.

1537 Klee, *Das Personenlexikon zum Dritten Reich*, p. 135

1538 Friedlander, *The Origins of Nazi Genocide*, p. 103

1539 de Mildt, *In the Name of the People*, p. 116.

lau, then to France, before finishing the war on the eastern front.[1540] He spent some time as a prisoner of war before, disguised as a civilian, he escaped captivity, and after a spell working at the State Mental Hospital at Hildesheim, returned to practicing medicine in Bettrum.

In 1966, together with Ullrich and Bunke, he stood in the dock of the Frankfurt court, accused, in Endruweit's case, of complicity in the murder of at least 2,250 individuals. All three men were acquitted on the grounds that an "unavoidable mistake of law" had led them to believe that their actions were legal.[1541] "...The general prohibition on killing had been curtailed by Hitler's directive. According to their interpretation, this directive had partially suspended the general prohibition on killing." In 1970, the Supreme Court reversed the decision and ordered a new trial; two years later Endruweit was declared medically unfit, although he was apparently healthy enough to continue to practice medicine. In 1986 proceedings against him were finally abandoned.[1542] On Endruweit's death the Lower Saxony General Medical Council published an obituary, which included the comment: "We will honour and remember him".[1543]

Willi Enke (1895–1974)[1544] was born in St.Gallen, Switzerland. After serving in the German army during the Great War he studied medicine in Leipzig, receiving his doctorate in 1923, and was subsequently employed at the psychiatric clinics in Tübingen and Marburg. A member of the NSDAP since April 1933 and thereafter of several other party organizations, in March 1934 he joined the Racial Policy Office of the NSDAP. In 1935 he was appointed a member of a Hereditary Health Court, and in the same year became ao professor in Marburg. On 1 January 1938 he was appointed director of the mental hospital at Bernburg, a position he continued to hold after September 1940, at which time the institution was divided in between the T4 section under Irmfried Eberl (q.v.) and Enke's "normal" section.

Enke was well aware of the "euthanasia" activities carried on at Bernburg (it could hardly be otherwise), and was responsible for completing *Meldebogen* in respect of his patients. He was arrested by the Americans in 1945, but was quickly released. Denazified in 1948, he returned for a short time to the University of Marburg before becoming head of the Hephata Psychiatric Clinic at Treysa, a position he held until his retirement.[1545]

Gottfried Ewald (1888–1963)[1546] was born in Leipzig, and in 1923 became a member of *Freikorps Oberland.* He had thus hoisted his political colours at an early date, yet his repeated attempts to join the NSDAP in the 1930s were consistently refused, despite his being considered to hold a "positive political attitude toward the Third *Reich.*" The reasons for this rejection are obscure. There is a suggestion that it was because Ewald had suffered amputation of a forearm as a result of an injury incurred during First World

1540 Klee, *Was sie taten–Was sie wurden*, p. 118.
1541 Bryant, *Confronting the "Good Death"*, p. 213.
1542 de Mildt, *In the Name of the People*, pp. 115–125.
1543 http://www.irren-offensive.de/kleespeech.htm (Accessed 4 December 2008).
1544 Klee, *Das Personenlexikon zum Dritten Reich*, p. 137.
1545 http://www.catalogus-professorum-halensis.de/enkewilli.html (Accessed 14 September 2009).
1546 Klee, *Das Personenlexikon zum Dritten Reich*, p. 141.

War military service, thus preventing his being an active member of the SA, although he did join the SA reserve.[1547] Alternatively, it is proposed that refusal to accept him as a member of the Nazi party arose because of his previous *Freikorps* membership.[1548] Neither explanation seems plausible. By no means all members of the NSDAP were expected to join the SA and a history of participation in a *Freikorps* did not prevent Werner Heyde (q.v.) and Friedrich Mauz (q.v.), for example, from being accepted as members of the party. The true reasons for Ewald's rejection are likely to remain unknown.

Ewald studied medicine at the University of Heidelberg from 1906 to 1911, obtaining his doctorate in 1912 and his approbation in 1913. Between 1913 and 1916 he worked as an assistant at the neurological clinics of the University of Rostock, the Berlin Charité and the University of Erlangen, where he habilitated in 1920.[1549] From 1922 to 1933 he worked at the mental hospital at Erlangen as a senior physician, until in 1933 he was appointed to the chair in Neurology and Psychiatry at the University of Greifswald, moving on in 1934 to occupy the chair in psychiatry at Göttingen, as well as heading Göttingen University Clinic and the local state hospital. He was therefore potentially a prize catch for T4. Yet although he accepted sterilisation and even "euthanasia" if the country faced "extreme conditions", such as food shortages in time of war, when in 1940 he attended a meeting called by Werner Heyde at which those present were invited to become *Gutachter* or *Obergutachter*, Ewald declined. "On principle I would not lend my hand to exterminate in this way patients entrusted to me", he stated.[1550]

Despite his concern, telling his wife, "Please be sensible about things, but be prepared for the possibility that, at any day from now on, I might be sent to a concentration camp," Ewald suffered no dire consequences as a result of his refusal to participate in the "euthanasia" programme, proof that it was indeed possible to refuse to become a perpetrator, even if one was sympathetic to the regime and its aims.[1551] However, it should be born in mind that Ewald's refusal to become a direct participant in mass murder did not extend to his preventing the transfer of his own handicapped patients.[1552] Moreover, at the end of the war he burned many of his files because, according to his wife, they "contained material against many people—they would lose their capacity to earn a living and could be sent to prison." He had also sheltered a substantial number of psychiatrists in his hospital, individuals who had not been as superficially ethical as himself concerning questions of "euthanasia". Finally, he became a sympathetic witness for the defence in early post-war trials, thus remaining one of the more enigmatic figures in this catalogue of the unsavoury.[1553]

1547 Lifton, *The Nazi Doctors*, p. 85.

1548 Friedlander, *The Origins of Nazi Genocide*, p. 324, note 108.

1549 http://www.egmed.uni-goettingen.de/geschichte/GottfriedEwald.html (Accessed 26 February 2009).

1550 Friedlander, *The Origins of Nazi Genocide*, p. 78; Lifton, *The Nazi Doctors*, pp. 82–83.

1551 Lifton, *The Nazi Doctors*, p. 86.

1552 Friedlander, *The Origins of Nazi Genocide*, p. 78

1553 Lifton, *The Nazi Doctors*, pp. 86–87.

Valentin Faltlhauser
Source: M.V. Cranach

Valentin Faltlhauser (1876–1961)[1554] was born in Wiesenfelden, Bavaria, and was renowned in the 1920s as a caring and innovative psychiatrist.[1555] He was the author of an important textbook for nurses, as well as numerous other publications.[1556] From 1929 he was director of the Kaufbeuren institution. An advocate of the eugenics movement and a keen supporter of "euthanasia", Faltlhauser joined the NSDAP in 1935, and was involved in the killing programme from an early stage.[1557] His name appears as a T4 *Gutachter* with effect from 6 September 1940, but he had been concerned with the implementation of the killing programme from at least February of that year.[1558] In 1945 Faltlhauser sought to rationalise and justify his actions in words his fellow criminals would have been all too ready to endorse:

> The euthanasia of the mentally ill was carried out on the grounds of a decree by the *Führer*. This decree was not only a specially binding condition, but was also a duty. The decree was the result of a hearing and was released with the agreement of the *Reich* Ministry of the Interior and the *Reich* Ministry of Justice. The decree had legal force. It was supported by a special law, which was unpublished, but was declared to be binding.
>
> I am a civil servant with 43 years service. As a civil servant I was educated to absolutely follow the prevailing orders and laws, therefore also to consider the euthanasia decree as a law. In each instance there was an order to perform euthanasia as a result of a conscientious examination of the special case by a specialist. Here I wish to clearly interject, that I, as nearly all German directors of psychiatric hospitals, had nothing to do with the first realisation of the decree. I always dealt with each case in good faith, following the dictates of humanity, and in the absolute conviction to do my duty in obeying the legal and lawful conditions. [...] My actions were not undertaken with the intention of committing a crime, but in contrast they were made with the consciousness to deal mercifully with the unhappy creatures, with the intention of freeing them from their suffering where there is no known method to save or to relieve, and therefore to act in good faith as a true and conscientious doctor. He who has experienced during a long service for the mentally ill, in hundreds and hundreds of cases, the terrible fate of individuals sinking to the level of an animal, only he really knows how to understand that euthanasia cannot be an offence against humanity, rather the opposite.[1559]

So starving children to death was not "an offence against humanity"? As has been illustrated, few of Faulthauser's comments bore any relationship to the truth. However, there were those who were rather more convinced by Faulthauser's self-righteous defence of his conduct. In July 1949 he was arraigned before the court in Augsburg. Under his stewardship, an estimated 1,200–1,600 patients had been murdered at Kaufbeuren between September 1941 and April 1945, at least 300 of them in the final year of the war. The court determined that killing on this scale was worthy of a term of imprison-

1554 Klee, *Das Personenlexikon zum Dritten Reich*, p. 144.
1555 Aly, Chroust, Pross, *Cleansing the Fatherland*, p. 160.
1556 McFarland-Icke, *Nurses in Nazi Germany*, p. 276, note 35.
1557 Friedlander, *The Origins of Nazi Genocide*, p. 51.
1558 Klee, "*Euthanasie*" *im NS-Staat*, p. 228, de Mildt, *In the Name of the People*, p. 346, note 13.
1559 http://www.phil.gu.se/sffp/reports/16.%20von%20Cranach.pdf (Accessed 14 November 2008).This quotation has been slightly modified in order for it to read more fluently.

ment of three years duration. Far from being a despicable murderer of the young and helpless, Faulthauser had acted out of "compassion" for his patients, "one of the noblest motives of human conduct."[1560] The court was possibly unaware of his role as a *Gutachter*—the subject was not raised at the trial.[1561] Whether knowledge of this aspect of Faulthauser's career might have influenced their verdict is open to question.

Thus, far from being a villain, Faulthauser was viewed as being almost heroic. It was a portent of things to come. After his release from jail, Faulthauser's license to practice medicine was rescinded.

Martha Fauser
Source: ZfP Südwürttemberg

Martha Fauser (1889–?) qualified as a physician in 1925. Her interest in psychiatry led to a career at a succession of mental institutions—Göppingen, Weissenau, and in September 1940, Zwiefalten, where she succeeded Alfons Stegmann (q.v.) as deputy director. A member of the National Socialist Women's League since 1934 and the NSDAP since 1937, she was fully aware of the fate of patients from Zwiefalten who had been transferred to Grafeneck; indeed she had on at least one occasion attended a gassing there at the invitation of her friend, Ernst Baumhardt (q.v.).

Fauser appeared in court at Tübingen in 1949 in connection with her involvement in the "euthanasia" killings, and received a sympathetic hearing. Her defence was the common one of remaining at her post in order to minimise the number of victims, a task at which she was singularly unsuccessful, but then as the court misogynistically observed "in the first place the case concerned a woman." Moreover, "taking charge of such a large institution as Zwiefalten and then particularly at such a time and under such circumstances without additional medical assistance would have been too much even for an experienced male director." The fact that she had personally administered fatal injections to three of her own patients failed to convince the court of any malevolent intentions on her behalf. She was sentenced to a jail term of one year and six months for the manslaughter of these three patients, but since her pre-trial confinement had exceeded that term, was immediately released.[1562]

Emil Gelny (1890–1961) was born in Vienna, and received his general medical licence in 1915, but did not receive accreditation as a psychiatrist until August 1943. He joined the illegal Austrian Nazi party and SA in 1932. After working for the similarly illegal Nazi intelligence service in Austria he was arrested in August 1934, but on his release continued to be active in the political manoeuvring leading up to the *Anschluss*. On 1 October 1943, Gelny was appointed provisional director of the Gugging state hospital in lower Austria. A dedicated Nazi, his appointment was intended to accelerate the so-called "wild euthanasia"—which it did.[1563]

1560 de Mildt, *In the Name of the People*, pp. 98–99.
1561 Ibid., p. 346, note 15.
1562 Ibid., pp. 166–167.
1563 Friedlander, *The Origins of Nazi Genocide*, pp. 159–160.

Dissatisfied with the conventional methods of medicalized murder, and with the electric chair as his model, Gelny introduced electrocution as a killing method in a process known as "electro-execution", which was described in the following terms:

> Once a patient became unconscious from the effects of electricity, the attendants then had to attach four electrodes to the patient's hands and feet. Dr. Gelny ran high voltage through [the electrodes] and after ten minutes at the most the death of the patient would set in.[1564]

It should be born in mind that these killings were not initially conducted under the aegis of T4; it was only in February 1944 that his unique contribution to "euthanasia" was recognized. In November of that year his domain was enlarged to include the institution at Mauer-Öhling, where he killed at least 39 people with drugs such as Veronal, Luminal and morphine. Overall, he is estimated to have personally killed almost 600 patients at Gugging and Mauer-Öhling.[1565] In February 1944, Gelny boasted in a letter to the *Gau* leader that thanks to his efforts "more than 400 incurably ill people who in the present situation are a serious burden on the state [have been] eliminated in the last four months."

Gelny fled from Vienna at the end of the war, moving first to Syria and then to Baghdad, where he practised as a doctor of some repute; he died peacefully in 1961.[1566]

Hans-Bodo Gorgass (1909–1993) [T4 pseudonym "Dr Kramer", post-war pseudonym "Dr Gerber"][1567] was the Leipzig born son of a railway official. Gorgass became interested in psychiatry during the course of his medical studies, and on graduation in 1936 was first employed at the Eichberg mental hospital. A protégé of Fritz Bernotat (q.v.), Gorgass moved to the Kalmenhof institution in 1938 where he was appointed physician-in-chief, a position he retained until drafted into the *Wehrmacht* on the outbreak of war for service as an army doctor. He had joined the SA in 1933 and the NSDAP in 1937.

In April 1941 he was suddenly declared "*uk-gestellt*" ("indispensable") and ordered to report to Bernotat, who in turn sent him to meet with Viktor Brack (q.v.) in Berlin. There, Gorgass was informed that he had been chosen to participate in the "euthanasia" programme. He was rapidly shipped off to Hartheim to be trained in the ways of killing by Rudolf Lonauer (q.v.), witnessing several gassings as part of his education. Although not averse to "euthanasia" as such, Gorgass claimed to inwardly reject the methods being used. However, he soon overcame any scruples he may have harboured. He visited Horst Schumann (q.v.) in Sonnenstein and observed another gassing. As described by Gorgass, "The death was a peaceful one. It is simply going to sleep in the true sense of the word. The people grew weary, lost all sense of the outside world, and went to sleep."[1568] Others were to describe the process in less comforting terms.

Thus suitably prepared, he reported to Friedrich Berner (q.v.) at Hadamar to begin his "euthanasia" career. Gorgass supervised the entire killing process, from the arrival

---

1564 Thomas Röder, Volker Kubillus, Anthony Burwell, *Psychiatrists—the Men Behind Hitler: The Architects of Horror* (Los Angeles: Freedom Publishing 1995), p. 208.

1565 http://en.doew.braintrust.at/m22sm110.html (Accessed 21 November 2008).

1566 http://www.memorialgugging.at/pdf/Czech_MedizinverbrechenGugging_engl.pdf (Accessed 21 November 2008).

1567 Klee, *Das Personenlexikon zum Dritten Reich*, p. 192.

1568 Burleigh, *Death and Deliverance*, p. 150.

of the unsuspecting patients to the removal of their corpses from the gas chamber. Gorgass admitted that between 18 June and the end of August 1941, about 2,000 victims had been gassed at Hadamar. He had personally been responsible for the majority of these murders.

With the completion of the first phase of "euthanasia", Gorgass was employed by *Sonderbehandlung* 14f13. Although he attempted to downplay his participation in that programme, claiming he had merely acted as an observer[1569], Friedrich Mennecke (q.v.) recorded that "Gorgass ... is said to have behaved dreadfully while at Buchenwald .... he supposedly acted more like a butcher than a doctor, thereby damaging the reputation of our whole operation."[1570] Together with other T4 personnel, in winter 1941/42 Gorgass was sent to the eastern front as part of the *Organisation Todt* mission.[1571]

Arrested in 1946 under the pseudonym of "Dr Gerber", in March 1947 Gorgass was put on trial in Frankfurt for his "euthanasia" activities. The court had no difficulty in dismissing Gorgass' rather puny defence, finding that he had acted maliciously in killing his victims, and was therefore guilty of murder. He was sentenced to death, a verdict upheld by the appeal court the following year, but with the abolition of the death penalty in the Federal Republic in 1949, the sentence was commuted to one of life imprisonment.[1572] Supposedly "repentant" and "physically broken",[1573] Gorgass was pardoned and released from custody in 1958 to take up employment with a pharmaceutical company.[1574]

Ernst-Robert von Grawitz (1899–1945)[1575] was born in Berlin, the son of a medical professor. He served in the German army during the Great War, and for a time became a prisoner-of-war of the British. On his release in 1919 he commenced medical studies at the University of Berlin. Although he obtained his license to practice and became a specialist in internal diseases in Berlin, he never achieved habilitation.[1576] This was largely because of his political activities. In 1920 he was involved in the Kapp *Putsch,* and subsequently joined a *Freikorps.* He claimed to have been a follower of Hitler from the earliest days of the NSDAP, although he never actually joined the party until 1932, by which time he had already been a member of the SS for a year.

In 1935 Himmler appointed Grawitz chief SS physician, the supreme authority for all SS medical affairs. Incongruously, Grawitz was appointed President of the German Red Cross in 1937, and two years later was made an honorary professor at the University of Graz.[1577] It was in his capacity as head of SS medical matters that Grawitz took a leading role in suggesting accommodating doctors to T4, thereby becoming heavily im-

---

1569 Friedlander, *The Origins of Nazi Genocide*, p. 348, note 56.

1570 Lifton, *The Nazi Doctors,* p. 141.

1571 Friedlander, *The Origins of Nazi Genocide*, p. 296.

1572 de Mildt, *In the Name of the People*, pp. 114–115.

1573 Burleigh, *Death and Deliverance*, p. 264.

1574 Klee, *Was sie taten–Was sie wurden*, p. 95.

1575 Klee, *Das Personenlexikon zum Dritten Reich*, p. 198.

1576 "Habilitation—a second dissertation or qualifying scholarly paper written after the Ph.D., dissertation allowing the candidate to formally teach at a German university. A *Habilitationschrift* is the thesis written to fulfil this requirement."[http://www.dictionaryofarthistorians.org/screens/foreign%20universities.htm (Accessed 13 October 2008)].

1577 Kater, *Doctors Under Hitler*, p. 131.

plicated in the "euthanasia" programme. In the summer of 1941 it was he who recommended gas chambers to Himmler as the most efficient method of mass killing. Together with Oswald Pohl, Grawitz was also responsible for the authorisation of horrific medical experiments on concentration camp prisoners.

On 24 April 1945 Grawitz saved the Allies the trouble of hanging him by the simple expedient of assembling his wife and two children and then detonating two hand grenades.[1578]

Heinrich Gross (1915–2005)[1579] was born in Vienna and joined the Hitler Youth in 1932, retaining his by then illegal membership of that organization between 1934 and 1938. He became a member of the SA in 1933, and the NSDAP in 1938. Gross qualified as a physician in 1939, and subsequently worked at the Mental Hospital of the City of Vienna during a time when patients were being transferred from that establishment to Hartheim for killing.

In 1940 he began working at the City of Vienna Young People's Welfare Institute – Am Spiegelgrund and its successor institutions. He was head of the children's ward at Am Spiegelgrund from 1941 to 1943, at which point he was drafted into the *Wehrmacht*, only to voluntarily return to the hospital in 1944 during a period of leave. His name and signature were later found on the death certificates of 238 Am Spiegelgrund child victims of "euthanasia". Gross was nicknamed "the Scythe", or "the Grim Reaper" by children who survived their hospitalization, one of whom recalled:

> There was a selection every 14 days or three weeks. There was Dr Gross. I remember him distinctly as a diligent, quiet young man. And he came up and pointed at some of us: he said, 'You, you, you, and you.' The children were taken from the group. The first children they selected were the bedwetters or harelips or the slow thinkers. That was the Nazi ideology; the doctors' instinct was that of an animal of prey. They selected the weakest out of a large herd. We did not dare ask where they were taken. We never saw them again.[1580]

Other survivors recalled children dying from exposure to cold, or from starvation or lethal medication. After their murder, parts of children's bodies, particularly brains, were retained for future study. It was only in 2002 that this material was buried at the Viennese Central Cemetery.

Gross was a Soviet prisoner of war until 1947. On his release he was charged by an Austrian court with treason (in respect of his illegal Nazi membership between 1934 and 1938) and murder. In 1950 he was acquitted on the treason charge. However, the court did find him guilty, not of murder, but of manslaughter, since at that time the law deemed that the definition of murder did not apply in the case of a mentally handicapped person because such individuals were incapable of reasoning.[1581] On appeal his case was ordered to be retried, but for reasons unknown no new trial occurred, and judicial proceedings against him were declared closed in 1951.

1578 Peter Padfield, *Himmler: Reichsführer-SS* (London: Papermac, 1990), p. 535.

1579 Klee, *Das Personenlexikon zum Dritten Reich*, p. 203

1580 Florian P Thomas, Alana Beres, and Michael I Sheveli, *"A Cold Wind Coming": Heinrich Gross and Child Euthanasia in Vienna* (Journal of Child Neurology, Vol.21, 2006), p. 344.

1581 Gabriel M Ronen, Brandon Meaney, Bernard Dan, Fritz Zimprich, Walter Stögmann, Wolfgang Neugebauer, *From Eugenic Euthanasia to Habilitation of "Disabled" Children: Andreas Rett's Contribution* (Journal of Child Neurology, Vol.24, No.1, 2009), p. 120.

Gross returned to the medical profession, in due course qualifying as a neurologist and psychiatrist. He pursued a highly successful career, and was regarded as a respected physician, in 1975 being awarded the Austrian Cross for Accomplishments in Science and the Arts. By the late 1970s, however, investigations into children's euthanasia at Am Spiegelgrund and Gross's role therein began to emerge. Despite clear evidence of his guilt, no criminal proceedings followed, and Gross remained a forensic expert to the Austrian courts until 1998. It was not until March 2000 that he finally appeared in court to account for his actions at Am Spiegelgrund. The case against him was closed after just one hour when the judge determined that Gross was suffering from dementia. Nevertheless, some punishment was meted out to him. The Austrian government rescinded his award of the Austrian Cross in March 2003.[1582]

Walter Gross

Source: Wikimedia Commons / Bundesarchiv, Bild 146-2007-0118 / CC-BY-SA 3.0 https://creativecommons.org/licenses/by-sa/3.0/de/deed.en

Walter Gross (1904–1945)[1583] a militant anti-Semite, was born in Kassel. He studied medicine in Göttingen, Tübingen, and Munich before obtaining his doctorate in 1928. Thereafter he worked for a time as a physician at a Brunswick hospital. He had joined the NSDAP in 1925, and became a member of the National Socialist Physicians' League in 1932. He headed the *Rassenpolitisches Amt* (*RPA*), the Race Political Office of the Nazi party from its inception in 1934, and in this capacity he was extensively involved in many of the regime's racial policies, including the Sterilisation Law, the sterilisation of the so-called *Rheinlandbastarde*, and the Nuremberg Laws.[1584] In 1938, Gross proclaimed that the work of the Race Political Office would not be complete until "the disappearance of the last Jew from our *Reich*." Three years later he went even further. Now a final solution would only be reached with the "complete removal of Jews from Europe."[1585]

Gross contributed a chapter entitled *National Socialist Racial Thought* to a 1938 English language book, *Germany Speaks*,[1586] in which he attempted to justify the sterilisation programme by arguing that the birth rate among the "unhealthy" was nine times greater than that of the "fit". He claimed that the Sterilisation Law was passed "to prevent the transmission of hereditary disease". He went on to suggest that

> Civilization is only possible through the individual becoming part of the whole and just as collective authority in the interests of all limits the egoism of the individual . . . , it similarly has the right to implement such measures for the benefit of the community as are scientifically proved expedient in the way of population policy or eugenics.

Gross continued with a familiar theme—Germany had suffered the loss of its racially purest in the Great War, and was only attempting to copy the discriminatory policies of

---

1582 Thomas, Beres, Sheveli, "*A Cold Wind Coming*", pp. 344–346.
1583 Klee, *Das Personenlexikon zum Dritten Reich*, p. 203.
1584 Proctor, *Racial Hygiene*, p. 87.
1585 Ibid, pp. 210–211.
1586 Claudia Koonz, *The Nazi Conscience*, Cambridge Massachusetts: Harvard University Press, 2003, p. 124.

other European nations and the United States in order to restore the balance—but saved his full rhetoric for his favourite subject. The Jews could not be tolerated, firstly because they were an alien race, secondly because they had too much financial power in Germany, and thirdly by their association with Communism. It was for these reasons that the Nuremberg Laws had been enacted. Jews and Germans were forbidden to intermarry, and illicit intercourse was subject "to punishment . . . designed primarily with a view to preventing the birth of further individuals of mixed blood . . ."

Gross continued to hold a variety of offices and academic positions throughout the years of Nazi rule. As his world collapsed around him, he committed suicide on 25 April 1945.

Käthe Gumbmann (1898–?) was born in Nuremberg, and after trying different jobs became a student nurse in 1919. She began working at Hadamar in 1932, and was enrolled into the "euthanasia" programme by Fritz Bernotat (q.v.) in November 1940, joining the NSDAP the following year. She claimed to have been working in the hospital kitchen and laundry throughout the first phase of "euthanasia" at Hadamar, but admitted to having participated in the killing of patients "once or twice" during the second phase. A defendant at the 1948 Frankfurt trial of nurses, Gumbmann was considered small fry within the greater scheme of things, and received a sentence of three years and one month's imprisonment.[1587]

Arthur Julius Gütt
Source: Wikimedia Commons / Public Domain

Arthur Julius Gütt (1891–1949)[1588] was born in Michelau, West Prussia. Commencing his medical studies in 1911, he served in the Great War as an army combat medical assistant before receiving his accreditation as a physician in 1918. He became involved in right-wing nationalist politics in the Weimar years, joining the NSDAP in 1932. On 1 May 1933, Leonardo Conti (q.v.) appointed Gütt head of the office responsible for public health affairs in the *Reich* Ministry of the Interior.[1589] Always an enthusiastic supporter of sterilisation, on 2 June 1933 Gütt became chairman of the newly formed "Expert Committee on Questions of Population and Racial Policy (*Sachverständigen-Beirat für Bevökerungsfragen und Rassenpolitik*), formed to pursue population issues initially raised a year earlier by the Prussian health office. Within a few weeks, the first results of the committee's deliberations—the Sterilisation Law—was enacted on 14 July 1933, with the important difference that unlike the Prussian draft law, sterilisation was not to be voluntary, but compulsory.[1590]

Gütt was co-author with Ernst Rüdin (q.v.) and Falk Ruttke (q.v.) of a commentary on the Sterilisation Law, which in 1937 led to a vociferous dispute with Gerhard Wag-

1587 de Mildt, *In the Name of the People*, p. 205.

1588 Klee, *Das Personenlexikon zum Dritten Reich*, p. 210

1589 Ian Kershaw, *Hitler 1889–36: Hubris* (London: Allen Lane, The Penguin Press, 1998), p. 487.

1590 Proctor, *Racial Hygiene*, pp. 95–96.

ner (q.v.), the *Reichsärzteführer*, over questions of interpretation of the legislation, resulting in some modification of the law in 1938.[1591] Gütt had been a member of the SS since November 1933, and in 1936 joined *Lebensborn* (see Chapter 10). He was appointed to a number of other important Nazi medical and administrative positions, and published numerous articles and books on matters of racial hygiene, including the Nuremberg laws, before in September 1939, following serious injuries incurred in a hunting accident, he resigned from his post at the Ministry of the Interior. Thereafter he was a person of minimal significance during the remaining years of the Third *Reich.*

Julius Hallervorden
Source: Wikimedia Commons / Public Domain

Julius Hallervorden (1882–1965)[1592] was born in Allenberg, East Prussia (today Chlebnikowo, Poland), the son of a psychiatrist. He studied medicine at the University of Königsberg, obtaining his doctorate in 1909. An initially distinguished career in neuropathology followed, in the course of which he worked with Professor Hugo Spatz; they jointly achieved fame through the identification of a congenital neurological condition known as "Hallervorden-Spatz Disease".

Through his friendship with Spatz, who had been appointed head of the Kaiser Wilhelm Institute for Brain Research, in 1937 Hallervorden was appointed head of the Department of Brain Histopathology in Berlin-Buch, whilst continuing anatomical activities at Brandenburg-Görden (a part of the aforementioned Kaiser Wilhelm Institute).[1593] Hallervorden joined the NSDAP in 1939, somewhat later than most of his medical contemporaries; in the same year Brandenburg-Görden became the "*Reich* Training Asylum" for children's "euthanasia", a place where, under the directorship of Hans Heinze (q.v.), many children were killed.[1594] Although it is highly likely that Heinze had informed him of T4 and its activities at a much earlier date, Hallervorden was only officially told of the existence of the organisation on 29 April 1940. Little more than two weeks later he received the first batch of children's brains from the Brandenburg killing centre, where the children had been gassed.[1595]

On 8 December 1942, Hallervorden had reported to the *Deutsche Forschungsgemeinschaft* (DFG–German Research Association) that, "during the course of the summer I was able to dissect 500 brains of imbeciles myself".[1596] In 1944, Hallervorden wrote to Paul Nitsche (q.v.), reporting that he had received a total of 697 brains, some of which

1591 Ibid., pp. 115–116.
1592 Klee, *Das Personenlexikon zum Dritten Reich*, p. 220.
1593 http://www.doew.at/thema/planck/planck1.html (Accessed 17 November 2008).
1594 Aly, Chroust, Pross, *Cleansing the Fatherland*, p. 220.
1595 Susanne Heim, Carola Sachse, Mark Walker, (eds.), *The Kaiser Wilhelm Society under National Socialism* (Cambridge: Cambridge University Press, 2009), p. 113.
1596 http://www.irren-offensive.de/kleespeech.htm (Accessed 17 November 2008).

he had personally removed at Brandenburg.[1597] The unanswered question remains of whether some or all these victims were killed at Hallervorden's request.[1598]

During his 1945 interrogation, while displaying his collection of murdered children's brains Hallervorden stated:

> I heard they were going to do that ["euthanasia"], and so I went up to them and told them "Look here now boys, if you are going to kill all these people, at least take the brains out so that the material could be utilized." They asked me: "How many can you examine?" and so I told them an unlimited number—"the more the better"...There was wonderful material among those brains, beautiful mental defectives, malformations, and early infantile disease. I accepted these brains, of course. Where they came from and how they came to me was really none of my business[1599]....[The patients] were selected from the various wards of the institutions according to an excessively simple and quick method. Most institutions did not have enough physicians, and what physicians there were either too busy or did not care, and they delegated the selection to the nurses and attendants. Whoever looked sick or was otherwise a problem was put on a list and was transported to the killing centre. The worst thing about this business was that it produced a certain brutalization of the nursing personnel. They got to simply picking out those whom they did not like, and the doctors had so many patients that they did not even know them, and put their names on the list.[1600]

After 1945, Hallervorden continued to work at what had now become the Max Planck Institute for Brain Research. Although the French viewed Hallervorden as a likely defendant in criminal proceedings, the Americans had no desire to pursue the matter, and in the event he escaped prosecution.[1601] Instead, in 1956 Hallervorden was awarded the *Grosses Verdienstkreuz*—The Order of Merit of the Federal Republic of Germany, awarded for outstanding achievements in political, economic, cultural, intellectual or voluntary work.

Benedikt Härtle (1901–?) was born in Peissenberg, a small town in Bavaria. He had originally intended to enter a monastery, but instead in 1927 decided on a career in nursing. He was employed at Hadamar from 1933 until the end of the war, becoming a member of the NSDAP in 1937. In November 1940 Ernst Baumhardt (q.v.) enrolled him as a participating member of the "euthanasia" operation. During the first phase of the programme, Härtle's duties consisted of accompanying transports from "feeder" institutions such as Eichberg and Galkhausen, as well as escorting victims to the gas chamber.[1602]

Härtle was another of the nurses tried at the second Hadamar trial at Frankfurt in 1947. He was found guilty of involvement in an unknown number of murders and re-

---

1597 Aly, Chroust, Pross, *Cleansing the Fatherland*, p. 228. Some 200 of the brains in Hallervorden's collection were those of Jewish typhus victims from Warsaw. (Weindling, *Nazi Medicine and the Nuremberg Trials*, p. 74.)

1598 Weindling, *Nazi Medicine and the Nuremberg Trials*, p. 96.

1599 Aly, Chroust, Pross, *Cleansing the Fatherland*, p. 224. The collection of brains continued to be studied at the Max Planck Institute for Brain Research in Frankfurt until 1990, at which time the material was buried in a Munich cemetery. (Ibid). It was also revealed that The Max Planck Institute of Psychiatry had in its collection brain specimens from children murdered in the child "euthanasia" programme. These specimens came from children murdered at Eglfing-Haar. [http://www.doew.at/thema/planck/planck1.html (Accessed 17 November 2008)].

1600 Leo Alexander, *Medical Science Under Dictatorship* (The New England Journal of Medicine 241, 1949), p. 40.

1601 Weindling, *Nazi Medicine and the Nuremberg Trials*, p. 96.

1602 de Mildt, *In the Name of the People*, pp. 197–198.

ceived a jail sentence of three years and six months.[1603] Although he had also been charged with participation in "second phase" killings, the court determined that there was insufficient evidence to convict in this respect.

Walter Heess (1901–1951?)[1604] was born in Ludwigsburg to an upper-middle-class family. He studied chemistry at the Technical College in Stuttgart, obtaining his doctorate in 1925 and his habilitation in 1937. Initially specialising in food chemistry, he turned instead to the criminal-technical division, where he was considered particularly industrious. Having joined the NSDAP and SA in 1933, in 1938 his professional and political diligence and enthusiasm received official recognition. In appointing Heess director of the Criminal Technology Institute (*Kriminaltechnisches Institut* – KTI) of the Security Police (*Reichskriminalpolizeiamt* – RKPA), Arthur Nebe, the head of *Kripo* noted that Heess was

> ...a purposeful personality full of character. As a result of his particular talents and his dogged hard work, he has achieved excellence in all areas of modern criminal technology. Under his direction the Criminal Technology Institute of the Security Police has come to be recognized as a leading institution.[1605]

Sometime immediately prior to *Barbarossa*, Heess had an informal discussion in Berlin with his assistant, Albert Widmann (q.v.), concerning the construction of a new kind of gas van. They concluded that rather than the bottled carbon monoxide which had been used up to that time to murder the disabled in some regions of Germany and Poland, a proposed modification would divert the van's exhaust fumes into an airtight compartment in the rear of the vehicle. This would achieve the desired result of killing the transportees, whilst at the same time overcoming the difficulties otherwise involved in conveying bottled gas to the Russian front. Heess was due to report on this subject the following day, probably to Heydrich.[1606] The head of the RSHA doubtless approved of the idea, for in due course a suitably modified vehicle was delivered to Sachsenhausen where, in the presence of Heess and others, a test gassing of about thirty Soviet prisoners was successfully completed in autumn 1941.[1607] Thereafter, gas vans of this type entered service in the East with the *Einsatzgruppen,* something not altogether to Widmann's liking, for whilst he was perfectly happy to use gas vans to kill the insane, he was less sanguine about employing them to exterminate "normal" individuals. Hees responded: "But you see, it is done anyway. Do you want to quit by any chance?" Widmann's conscience was quickly salved by a combination of Hees' nonchalant reply and the promotion that followed soon after.[1608]

The deputy director of the KTI, Walter Schade, testified that in his presence Heess and Widmann had openly discussed the technique of killing people through the use of gas vans; further, Heess claimed to have witnessed the gassing of patients at Pirna-Sonnenstein. Schade alleged Heess had commented that, given a state of total war, it

1603 Ibid, p. 371, note 96.
1604 Klee, *Das Personenlexikon zum Dritten Reich,* p. 236.
1605 Wildt, *An Uncompromising Generation*, pp. 183–184.
1606 Eugen Kogon, Hermann Langbein, Adalbert Rückerl (eds.), *Nazi Mass Murder: A Documentary History of the Use of Poison Gas* (New Haven: Yale University Press, 1993), p. 38.
1607 Wildt, *An Uncompromising Generation*, p. 186.
1608 Hilberg, *The Destruction of European Jews*, p. 345.

was no longer possible "to continue feeding people with incurable mental illnesses." Moreover, Heess concluded, there was an acute requirement for the nursing staff involved in caring for such patients to be transferred to military hospitals.[1609]

Having first poisoned their children, Hees' wife committed suicide at the end of April 1945. Like many other Nazi criminals Heess then disappeared from view, the question of his death or survival left unanswered. Any remaining doubts were not altogether extinguished when in 1951 his sister had him officially declared dead.[1610]

Hans Heinze (1895–1983)[1611] was born in Elsterberg, Saxony. In a somewhat odd misreference, he is referred to as "Carl Hans Heinze Sennhenn" on a number of websites. The apparent source of this error is a Federal Bureau of Investigation list of files handed over to the National Archives and Records Administration in 2002, many of which relate to possible German and Japanese war criminals. Included on the list are the names of a "Karl Heinz" and a "Dr Carl Hans Heinz Sennhenn."[1612] Dr Karl Heinz Sennhenn was an SS-*Obersturmführer* with no evident connection to T4 or "euthanasia."[1613] Who Karl Heinz may have been is unknown. The career of Hans-Heinz Schütt, who was extensively involved in both T4 and *Aktion Reinhard*, is set out below. Hans Heinze, a T4 *Gutachter* from 17 November 1939, was a much more senior cog in the killing programme.[1614] This kind of confusion regarding identities in the aftermath of war may go some way toward explaining how easy it was for so many to escape pursuit and prosecution. A classic example of such a misunderstanding was the frequent inability of the Allies to distinguish between the activities of Karl Brandt (q.v.) and Rudolf Brandt, Himmler's personal adjutant, a fact that did neither man any favours.[1615]

Unlike most of his generation, Heinze did not serve in the military during the Great War, instead becoming a member of the Red Cross. This probably encouraged him to study medicine, which he did in Leipzig. A specialist in psychiatry and neurology, Heinze was made director of the mental hospital at Potsdam in 1934 before taking up the position of director of the Brandenburg-Görden asylum in 1939.[1616] He had joined the Nazi party in 1933 and had been recommended to the group concerned with the planning of children's "euthanasia" by Herbert Linden (q.v.).[1617] It was under Heinze's supervision that the first children's killing ward had been opened at Brandenburg-Görden in October 1939. Heinze's contribution to the activities of T4 have already been described.

In March 1946, Heinze was sentenced to seven year's imprisonment by a Soviet military tribunal. Released in 1952, he made his way to West Germany to resume his medical career, and became director of the Wunstorf clinic for juvenile psychiatry in

1609 Wildt, *An Uncompromising Generation*, p. 187.
1610 Ibid, p. 363.
1611 Klee, *Das Personenlexikon zum Dritten Reich*, p. 240.
1612 http://www.fas.org/sgp/othergov/fbi-nwcda-091002.htm (Accessed 7 November 2008).
1613 Klee, *Das Personenlexikon zum Dritten Reich*, p. 579.
1614 Klee, *"Euthanasie" im NS-Staat*, p. 228.
1615 Schmidt, *Karl Brandt*, pp. 328–329.
1616 Burleigh, *Death and Deliverance*, p. 104.
1617 Friedlander, *The Origins of Nazi Genocide*, p. 44.

1954.[1618] Investigations into his Nazi past were commenced in 1962. Not surprisingly he "clearly displayed a hyper-sensitivity towards questions dealing with 'euthanasia'". Proceedings against him were abandoned in 1966[1619] after the court ruled he was "a psychological wreck" unable to withstand the tribulations of another trial. The "psychological wreck" survived for a further seventeen years. On his death, his colleagues published an obituary declaring: "We will honour his memory."[1620]

Günther Hennecke (1912–1943) [T4 pseudonym "Dr Fleck"][1621] was born in Halle, and joined both the NSDAP and SA in 1933. Having obtained his medical licence in 1939, from 25 April 1940 he was the deputy doctor in charge of gassing at Grafeneck, moving to Hadamar in January 1941 to carry out similar duties there until June of that year. A "comfort" letter written by Hennecke from Hadamar in March 1941 has survived.[1622]

In similar fashion to Ernst Baumhard (q.v.), Hennecke apparently fell out with Viktor Brack (q.v.) in the summer of 1941 (probably for the same reasons, or out of a comparable desire to escape T4) and, again like Baumhard, enlisted in the navy to serve as a U-boat medical officer. He was killed when U 538 was sunk on 21 November 1943.[1623]

Werner Heyde (1902–1964) [post-T4 pseudonym "Dr Fritz Sawade"][1624] was born in Forst, Brandenburg. He studied medicine in Berlin, Freiburg, Marburg, Rostock, and Würzburg, finally receiving his medical licence in 1926, and his accreditation as a specialist in psychiatry and neurology in 1929. He was appointed a physician at the psychiatric clinic of the University of Würzburg, where in 1933 he met Theodor Eicke, who was to go on to become commandant of Dachau and creator of the brutal concentration camp regime applied there and at all other camps. Eicke, who had justifiably been committed to the psychiatric clinic for observation, was eventually to become inspector of the entire concentration camp universe. Heyde and Eicke developed a friendship (both were apparently homosexual), and through Heyde's favourable diagnosis and intervention with Himmler, Eicke was discharged from the clinic to pursue his SS career. Although Heyde had been a *Freikorps* member during the Weimar era, fighting first in Estonia and then in Cottbus, it was only after he had met Eicke that he became a member of the Nazi party in May 1933.[1625]

One good turn deserved another. In 1934 Heyde was promoted to senior physician at Würzburg. He became a lecturer in so-called "hereditary diseases", served as an expert witness in court cases, was a member of the local sterilisation court, and was appointed head of the Würzburg Nazi Party race office. Not least, through Eicke's influ-

---

1618 Burleigh, *Death and Deliverance*, p. 273.
1619 de Mildt, *In the Name of the People*, p. 127.
1620 Klee, *Was sie taten–Was sie wurden*, pp. 138–139.
1621 Friedlander, *The Origins of Nazi Genocide*, p. 103 suggests an alternative pseudonym for Hennecke of "Dr Ott".
1622 Burleigh, *Death and Deliverance*, pp. 148–149.
1623 Klee, *Das Personenlexikon zum Dritten Reich*, p. 245.
1624 Ibid., p. 252.
1625 Burleigh, *Death and Deliverance*, p. 116.

ence he became adviser on mental health matters to the *Gestapo*, and assumed responsibility for the investigation of psychiatric cases and supposed hereditary diseases among concentration camp prisoners (with a personal history of issuing recommendations for compulsory sterilisation in many instances).[1626] In June 1936 he had joined the SS, and in 1938, was made chief of staff of the medical department in the SS-*Hauptamt* (headquarters). Inevitably this meteoric rise led to Heyde being appointed to the chair of psychiatry in Würzburg in 1939, and made him a natural choice for the medical leadership of the "euthanasia" programme from July of that year.[1627]

The importance of Heyde's contribution to so many facets of Nazi eugenic policy cannot be overstated. His name recurs time and again in connection with sterilisation, "euthanasia", 14f13, as head of the T4 medical office and the recruitment of physicians, as *Obergutachter*—his endorsement of and dedication to the Third *Reich*'s racial hygiene programme is indisputable. In the words of Franz Suchomel (q.v.):

> [Heyde] was the head of the whole thing, he developed it...[He] had a flat at Tiergartenstrasse 4, next to my office. He was the top expert in the mercy-killing business. He only stayed at his flat when he had official business in Berlin. He was, I was told, an authority in his field...I know that there was a research institute into mental illness in Strasbourg; he may have run that. That's where the brains of selected mental patients were sent for research purposes.[1628]

In December 1941 Heyde was replaced as chief of T4's medical office by his deputy, Paul Nitsche (q.v.) allegedly on the grounds of Heyde's homosexual proclivities.[1629] Thereafter Heyde continued his professorship at Würzburg and became head of the SS hospital for psychiatric and neurological injuries affiliated to the Würzburg clinic. Because of Allied bombing, in 1945 the SS hospital was relocated to Denmark, and it was here that Heyde was arrested by the British. He was returned to Germany to give evidence at both the Hadamar and Nuremberg medical trials. Sentenced to death *in absentia* by a German court,[1630] whilst under arrest in July 1947 Heyde escaped from American custody by the simple expedient of jumping off the back of a lorry in Würzburg en route to Frankfurt. He made his way to Schleswig-Holstein, on the way effecting some minor changes to his appearance, and acquiring false identity papers in the name of "Fritz Sawade".

After working for a time as a gardener, Heyde/Sawade successfully applied for a job as a sports physician in Flensburg. He was soon acting as a medical referee for public insurance offices and in the local courts. His true identity was common knowledge. It could hardly be otherwise with a wife who, taking umbrage at being treated with a certain disdain by the wives of professors because her husband was only a sports doctor, would blurt out: "Don't you know who I am? I am not just a mere Frau Doctor, but, just so you know, Frau Professor Heyde!" Erika Hyde had been claiming a widow's pension since 1951, an offence for which she received a sentence of one year's imprisonment in 1962.

1626 Aly, Chroust, Pross, *Cleansing the Fatherland*, p. 37.

1627 Friedlander, *The Origins of Nazi Genocide*, pp. 65–66.

1628 Sereny, *Into That Darkness*, p. 57.

1629 Burleigh, *Death and Deliverance*, p. 117.

1630 Wistrich, *Who's Who in Nazi Germany*, p. 107.

Given such circumstances, it was only a matter of time before Heyde's conspicuous behaviour ended in his arrest, which it did in November 1959. He was imprisoned in Limburg, awaiting trial for complicity in the murder of "about 100,000" people. However, it was not the end of a life which, despite its involvement in the horrors of mass murder, had degenerated into farce. Having (unusually for the time) failed in an attempt to avoid legal proceedings on health grounds, in a plot twist that would be considered improbable in a work of fiction, Heyde hatched an escape plan involving duplicate keys, a two-way radio, and a beautiful blonde, but was moved to another prison in Butzbach before the plan could be enacted. On 12 February 1964, Friedrich Tillmann (q.v.) fell from the eighth floor of a Cologne building. The next day Heyde was found dead in his cell, having apparently committed suicide by hanging himself with his own belt. The two deaths were fortuitous for many in the German medical and legal professions, to say the least.

Heyde's obituary contained the following: "After an active life and after a long, troubling denial of liberty, Herr Prof. Dr med. Werner Heyde has been summoned at the age of 62 before his heavenly judge."[1631] What that judgement may have been will forever remain a mystery; the judgement of history is somewhat clearer.

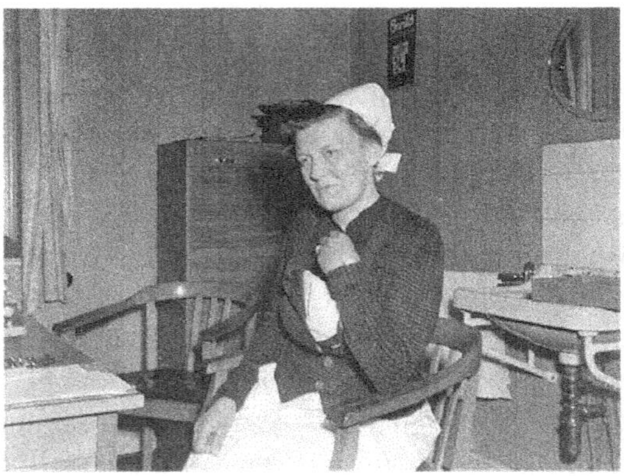
Irmgard Huber
Source: USHMM / Signature #05472

Irmgard Huber (1901–1983) initially worked as a domestic servant before training as a psychiatric nurse.[1632] She commenced employment at Hadamar in 1932, and remained there until July 1945. Although a member of the German Labour Front (DAF) and National Socialist People's Welfare (NSV), she was not a Nazi party member.[1633] Charges were later only brought against Huber for participation in the "second phase" of "euthanasia", but given her presence throughout the entire period Hadamar served as a killing centre, it seems probable that she was at the very least aware of events from 1940 onwards.[1634] The finding of the American military court at the first Hadamar trial held at Wiesbaden in October 1945 read in part:

> Irmgard Huber was the chief female nurse at the institution, carrying out the orders of Dr. Wahlmann (q.v.) and overseeing the duties of the seven other female nurses. She knew beforehand of the arrival of the first transport and made preparations for housing the victims. There was some evidence that the female nurses actually gave injections. It was at least well established that the accused Huber took part in daily morning conferences at which Wahlmann signed death certificates. She obtained narcotics from the pharmacy in Dr. Wahlmann's office...and she was actually present on at least one occasion when fatal injections or dosages were given to patients, and when false death certificates were made out.

1631 Burleigh, *Death and Deliverance*, pp. 270–273.
1632 Ibid., pp. 239–240.
1633 Ibid., p. 195.
1634 Ibid, p. 370, note 87.

A sentence of twenty-five years' imprisonment was imposed on Huber.[1635] Two years later, Huber was in the dock again, this time in Frankfurt for the second Hadamar trial. The court determined that Huber was guilty of participation in "at least" 120 cases of murder, and sentenced her to a further eight years' imprisonment.[1636] However, she was another to benefit from a dramatic change in the political climate. Despite her twice proven guilt, Huber was released in 1952.[1637]

Murad Ibrahim

Murad Jussuf Bey Ibrahim (1877–1953)[1638] was born in Cairo, the son of an Egyptian doctor father and a German mother. After commencing his medical studies in Cairo, he left for Munich, where he graduated as a doctor. From 1902 to 1906 he was assistant physician at the Heidelberg paediatric clinic, qualifying in 1905 as a paediatric specialist in Berlin. In 1907 he returned to Munich as director of the Gisela *Kinderspital* (children's hospital), and in 1916 was appointed ao.Professor in Würzburg. In 1917 he became o. professor[1639] at the Friedrich Schiller University in Jena, where he was head of the children's clinic until his death in 1953, following which the clinic was named the *Kinderklinik Jussuf Ibrahim*.[1640]

On the occasion of his 70th birthday, in 1947 Ibrahim had been made an honorary citizen of Jena for his distinguished work in connection with the care of the young, but in 1985 evidence began to emerge of his involvement in children's "euthanasia". In 2000 after intensive examination of all available documentation, "The Commission on the 'Jussuf Ibrahim' Children's Hospital"—consisting of 6 independent scientists of the Friedrich Schiller University—came to the conclusion that there was no doubt that Ibrahim had been actively involved in children's "euthanasia" between 1941 and 1945. "Incurable" children had been sent by Ibrahim to Gerhard Kloos (q.v.) at the Stadtrota clinic for killing.[1641] The commission reported that Ibrahim not only endorsed the killing of handicapped children, but directly contributed towards their death. In the light of these findings, the clinic was immediately re-named *Die Universitätsklinik für Kinder- und Jugendmedizin Jena* ("The Jena University Hospital for Children and Adolescent Medicine").

Ernst Illing (1904–1946)[1642] was born in Leipzig. He received his licence to practice medicine in 1930, became an accredited specialist in psychiatry and neurology in 1937,

---

1635 http://www.ess.uwe.ac.uk/WCC/hadamar.htm#Huber (Accessed 7 January 2009).

1636 de Mildt, *In the Name of the People*, p. 371, note 96.

1637 http://www.ushmm.org/wlc/article.php?lang=en&ModuleId=10007146 (Accessed 7 January 2009).

1638 Klee, *Das Personenlexikon zum Dritten Reich*, p. 277

1639 Professor *extraordinarius* (*ausserordentlicher* Professor, ao. Prof.): a professor without a chair, or one subordinated to a professor with a chair. Professor *ordinarius* (*ordentlicher* Professor, o. Prof.): a professor with a chair.

1640 http://www.whonamedit.com/doctor.cfm/1518.html (Accessed 4 September 2008).

1641 Mark Walker (ed), *Science and Ideology—A Comparative History* (London: Routledge, 2003), p. 202.

1642 Klee, *Das Personenlexikon zum Dritten Reich*, p. 278.

and completed his habilitation in 1942.[1643] He joined the Nazi party in 1933, and served his apprenticeship as a murderer in the children's killing ward at Brandenburg-Görden, where he held the position of assistant to Hans Heinze (q.v.).

Situated within the Am Steinhof institution in Vienna, the Am Spiegelgrund children's sanatorium opened in July 1940. It was to serve as the principal killing centre for children's "euthanasia" in Austria. In July 1942, Illing succeeded Erwin Jekelius (q.v.) as director of Am Spiegelgrund. His responsibilities were quite clear—to continue killing as he had done at Görden.[1644]

At the conclusion of the war, Illing was tried by the People's Court in Vienna and found guilty of an estimated 200 cases of murder. He was hanged on 20 November 1946.

Erwin Jekelius (1905–1952)[1645] was born in Hermannstadt (today Sibiu, Romania). He joined the then illegal Austrian NSDAP in 1933. After becoming an accredited specialist in neurology in 1938, he was appointed leader of the psychiatric care department of the Vienna public health office in 1939.[1646] He became the first medical director of the Am Spiegelgrund institution in July 1940, a position he relinquished to Ernst Illing (q.v.) in 1942 when Jekelius was called up for army service.[1647] By then, Jekelius had played a leading role in the implementation of the "euthanasia" programme in Vienna.[1648] It was Jekelius (a *Gutachter* since 14 October 1940) who first experimented with electrocution as a killing technique before passing the method into the hands of Emil Gelny (q.v.)[1649]

Jekelius died from cancer of the bladder in a Soviet jail in 1952 whilst serving a 25 year sentence for his part in the killing of an estimated 4,000 victims of "euthanasia." In recently released transcriptions of interrogations conducted by his captors, Jekelius justified his actions by saying he totally agreed with Hitler's policy of exterminating mentally ill people. He also said that his personal and professional view was that such people were a burden to their families and to society.

He went on to describe how six to 10 children a month were killed at Am Spiegelgrund, and how his doctors created credible, if wholly fictitious, causes of death to explain the often sudden demise of a child to his or her grieving parents. It has been suggested that the reason Jekelius fell into Soviet hands was because he had become engaged to Hitler's sister Paula, and that on hearing the news the irate *Führer* ordered Paula's lover be shipped off to the eastern front.[1650] The source of this suggestion (other than the alleged memoirs of Paula Hitler), would appear to be a note made by Himmler of a telephone conversation with Heydrich that took place on 30 November 1941, and which includes the brief phrase: "Arrest Dr Jekelius." These three enigmatic words have given rise to an abundance of interpretations, but in truth could mean almost anything.

1643 Friedlander, *The Origins of Nazi Genocide*, p. 127.

1644 Ibid., p. 49.

1645 Klee, *Das Personenlexikon zum Dritten Reich*, p. 286.

1646 Marius Turda, and Paul J Weindling, (eds.) *"Blood and Homeland". Eugenics and Racial Nationalism in Central and Southeast Europe, 1900–1940*, (Budapest: Central European University Press, 2007), p. 329.

1647 Friedlander, *The Origins of Nazi Genocide*, p. 49.

1648 Turda and Weindling, "*Blood and Homeland*", p. 323.

1649 Friedlander, *The Origins of Nazi Genocide*, p. 169.

1650 http://www.shoah.dk/Paulahitler/page_3.htm (Accessed 25 November 2008).

Eva Justin (1909–1966)[1651] was born in Dresden, and having trained as a nurse was employed by Robert Ritter (q.v.) in his eugenic research department at Tübingen University, where she was described as Ritter's "right hand." She moved to Berlin with Ritter in 1938, acting as his research assistant with regard to Gypsy matters. Despite her lack of a university degree, she received a doctorate in anthropology in 1943; her sponsors and/or referees were Eugen Fischer (q.v.), Hans Reiter (q.v.), Herbert Linden (q.v.), Paul Werner (q.v.), and Richard Thurnwald, the latter a founder member of Alfred Ploetz's (q.v.) Society for Racial Hygiene.[1652] Justin's dissertation was the result of her study of Gypsy children in the Mulfingen home; these children were subsequently deported to Auschwitz, where most were murdered.[1653]

In 1948 Justin was employed as an expert psychologist on Gypsy matters by the city of Frankfurt am Main, where her superior was none other than Robert Ritter.[1654] Ten years later her case was investigated by the authorities in Frankfurt, who concluded that although Justin was involved in the eugenic assessment of Gipsy children during the Nazi epoch, there was no evidence that she had been aware of the children's subsequent extermination. Other issues relating to her participation in sterilisation matters were by that date statute barred. Proceedings against her were therefore terminated in December 1960.

Helmut Kallmeyer (1910–2006)[1655] was born in Hamburg. Unusually, although a member of the SA he never joined the NSDAP, something that may subsequently have served him well in post-war investigations into his activities during the Nazi era. He studied chemistry at a number of universities before receiving his certification from the Berlin Technical Institute in 1939 and his doctorate in 1940. Between September 1939 and September 1941 he served in the German navy, at which point Viktor Brack (q.v.) secured his release from military service. Thereafter Kallmeyer was employed, first by the KdF and then by the KTI (the Criminal Technology Institute of the Security Police) for the rest of the war.

Although by the time of Kallmeyer's recruitment the "stop order" had been issued, Brack needed another expert chemist for the larger ventures in which T4 was about to become engaged. August Becker (q.v.) was busy helping the *Einsatzgruppen* kill Jews in gas vans, and Albert Widmann (q.v.) was with Arthur Nebe in Byelorussia, experimenting with other methods of mass murder. Having apparently done little or nothing for several months, in early 1942 Kallmeyer was dispatched to Lublin. Precisely what he did there is unknown. The Belzec extermination camp was in the course of construction, and it seems inconceivable that he was not involved in some aspect of its intended function. Kallmeyer was less than helpful in this regard. In his post-war testimony he could recall neither the name of the agency to which he reported in Lublin, nor the name of the man he met there. After a week in Lublin he claimed to have returned to Berlin, where he was admitted to hospital suffering from typhus, supposedly on 28 Feb-

---

1651 Klee, *Das Personenlexikon zum Dritten Reich,* p. 294

1652 Proctor, *Racial Hygiene,* p. 17.

1653 Friedlander, *The Origins of Nazi Genocide,* pp. 250–251.

1654 http://www.hagalil.com/01/de/Antisemitismus.php?itemid=1192 (Accessed 26 November 2008).

1655 Klee, *Das Personenlexikon zum Dritten Reich,* p. 297

ruary 1942. On his recovery he was transferred from the KdF to the KDI, where he was one of those responsible for delivering lethal substances to T4.

There was further documentary evidence of Kallmeyer's apparent expertise in gassing technique, even if in this case it was not utilised. A letter from Alfred Wetzel of the *Reich* Ministry of the Occupied Eastern Territories to Heinrich Lohse, the *Reich* Commissar for the Ostland, dated 25 October 1941 included the following:

> ...*Oberdienstleiter* Brack of the Führer Chancellery has agreed to collaborate in the production of the required shelters and gassing devices. At this time, the envisaged devices are not available in sufficient quantity; they will first have to be manufactured. Since in Brack's opinion, the manufacture of the devices in the *Reich* will cause much greater difficulties than doing it on the spot, Brack considers it most expedient to send his people to Riga, especially his chemist Dr. Kallmeyer, who will effect all further steps there...In these circumstances, I request that you address yourself to *Oberdienstleiter* Brack in the *Führer* Chancellery through your Higher SS and Police Leader and request the dispatch of the chemist Kallmeyer and other assistants... As things now are, there are no objections if the Jews who are not capable of work, are eliminated with the Brackian remedy.[1656]

Kallmeyer naturally denied all knowledge of this letter, and there is no evidence that he ever actually visited Riga. However, there was one other undeniable connection between himself and T4. In December 1940, Kallmeyer married Gertrud Fröse, who had been Brack's secretary at the KdF. During 1940 she had been posted to Grafeneck, allegedly prior to the time that killing began there. She later revised her testimony; she could have left her job with Brack later than originally claimed, she might have been at Grafeneck after the killing started, and perhaps she had met Horst Schumann (q.v.) there, all matters that she had previously denied.

Even when added to the fact that he and his wife had met with Schumann in Ghana after the war, there was considered insufficient evidence to directly implicate Kallmeyer in any criminal activity, and he was never indicted.[1657] Post-war, he was employed by the national statistical office of Schleswig-Holstein, before working for the United Nations in Cuba and Ghana in the 1960s.

Anna Katschenka (1905–?) was born in Vienna, and graduated from nursing school in 1927. From that time until 1941 she was employed by three separate Viennese hospitals, and was considered an asset in each of them. She was never a member of the Nazi party or any of its related organizations; in fact, her political inclinations were, if anything, left wing. Katschenka married a Jewish medical student in 1929, but was divorced the following year. In a state of depression after the *Anschluss,* her hospital sent her to Erwin Jekelius (q.v.) for treatment. He helped her to overcome her problem and, not unusually, she formed an attachment to him. In 1941 she heard that the children's hospital at Am Spiegelgrund, of which Jekelius was head, needed nurses, and she volunteered to join him.

Jekelius informed Katschenka of the children's "euthanasia" programme, and asked her if she was prepared to administer the necessary lethal medication. She agreed to do so, stating after the war that she had never believed children's "euthanasia" to be unlawful, since Jekelius had assured her that it was only hopeless cases with no prospect

1656 http://www.holocaust-history.org/19411025-wetzel-no365/ (Accessed 27 November 2008).

1657 Friedlander, *The Origins of Nazi Genocide*, pp. 211–214.

of recovery that were to be treated, and in any event it was her understanding that a binding law existed which permitted "euthanasia".[1658] The similarities to the case of Heléne Schürg (q.v.) are evident. When Jekelius departed for the *Wehrmacht*, Katschenka continued killing under his successor, Ernst Illing (q.v.).[1659]

Katschenka was tried in Vienna in 1948 and sentenced to eight years' imprisonment. She was released in 1951, to continue working as a children's nurse.[1660]

Berthold Kihn (1895–1964)[1661] born in Schöllkrippen in Bavaria, was another paediatrician. He joined the SA in 1933 and was appointed an ao. Professor at Erlangen University in 1934. From 1938 he was a judge at the Hereditary Health Court in Jena, and in the same year was appointed Professor of psychiatry and leader of the psychiatric, mental health, and outpatient's clinic at the University of Jena. In 1939 he became an o.Professor. He is known to have co-operated with Jussuf Ibrahim (q.v.) and Gerhard Kloos (q.v.) in children's "euthanasia" at the Stadtroda clinic.[1662] He appears on a list of T4 personnel as a *Gutachter*, effective 5 June 1940, but had been involved in the planning of adult "euthanasia" from a much earlier date.[1663]

In 1944 he was appointed Dean of Erlangen University, but was dismissed from this position in September 1945. By 1952 he was again an honorary professor at the university. He also operated his own private sanatorium. In 1963 preliminary proceedings against him in connection with his activities during the Nazi era were abandoned.

Gerhard Kloos (1906–1988)[1664] was a psychiatrist born in Sächsisch-Regen (today Reghin in Romania). At the time of Kloos' birth the city was within the Habsburg Empire, but during the first half of the twentieth century it became successively Romanian (in 1920), then Hungarian (in 1941), before being restored once more to Romania (in 1945). There had been an ethnic German population in Sächsisch-Regen for centuries; in 1910, 41 percent of the city's population were of German descent, but by 2002 less than 1 percent of the inhabitants claimed such ancestry.

In 1933 Kloos joined both the NSDAP and the SA, and in July 1939 was appointed head of the Thuringian State Hospital at Stadtroda and its children's ward. Kloos was an enthusiastic proponent of "euthanasia", as a consequence of which it has been estimated that there were approximately 900 "euthanasia" victims at Stadtroda, among them many children. In a 1939 Nazi medical journal, Stadtroda was commended for being "a symbol of the definitive break with a weak, humanitarian past—not only in Germany, but rather in the entire cultural world."[1665]

As the war drew to a close, Kloos fled before the advancing Red Army to pursue a successful career in West Germany, where amongst other positions he was made direc-

1658 http://www.doew.at/frames.php?/thema/thema_alt/justiz/euthjustiz/gross.html (Accessed 3 January 2009).
1659 Friedlander, *The Origins of Nazi Genocide*, p. 232.
1660 http://www.oegkv.at/uploads/media/fuerstler_01.pdf (Accessed 3 January 2009).
1661 Klee, *Das Personenlexikon zum Dritten Reich*, p. 308
1662 Walker, *Science and Ideology*, p. 202.
1663 Friedlander, *The Origins of Nazi Genocide*, p. 66.
1664 Klee, *Das Personenlexikon zum Dritten Reich*, p. 317.
1665 Walker, *Science and Ideology*, p. 202.

tor of the state hospital in Göttingen. An investigation into his Nazi era activities was abandoned in the early 1960s due to lack of evidence. In 1985, Professor Dr Dr Kloos (he held not only a medical doctorate but was also a PhD) unsuccessfully attempted to bring an action against a young physician, Helmut Becker, at that time Vice-President of the Berlin Medical Council, after being described by Becker as making a significant contribution towards the "euthanasia" of children during the Nazi regime. Kloos was forced to reluctantly admit his acceptance of the notion that in certain circumstances the killing of children was not only permissible, but desirable. However, he himself had of course never practised such a thing. "I am a meat eater but that does not make me a slaughterer," was how he chose to describe his position on the subject.[1666]

Pauline Kneissler (1900–?) is of particular interest, not only for her very extensive "euthanasia" activities, but because of the light she shed on the frequently mentioned *Organisation Todt* expedition to the eastern front in 1941/1942.

Kneissler, the daughter of a bourgeois *Volksdeutsche* farmer, was born in the Ukraine. In the wake of the Russian revolution she and her family lost their property and in 1918 fled to Germany, settling in Westphalia, where her father once again took up farming. In 1920 Kneissler began her training as a nurse, and in 1925 was employed at the Berlin-Buch mental institution, where she remained until in 1940 she was enrolled into T4, probably by Dr Wilhelm Bender, the director of the institution, and one of those involved in the planning of "euthanasia".[1667] She joined the Nazi party in 1937, and was also a member of the National Socialist's Women's League.

In early 1940 she was ordered to report to T4 headquarters, where she and a number of other nurses were informed by Werner Blankenburg (q.v.) that they had been chosen for the top secret "euthanasia" programme. They were given the option of whether or not to participate. None declined to do so, and therefore all were sent to Grafeneck. There Kneissler accompanied patients during their transfer to the killing centre, helped them to disrobe, assisted in their medical "examination", then led the victims to the gas chamber. As she testified:

> The patients we evacuated were not necessarily serious cases; they were indeed mentally ill but very often in good physical condition. Each transport consisted of about 70 people and we carried out such transports almost every day...In most cases the patients were killed in Grafeneck within 24 hours of their arrival. I have spent almost a year at Grafeneck and I know of only a few cases in which patients were not gassed.[1668]

When Grafeneck closed she was transferred to Hadamar, where she performed a similar function until the "stop order" of August 1941. Shortly afterwards she was sent to the East, where she spent several months.

Kneissler's post-war evidence that some T4 medical orderlies gave lethal injections severely wounded soldiers—the true purpose for the dispatch of so many T4 operatives to the Soviet Union—is believed to be accurate. Those participating in the *Organisation Todt* mission included numerous male and female T4 nurses, as well as SS-men

---

1666 http://www.zeit.de/1985/11/Der-Fleischesser-schlachtet-nicht-selbst (Accessed 27 February 2009).
1667 de Mildt, *In the Name of the People*, p. 372, note 107.
1668 Ibid., p. 372, note 109.

who had previously been employed as *Brenner* in the T4 killing centres. Kneissler stated that she and her unit administered such injections to blinded, brain-damaged, and mutilated troops, as well as to amputees; that is to say those who were so damaged there was no possibility of their being recycled back to the *Wehrmacht*. What possible reason could she have for fabricating such testimony?

Following her arrival back in Germany, Kneissler worked for a short time at the Weilmünster hospital before returning to Hadamar in September 1942, where she quickly became directly involved in the murder of patients through the administration of lethal medication. In May 1943 she became unwell and was excused from killing activities, until in April 1944 the by now very experienced murderess found new employment at Kaufbeuren-Irsee, where she continued to dispatch victims until the end of the war. As Kneissler declared: "Since the year 1940, I have killed thousands of mentally ill Germans and foreigners...in the mental institutions of Grafeneck, Hadamar and Kaufbeuren. I am unable to give an exact figure of the number killed since it concerned too many persons."

Together with five other nurses (two of whom were acquitted for lack of evidence), Kneissler was tried by the Frankfurt court in 1948. For her involvement in the murder of an unknown number of victims she was sentenced to imprisonment for three years and one month. She was paroled one year later.[1669] By 1950 Kneissler was working as a nurse again; she retired in 1963.[1670]

Edith Korsch (1914–?) was born in West Prussia, a region ceded to Poland at the end of the Second World War. Her first job was as a household maid; she then worked for her mother for a time before entering the Neuruppin mental hospital as a student nurse in 1938. She had joined the National Socialist Women's League in 1936, but never became a member of the Nazi party.

Like Pauline Kneissler (q.v.) (whose career her own closely mirrored) and others, Korsch was ordered to report to T4 headquarters in Berlin in January 1940, where she was enrolled in the "euthanasia" programme and dispatched to Grafeneck. When that killing centre closed in December 1940, she was first transferred to Hadamar, then briefly to Eichberg, before joining the *Organisation Todt* expedition to the eastern front. On her return she was again sent to Neuruppin (where she married in 1942), prior to being assigned briefly to Bernburg, and finally to Hadamar once more.[1671] Although questioning the legality of killing patients, like many others she was prepared to accept that the orders to do so were binding on her. She was thus involved in both phases of the "euthanasia" operation, only leaving T4 in May 1944 after becoming pregnant.

Together with other nurses, Korsch was tried in Frankfurt in January 1948. She was found guilty of being an accessory to murder "in an unknown number of cases" during the first phase of "euthanasia" as well as participating "in at least 20 cases" during the second phase, and was sentenced to a jail term of three years and four months.[1672]

1669 Burleigh, *Death and Deliverance*, p. 154.
1670 de Mildt, *In the Name of the People*, pp. 200–203.
1671 Ibid., p. 372, note 123.
1672 Ibid., pp. 204–205.

Fritz Kühnke (1911–?)[1673] was born in Stettin, today Szczecin, Poland. On leaving school in 1932 he initially intended to study music, but instead was persuaded by Ernst Wentzler (q.v.), both a distant relation and a mentor, to become a physician. He attended the universities of Tübingen, Königsberg, Munich and Kiel, graduating from the latter in December 1937. In January 1933 he had joined the *Stahlhelm*, a paramilitary nationalist organization subsequently integrated with the SA. He became a member of the Nazi party in 1937.

Kühnke qualified as a doctor in January 1939 and was employed by a Berlin children's clinic, until in June 1940 he was conscripted into the *Wehrmacht*. Unhappy with his lot, in September of that year he met with Wentzler, who, as a T4 *Gutachter*, suggested that Kühnke apply to have his military service suspended and instead join the "euthanasia" programme. Kühnke hesitated, unsure of the legality of the T4's activities. Wentzler assured him that everything was above board, and that a legally binding *Führer* decree existed, but had to remain secret.

Kühnke procrastinated for four weeks before agreeing to Wentzler's proposal. Although he still retained lingering doubts, he felt that a doctor "had, under all circumstances, the duty to help pitiable and joyless pointlessly vegetating living creatures". Released from the military through Wentzler's intervention, in November 1940 Kühnke arrived at Eglfing-Haar, where he was to remain until December 1941. During that 13 month period he also occasionally acted as Richard von Hegener's (q.v.) locum at the KdF. At his post-war trial it was determined that in cooperation with the head of Eglfing-Haar, Hermann Pfannmüller (q.v.), Kühnke had been guilty of killing 19 children. Furthermore, he had personally killed another 8 children at the Wiesloch mental institution when on temporary assignment there.

By February 1942 Kühnke had rejoined the army, serving as a doctor in Norway, Poland, Russia, and France. He was temporarily a British prisoner of war, but was released from custody in December 1945. Shortly thereafter he received recognition as a paediatrician, and in April 1947 began to practice as such in Hamburg.

In May 1968 proceeding against Kühnke for his participation in "euthanasia" began in Munich. The court went to extraordinary lengths to excuse what were plainly cases of murder, accepting without reservation the defendant's claim that he had acted solely out of a sense of compassion for his victims and never on racial or eugenic grounds. Consequently the court acquitted Kühnke of the charge of murder, instead finding him guilty of 27 cases of manslaughter. However, since the statute of limitations had become operable with regard to such crimes, Kühnke walked from the courtroom a free man, to resume his paediatric practice.[1674]

Alfred Leu (1900–?)[1675] was born in Schwerin, and originally trained as a locksmith before deciding upon a career in medicine. He was a student at the universities of Vienna, Rostock, and Hamburg, graduating from the latter in 1925. Four years later he commenced part-time employment at the Sachsenberg mental hospital; the rest of his time was devoted to his private psychiatric practice. He joined the NSDAP in 1933, and like

1673 Klee, *Das Personenlexikon zum Dritten Reich*, p. 348.

1674 de Mildt, *In the Name of the People*, pp. 140–145.

1675 Klee, *Das Personenlexikon zum Dritten Reich*, p. 369

his Sachsenberg colleagues became an expert advisor to one of the Hereditary Health Courts sitting in judgement in cases of compulsory sterilisation. In September 1941, Hans Hefelmannn (q.v.) and Richard von Hegener (q.v.) interviewed Leu at the KdF and invited him to take charge of a children's killing ward at Sachsenberg, which, despite his claimed moral repugnance at participating in such reprehensible acts, he did.[1676]

Leu, who remained at Sachsenberg until 1945, freely admitted that he had completed patient's *Meldebogen* in full knowledge of their purpose, but had only done so as the lesser of two evils; this was to become the familiar *ex post facto* defence of sacrificing some in order to rescue others. He had ordered the killing of about 100 children and adults by lethal injection, but only so as to enable him to save an equivalent number of potential victims. In October 1951, a Cologne court found Leu's defence convincing, and acquitted him. On appeal, the Supreme Court reversed the decision and ordered a retrial.

Leu's defence was not supported by the evidence of others. Witnesses claimed that he was a dedicated Nazi, and a wholehearted supporter of "euthanasia", a man who had killed 200–300 adults with overdoses of medication. So efficient was Leu at murder, that in 1942 an additional children's ward was added at Sachsenberg. Although it was apparent that such a ward had existed solely for the purpose of killing, the court found that if children had died there, it was not because of Leu, but due to "natural causes". Needless to say, Leu claimed that his party membership and apparent enthusiasm for "euthanasia" was all a sham; he had to pretend to support the regime and its murderous policies in order to act like a medical Scarlet Pimpernel. Unlike almost everybody else, the court was totally convinced of Leu's innocence, concluding:

> Contrary to his inner disapproving attitude towards euthanasia, the defendant...considered it imperative to camouflage his sabotaging measures by pretending to the outside world to be in agreement with the *Aktion* ordered by the state authority. To his colleagues as well as towards the nursing personnel he therefore repeatedly expressed himself in a way showing that he agreed with the euthanasia *Aktion* and considered it justified. Towards...relatives of mentally ill patients he made remarks of a similar content.[1677]

Despite the fact that Leu had not only knowingly arranged for the transport of his patients to their death, but had personally killed others, in 1953 he was acquitted "for lack of evidence."[1678] One wonders what kind of evidence would have been necessary to convince that particular court to convict anybody of the crimes under consideration.

Herbert Linden (1899–1945)[1679] was born in either Konstanz, Baden[1680] or Berlin,[1681] (sources differ) and qualified as a doctor in 1925. He joined the Nazi party in the same year. Apparently he was never a general practitioner, nor did he ever specialize in a specific medical field. Instead he worked as an assistant physician in Heidelberg until in 1931 he moved to the *Reich* Ministry of Health, remaining with that agency when it became a division of the

1676 Friedlander, *The Origins of Nazi Genocide*, p. 53.
1677 de Mildt, *In the Name of the People*, pp. 176–179.
1678 Bryant, *Confronting the "Good Death"*, pp. 198–203.
1679 Klee, *Das Personenlexikon zum Dritten Reich*, p. 373.
1680 Friedlander, *The Origins of Nazi Genocide*, p. 43.
1681 Klee, *Das Personenlexikon zum Dritten Reich*, p. 373.

Reich Ministry of the Interior in November 1933. Clearly having decided that practicing medicine was not for him, Linden instead elected for the life of a *Schreibtischtäter*, and was prominent in virtually every aspect of eugenic affairs throughout the history of the Third *Reich*. He was instrumental in the initiation of the sterilisation and marriage laws, and in 1936 was a co-author with Arthur Gütt (q.v.) and Franz Massfeller of a commentary on the "Blood Protection and Marriage Health Laws".[1682] Working under Fritz Cropp (q.v.), he also became one of the principle architects of the "euthanasia" programme. As Viktor Brack (q.v.) commented: "He [Linden] was always there for the entire affair."[1683]

Linden's job was to ensure efficient co-operation between state health officials and T4 functionaries. On 23 October 1941, Linden was appointed *Reich* Commissioner for Mental Hospitals and Nursing Homes, promoting him to a position above Werner Heyde (q.v.) and Paul Nitsche (q.v.) in the T4 pecking order, and thereby making the entire "euthanasia" organization a division of Department IV of the Ministry of the Interior.[1684] When the "euthanasia" programme entered its second phase in late 1942, it was Linden who provided the vital connection between the Ministry and T4.[1685]

The so-called "Gerstein Report" mentions the presence of Linden at Belzec in August 1942, although there is no other evidence of a direct connection between Linden and *Aktion Reinhard*. Such a connection is within the realm of possibility. From July 1942 to 1944 or 1945 Linden was an honorary judge at the *Volksgerichtshof* (the People's Court). What qualified him to hold such a post other than a long term devotion to Nazism is difficult to surmise.

Linden committed suicide by shooting himself on 27 April 1945.

Lothar Loeffler [or Löffler] (1901–1983)[1686] a prominent racial biologist, was born in Erfurt. From 1927 he worked at the Kaiser Wilhelm Institute for Anthropology, and went on to hold a number of other academic positions at the universities of Kiel, Königsberg, and Vienna, where in 1942 he became the director of the Office of Hereditary and Racial Biology. He had joined the NSDAP and SA in 1932, and as an associate of Walter Gross (q.v.) of the Office of Racial Policy, was extensively involved in many aspects of that department's activities.[1687] His connection to children's and adult "euthanasia" is deserving of further research.

Loeffler's academic and professional career continued uninterrupted after the war, as he was appointed to a variety of successively important posts.

Rudolf Lonauer (1907–1945)[1688] was born in Linz, the son of an Austrian civil servant who had been an early member of the Nazi party. Lonauer followed his father into the then illegal NSDAP in 1931, and also joined the SS in 1933. He qualified as a neurologist at the University of Graz, and after the *Anschluss* in 1938, was promoted to the position of director of the Niedernhart sanatorium. He went on to head the department of neu-

1682 Friedlander, *The Origins of Nazi Genocide*, p. 43.
1683 Ibid., p. 200.
1684 Aly, Chroust, Pross, *Cleansing the Fatherland*, p. 46.
1685 Schmidt, *Brandt*, p. 215.
1686 Klee, *Das Personenlexikon zum Dritten Reich*, p. 376.
1687 Proctor, *Racial Hygiene*, p. 87.
1688 Klee, *Das Personenlexikon zum Dritten Reich*, p. 379

rological diseases at Linz general public hospital, as well as having a private practice in Linz itself.

From spring 1940 Lonauer became physician-in-charge at Hartheim whilst retaining his position at Niedernhart. This unusual relationship between the two institutions meant that Niedernhart served as a kind of waiting room for Hartheim. Patients selected for immediate killing went straight to Hartheim; those where a delay was considered appropriate before gassing occurred were first sent to Niedernhart.[1689] There is circumstantial evidence that Lonauer was also involved in killings in at least one other Austrian institution. A doctor at the Gugging state hospital testified that Lonauer and two colleagues were resident there for a short time in 1943. The visitors claimed that there was a "typhus epidemic" raging in the infectious diseases ward, entry to which was then forbidden to the witness for two or three weeks. He continued:

> During this period, the mortality rate in the infectious diseases ward rose sharply and many patients from other departments were also transferred to the ward. In the period between 28 March and 8 April 1943...a total of 112 patients died. Dr. Lonauer was at the hospital the entire time and also lived there.[1690]

Lonauer is yet another whose name appears on the surviving list as a *Gutachter* with effect from 1 April 1940.[1691]

In early 1941, Lonauer became one of the doctors assigned to *Sonderbehandlung* 14f13, visiting Mauthausen, Gusen, Dachau, and Buchenwald to select prisoners for the Hartheim gas chamber. He thus not only chose the victims as a 14f13 physician, but also supervised the murder of some of them as head of Hartheim.[1692] Lonauer joined the Prinz Eugen division of the *Waffen*-SS in September 1943. Quite what his duties there involved is unknown, but in November 1944 he returned to once again to become head of Niedernhart and Hartheim, taking over from his assistant Georg Renno, who had been in charge in Lonauer's absence.

Married to a woman who was reputedly an even more enthusiastic Nazi than Lonauer himself, on 5 May 1945, having first killed their two daughters, the husband and wife jointly committed suicide by taking poison. The number of deaths that can be laid at Lonauer's door in his capacity as a leading practitioner of "euthanasia" is unquantifiable.

Wilhelm Lückoff (1909–?) was born in Wissenbach. He joined the NSDAP in 1928 and the SA in 1930. An unskilled worker, he suffered lengthy periods of unemployment before in 1937 he joined the Herborn mental institution as a student nurse. He was transferred to Hadamar in 1940 and rapidly enrolled into T4. Although he claimed not to have been involved in the first phase of "euthanasia", working instead at a variety of odd jobs, he was yet another sent on the *Organisation Todt* mission to the East in winter 1941/1942. He spent a brief period at Weilmünster on his return before transferring to

---

1689 Friedlander, *The Origins of Nazi Genocide*, p. 109.

1690 http://www.memorialgugging.at/pdf/Czech_MedizinverbrechenGugging_engl.pdf (Accessed 18 September 2009.)

1691 Klee, "*Euthanasie*" *im NS-Staat*, p. 229.

1692 Friedlander, *The Origins of Nazi Genocide*, p. 222.

Hadamar once more to work as a night-watchman. It was in this capacity that he administered lethal medication to patients.[1693]

A defendant at the second Hadamar trial in 1947, Lückoff was found guilty of complicity in at least 8 cases of murder, and was sentenced to three years and one month in jail.[1694]

Heinrich Arthur Matthes (1902–?) was born in Wermsdorf, a small town in Saxony. He began his working life as a tailor, but in 1924 changed careers and became a student nurse at Sonnenstein. After passing his nursing examinations there, he worked at the hospitals in Arnsdorf and Bräunsdorf, before he was conscripted by the *Wehrmacht* in 1939. He served in the army in Poland and France until September 1941, when he was summoned to the KdF, his military service terminated. He had joined the SA in 1934, and applied for membership of the NSDAP in 1937, but it is unclear whether his application was ever accepted. At T4 headquarters he worked in the photographic laboratory for a time, until in winter1941/42 he was sent eastwards with the *Organisation Todt* mission, serving as a male nurse in the Minsk and Smolensk area. In February/March 1942 he returned from the Soviet Union to take up his previous post at T4.[1695]

In late August 1942, Matthes was assigned to Odilo Globocnik's organisation in Lublin, where he was given the usual SS-*Scharführer* rank and sent to Treblinka. There he was put in charge of Camp II where the gas chambers were located. In addition to supervising the resident Jewish workforce, he actively participated in the entire extermination process, from herding prisoners into the gas chambers to the transport of victims' corpses. He did all this with great brutality, seemingly deriving enormous pleasure from the cruelty his unlimited power gave him over those unfortunate enough to pass through his domain.[1696] Matthes appears to have had an obsession with cleanliness; he once shot two prisoners because at the end of the working day he was not satisfied with the state of the stretcher which they had used to carry bodies. A survivor recalled of Matthes:

> He used to beat the prisoners with a completely expressionless, apathetic look on his face, as if the beatings were part of his daily routine. He saw to it that the roll-call area would always be extremely clean. One of the prisoners had to rake the sand in the square all day long, and he had to do it with Prussian exactness.[1697]

Matthes was in Treblinka at the time of the uprising in that camp on 2 August 1943.[1698] The following month he was transferred to Sobibor; since he remained there until early December of that year, he was certainly a staff member at the time of the uprising in Sobibor on 14 October 1943, as well as being present during the subsequent liquidation of the camp. After a short spell in Berlin, Matthes joined his former *Aktion Reinhard* comrades in Italy for anti-partisan activities and the continued extermination

---

1693 de Mildt, *In the Name of the People*, p. 198.

1694 Ibid., p. 371, note 96.

1695 Friedlander, *The Origins of Nazi Genocide*, p. 242.

1696 de Mildt, *In the Name of the People*, p. 260.

1697 Arad, *Belzec, Sobibor, Treblinka*, p. 194.

1698 Sereny, *Into That Darkness*, pp. 240–241.

of Jews.[1699] He was briefly an American prisoner of war in 1945, but it was not long before he was once again working as a nurse in a succession of mental hospitals.

On 12 October 1964, the trial of 10 men accused of mass murder at Treblinka opened in Düsseldorf.[1700] One of the defendants was Matthes, who was found guilty of the murder of "at least 100,000 people", plus eight specific charges of homicide, which given his length of service in the camp seems a remarkably modest total. The court imposed 5 life sentences on him.[1701]

Otto Mauthe

Source: Landesarchiv Baden-Württemberg / StAL F215 Bü 184 / © 2017

Otto Mauthe (1892–1974)[1702] was born in Derdingen. He studied medicine in Tübingen and Kiel before serving as a doctor during the Great War. After qualifying as a physician in 1920 he worked in a women's clinic in Tübingen, before in 1927 he commenced practising as a doctor in Herrenberg. He joined the SA in 1933 and the NSDAP the following year. In 1936 he was appointed deputy to the chief medical officer of the health department of the Württemberg Ministry of the Interior, Eugen Stähle (q.v.), with responsibility for the 48 mental hospitals within the state. In this capacity Mauthe ensured the smooth operation of the "euthanasia" programme in Württemberg, including the passing on of directives and transport lists to the institutions under his control, regulating the discharge of patients, withholding information concerning patient's whereabouts from their relatives, completing registration forms, and the like.

In late November 1939, Mauthe began the process by which patients marked for murder were transferred to Grafeneck, then after the closure of that killing site, to Hadamar. Mauthe personally involved himself in every aspect of the programme, even visiting Grafeneck to witness a gassing of patients.[1703] Apart from the murders at Grafeneck and Hadamar, he also authorised the transfer of at least 93 children to killing wards outside Württemberg. In short, if not a murderer in the sense of directly killing anybody with his own hands (so far as is known), Mauthe was the classic example of the *Schreibtischtäter*, on one occasion offering as a defence for his actions: "One lived in an authoritarian state and had to comply with the orders given without asking what was right."[1704]

In July 1949, proceedings against Mauthe began in Tübingen. Despite an overwhelming amount of documentary evidence proving Mauthe's guilt, rejecting the defence that he had acted as he did on the "surrender few, save many" principle which was to become such a familiar feature of post-war German trials, and accepting that he had willingly colluded with the "euthanasia" programme to an extent far beyond that which might be considered the norm, the court sentenced Mauthe to a derisory 5 years'

1699 de Mildt, *In the Name of the People*, p. 395, note 49.

1700 An eleventh defendant, Kurt Küttner, died before proceedings commenced.

1701 de Mildt, *In the Name of the People*, p. 260.

1702 Klee, *Das Personenlexikon zum Dritten Reich*, p. 396.

1703 de Mildt, *In the Name of the People*, pp. 106–107.

1704 Ibid., p. 348, note 48.

imprisonment for participation in the murder of an estimated 4,000 people. Shortly afterwards the sentence was suspended and Mauthe was released.[1705]

Robert Friedrich Mauz (1900–1979)[1706] was born in Esslingen, near Stuttgart. In 1922 he joined the *Freikorps* Epp, a right-wing, paramilitary formation. He became a lecturer at the University of Marburg in 1928, and was appointed ao. Professor there in 1934. He joined the SA in the same year and the NSDAP, together with several other Nazi organisations including the National Socialist Physicians' League, in 1937. Between 1939 and 1945 he was the director of the psychiatric clinic at the University of Königsberg. Mauz was involved in the development of "euthanasia" from inception, being one of forty-eight physicians asked to evaluate potential victims of the programme in October 1939.[1707] His name appears as a *Gutachter* on a list of T4 personnel, apparently appointed on 2 September 1940.[1708]

Mauz enjoyed a highly successful post-war career, in 1945 becoming director of the mental hospital at Ochsenzoll, Hamburg, and in 1953 was appointed director of the psychiatric clinic at the University of Münster. Few seemed concerned about his Nazi past. In 1948, he had been an official delegate to the Third International Congress of Mental Hygiene in London. Ten years later he was president of the German Society of Psychiatrists and Neurologists. The "*Münsterische Zeitung*" of 17 May 1980 reported:

> The medical faculty of the University of Münster commemorates today their member of long-standing and director of the University Clinic, Prof. Dr. Robert Friedrich Mauz...He has helped, as hardly any other psychiatrist of his generation has, to shape the thoughts and actions of German medicine...(Let us) then pay tribute to the special accomplishments of Friedrich Mauz.[1709]

Friedrich Mennecke
Source: Vernetztes Erinnern Hildesheim

Friedrich Wilhelm Heinrich Mennecke (1904–1947)[1710] was born in Gross-Freden, near Hannover, the son of a stonecutter and mason. His father served in the Great War, returning a shell-shocked invalid to die in 1923. His father's disability may well have contributed to Mennecke's desire to enter the medical profession. But for someone with his background and in the prevailing economic conditions, this was far from easy. After graduating from the *Gymnasium* in 1923, Mennecke became a commercial apprentice at a company in Freden, eventually obtaining the position of export salesman. In 1927 he left that employment to commence his medical studies in Göttingen, qualifying as a doctor in 1934. He had joined both the Nazi party and the SS in 1932.

On 1 July 1935, he elected for a specialization study in the Department of Surgical Gynaecology at the Bad Homburg District Hospital. Unhappy there, two months later he transferred to the Eichberg State Mental Hospital, thereby electing instead for a career

1705 Bryant, *Confronting the "Good Death"*, pp. 164–167.
1706 Klee, *Das Personenlexikon zum Dritten Reich*, p. 396
1707 Proctor, *Racial Hygiene*, p. 189.
1708 Klee. *"Euthanasie" im NS-Staat*, p. 228.
1709 Röder, Kubillus, Burwell, *Psychiatrists—the Men Behind Hitler*, p. 104.
1710 Klee, *Das Personenlexikon zum Dritten Reich*, p. 403.

in psychiatry.[1711] Within two years Mennecke had risen to the post of chief physician, and on 30 January 1939 he was appointed director of the hospital. In the interim, he had married Eva Wehlan, a medical technician.[1712] Ambitious and in tune with National Socialist ideology, he rapidly gained promotion within the SS, attaining the rank of *Hauptsturmführer* by October 1940.

Although appointed troop physician at the beginning of the Second World War, Mennecke's military career was terminated in January 1940 by special order of the KdF. After attending a meeting at T4 headquarters one month later where, he claimed, the presence of so many distinguished medical professionals was proof of the validity of the "euthanasia" programme, he agreed to participate in it.[1713] His duties initially consisted of visiting various psychiatric institutions with a view to selecting inmates for extermination. Later he was extensively involved in assessing concentration camp inmates within the framework of *Sonderbehandlung* 14f13, being responsible for condemning at least 2,500 such prisoners. He also served as a *Gutachter*, reviewing nearly 7,000 forms on behalf of T4. Mennecke was resolute in his work; when a pregnant woman was sent from Eichberg to a killing centre for gassing but was returned the same evening, he ordered that she be sent back with another transport the following day. The woman did not return a second time.[1714] Because of his broader involvement in "euthanasia", Mennecke was unable to personally fully implement his plans for a children's killing section at Eichberg, leaving that to his deputy, Walter Schmidt (q.v.). However, he was able to ensure that Carl Schneider (q.v.) at the University of Heidelberg received a steady supply of brains removed from the children murdered at Eichberg.[1715]

Having incurred the enmity of Fritz Bernotat (q.v.), in 1943 Mennecke was once more appointed troop physician, at first on the French Channel coast and then on the eastern front. In August 1943, suffering from goitre, he was transferred back to the *Reich*, but was not permitted to return to Eichberg. Instead, still in the capacity of a military doctor, he was sent to a hospital in Bühl, although he maintained close connections with T4, participating in conferences with other experts in matters of mass murder. In 1944 he was considered as a possible director of the "euthanasia" institutions at either Meseritz-Obrawalde, Bernburg-Saale, Graz or Plagwitz, at the last of which a large number of children had been assembled for killing. Instead, during a routine x-ray examination, tuberculosis was discovered in Mennecke's lungs. He ended the war as a patient, moving with his wife, Eva, to Northeim in autumn 1945. Eva was briefly arrested on suspicion of involvement in "euthanasia" matters, but was quickly released. It was only when Mennecke attempted to begin working at a refugee camp that he was arrested in April 1946, subsequently appearing as a witness in the Medical Trial at Nuremberg in January 1947.[1716]

1711 Burleigh, *Death and Deliverance*, p. 217.

1712 Friedlander, *The Origins of Nazi Genocide*, p. 229.

1713 Ibid.

1714 Ibid., p. 168.

1715 Ibid., p. 229.

1716 http://www.lernen-aus-der geschichte.de/resmedia/document/document/A028T07E.PDF (Accessed 19 May 2008).

By then proceedings against him had already begun in Frankfurt am Main (the so-called "Eichberg Trial"). Among the most damning evidence produced were the hundreds of letters he had written to his wife whilst he was absent on his "selection tours" (some 2,500 of an estimated 8,000 pages of Mennecke's correspondence with his wife, relatives, and colleagues have survived). Seemingly compiled on an almost hourly basis, they set out in extraordinary detail the trivia of Mennecke's everyday existence alongside comprehensive details and statistics regarding the selection of his victims. The letters proved beyond any doubt that Mennecke was an enthusiastic and dedicated proponent of "euthanasia." His wife was apparently no less committed to the cause, sometimes helping her husband with the registration of patients. The verdict of the court on Mennecke was withering:

> Mennecke...was not a man of great convictions. [He] was driven by unrestrained ambition and boundless assertiveness. He aspired to positions which did not match his capabilities and, once achieved, he wanted to retain them under all circumstances. It flattered his vanity to belong to a circle of renowned men, about whose recognition he was delighted in a childlike fashion...The court considers it proven that the defendant accommodated his past professional career solely in accordance with his personal advantage and that, in order to achieve this goal, he unhesitatingly threw everything overboard which is considered as right—moral and professional ethics, honour and decency—and that this eventually led him to take part unscrupulously in mass murder.

Mennecke was found guilty of the murder of at least 2,500 individuals, and was sentenced to death. However, he cheated the hangman. The death sentence would have almost certainly been commuted to a lengthy prison sentence on appeal, as occurred in similar cases, but on 28 January 1947 Mennecke died in prison, allegedly from acute tuberculosis, although there is reason to suspect that a visit from Eva two days before his death was not entirely unconnected with his demise.[1717]

Erich Karl Friedrich Moos (1904–?) had been unemployed for years prior to enrolling as a student nurse at the Weilmünster mental hospital in 1936. A member of the NSDAP since February 1933, he was transferred to Hadamar in July 1941, allegedly against his will, and recruited for the "euthanasia" programme. His responsibilities consisted of accompanying transports of patients from other institutions to Hadamar, then conducting them to the gas chamber. When the first phase of the killings stopped he returned for a time to Weilmünster, but was soon back at Hadamar, where he worked as a gardener. However, he also sometimes performed nursing duties, which at Hadamar inevitably came to include the administration of lethal medication to patients.[1718]

Moos was another of the nurses tried at the second Hadamar trial in Frankfurt. He was found guilty of complicity in an unknown number of murders and was sentenced to 4 years' imprisonment.[1719]

Theophil Mootz (1872–?)[1720] was a psychiatrist born in Fischau (today Fiszewo, Poland). A member of the Nazi party and the SA since 1937, he was brought out of retire-

---

1717 Aly, Chroust, Pross, *Cleansing the Fatherland*, pp. 238–295; de Mildt, *In the Name of the People* pp. 108–112; Bryant, *Confronting the "Good Death"*, pp. 121–126.

1718 de Mildt, *In the Name of the People*, p. 197.

1719 Ibid., p. 371, note 6.

1720 Klee, *Das Personenlexikon zum Dritten Reich*, p. 415.

ment and appointed medical director of Meseritz-Obrawalde in March 1942. During his time in charge, an estimated 18,000 individuals were murdered at Obrawalde. He died in captivity at Waldheim, East Germany on an unknown date.

There was little doubting Mootz's enthusiasm for "euthanasia". A number of ex-nurses from Obrawalde testified to this effect; he had told one of them that the programme had been ordered "from above", and there was nothing to do but comply with those orders. The killings were legal, and he would take full responsibility for them. Another nurse who questioned the legality of proceedings was simply told to do her duty. A third nurse initially refusing to participate was told by Mootz that it was her "legal duty" as a civil servant to do so. Cowered, the nurse went on to become involved in the murder of at least 120 patients.[1721]

Maria Müller (?-?) was a nurse at the Kalmenhof mental institution, where she served under Mathilde Weber (q.v.) and was responsible for the murder of an unknown number of children through the addition of Luminal to their food. In return for such services, Müller received 50 percent of a monthly stipend of 30 *Reichsmarks* provided by T4, plus 50 percent of the 5 *Reichsmarks* "per death" paid by the institute's director, Wilhelm Grossman (q.v.), Müller's co-beneficiary in both cases being Anna Wrona (q.v.).

Arrested by the Americans at the end of the war, Müller escaped from custody in October 1945 and was never traced. This naturally made it easier for others deeply implicated in the killings to pass all of the blame for these crimes on to her.[1722]

Robert Müller (1886–1945)[1723] was a psychiatrist born in Schwientochlowitz, Silesia (today Swietochlwice, Poland) employed at the Königslutter state hospital and sanatorium. Müller was an enthusiastic eugenicist, and apart from the known transfers of patients from his institution to other killing centres, the hospital's abnormally high death rate for both children and adults attests to the fact that so-called "wild euthanasia" was carried out at Königslutter itself.[1724]

As a holder of the Golden Party Badge, Müller had been recognized by the regime as a long-standing and loyal member of the NSDAP. A list of *Gutachter* indicates that he was one of the doctors employed at T4 central office in Berlin from 8 November 1940. In 1941 Müller was assigned to *Sonderbehandlung* 14f13, assessing concentration camp inmates for gassing. His name appears several times in letters written to his (Mennecke's) wife by Friedrich Mennecke (q.v.) from various camps the two men were visiting together for this purpose.

Müller was another to commit suicide at the end of the war, in his case in Königslutter on 2 June 1945

Hermann Paul Nitsche (1876–1948)[1725] was born in Colditz, the son of a physician. He studied medicine at the universities of Leipzig and Göttingen, from where he received

1721 de Mildt, *In the Name of the People*, p. 366, note 96.
1722 Ibid., p. 130, p. 186, p. 352, note 119.
1723 Klee, *Das Personenlexikon zum Dritten Reich*, p. 422.
1724 http://www.pk.lueneburg.de/gedenkstaette/NS-Psychiatrie-englisch.pdf (Accessed 2 December 2008).
1725 Klee, *Das Personenlexikon zum Dritten Reich*, p. 437.

his doctorate in 1902. Between 1904 and 1907 he was an assistant to the eminent psychiatrist Emil Kraepelin.[1726] Nitsche worked at the Frankfurt asylum, as well as at the university hospitals of Heidelberg and Munich. In 1911, together with Alfred Ploetz (q.v.) with whom he was acquainted, he participated in an exhibition in Dresden dedicated to the subject of racial hygiene, before in 1913 he became the senior physician at the Dresden mental hospital. From 1914 until early 1918 he was the acting head at the Pirna-Sonnenstein institution.

In April 1918 Nitsche became director of the Leipzig-Dösen asylum. He achieved the title of professor in 1925, and in 1927 was made a consultant psychiatrist to the government of Saxony. On 1 August 1928 he returned to Pirna-Sonnenstein, this time as director, a position he was to occupy until May 1939. Between January and April 1940 he went back to Leipzig-Dösen to conduct killing experiments on patients using barbiturates.[1727] Nitsche was thus involved in "euthanasia" from a very early stage. Although he only joined the Nazi party in 1933, somewhat later than many, he had been a supporter of eugenics in all of its manifestations for many years before acting first as a Hereditary Health Court judge, then as a T4 *Gutachter* (officially from 28 February 1940) and *Obergutachter*. He succeeded Werner Heyde (q.v.) as head of the T4 Medical Office in December 1941, functioned as an expert assessor for *Sonderbehandlung* 14f13, and enthusiastically endorsed "wild euthanasia" as well as a proposed official post-war recommencement of the killing programme.[1728] Nitsche's overall contribution to T4 was immense, as has been indicated. As he stated at his trial, he considered his participation in "euthanasia" as "a service to humanity".[1729]

In his pre-National Socialist days, Nitsche had been a highly regarded psychiatrist, writing somewhat ironically in 1929:

> The nursing staff must be warned expressly against the dangers of incorrect attitudes; they must be made aware that a lack of skill and kindliness can brutalize psychiatric nursing. The patient should always feel that the intention is kind and benevolent.[1730]

Nitsche was arrested in spring 1945 by the Soviets, tried in 1947 in Dresden by an East German court, sentenced to death, and guillotined in 1948, one of the few senior T4 medical staff to receive a sentence commensurate with his crimes.

Friedrich Walter Creutz Panse (1899–1973)[1731] was born in Essen. Like his friend Kurt Pohlisch (q.v.), he had served in the military during the First World War, then studied medicine at the University of Berlin. After graduating in 1924, he worked as an assistant psychiatrist at the Wittenau mental hospital in Berlin, where he first met Werner Heyde (q.v.). Because of his friendship with Pohlisch, who was a rising star in the Nazi psychiatric firmament, in 1936 Panse was appointed chief physician at the

---

1726 In 1899 Kraepelin had suggested that heredity was "perhaps the strongest cause of mental illness," and that it was the duty of the state to incarcerate the mentally ill for the sole purpose of preventing their reproduction. (Weikart, *From Darwin to Hitler*, p. 137).

1727 Friedlander, *The Origins of Nazi Genocide*, p. 91.

1728 Klee, *"Euthanasie" im NS-Staat*, p. 228.

1729 Schmidt, *Karl Brandt*, p. 136.

1730 Aly, Chroust, Pross, *Cleansing the Fatherland*, p. 158.

1731 Klee, *Das Personenlexikon zum Dritten Reich*, p. 449.

Provincial Institute for Psychiatric and Neurological Genetic Research in Bonn. The following year he joined the NSDAP.

Panse was well respected by his colleagues, who testified that he had been opposed to "euthanasia", instead being dedicated to purely scientific research. Unfortunately this endorsement of Panse's virtues ignored the fact that, like his friend Pohlisch, during the 1930s he had been actively involved in a project promoted by the Central Office for the Hereditary Biological Inventory, which involved compiling a nation-wide data bank of genetic and racial information. Many of those contributing to this undertaking were later to reappear as leading members of T4. Furthermore, Pohlisch and Panse's joint project on epilepsy was sponsored by Herbert Linden (q.v.), a man immersed in Nazi eugenics. Admittedly this was all guilt by association, but it is also true that a man is (or should be) judged by the company he keeps.

In April 1940, Panse was summoned to T4 headquarters in Berlin together with 50–60 other psychiatric experts, including Pohlisch. Those present were invited by Viktor Brack (q.v.) to become involved in the "euthanasia" programme. Although it was subsequently asserted that there was general resistance to any such co-operation, in fact all present agreed to participate in the preparation of a modified "euthanasia" registration form. At the conclusion of the meeting Linden asked Panse and Pohlisch to become *Gutachter.* Panse testified that he refused, even though his name appears on the surviving list of *Gutachter*, effective 14 May 1940 until 16 December 1940.[1732] In any event, between May and September 1940, Panse received hundreds of completed forms on which to pass his expert opinion. He claimed to have done so in only 15 cases. Thereafter there is no further evidence of any activity on his behalf as a *Gutachter.*[1733]

Panse continued to work as a psychiatrist throughout the war, in 1942 being appointed Professor of Psychiatry, Neurology, and Racial Hygiene at Bonn University. In addition he acted as physician-in-chief at the university's mental clinic, as well as director of the provincial mental hospital. In 1948, he appeared in the dock in Düsseldorf alongside Pohlisch, accused of participation in the "euthanasia" programme. The two men presented a similar, and familiar, defence. They had objected to "euthanasia", and had done everything in their power to preserve the lives of their patients. "Professor Pohlisch and myself were the leaders of the opposition," Panse declared.[1734] Those patients they had selected were terminally ill; none had died solely as a result of their actions. The court unhesitatingly accepted their evidence, and acquitted them on the grounds of "proven innocence". The verdict was annulled by the Supreme Court of the British Zone and a retrial ordered, which took place in January 1950, again resulting in acquittal. Not only were they innocent, but as in the case of Walter Creutz (q.v.) the court deemed their behaviour as verging on the heroic.[1735]

Panse went on to enjoy a distinguished post-war academic career.

1732 Klee, *"Euthanasie" im NS-Staat*, p. 228.

1733 de Mildt, *In the Name of the People*, pp. 158–159.

1734 Ibid., p. 360, note 57.

1735 Bryant, *Confronting the "Good Death"*, pp. 160–163.

Eduard Pernkopf
Source: Josephinum, Ethik, Sammlungen und Geschichte der Medizin, MedUniWien / Josephinum, Ethics, Collections and History of Medicine, MedUni Vienna

Eduard Pernkopf (1888–1955)[1736] was born in Rappottenstein, Austria. He graduated from the University of Vienna in 1912, and in 1920 became an assistant to Ferdinand Hochstetten at the university's Anatomical Institute. Appointed a professor the following year, in 1933 he succeeded Hochstetten as head of the institute. Pernkopf joined the Nazi party the same year and the SA in 1934. Between 1938 and 1943 he was the Dean of the university, and from 1943 to 1945 its Rector.

Following the *Anschluss*, at his first lecture as Dean on 6 April 1938 Pernkopf left no-one in any doubt as to where his loyalties lay. Racial hygiene was to be the dominant factor in medicine. Hitler was "the greatest son of our home country" and enjoyed the "joyful devotion and loyalty" of all, Students would be "trained as National Socialist doctors, who are to put the medical profession at the service of National Socialism." In July of that year he affirmed the importance and necessity of an Institute of Genetics and Racial Biology in Vienna, and such a department was created at the university in October 1938, with the intention of it becoming operational on 1 November 1939. In fact it was to take four years before Lothar Loeffler (q.v.) assumed control of a fully functioning Institute.[1737]

If Pernkopf is remembered at all today it is for his *Topographical Anatomy of Man*, a text which has been a widely used as the standard work of anatomy for over sixty years. The book continues to be published under the imprint of the original publisher. However, the appearance of one of the subjects raised questions as to whether the real-life model may have been a prisoner of the Nazis. In March 1995, Yad Vashem formally requested that the University of Vienna undertake an independent inquiry into the backgrounds of the subjects in *Pernkopf's Anatomy*. In February 1997, the then Rector of the university announced a commission of investigation and a report was issued on 1 October 1998, which revealed that the Institute of Anatomy had received almost 1,400 cadavers from the *Gestapo* execution chamber at the Vienna Regional Court. While the Anatomical Institute and its collection were destroyed by bombing near the end of the war, the investigation did identify approximately 200 Institute specimens from the Nazi era that were still in the collections of other universities. The possible use of these bodies as models cannot be excluded for up to half of the approximately 800 plates in Pernkopf's book.[1738]

Pernkopf was dismissed from his academic post at the end of the war, and spent three years in an internment camp. On his release he continued to work on his anatomical atlas until his death.

1736 Klee, *Das Personenlexikon zum Dritten Reich*, p. 453.

1737 http://www.doew.at/information/mitarbeiter/beitraege/rachyg.html (Accessed 5 December 2008).

1738 http://www.jewishvirtuallibrary.org/jsource/Holocaust/medmurder.html (Accessed 5 December 2008).

Hermann Pfannmüller (1886–1961)[1739] was born in Munich. He qualified as a doctor in 1913, and as a specialist in psychiatry in 1918.[1740] During the First World War he worked at the Homburg mental hospital, where the sight of so many patients dying of malnutrition is likely to have had a profound influence on him. He certainly became an enthusiastic, not to say fanatical exponent of "euthanasia", a belief which was undoubtedly bolstered by his joining the NSDAP in 1922. He left the party in 1925 after a dispute with Julius Streicher, but rejoined after the *Machtergreifung*. In 1934 Pfannmüller had briefly been at the Kaufbeuren asylum, before In 1936 he was appointed chief physician at the Augsburg Consultation Office for Genetic and Racial Care. He became director of the Eglfing-Haar mental institution in 1937.[1741] In November 1939, and already a *Gutachter*,[1742] he set out his philosophy in a report to the Ministry of Justice:

> As a confessionally unattached and convinced National Socialist director of an institution, I consider it my duty to present a genuine economization that is suited to influence favourably the financial position of the institutions. In this respect I consider it appropriate openly and in all clarity to point to the necessity that, with regard to the medical treatment of life-unworthy life, we as doctors draw the final conclusion in the sense of extermination...Those patients who are in themselves to be pitied [but] who have become totally useless for social integration into human society...[and] who have become a burden and a pain to themselves, their relatives, and their environment, should be subjected to intensified extermination...[1743]

Pfannmüller practiced what he preached. Almost 1,000 of his patients were transported to other killing centres for gassing, whilst under his supervision a minimum of 120 children were murdered at Eglfing-Haar. Pfannmüller's dedication to "euthanasia" has been illustrated elsewhere in this work, and although there were those, like Gerhard Schmidt, Pfannmüller's successor as director of Eglfing-Haar near the end of the war, who were charitable enough to believe that Pfannmüller was "a simple man [who] was strongly convinced that [the "euthanasia" programme] was urgently necessary", and who ordinarily "could not hurt a fly," the record suggests otherwise.[1744] Viktor Brack (q.v.), for example, considered him one of the leading lights of "euthanasia."

Having appeared as a witness at the Nuremberg Medical Trial, in 1949 Pfannmüller was arraigned before a Munich court in connection with his own "euthanasia" activities. His defence—essentially that the legality or otherwise of his actions could be reduced to a matter of opinion—was, not surprisingly, rejected by the court. That he was well aware of the illegal nature of the programme was illustrated shortly before his arrest by American troops, when he destroyed all records of his activities at Eglfing-Haar. Nonetheless, the court determined that it was "not proven beyond doubt" that Pfannmüller was fully aware of all aspects of T4's murderous programme, and in March 1951 it sentenced him to five years' imprisonment for complicity in manslaughter,[1745] a verdict

---

1739 Klee, *Das Personenlexikon zum Dritten Reich*, p. 458.

1740 Friedlander, *The Origins of Nazi Genocide*, p. 49.

1741 Michael Burleigh, *Ethics and extermination: Reflections on Nazi genocide*, (Cambridge: Cambridge University Press, 1997), p. 245, note 23.

1742 Klee, *"Euthanasie" im NS-Staat*, p. 228.

1743 de Mildt, *In the Name of the People*, p. 102.

1744 Lifton, *The Nazi Doctors*, p. 120.

1745 de Mildt, *In the Name of the People*, pp. 101–103.

now considered "a travesty of justice".[1746] After deduction of time spent in pre-trial confinement, Pfannmüller was released after two years in jail.[1747]

Kurt Pohlisch (1893–1955)[1748] was born in Remscheid. His life and career largely replicated that of Friedrich Panse (q.v.), as has been described. Although he did not join the Nazi party until 1937, he indicated his allegiance to the cause in other ways. He was a prominent official in the Hitler Youth, a member of National Socialist Public Welfare, the National Socialist Lecturers' League, and the *Reich* Air Defence League, as well as a sponsoring member of the SS. All of this undoubtedly helped to influence his appointment in 1934 as Professor of Psychiatry and Neurology at the University of Bonn, where he also became physician-in-chief at the university mental clinic, and director of the regional mental institution.

Pohlisch appears on the list of T4 *Gutachter* with effect from 30 April 1940 to 6 January 1941. Like Panse, he claimed to have declined Herbert Linden's (q.v.) invitation to participate in the "euthanasia" programme, yet he still completed a few of the *Meldebogen* sent to him (in his case 10 "at the most"). Pohlisch's protestations of innocence seem somewhat compromised by Werner Heyde's (q.v.) assertion in 1961 that

> ...the testimony of Professor Pohlisch is false in all essential points. This is already shown by the fact that during the entire time of my own involvement in the euthanasia measures, he was uninterruptedly active as a *Gutachter*...On no occasion did Pohlisch express a principled disapproving attitude towards the euthanasia measures, either in words or in behaviour; the opposite is true.

Heyde may have had his own reasons for implicating Pohlisch so thoroughly in T4 activities, but it is nonetheless undeniable that in February 1941 (that is after he apparently had ceased to be a *Gutachter*), Pohlisch was involved in the selection of patients at the Bethel mental home. Moreover, this dedicated opponent of "mercy killing", who according to his own testimony had, through his intervention, rescued "a great many thousands of people", and had even "intentionally sabotaged the *Aktion*, even risking my life", was also a participant in the drafting of a proposed "euthanasia" law in 1940.[1749]

In fairness, at his trials in 1948 and 1950 the court was probably unaware of this evidence. As with Panse, Pohlisch was duly acquitted, and successfully continued with his psychiatric career in the Federal Republic.[1750]

Amanda Ratajczak (?-1945) was head nurse at Meseritz-Obrawalde. She escaped from the institution before the arrival of the Red Army on 29 January 1945, but was captured that March, and tried before a Soviet tribunal. She admitted to murdering more than 1,500 patients, the last on the day before the liberation of Obrawalde. Sentenced to death, she was shot on 10 May 1945.[1751]

1746 Bryant, *Confronting the "Good Death"*, p. 197.

1747 de Mildt, *In the Name of the People*, p. 346, note 33.

1748 Klee, *Das Personenlexikon zum Dritten Reich*, p. 467.

1749 de Mildt, *In the Name of the People*, p. 360, note 57.

1750 Ibid., pp. 156–160.

1751 Susan Benedict, Arthur Caplan, Traute Lafrenz Page, *Duty and 'Euthanasia': the Nurses of Meseritz-Obrawalde* (Nursing Ethics, Vol. 14(6), 2007), p. 784.

Hans Reiter
Source: Wikimedia Commons / CC-BY-SA 4.0 https://creativec ommons.org/licenses/by-sa/4.0/deed.en

Hans Conrad Julius Reiter (1881–1969)[1752] was born in Reudnitz, now incorporated as part of Leipzig. He studied medicine in Leipzig, Breslau, and Tübingen, where he received his doctorate. Subsequently he continued his studies in Berlin, Paris, and London. He was a military physician during the Great War, and in 1919 was appointed ao. Professor of Social Hygiene at the University of Rostock. A committed Nazi, Reiter joined the party in 1931 and the SA ten years later.

In 1933 Reiter was appointed head of the *Reich* Health Office.[1753] At his post-war interrogation he was rather economical with the truth, insisting that his Office had no connection with "euthanasia", or other eugenic matters.[1754] In fact, Robert Ritter's (q.v.) Eugenic and Population Biological and Research Station operated as a department of the *Reich* Health Office.[1755] Reiter had been an enthusiastic supporter of both the sterilisation and "euthanasia" policies, and was involved in a study that inoculated concentration camp internees at Buchenwald with an experimental typhus vaccine, thereby causing the death of many of them.

Although he was interned by the Americans between 1945 and 1947, by 1949 Reiter was working again as a doctor in Kassel.

Georg Renno
Source: Hessisches Hauptstaatsarchiv / Signature Abt. 631 a, Nr. 2367

Georg Renno[1756] (1907–1997) [T4 Hartheim pseudonym, "Dr Steinert",[1757] post-war pseudonym, "Dr Georg Reinig"][1758] was born in Strasbourg, Alsace, a region that has repeatedly switched hands between France and Germany. At the time of Renno's birth, Alsace was part of Germany, but the region was returned to France at the end of the Great War. Of German descent, and therefore considered of questionable loyalty to France, in what today would be termed an act of "ethnic cleansing", the Renno family was expelled from Strasbourg in 1919 to settle in Ludwigshafen, where Renno received his basic education. After his matriculation in 1926, he commenced studying medicine at hospitals in Heidelberg and Frankenthal. He graduated as a physician in 1933, and took up his first medical position at the children's hospital for tuberculosis in Scheidegg.[1759] By that time he was already a member of the Nazi party, having joined

1752 Klee, *Das Personenlexikon zum Dritten Reich,* p. 490.

1753 Weindling, *Nazi Medicine and the Nuremberg Trials,* p. 159.

1754 Ibid., p. 236.

1755 Friedlander, *The Origins of Nazi Genocide,* p. 250.

1756 Much of the following biographical material has been derived from Mireille Horsinga-Renno, *Cher oncle Georg: La bouleversante enquête d'une femme sur un médecin de la mort impuni* (Strasbourg : La Nuée Bleue, 2006).

1757 Horsinga, *Cher oncle Georg,* p. 89.

1758 Ibid., p. 158.

1759 Klee, *Das Personenlexikon zum Dritten Reich,* p. 491.

in August 1930 after becoming a member of the NS-*Studentenbund*, the Nazi Students' League, one year earlier.[1760] An enthusiastic flautist, he became a member of the SS in 1931, ostensibly to play military music, but also to act as physician to his unit.[1761]

In November 1935, Renno began specializing in neurology at the psychiatric institution at Leipzig-Dösen,[1762] becoming a consulting physician one year later. He spent another year at, first, the Kaiser-Wilhelm Institute for Anthropology in Berlin, then at the German Research Institute for Psychiatry in Munich, before becoming medical advisor to the State of Saxony. He was considered to be an expert in genetics and "racial science".[1763] Of his convinced Nazism and unquestioning obedience and loyalty to his *Führer* there was no doubt.

In 1939 he temporarily left public service for private practice, but the outbreak of war put an end to that. In March 1940 he was recruited for T4 by Paul Nitsche (q.v.), head of the institution at Leipzig-Dösen, whom he had already assisted in the killing of patients at that hospital.[1764] According to Renno, Nitsche simply sat him down and casually began their meeting with the words: "Let's see, what do you think of euthanasia?" Ambitious young Dr Renno was not about to curb his enthusiasm for the concept.[1765] After all, Himmler had pronounced that the SS physician was a soldier first and a doctor second, and Renno was nothing if not an obedient SS-man.[1766]

In May 1940 Renno was appointed a *Gutachter*, and was posted to Niedernhart and Hartheim as assistant to Rudolf Lonauer (q.v.), the physician-in-charge at those establishments.[1767] Between the summers of 1940 and 1941, Renno visited around fifty institutions within the surrounding region, meticulously selecting those mental patients he considered suitable for transfer to Hartheim and gassing.[1768] At the suggestion of Werner Heyde (q.v.), in spring 1941 Renno also participated in the making of a "euthanasia" propaganda film, *Dasein ohne Leben* ("Existence Without Life"). The film was shot at Sonnenstein, where the camera accompanied the elegant Dr Renno as he made his rounds and consulted with his patients.[1769] Of course, the film did not show the consequences of Renno's diagnoses.

Renno considered murder of an impersonal nature to be somewhat beneath him, informing Heyde, "I did not study medicine to operate a gas valve,"[1770] although doing so "was nothing special."[1771] He clearly preferred a more "hands-on" approach, an opportunity for which arose when in May 1941 a new children's killing ward was proposed

---

1760 Ibid.

1761 Friedlander, *The Origins of Nazi Genocide*, p. 224.

1762 However, he was never certified as a specialist in either neurology or psychiatry.(Friedlander, *The Origins of Nazi Genocide*, p. 58).

1763 Horsinga, *Cher oncle Georg*, p. 54.

1764 Friedlander, *The Origins of Nazi Genocide*, p. 225.

1765 http://www.werkblatt.at/fallend/kohl.htm (Accessed 22 August 2008). Nitsche was experimenting with the use of lethal injections as a killing method, and chose two junior physicians to assist him, Renno being one of them. (Friedlander, *The Origins of Nazi Genocide*, p. 224).

1766 Horsinga, *Cher oncle Georg*, p. 25.

1767 Friedlander, *The Origins of Nazi Genocide* , p. 102.

1768 Horsinga, *Cher oncle Georg*, p. 76.

1769 Ibid., p. 97.

1770 Friedlander, *The Origins of Nazi Genocide*, p. 218.

1771 Klee, *Das Personenlexikon zum Dritten Reich*, p. 491.

for the institution at Waldniel, near Andernach. Renno was appointed as its head in October 1941,[1772] but two months later he developed tuberculosis, and was forced to resign his position.[1773] He left Waldniel in February 1942 for several months' treatment in Leipzig, Sankt Blasien, and Davos, before in 1943 he returned to Hartheim to again take up his chosen career of mass murderer.[1774]

In September 1943, Lonauer was called-up for service in the *Waffen*-SS; Renno replaced him as physician-in-chief, and was solely responsible for the continued gassing of concentration camp inmates at Hartheim under the aegis of 14f13. Lonauer returned from the military in November 1944, by which time Hartheim was about to cease operations. In mid-December Renno and his family left for the rest home that had been established by T4 at Lake Attersee. There was a recurrence of Renno's tuberculosis requiring a further sojourn at Davos, but in April 1945 he returned to Attersee, where he remained until the end of the war. Bearing false identity papers, Renno settled first in Aschaffenburg, then Bockenheim, where he was to live for the remainder of his life. In 1948 he commenced employment as a researcher with the pharmaceutical company Schering AG.[1775]

The next thirteen years passed peacefully enough, until October 1961, when Renno, by that time again having resumed his true identity, was arrested in connection with his T4 activities. Three months later he was released from custody; it required more than another seven years before his trial could commence in August 1969. In March 1970, Renno entered hospital for the removal of his appendix. Although suffering from no recognizable symptoms, Renno's condition deteriorated to the point where he was declared unfit to stand trial. The case against him was abandoned in December 1975.[1776] Renno then staged a remarkable recovery, extending his lifespan by almost a further twenty two years.

For his involvement in an unknown number of murders of both adults and children, Renno thus received no judicial punishment whatsoever, and was able to arrogantly boast to the end of his life, "I am not guilty."[1777] By that time he had probably come to believe this kind of lie, as well as those he told to his great niece when stating: "It was the Americans who circulated rumours about the gas chambers to discredit the Nazi regime. Believe me, such things never existed."[1778]

Paul Reuter (1907–?) was born in Wolfenhausen, and began his working life as a gardener, but in common with many others was unemployed for long periods. When he did find employment it was as a farm labourer, or on relief projects, until in 1936 he became a student nurse at the Weilmünster mental hospital. He had joined the NSDAP in 1930, left after a few months membership, but re-joined the party in December 1933.

1772 Friedlander, *The Origins of Nazi Genocide*, pp. 52–53. The first child victims arrived at the end of 1941 or the beginning of 1942. (Horsinga, *Cher oncle Georg*, p. 100).

1773 de Mildt, *In the Name of the People*, p. 133, p. 152.

1774 Horsinga, *Cher oncle Georg*, p. 92.

1775 Ibid., pp. 201–202.

1776 de Mildt, *In the Name of the People*, p. 83.

1777 Horsinga, *Cher oncle Georg*, p. 168.

1778 Ibid., p. 86.

In July 1941 he was transferred to Hadamar and inducted into the "euthanasia" programme. He performed the usual duties of accompanying transports from other institutions to Hadamar, assisting new arrivals to disrobe, and guiding them to the gas chamber. He also witnessed gassing operations. He claimed to have agreed to participate in the programme only when he was assured that the patients to be killed were "life unworthy of life".[1779] However, at his trial he also accepted that among the victims were patients who appeared healthy, could undress themselves, and were quite capable of communicating with him. Reuter was yet another sent eastwards with the *Organisation Todt* mission in winter 1941/1942, returning to Germany in March 1942. He was briefly at Eichberg before returning to Hadamar to continue his nursing duties, which included administering lethal injections to 10–20 patients.[1780]

One of the defendants at the second Hadamar trial held at Frankfurt in February 1947, Reuter was found guilty of involvement in the murder of an unknown number of victims, and sentenced to four years and six months' imprisonment.[1781]

Robert Ritter (1901–1951)[1782] was born in Aachen. Too young to serve in the Great War, he became a *Freikorps* member in Oberschlesien at the age of 17 before becoming active in a youth group opposing French occupation forces in the Rhineland.[1783] After studying at a variety of universities, he received his doctorate in educational psychology from the University of Munich in 1927, and went on to obtain his medical doctorate from the University of Heidelberg in 1930. He was granted his medical license in the same year, and gained accreditation as a child psychiatrist in 1934. Ritter worked at hospitals in Paris, Zurich, Berlin and Tübingen, receiving his habilitation from the latter university in 1936. His principal area of research was juvenile antisocial behaviour.

Ritter appears to have been a National Socialist in spirit, if not in fact, for despite his obvious political inclinations, there is no evidence that he joined the Nazi party or any of its associated organizations, although he was evidently a child psychiatrist for the Hitler Youth.[1784] As already described, in 1936 he was appointed head of the newly created Eugenic and Population Biological Research Station of the *Reich* Health Office, under Hans Reiter (q.v.). Together with his assistants, Ritter investigated the family histories of certain petty criminals, particularly "those of alien race", most significantly Gypsies, who were categorized as being either "troublemaker, loafer, sponger, talentless, violent criminal, knave, crook, [or] hereditarily insane".[1785] Ritter hoped to prove a direct link between heredity and criminality in the Gypsy population, a subject completely without any scientific validity. His attitude towards the key objects of his studies can be summarised in a report he wrote in 1940:

> ...We have been able to establish that more than 90% of so-called native Gypsies are of mixed blood...Further results of our investigations have allowed us to characterise the Gypsies as being a people

1779 Sylvia Anne Hoskins, *Nurses and National Socialism—a Moral Dilemma: One Historical Example of a Route to Euthanasia* (Nursing Ethics, 12 (1), 2005) 79–91, p. 86.

1780 de Mildt, *In the Name of the People*, p. 197.

1781 Ibid., p. 371, note 96.

1782 Klee, *Das Personenlexikon zum Dritten Reich*, p. 499.

1783 Wildt, *An Uncompromising Generation*, p. 179.

1784 Friedlander, *The Origins of Nazi Genocide*, p. 371, note 17.

1785 Wildt, *An Uncompromising Generation*, p. 180.

> of entirely primitive ethnological origins, whose mental backwardness makes them incapable of real social adaptation....The Gypsy question can only be solved when the main body of a social and good-for-nothing Gypsy individuals of mixed blood is collected together in large camps and kept working there, and when the further breeding of this population of mixed blood is stopped once and for all.[1786]

As the persecution of the Gypsies intensified, in 1941 Ritter was appointed head of another new department—the Criminal Biological Institute of the Security Police, a division of the RSHA—where his identification of Sinti and Roma, thereby enabling their arrest and deportation, contributed in no small measure to the annihilation of the Gypsies of the *Reich*. In winter 1941–1942, Ritter attended a conference at which it was proposed that Germany's 30,000 Gypsies be eliminated by loading them onto ships, which would then be sunk by gunfire in the Mediterranean.[1787] Whether Ritter approved of such a measure is unknown, but he was certainly well aware of the fate that awaited any Gypsies marked for deportation.

After the war, Ritter was employed by the Frankfurt Health Office as a children's physician. Investigation into his and his associates' involvement in the extermination of German and other Gypsies was abandoned with his death.[1788]

Curd Runckel (1913–?)[1789] was born in Berlin, and was one of the doctors who operated out of T4 headquarters, functioning as a *Gutachter* from 1 September 1941.[1790] In the summer of 1944, Runckel, acting as Paul Nitsche's (q.v.) roving reporter, commented in unflattering terms on the mind-set of German psychiatrists *vis-à-vis* "euthanasia":

> I always find the attitude of numerous asylum directors particularly sad as far as the therapy and the problems tackled by the *Reich* committee are concerned...I also saw time and time again in a number of asylums that chronically severe mentally ill patients are kept alive through all kinds of substances. It is an attitude for which it is especially difficult to have empathy during a war! I always ask the senior physicians in the asylums about therapy and also about the euthanasia problem, and have so far found no enthusiasm for this, except in those institutions with which we collaborate.

Runckel frequently found that doctors refused to discuss the killing of patients through the use of medication, yet were prepared to allow their charges to starve to death.[1791] Whether these doctors were preparing a post-war defence for themselves—for example that starvation of patients was inevitable in wartime conditions, as had been proven in 1914–1918—whilst in fact cooperating with the killing programme, is a matter for debate. Certainly by this late stage of the war, there was much thought being given to "plausible deniability". That was, after all, the basis upon which the Third *Reich* was ruled.

Following a meeting with Karl Brandt (q.v.), on 24 July 1944 Runckel informed Nitsche that Brandt had queried whether it would be possible to "inconspicuously prepare the activation of our specific therapy."[1792] The meaning of this coded message was clear. In the convoluted language of the Nazi regime, consent was being given at the

---

1786 Angus Fraser, *The Gypsies* (Malden: Blackwell Publishing Ltd., 2003), p. 259.
1787 Proctor, *Racial Hygiene*, p. 215.
1788 Friedlander, *The Origins of Nazi Genocide*, pp. 249–262.
1789 Klee, *Das Personenlexikon zum Dritten Reich*, p. 515
1790 Klee, "*Euthanasie*" *im NS-Staat*, p. 229.
1791 Schmidt, *Karl Brandt*, p. 246.
1792 Aly, Chroust, Pross, *Cleansing the Fatherland*, pp. 83–84.

highest levels of government for the formal expansion of "wild euthanasia", for it could be assumed that anything emanating from Brandt had been approved by Hitler.[1793]

After the war, Runckel pursued a successful medical career in Switzerland. Although a warrant for his arrest in connection with the murder of "at least 10,000 people" was issued in August 1961, proceedings against him were delayed, before eventually being abandoned eight years later.[1794]

Josef Schicker (1879–1949)[1795] a psychiatrist born in Summerau, Austria, joined the illegal Austrian Nazi party in 1932. With the reputation of a "yes man", he was appointed director of the Gugging mental institution located in the outskirts of Vienna. Although subsequently cleared of any complicity in "euthanasia", he was involved in the compulsory sterilisation of at least 102 men and women. At his post-war trial, Schicker stated that in 1940 he had been informed by his medical superior that "measures to empty the wards" were to be taken. In the same year, by order of Erwin Jekelius (q.v.) nine medical students appeared at Gugging to study patient's case histories and complete *Meldebogen* where appropriate. Thereafter, 675 inmates were transferred to Hartheim for gassing in the first phase of "euthanasia". With the advent of "decentralised euthanasia", the sinister Emil Gelny (q.v.) became responsible for the killing of patients at Gugging.[1796]

Curt Schmalenbach (1910–1944)[1797] [T4 pseudonym "Dr Palm"][1798] was born in Wuppertal. He joined the SS in 1931/32 and also briefly became a member of the NSDAP for the first time in 1932, subsequently being excluded from membership of the party when he was unable to pay his subscription. He rejoined the party in late 1940. After qualifying as a physician, he worked as Paul Nitsche's (q.v.) assistant at Sonnenstein, it probably being Nitsche who recommended Schmalenbach to T4.[1799]

Schmalenbach appears on a T4 personnel listing as "serving at headquarters" with effect from 26 June 1940.[1800] He began his "euthanasia" duties at Sonnenstein as Horst Schumann's (q.v.) assistant in autumn 1940, remaining at that institution until December of that year; amongst other duties, in Schumann's absence Schmalenbach was in charge of the gassing of patients. He then became the point of contact between T4 in Berlin and the killing centres, acting for a number of hospitals as the *Gutachter* responsible for the selection of patients to be murdered.

Schmalenbach was one of the doctors involved in *Sonderbehandlung* 14f13. Unfavourable reference was made to him in a letter Friedrich Mennecke (q.v.) wrote to his wife from Fürstenberg on 21 November 1941, in which Mennecke complained that "...Mr Schmalenbach wreaked havoc with the order I had created yesterday for the

---

1793 Schmidt, *Karl Brandt*, pp. 247–249.

1794 Klee, *Was sie taten–Was sie wurden*, pp. 184–185.

1795 Klee, *Das Personenlexikon zum Dritten Reich*, p. 533

1796 http://www.memorialgugging.at/pdf/Czech_MedizinverbrechenGugging_engl.pdf (Accessed 18 September 2009.)

1797 Klee, *Das Personenlexikon zum Dritten Reich*, p. 541.

1798 Friedlander, *The Origins of Nazi Genocide*, p. 103 suggests a pseudonym for Schmalenbach of "Dr Blume".

1799 Friedlander, *The Origins of Nazi Genocide*, p. 222.

1800 Klee, *"Euthanasie" im NS-Staat*, p. 229.

work we were doing." The "work" was being conducted at Ravensbrück. Four days later, the medical party was in Buchenwald, where Mennecke "had to 're-examine' the files that Schmalenbach and I myself had prepared."[1801]

In late 1941 Schmalenbach was appointed head of Hadamar; shortly thereafter he joined the mysterious T4 mission to the eastern front. After his return, the decision was taken to cease gassing operations at Hadamar. Under Schmalenbach's supervision the gas chamber was dismantled. Thereafter he ostensibly became a hospital surgeon, but as Viktor Brack's (q.v.) adjutant he visited Paris, Warsaw, Prague and other cities, carrying with him an authorization from Philip Bouhler (q.v.) allowing him to refuse to disclose to any curious interrogator information about his activities and the purpose of his trip.

Schmalenbach died in an airplane crash over Lake Como on 15 June 1944. Whether he was in Italy as part of the transfer of *Aktion Reinhard* staff to that region is speculation, but seems possible.

Walter Eugen Schmidt (1910–1970)[1802] was born in Wiesbaden, and studied medicine at the University of Frankfurt, from where he graduated in November 1937. He was a devoted Nazi, joining the Hitler Youth at 16, the NSDAP in 1930, and the SS two years later. In May 1939 he was appointed assistant to Friedrich Mennecke (q.v.) at Eichberg. On the outbreak of war, Schmidt was conscripted into the *Wehmacht*, but when in early 1941 it was decided to establish a children's killing ward at Eichberg under Mennecke, the latter recommended Schmidt as his lieutenant. Hans Hefelmannn (q.v.) and Richard von Hegener (q.v.) agreed, and arranged for the suspension of Schmidt's military service.

In the summer of 1941, Schmidt accompanied Mennecke to a conference at the KdF together with 30–40 other physicians. All were invited to participate in the "euthanasia" programme. None declined. On his return to Eichberg Schmidt was promoted to physician-in-chief, Mennecke being otherwise engaged in killing activities. However, Schmidt's name appears on the list of T4 *Gutachter*, effective 2 September 1940, which suggests a much earlier involvement in "euthanasia".[1803] His attendance at the Berlin conference of medical dignitaries in February 1940 that so impressed Mennecke would appear to confirm this.[1804] Certainly he had no qualms about participating in the programme, commenting as he later did: "For me there was only the possibility of 'treating' the children concerned at the right moment, they were to be killed in any case."[1805]

At his December 1946 trial in Frankfurt, where he, Friedrich Mennecke (q.v.), Heléne Schürg (q.v), and Andreas Senft (q.v.) were co-defendants, Schmidt admitted to killing a minimum of seventy children, plus an unspecified number of adults. In at least thirty documented cases of child murder he had personally administered lethal doses of medication.[1806] A nurse employed at Eichberg remarked: "It was generally known

1801 Kogon, Langbein, Rückerl, *Nazi Mass Murder*, pp. 41–42.

1802 Klee, *Das Personenlexikon zum Dritten Reich*, p. 546.

1803 Klee, *"Euthanasie" im NS-Staat*, p. 228.

1804 Alexander Mitscherlich, and Fred Mielke, *Doctors of Infamy: The Story of the Nazi Medical Crimes* (New York: Henry Schuman, 1949), p. 93.

1805 de Mildt, *In the Name of the People*, p. 128.

1806 Bryant, *Confronting the "Good Death"*, p. 125.

among the nurses and throughout the Eichberg [institution] that Schmidt killed patients. We just called him the 'mass murderer'".[1807] Although there was a substantial body of evidence to the effect that Schmidt had ruled his kingdom like a petty despot,[1808] the court credited him with his dedication to healing those patients he had not killed (!), noting in particular that in order to acquire better diagnostic data "he dissected the brains of the children killed or deceased and sent them for scientific research to the psychiatric university at Heidelberg." It would be charitable to assume that the court was unaware of the activities of Carl Schneider (q.v.) at Heidelberg and his involvement in so many aspects of T4. In fact, Eichberg was one of the most important sources of material for Schneider's "research".[1809]

Whilst adjudging Schmidt guilty of murder, the court managed to find reasons not to impose the death penalty, and instead sentenced Schmidt to life imprisonment. The Regional Appeal Court overruled that verdict, and in 1947 did sentence Schmidt to death, but the sentence was never carried out. Instead, in 1953, in a rather different political climate, and with his supporters even claiming that he had found a cure for Multiple Sclerosis, Schmidt was released from custody.[1810] Despite having had his medical license revoked, he reputedly continued to practice as a doctor. By the mid-1950s none of the principal practitioners of child murder at Eichberg remained in custody.[1811]

Carl Schneider (1891–1946)[1812] was born in Gembitz in the province of Posen (now Gebice, Poland).[1813] Of modest origins, he won a scholarship to an elite school and went on to study medicine at Würzburg, before becoming a medical orderly during the Great War. He graduated as a physician in 1919, and for a short period worked as assistant to the eminent psychiatrist Oswald Bumke at Leipzig University. In 1922 he was appointed physician at the Arnsdorf mental hospital, where he remained for the next eight years. In 1930, the same year he worked with Paul Nitsche (q.v.) in creating the International Hygiene Exhibition in Dresden, he became the first medical director of the Bodelschwingh asylum in Bethel. He joined the NSDAP in 1932, and the following year left Bethel to take up the chair of psychiatry and neurology at Heidelberg.[1814] Although it has been suggested that this was not a political appointment, Schneider's predecessor had been removed because of an apparent lack of enthusiasm for the regime, and it seems likely that almost every academic appointment would have been political by October 1933.[1815] Schneider's membership of the party was certainly no impediment to his burgeoning career, for he also became head of the Racial Political Office in Baden.

Having advanced to the position of head of the University Clinic at Heidelberg, a post he held until the end of the war, Schneider became one of the major organizers of "euthanasia"; a *Gutachter* from April 1940 and a member of the *Reich* Committee, he was

---

1807 McFarland-Icke, *Nurses in Nazi Germany*, p. 231.

1808 Friedlander, *The Origins of Nazi Genocide*, pp. 179–181.

1809 de Mildt, *In the Name of the People*, p. 352, note 109.

1810 Ibid., pp. 128–129.

1811 http://www.mdpi.com/2075-4698/2/3/157/htm (Accessed 29 October 2016)

1812 Klee, *Das Personenlexikon zum Dritten Reich*, p. 551.

1813 At least one source incorrectly describes Schneider as Austrian (Lifton, *The Nazi Doctors*, p. 35).

1814 Burleigh, *Death and Deliverance*, pp. 117–118.

1815 de Mildt, *In the Name of the People*, p. 401, note 6.

also considered T4's leading researcher. His overnight conversion from caring, principled doctor to ruthless, dedicated murderer has been ascribed to overweening ambition; certainly that must have made an important contribution to both his attitude and his actions, although his enthusiasm for killing may also be seen, at least in part, as yet another example of the ability of power to corrupt.[1816] In 1941 Schneider documented the way forward:

> All current measures to relieve economic pressure on our people resulting from expenditures for useless institutional inmates [meaning "euthanasia"], and all eugenic measures in the broadest sense [meaning "euthanasia"], are long term measures. It will take centuries [to eliminate mental disease]... But it is still possible to relieve our people of a great number of congenital, hereditary, or acquired chronic mental infirmities in other ways [meaning "euthanasia"].

At least 211 former patients of the Heidelberg Psychiatric Hospital were murdered during the various phases of "euthanasia". Another 21 children were killed at Eichberg for the purposes of Schneider's Heidelberg research. After many years, a memorial commemorating the death of these children was dedicated at the front of the Heidelberg Hospital. Under Schneider, Heidelberg's clinic became the destination for the brains of an unknown number of "euthanasia" victims. As he journeyed around various institutions, Schneider was constantly looking out for suitable subjects. The fate of those he discovered can be imagined.[1817]

Schneider had ambitious (and expensive) plans to make psychiatry the cornerstone of German medical research. To this end he set up a new department employing a number of specialist physicians in spring 1942, initially sited at the Heidelberg University Psychiatric Clinic, but subsequently moving to the Wiesloch asylum.[1818] But all of Schneider's grandiose visions vanished with the collapse of the Third *Reich*. He was briefly an American prisoner before being released to become a patient in a psychiatric hospital. Rearrested as a potential witness in the Eichberg trial, and faced with the inevitable consequences of his activities, on 11 December 1946, whilst in American captivity, he was another to hang himself in his cell.[1819]

Agnes Schrankel (1907–?), born in Bochum, was a nurse at Hadamar, where she had worked since 1937. She never joined the Nazi party, but in 1940 was recruited by T4, even though she claimed she had not participated in the mass gassings that occurred at Hadamar in the first phase of "euthanasia". Instead, she testified, she had been working in the kitchen. It was impossible to prove otherwise, but it was established that she had assisted her superior Minna Zachow (q.v.) in the killing of some 20–25 patients during the programme's second phase.[1820]

1816 Ibid., pp. 303–304.

1817 Friedlander, *The Origins of Nazi Genocide*, pp. 130–131.

1818 Aly, Chroust, Pross, *Cleansing the Fatherland*, pp. 202–219.

1819 http://www.chgs.umn.edu/histories/documentary/hadamar/research.html (Accessed 13 December 2008).

1820 de Mildt, *In the Name of the People*, p. 196.

Another to be tried at the second Hadamar trial in Frankfurt, Schrankel received a jail sentence of three years and six months for participation in at least 20 cases of murder.[1821]

Josef Artur Schreck (1878–1963)[1822] was born in Baden-Baden, a man who on taking up psychiatry in 1913 as the branch of medicine in which he intended to specialise, did so "because there, the doctor could mean more to his patients." Indeed, Schreck enjoyed a high reputation as a "fatherly benign doctor who sacrificed himself on behalf of his patients." But there was a worm in the bud. Schreck had been hugely influenced by his dreadful experiences at the Illenau asylum during the First World War, and then by the writings of Binding and Hoche. He believed that "euthanasia" was acceptable, if not essential, in the case of "full idiots" or those who had become "feebleminded" due to accident or illness. A merciful death was something he himself would desire if he were ever unfortunate enough to fall into such a condition.[1823] Schreck clung to these convictions to the end, although as he admitted at his trial, he was fully aware that Nazi "euthanasia" was a perversion of Binding and Hoche's vision of legalized "mercy killing". Nonetheless, given the exigencies of wartime, in his view killing on such a scale was justified on purely economic grounds.

Given such opinions, and his 1933 membership of the Nazi party, T4 had no difficulty in recruiting Schreck for the "euthanasia" programme. He served at three Baden mental institutions—Illenau, Rastatt, where he was appointed director in 1934, and Wiesloch, where he assumed a similar position on the dissolution of Rastatt in 1940. In the same year his name appears on a surviving list as a *Gutachter*.[1824] He personally witnessed a gassing at Grafeneck, and had no hesitation in knowingly sending his patients to their death there.

1n 1948 Schreck appeared in court at Freiburg to account for his crimes. He was unapologetic. Unlike many of his accomplices he did not seek to avoid responsibility for his actions, clearly considering himself nothing less than a martyr to the cause of euthanasia, with little or no comprehension of the criminality of his activities. He was sentenced to life imprisonment, plus a further ten years. In 1950, the overall sentence was reduced to one of 12 years. Twelve months later he was a free man, permitted to again practice as a physician.[1825]

Horst Schumann (1906–1983) [T4 pseudonym "Dr Klein" or "Dr Blume"].[1826] Even among this company of what might be termed "anti-physicians", Schumann can be considered exceptional. Few, if any, could match the extent of his participation in so many different aspects of T4, nor his sickening record of cruelty and murder.

1821 Ibid., p. 371, note 96.

1822 Klee, *Das Personenlexikon zum Dritten Reich*, p. 559.

1823 Bryant, *Confronting the "Good Death"*, pp. 172–176.

1824 Klee, *"Euthanasie" im NS-Staat*, p. 228.

1825 de Mildt, *In the Name of the People*, pp. 99–101.

1826 Klee, *Das Personenlexikon zum Dritten Reich*, p. 570. Friedlander, *The Origins of Nazi Genocide*, p. 103 suggests a pseudonym of "Dr Keim" for Schumann, and attributes the pseudonym of "Dr Blume" to Schmalenbach.

He was born in Halle, the son of a doctor who had distinctly right-wing and nationalistic political views. This was undoubtedly a profound influence on Schumann, who as early as 1923/1924 became a member of a group of National Socialist sympathisers. He joined the party itself in 1930, and the SA in 1932. By this time he was a medical student, graduating in 1933 to work in the public health office in Halle. There he remained until shortly before the outbreak of war, when he was conscripted to serve as a medical officer in the *Luftwaffe*. His military service was brief, for in early October 1939, Schumann was summoned to the KdF, informed of the "euthanasia" programme, and invited by Viktor Brack (q.v.) to become physician-in-charge at Grafeneck. Schumann, who had no knowledge of psychiatry, but presumably had other credentials that appealed to Brack, such as a complete lack of scruples so far as murder was concerned, willingly accepted.[1827] He had not been Brack's first choice for the position, but the man who was—an SS doctor named Werner Kirchert—refused Brack's offer, and instead recommended his friend and fellow former citizen of Halle, Schumann.

Schumann arrived in Grafeneck in January 1940, remaining at that institution until late April or early May of that year, when he was transferred to Sonnenstein to supervise the construction of the new killing centre there.[1828] Whilst still officially employed at Sonnenstein, he became a member of the *Sonderbehandlung* 14f13 doctors' commission that visited concentration camps to select prisoners for killing. This led to Schumann's first visit to Auschwitz, where on 28 July 1941 he selected 575 prisoners for subsequent transfer to Sonnenstein and gassing. Like Rudolf Lonauer (q.v.), Schumann thus not only selected those who were to die, but oversaw the actual murder of certain of them.[1829]

In November 1942, Schumann returned to Auschwitz, this time to conduct sterilisation experiments on male and female inmates. His brutal and callous methods have been described above. Schumann left Auschwitz in spring 1944. His subsequent actions are uncertain, but it is known that he conducted sterilisation experiments at Ravenbrück similar to those he had initiated at Auschwitz, and it is likely that he continued to select victims for gassing. According to Schumann he served as a doctor in the *Wehrmacht* from January 1945, but the war over, he surfaced in October of that year in Gladbeck, where he found employment as a sports doctor. He made no attempt to hide his identity, and in 1949 began working as a general practitioner. He might have continued to lead an untroubled existence if he had not applied for a hunting licence. He had been on the "wanted" list since the time of the Nuremberg Medical Trial, but it was only now that his whereabouts were discovered. When on 26 February 1951 police officers arrived to arrest him he was gone, probably warned by somebody in authority.

Thereafter he led a strange existence. He claimed to have been a ship's doctor for the next three years; since he had no German passport, he had applied for and obtained one in his own name in Japan in 1954. One year later he turned up in Egypt, then he and his wife settled in the town of Li Jubu in the Sudan, where he became the director of a hospital for lepers. Unfortunately for Schumann, a German journalist published an article about this latter-day Albert Schweitzer, which immediately aroused the interest of

---

1827 Friedlander, *The Origins of Nazi Genocide*, p. 220.
1828 Ibid., p. 92.
1829 Ibid., p. 148.

the German authorities. In summer 1959 a request for Schumann's extradition was filed with the Sudanese government, but he had already fled the country, this time for Ghana, where he lived under the protection of Kwame Nkrumah, that country's leader. There Schumann could meet with friends like Helmut Kallmeyer (q.v.) and relive the good old days. But when Nkrumah was ousted from power in 1966, Schumann was again vulnerable. This time an extradition application was successful, and in November of that year he was back in Germany.[1830]

It was to take almost another four years before proceedings against him finally began in September 1970. They did not last long. His cellmate Hans-Joachim Becker (q.v.) testified that Schumann simulated a haemorrhage by swallowing his own blood then violently regurgitating it. The trial of the ex-doctor (he had been struck off in 1961) was suspended in April 1971 because of his "ill health", and was terminated in July 1972. Schumann was released from custody.[1831] Despite his "weak heart", he managed to survive for a further eleven years.

Heléne Schürg (1904–?) was born in Karlsruhe and became a student nurse in 1924. She was employed at the Weilmünster mental hospital from 1933 to October 1937, when she was transferred to Eichberg to take up the post of head nurse, a position she retained until her dismissal in July 1945. She joined the NSDAP in 1940, and was a sponsoring member of the SS.

In early 1941 a children's killing ward was established at Eichberg, supervised by Walter Schmidt (q.v.). Schürg, enamoured of her superior, assisted Schmidt in the killing of an estimated fifty children, for which service she received bonus payments from time to time from T4. She was a defendant at the Eichberg trial held in Frankfurt in December 1946, and made no attempt to deny her guilt. Rather, she had committed these crimes because of her infatuation with Schmidt, a defence the court was prepared to accept, stating: "She had so divested herself of her own will that she placed all her fortitude at Schmidt's disposal, seeking to translate his will into reality and doing precisely that."[1832] Playing Trilby to Schmidt's Svengali stood Schürg in good stead, for the court determined that she had not been a murderess, but had behaved as she did from "a wrongly understood duty of obedience," and imposed a sentence of eight years' imprisonment. Time spent in pre-trial confinement was deducted from the sentence.

Andreas Senft (1883–?) was another Eichberg nurse. He had trained as a cartwright, but had been a nurse at the institution since 1906. He joined the Nazi party in 1940, and never attempted to deny his involvement in the murder of an estimated twenty patients by means of lethal injections. Another under the baleful influence of Walter Schmidt (q.v.), when asked if he had any reservations about these killings, he replied:

> I was close to 40 years in the institution and I have always just obeyed the Herr Director and the doctors...Well, I thought, this Dr Schmidt, he surely could not do this without any foundation, there must definitely be an instruction...Certainly I questioned my conscience on this from time to time. I have simply

1830 de Mildt, *In the Name of the People*, pp. 82–83.
1831 Burleigh, *Death and Deliverance*, p. 269.
1832 Bryant, *Confronting the "Good Death"*, p. 126.

only obeyed, nothing more. If, for instance, Dr Schmidt had ordered me to steal something, that is of course something different, I would not have done that.

So murder was acceptable, but stealing was taboo. The Frankfurt court considered that Senft had acted as he did "out of a certain naivety", which seems something of an understatement, and sentenced him to four years' imprisonment.[1833]

Ludwig Sprauer (1884–1962)[1834] was born in Heidelberg, and studied medicine at Freiburg, Strasbourg, and Berlin, before qualifying as a doctor in 1907. Having spent some time gaining experience in various institutions, he became a general practitioner in 1910, and after serving in the army as a physician for the duration of the Great War, then spending a year as a prison doctor in Mannheim, returned to general practice in 1920 in the small town of Staufen im Breisgau. Despite no apparent previous right-wing political sympathies, he joined the NSDAP in February 1933. It was a good career move, for shortly thereafter he was appointed head of the Health Department in the Baden Ministry of the Interior.[1835] In this capacity he was responsible for twelve mental hospitals in the Baden district.

In October 1939 Sprauer received a summons from Herbert Linden (q.v.) at the *Reich* Ministry of the Interior. Linden first swore Sprauer to secrecy, then informed him of the impending "euthanasia" programme. In the region for which he was responsible, it was to be Sprauer's responsibility to arrange the registration and transportation of patients deemed incurable.[1836] Although he claimed to have been "devastated" by the sight of patients being gassed at Grafeneck, and indeed to have felt distinctly generally uneasy about carrying out his instructions, he managed to overcome his misgivings.[1837] In fact, Sprauer was a good deal more enthusiastic about killing the patients in his care than most of the physicians at the Baden institutions, on occasion even going so far as to threaten with arrest those in positions of authority who refused to cooperate.[1838] In addition he had been instrumental in establishing a children's killing ward at the Wiesloch asylum.[1839] If no fanatical eugenicist, Sprauer was a faithful follower of orders and thus, from T4's point of view, the ideal man to have in an important position.

In 1948, Sprauer was tried alongside Artur Schreck (q.v.) in Freiburg. The court found a catalogue of extenuating circumstances for Sprauer's behaviour; he possessed "a false bureaucratic ambition and submissiveness to authority", "did not want to lose his job in the ministry", and was merely a "typical accomplice" of the true culprits, the Nazi leadership, most of whom were dead.[1840] Nonetheless, the court could not ignore Sprauer's complicity in the death of at least 3,000 patients. He was sentenced to life imprisonment, subsequently reduced to a term of eleven years. A few months later he was not only released, but awarded a pension by the government of Baden-Württemberg for services rendered.

1833 de Mildt, *In the Name of the People*, pp. 183–184.
1834 Klee, *Das Personenlexikon zum Dritten Reich*, p. 592.
1835 de Mildt, *In the Name of the People*, pp. 103–104.
1836 Bryant, *Confronting the "Good Death"*, p. 170.
1837 de Mildt, *In the Name of the People*, p. 62.
1838 Ibid., pp. 104–105, page 347, note 40.
1839 Friedlander, *The Origins of Nazi Genocide*, p. 52.
1840 Bryant, *Confronting the "Good Death"*, p. 171.

Eugen Stähle
Source: SPD Calw / Public Domain

Eugen Stähle (1890–1948)[1841] was born in Stuttgart. Having completed his medical degree in Tübingen, he served as a military doctor during the Great War, suffering a severe gassing. He quickly showed his fascist sympathies by becoming a member of the *Freikorps* Epp in 1919, one of those responsible for suppressing the newly declared Bavarian Soviet Republic in Munich. He was an early member of the Nazi party, joining first in 1923, and then after the party was banned for a time following the Munich *Putsch*, rejoining in 1927. In 1930 he was appointed the Nazi Physician's League deputy for Württemberg.[1842] Stähl served as an NSDAP member of the *Reichstag* for a few months in 1933, and in the same year became the chief medical officer in the health department of the Württemberg Ministry of the Interior. In this capacity he was deeply involved in the "euthanasia" of both adults and children. It was he who in 1939 offered Herbert Linden (q.v.) Grafeneck as a killing centre.[1843] Subsequently, Stähl instructed the directors of the institutions under his control as to the part they were to play in the programme, as well as personally being present at a number of gassings.[1844]

In the mid 1930s, long before "euthanasia" was even officially contemplated, Stähl was asking the directors of his institutions their hypothetical reaction "if during the war, the state considered the extermination of certain categories of patients under the circumstances that, due to the drying up of imports, food supplies no longer sufficed to feed the population."[1845] When in 1940 he was reproached by a member of the clergy about the killing of persons allegedly "unfit to live", Stähl responded: "The fifth commandment, 'thou shalt not kill' is not a commandment from God, but merely a Jewish invention."[1846]

A holder of the Nazi's Golden Party badge, on 13 November 1948 whilst a prisoner on remand, Stähl died in the district hospital at Münsingen.[1847]

Alfons Stegmann (1908–?) joined the Nazi party in 1930, and between 1935 and 1939 was a medical student. Following the grant of his degree he worked for a short period as an assistant physician at the Winnental asylum. Although conscripted at the beginning of the war, his military service was equally brief, for in November 1939 he was appointed deputy director of the Zwiefalten mental hospital. It was in this capacity that he attended a meeting of the directors of Württemberg institutions called by Eugen Stähl (q.v.) in February 1940 at which those present were informed of the "euthanasia" programme, and the part they were to play in it. Stegmann apparently had no problem with the concept or its implementation, for he knowingly sent hundreds of his patients to their death at Grafeneck, whose director, Ernst Baumhardt (q.v.) was a friend.

1841 Klee, *Das Personenlexikon zum Dritten Reich*, p. 594

1842 Kater, *Doctors Under Hitler*, p. 61.

1843 Friedlander, *The Origins of Nazi Genocide*, p. 201.

1844 de Mildt, *In the Name of the People*, p. 107, p. 113.

1845 Ibid., p. 52.

1846 http://migs.concordia.ca/occpapers/brutality.htm (Accessed 19 December 2008).

1847 Klee, *Was sie taten–Was sie wurden*, p. 85.

In late 1940, Stegmann was suspended from his position, supposedly for disciplinary reasons. In July 1943 he was convicted of having performed illegal abortions (the seriousness of which in Nazi Germany, so far as "Aryan" women were concerned, has been demonstrated above). He was sentenced to two years' imprisonment together with suspension of his medical licence, although the sentence was subsequently held over.

In July 1949, Stegmann appeared in the court at Tübingen alongside several defendants, including Otto Mauthe (q.v.), protesting his abhorrence of "euthanasia" like so many others. The court found his defence unconvincing, since he had not only visited Grafeneck on several occasions, but in some cases had personally selected patients for transportation there. However, once again the court managed to find grounds for excusing his actions; he had not personally designed the killing programme (!), he was a young man easily influenced by his elders, and his involvement was "relatively small".[1848] Thus a sentence of two years' imprisonment was considered an "appropriate but also adequate atonement".

Stegmann did not need to have any cause for concern. He was immediately pardoned by the prime minister of Baden- Württemberg and returned to general practice.[1849]

Theodor Steinmeyer (1897–1945)[1850] was born in Öttingen, Bavaria. After serving in the military during the First World War, he became a medical student in Erlangen, and having obtained his degree, by 1925 was practising medicine in Nuremberg. He joined the NSDAP and SA in 1929, the same year he began training as a psychiatrist. In due course was appointed deputy director of the Wehnen mental hospital, and in 1934 was promoted to full directorship of the Ellen institution near Bremen.[1851] From 1939 he became physician-in-chief and paediatric psychiatrist at the Marsberg and Niedermarsberg institutions, where the killing of children was to occur.

Steinmeyer was involved with T4 from an early stage, acting on occasions as deputy to Irmfried Eberl (q.v.) and Heinrich Bunke (q.v.) at Bernburg. He had been a T4 *Gutachter* since 28 February 1940, and was a member of the Doctors' Committee.[1852] From 1941 to 1943 he worked out of T4 headquarters in Berlin, and was a participant in *Sonderbehandlung* 14f13 together with other doctors, including Friedrich Mennecke (q.v.), who mentions Steinmeyer in one of his many letters to his wife. For a time in 1942, while victims of 14f13 were being murdered there, Steinmeyer again acted as deputy head at Bernburg, before he moved on to become director of the Warstein psychiatric hospital.

In 1943 he became head of the institution at Pfaffenrode near Mühlhausen, where murder was perpetrated on a level that later enabled him to write enthusiastically to his friend Mennecke: "The mortality is fantastic." In another letter to Mennecke,

---

1848 Bryant, *Confronting the "Good Death"*, p. 167.

1849 de Mildt, *In the Name of the People*, pp. 112–113.

1850 Klee, *Das Personenlexikon zum Dritten Reich*, p. 601.

1851 Francis R Nicosia, and Jonathan Huener (eds.), *Medicine and Medical Ethics in Nazi Germany: Origins, Practices, Legacies* (New York: Berghahn Books, 2002), p. 82.

1852 Klee, *"Euthanasie" im NS-Staat*, p. 228.

Steinmeyer referred to the forthcoming installation of a crematorium at Pfaffenrode: "...I am supposed to be getting the same building as Faltlhauser" (q.v.) [director of Kaufbeuren, where a crematorium had recently become operational]. "You know what I mean."[1853]

Steinmeyer committed suicide on 26 May 1945 while a prisoner in Mühlhausen,

Adolf Thiel (1905–?) was tried together with Karl Todt (q.v.) in Koblenz in 1948. Physician-in-charge at the Scheuern mental hospital where Todt was director, Thiel had joined and left the NSDAP twice before the *Machtergreifung*, renewing his membership for a third time in 1933, although the court considered him no fanatic. Like Todt, he was apparently a devout Christian, respected and admired by all, who had taken up his position at Scheuern in 1938.

In the summer of 1940, Todt and Thiel had been party to the completion of some 500–600 "euthanasia" registration forms, supposedly without being aware of their purpose. The first patients designated for killing left Scheuern in March 1941 for unknown destinations. When notification of the death of those transferred arrived shortly afterwards, Todt and Thiel realized just what they had become involved in. It was also in March 1941 that Scheuern was designated as a transit centre for Hadamar. Despite their alleged distress at their participation in the programme, between that time and September 1944, Todt and Thiel were responsible for the transfer of approximately 1,640 patients from Scheuern, of whom about 1,000 were murdered.

Todt and Thiel advanced the defence that they had remained in their posts despite certain knowledge of the fate of those they transported in order to save lives; if they had not sacrificed some patients, other more zealous militants would have replaced them and ensured that the programme devoured countless further victims. It was for this reason they had not resigned, despite their abhorrence of "euthanasia."[1854] The court found this argument convincing, and acquitted both men.[1855]

Lydia Thomas (1910–?) was born in Aumetz, Lothringen (Lorraine), and had entered nursing at the Herborn mental hospital in 1928. She was briefly employed at Weilmünster, before in July 1941 she was transferred to Hadamar, allegedly against her will. A member of the Nazi party since 1933, as well as the National Socialist Women's League and the German Labour Front, on arrival at Hadamar she was enrolled into T4 by Fritz Bernotat (q.v.).

A defendant at the second Hadamar trial, Thomas admitted to having participated in both phases of the "euthanasia" programme. During the first phase she had attended victims up to the point at which they entered the gas chamber; in the second phase, un-

1853 Aly, Chroust, Pross, *Cleansing the Fatherland*, p. 91.

1854 This reasoning is remarkably similar to that advanced by certain leaders of the *Judenrate* in the ghettos of eastern Europe, justifying the sacrifice of part of the community in order to save the remainder. See, for example, Chaim Rumkowski in Lodz, Jacob Gens in Vilna, and Mosze Meryn in Sosnowiec. Of course, there were others like Adam Czerniakow in Warsaw, or Elchanan Elkes in Kovno, who either chose suicide rather than condemn their fellow Jews, or steadfastly refused to cooperate with the murderers despite the risks involved. http://www.jewishgen.org/yizkor/terrible_choice/Terrible_Choice.html#TOC (Accessed 22 December 2008)].

1855 Bryant, *Confronting the "Good Death"*, pp. 148–149.

der the supervision of Irmgard Huber (q.v.) she had assisted in the murder of about 25 patients.[1856] Found guilty of participation in an unknown number of cases of murder, Thomas received a sentence of five years' imprisonment.[1857]

Karl Todt (1886–?), whose doctorate was in education, not medicine, succeeded his father as director of the Scheuern mental hospital in 1921. He joined the Nazi party in 1937 at the insistence of the notorious National Socialist, Fritz Bernotat (q.v.). It was the misfortune of Todt and his chief physician, Adolf Thiel (q.v.), that their institution fell under the bailiwick of Bernotat, a fervent advocate of "euthanasia". On the outbreak of war Todt served for a time in the *Wehrmacht*, but in September 1940 was discharged because of his age and returned to Scheuern (Thiel had been deemed medically unfit for military service). As has been described, the two men were jointly responsible for the enactment of the "euthanasia" programme at Scheuern over a period of years.

At the trial and acquittal of Todt and Thiel in 1948, the court determined that in condemning 1,000 patients, the duo had saved the lives of 250 others. This was regarded as a significant success; it may be considered doubtful that the dead and their families concurred with this judgement. In support of their verdict, the court quoted with approval the criminal law expert Professor von Weber:

> One should acknowledge that remaining in position and cooperating while simultaneously hampering the implementation of orders, often calls for the greater moral courage and that, because of this proven behaviour of men conscious of their responsibilities, much harm was prevented during the time of the National Socialist regime.

This may well have been true; alternatively it provided a perfect excuse for inertia—or worse. In any event, both von Weber and the court were prepared to leave the final judgement to a higher authority. "The solution to such conflicts can only be found in the absolute values of the conscience; it is up to the individual with his God," stated von Weber. The court agreed: "Whether the defendants can feel themselves free of all guilt is a matter for their personal conscience." Thus the defendants were acquitted, not because of their innocence, but rather on the grounds that their acts could be considered "excused, even if not justified."[1858]

Aquilin Ullrich (1914–2001) [T4 pseudonym "Dr Schmitt"][1859] was born in Dillingen an der Donau, Bavaria. He came from a devoutly Roman Catholic family, and initially enrolled as a theology student at Munich University before transferring to the study of medicine. He joined both the National Socialist Students' Association and the SA in 1934, and like his friend Klaus Endruweit (q.v.), devoted a year to voluntary service with the *Reichswehr* before taking up his studies again at the universities of Würzburg and Freiburg. It was at the latter that he first met Heinrich Bunke (q.v.), an association that was to have disastrous consequences for humanity. In May 1937 Ullrich joined the Nazi party.

1856 de Mildt, *In the Name of the People*, p. 196.
1857 Ibid., p. 371, note 96.
1858 Ibid., pp. 145–149.
1859 Klee, *Das Personenlexikon zum Dritten Reich*, p. 635.

Whilst still at Freiburg, Ullrich attended several lectures delivered by the man who was destined to become the medical head of T4, Werner Heyde (q.v.), at that time the chief physician of the Würzburg psychiatric clinic. As already described, Endruweit and Ullrich were among a party of students who undertook a pseudo-scientific, racially orientated expedition to Bessarabia in summer 1938. As a consequence of the report this "research" generated, in July 1939 Ullrich met not only Joseph Goebbels, but Hitler himself. The extent to which all of these events influenced Ullrich's attitudes is difficult to evaluate, but it is reasonable to assume that they had some bearing on his subsequent conduct.

The coming of war not only saw Ullrich conscripted, but prevented from completing his degree. Instead, in November 1939 he was another to be granted *Notapprobation* status as a physician. Unhappy with life as an army doctor, he was advised by a friend to contact Heyde, who was supposedly looking for young doctors for a special assignment. Ullrich spoke to Heyde, who, without going into specifics, informed him in general terms of the secret "euthanasia" programme. Ullrich could join it if he so wished—the decision was entirely his. It did not take Ullrich long to make up his mind. Blinded by the eminence of those involved, and gratified to be a member of such exalted company, Ullrich quickly agreed to become a participant. Two weeks later he reported to the KdF, where Hans Hefelmann (q.v.) again outlined the nature of the programme, and informed him that he was to report to Brandenburg as assistant to Irmfried Eberl (q.v.).

At Ullrich's request, Werner Blankenburg (q.v.) escorted him to Brandenburg to inspect his new place of employment. Although no gassings were taking place at the time, it was immediately apparent to Ullrich that this was a killing centre. In fact, it was confirmed during the course of his visit that individuals were being gassed at Brandenburg with carbon monoxide. If he harboured any reservations about his new job, Ullrich was certainly not about to go back to the army. He served at Brandenburg from early April until the beginning of August 1940, supervising the entire extermination process, from the arrival of patients to the issuing of false death certificates and letters of condolence to the victims' relatives. He then managed to get himself transferred to Bessarabia for a few months, and on his return in November 1940, informed Heyde that he was no longer prepared to work in a killing centre. Instead, he was employed in the planning department at T4 headquarters in Berlin, before in January 1942 he joined the *Organisation Todt* mission to the Soviet Union. Returning to Germany three months later, he secured his release from T4, and thereafter worked at the Pathological Institute of Munich University before rejoining the *Wehrmacht*.[1860]

After the war Ullrich successfully pursued a career as a gynaecologist in Stuttgart, until in 1966 he appeared in the dock in Frankfurt alongside Bunke and Endruweit, accused, in Ullrich's case, of participation in the murder of "at least 1,815" individuals. All three men were acquitted on the grounds that they had believed their actions to be legal. They were "without culpability."[1861] Although the German Supreme Court reversed the decision in 1970 and ordered a retrial, the new proceedings were suspended be-

1860 de Mildt, *In the Name of the People*, pp. 116–118.

1861 Ibid., pp. 122–123.

cause of the men's alleged ill health.[1862] It was not until 1986 that the three men appeared once more in the Frankfurt courtroom. Endruweit was quickly released because of his continuing health problems, but Ullrich was found guilty of involvement in the murder of "at least 4,500 patients" and sentenced to four years' imprisonment. Bunke received a similar sentence for his participation in the killing of a rather greater number of people. In December 1988 the Federal High Court reduced the number of Ullrich's victims to 2,340, and lowered his term of imprisonment to three years. Why the severity of a sentence should be determined on the basis of the number of victims murdered is difficult to understand.

Ullrich was released after having served twenty months in prison.[1863]

Hellmuth Unger (1891–1953).[1864] was an ophthalmologist born in Nordhausen. Having studied medicine in Würzburg, Rostock, Halle, and Leipzig, in 1928 he became press officer to the *Hartmannbund*, the German Doctors' Association, and five years later was appointed press officer to Gerhard Wagner (q.v.) himself. Unger was a prolific author, producing fifty plays, non-fiction works, and novels. Among these was *Sendung und Gewissen* ("Mission and Conscience") published in 1936, which was to be adapted for the screen as *Ich klage an* ("I Accuse") in 1941.[1865] He had also been co-author of the screenplay of the "euthanasia" propaganda film *Erbkrank* ("Hereditarily Ill") in 1936. Astonishingly, although Unger had applied to join the NSDAP, he was never accepted for membership, nor was he ever a member of the National Socialist Physician's League. This was probably because of his association with freemasonry, where he was a lodge master. It was doubtless for the same reason that a proposal to appoint him a professor was rejected in 1944.

Since Unger had never disguised his enthusiasm for its introduction, it was only natural that he became a member of the committee responsible for children's "euthanasia". He had been suggested to Hans Hefelmann (q.v.) by Richard von Hegener's (q.v.) sister as a potential recruit.[1866] Unger went on to become involved in the planning of the extension of the killing programme to the adult population.[1867] After the war, Unger worked again as an ophthalmologist in Bad Harzburg, until ceasing to practice medicine and moving to Freiburg, where he died in 1953.

Werner Villinger (1887–1961)[1868] a psychiatrist and neurologist dedicated to the principles of eugenics, was born in Besigheim. Having been appointed head of the Youth Authority in Hamburg in 1926, in 1934 he became physician-in-chief at the Evangelical Inner Mission's mental hospital at Bethel. By political inclination a conservative nationalist, he joined the Nazi party in 1937, as well as the National Socialist People's Welfare, the National Socialist Physician's League, and the paramilitary National Socialist Flying Corps.

1862 Bryant, *Confronting the "Good Death"*, p. 213.
1863 de Mildt, *In the Name of the People*, pp. 125–126.
1864 Klee, *Das Personenlexikon zum Dritten Reich*, p. 636.
1865 Burleigh, *Death and Deliverance*, p. 200.
1866 Friedlander, *The Origins of Nazi Genocide*, p. 44.
1867 Ibid., p. 66.
1868 Klee, *Das Personenlexikon zum Dritten Reich*, p. 641

Although during the lifetime of the Weimar Republic he had opposed enforced sterilisation, Villinger was quick to take advantage of the Nazi Sterilisation Law, registering some 1,700 of his 3,000 patients for such treatment. He reported to his colleagues that one way of persuading patients to undergo the operation was by convincing them that to do so was "a patriotic sacrifice." So successful was this strategy, that in July 1934 he further reported: "At the moment, our hospital is unable to cope with the sterilisations. We have sterilisation-day only once weekly, and then only a limited number can be carried out. It is absolute chaos then."[1869] He also informed the Inner Mission that he was prepared to sterilise foreigners, with one exception; an Austrian boy in his charge came from Braunau, the birthplace of Adolf Hitler.

In 1937 Villinger became a judge in the Higher Hereditary Health Court at Hamm, and in 1940 professor of psychiatry and neurology at the University of Breslau, where he also became director of the clinic for nervous diseases. Villinger appears as a T4 *Gutachter* with effect from 28 March 1941.[1870] His attitude towards "euthanasia" seems ambivalent, Paul Nitsche (q.v.) and others considering him too lenient in his assessments.[1871] However, Villinger was prepared to become involved in medical experiments on mental patients in his care. Kurt Gutzeit, professor of internal medicine at Breslau, asked him to agree to infect a number of them with hepatitis without first obtaining their consent, which Villinger did. Gutzeit was researching the causes of and a possible cure for the infection.[1872]

1946 found Villinger a professor at the University of Marburg. He enjoyed a distinguished post-war academic career, amongst other things being elected president of the German Association for Child and Youth Psychiatry in 1952, and in 1954 becoming head of the medical department at Marburg, before being appointed Rector of the university in 1955. He was even decorated with the Federal Cross of Merit. When investigations into his past began in 1960, he declared: "I never worked as a *Gutachter*...I was known as an opponent of the racial policies of the Nazi party."

As the evidence concerning his Nazi associations began to emerge, Villinger decided to put an end to any further speculation. On 8 August 1961, he committed suicide by throwing himself off of a mountain near Innsbruck.[1873]

1869 de Mildt, *In the Name of the People*, p. 50.
1870 Klee, *"Euthanasie" im NS-Staat*, p. 228.
1871 Schmidt, *Karl Brandt*, p. 238.
1872 Ibid., p. 265.
1873 http://www.beratungszentrum-alsterdorf.de/cont/ErnstKlee(3).pdf (Accessed 23 December 2008).

Gustav Adolf Waetzoldt (1890–1945) was head of the Wittenau mental hospital in Berlin (today the Karl Bonhoeffer Clinic of Neurology). A transit institution during the first phase of "euthanasia", Wittenau later became a killing centre in its own right. The brains of 106 children murdered there were sent to Berthold Ostertag, a colleague of Julius Hallervorden (q.v.), for "research purposes".[1874] Waetzoldt's sympathies can be gleaned from the title of his work "*Hereditary Improvement through Eradication*". It is probable that it was Waetzoldt who was approached by the bacteriologist Arnold Dohmen in spring 1943, when the latter wished to conduct medical experiments on asylum inmates in the search for the cause and treatment of viral hepatitis. In the event, although Waetzoldt appears to have had no objection to this blatant violation of both law and medical ethics, Karl Brandt (q.v.) prohibited the use of mental patients for this purpose. Instead, a group of young Jewish boys from Auschwitz was chosen, equally illegally, and equally in contravention of every medical principle.[1875] The consent of those experimented upon by the Nazis was never sought, nor obtained.

Gerhard Wagner
Source: USHMM / Courtesy of Geoffrey Giles / Signature #45278

Gerhard Wagner (1888–1939).[1876] was born in Neu-Heiduk, Upper Silesia (now part of Chorzow, Poland), the son of a professor of surgery. He was a medical student in Munich, and served as a frontline doctor during the Great War, being awarded the Iron Cross first class, among other decorations. At the war's end Wagner became a general practitioner in Munich, and in the years 1921–23 was a member of *Freikorps* Epp and Oberland (two right-wing paramilitary organizations). He joined the Nazi party in 1929, and in the same year was one of the founders of the National Socialist Physicians' League (NS-*Ärztebund*), an organization intended to "purify" German medicine of the taint of "Jewish Bolshevism".[1877] In late 1932 he replaced Ludwig Liebl as head of the *Hartmannbund*, the German Doctors' Association.[1878]

With the *Machtergreifung*, Wagner's career rapidly advanced. By 1933 he was leader of the Main Office for National Health; one year later he was *Reich* Doctors' Leader and "The *Führer's* Commissioner for National Health". In December 1935, Wagner became leader of the *Reich* Physicians' Chamber, and in 1936 he was appointed head of the German Panel Fund Physicians Union. Eventually he was to head virtually every significant German medical organization. Since all doctors working in public health had to belong to the German Panel Fund Physicians Union, and every practising doctor to

1874 Nicosia and Huener, *Medicine and Medical Ethics in Nazi Germany*, p. 100.
1875 Schmidt, *Brandt*, pp. 265–276.
1876 Klee, *Das Personenlexikon zum Dritten Reich*, p. 649.
1877 Eric Joseph Epstein and Philip Rosen, *Dictionary Of The Holocaust* (Westport: Greenwood Press, 1997), p. 206.
1878 Kater, *Doctors Under Hitler*, p. 20.

the *Reich* Physicians' Chamber, Wagner was now chief physician of the *Reich*, the *Reichsärzteführer*, lord of all matters pertaining to issues of health.

As a venomously anti-Semitic proponent of racial hygiene, and personal physician to Rudolf Hess, from a Nazi perspective Wagner was the right man in the right job at the right time. His influence on all medical matters was comprehensive, and his role in the establishment of Nazi racial policy indisputable. He was considered by many to have been the 'godfather of "euthanasia"',[1879] declaring as he did in 1933:

> Knowledge of racial hygiene and genetics has become, by a purely scientific path, the knowledge of an extraordinary number of German doctors. It has influenced to a substantial degree the basic world view of the State, and indeed may even be said to embody the very foundations of the present state.[1880]

In Wagner's view, the Sterilisation Law was not sufficiently racial;[1881] something more radical was required. Quoting spurious statistics to the 1935 Nuremberg Party Congress, Wagner concluded that Jews were "a diseased race", and Judaism was "disease incarnate."[1882] To add to the mix, criminality was higher among Jews than among gentiles; Jews were thus both sick and criminal—what other reason was needed for their elimination?[1883]

It was in 1936 that Wagner conducted an exchange of views with like-minded friends, that is to say high-ranking officials and doctors, about the killing of "idiot children" and mentally ill adults. When Hermann Boehm, head of teaching at the Nazi Physicians' School expressed concern about the implications of such a programme, Wagner put Boehm's mind at ease. Eliminating the mentally and physically incapacitated was always easier in time of war because then life was cheap, and Hitler had made clear his intention to join "euthanasia" to the outbreak of war. In the meantime, a massive propaganda campaign was about to begin with the aim of educating the public about the many benefits the proposed programme promised.[1884] In order to show the misery of patients' lives, it was intended to produce documentary films shot in asylums and other institutions, as indeed occurred.[1885]

The following year Wagner became an SA-*Obergruppenführer*. Two years after that he was dead from an undisclosed ailment, and was succeeded by Leonardo Conti (q.v.), although it was Karl Brandt (q.v.) who was to go on to become the dominant figure in Nazi medical matters. Even though Wagner died in March 1939, well before the official commencement of "euthanasia", his malign influence on this, as on so many other aspects of Nazi medical practice, cannot be overstated.[1886]

1879 Lifton, *The Nazi Doctors*, p. 35.

1880 Proctor, *Racial Hygiene*, p. 45.

1881 Weindling, *Nazi Medicine and the Nuremberg Trials*, p. 251.

1882 Proctor, *Racial Hygiene*, pp. 195–196.

1883 Ibid., pp. 204–205.

1884 Schmidt, *Karl Brandt*, p. 80.

1885 Lifton, *The Nazi Doctors*, p. 50.

1886 Wolfgang, Weyers, A. Bernard Ackerman, (ed), *Death of medicine in Nazi Germany: Dermatology and Dermatopathology Under the Swastika* (Philadelphia: Lippincott-Raven Publishers, 1998)—passim.

Adolf Wahlmann (1876–1956)[1887] was born in Koblenz. Following the pattern common at the time, he studied medicine at several different universities before obtaining his degree in 1903. Two years later he received accreditation as a specialist in psychiatry. Thereafter, apart from a period served in the military during the Great War, he worked at a number of different mental institutions, being appointed physician-in-chief at both Hadamar (twice) and Eichberg, until his retirement in 1937. He joined the NSDAP in 1933, although he was never active in party affairs, since he devoted most of his spare time to directing and conducting his church choir, much to the displeasure of party officials. He considered himself as much a musician as a physician.[1888]

On the outbreak of war, Wahlmann was called out of retirement, and in June 1940 appointed chief doctor at the Weilmünster mental hospital, before in August 1942, Fritz Bernotat (q.v.) made him physician-in-chief at Hadamar once again.[1889] By that time the gas chamber at Hadamar had been dismantled; killing was now to be accomplished through the use of drugs. Wahlmann claimed that at the time of this appointment he was not aware that Hadamar had been a killing centre, and was programmed to continue to be one in the future. If that was the case, he was quickly disabused; subsequently he certainly appeared to have no compunction about murdering his patients. Thus, when in 1944, Polish and Russian workers already diagnosed as "tubercular" arrived at Hadamar, they were killed without further ado. The only medical examination performed on these workers occurred *after* they had received their lethal injections or tablets, and was solely for the purpose of confirming that they were in fact dead.[1890]

On 8 October 1945, Wahlmann was one of seven defendants in the first post-war "euthanasia" trial, conducted by the United States occupation authorities in Wiesbaden. The charges related, not to German citizens, but to the Polish and Russian forced labourers who had been killed in the final months of the war at Hadamar. Wahlmann, found guilty of the murder of an unspecified number of victims, was sentenced to life imprisonment.[1891] Less than eighteen months later he was in the dock again, this time before a German court for his part in the murder of German citizens. The second trial took place in Frankfurt in March 1947. Wahlmann did not attempt to disguise his involvement in the killings, freely admitting to having personally ordered the death of almost one thousand victims. Moreover, he was fully aware that these homicides had nothing to do with the commonly accepted definition of a "mercy death". Not that this had troubled him, for he continued to organize killing at Hadamar to the end of the war.

The verdict of the court was never in doubt. Wahlmann was sentenced to death, but with the abolition of the death penalty in West Germany in 1949, the sentence was commuted to one of life imprisonment. Wahlmann was released in 1953.[1892]

Mathilde Weber (1909–1996)[1893] was born in Dinslaken and studied medicine at Bonn University. In June 1939 she became assistant to Hans-Bodo Gorgass (q.v.) at the Kal-

---

1887 Klee, *Das Personenlexikon zum Dritten Reich,* p. 652

1888 Klee, *"Euthanasie" im NS-Staat,* p. 170.

1889 Bryant, *Confronting the "Good Death"*, p. 129.

1890 Lifton, *The Nazi Doctors,* pp. 100–101.

1891 Bryant, *Confronting the "Good Death"*, pp. 76–90.

1892 de Mildt, *In the Name of the People,* pp. 97–98.

1893 Klee, *Das Personenlexikon zum Dritten Reich,* p. 658.

menhof psychiatric hospital. Although she never joined the NSDAP, she was a member of the National Socialist Women's Organisation. Her involvement in adult euthanasia appears to have been peripheral; she did register approximately 1,000 patients with T4 in 1940, of whom 232 were gassed at Hadamar in early 1941, but during the initial phase of "euthanasia" Kalmenhof served as a transit institution rather than a killing centre, and Weber was adjudged to have had no knowledge of the fate of patients transported from there to Hadamar.

Whether or not that was the case, there was no doubt about her involvement in children's "euthanasia". In 1942, transports of children from the western *Reich* arrived in Kalmenhof for extermination. The children, who had been pre-selected for killing, were neither examined nor treated, except to the extent of the administratiom of lethal doses of Luminal by the nurse Maria Müller (q.v.), a process supervised by Weber. It was impossible to arrive at a definitive number of child victims at Kalmenhof, but it was estimated to be a minimum of many hundreds. When Weber and Müller were on sick-leave between late September and early November 1943, not a single child died at Kalmenhof.[1894] When they were both present, "there were many days when 3–6 children died." In May 1944, Weber was replaced as physician-in-chief by Hermann Wesse (q.v.). According to Weber: "I went because *Landesrat* Bernotat [q.v.] threw me out as he found that the children did not disappear fast enough..."[1895]

The trial of Weber was held in Frankfurt in January 1949. Despite her ludicrous assertion that (a) the deaths of children on her ward were attributable to natural causes, and (b) if they weren't, she knew nothing about it, since she had not resided in the hospital, nor was she present when the overdoses were administered, the court had no difficulty in establishing her guilt. In summer 1942, along with Friedrich Mennecke (q.v.), she had attended a training course organized by Carl Schneider (q.v.), one of the leading T4 medical experts. It was not difficult to surmise what the subject of the course had been. Moreover, she had received bonuses from the *Reich* committee for unspecified services rendered, the nature of which it was also easy to imagine. She was sentenced to death.[1896]

Twelve months later the sentence was annulled on appeal, and a retrial ordered. In February 1949, a rather more lenient sentence was returned. Weber had not been a perpetrator after all, merely an "accomplice". She had "a weak and unstable personality", as a result of which she had "closed her eyes to what went on around her". Therefore, rather than death, a sentence of three years and six months in jail was considered appropriate. However, because of Weber's "poor health", the sentence was suspended, and she was released. Having had her request for the reimbursement of her legal costs denied, Weber was finally considered fit enough to complete her sentence on 11 October 1954. On 16 November, having been in jail for little more than a month, she was pa-

---

1894 Weber and Müller were both infected with tuberculosis as a result of dealing with a shipment of children from Scheuern. They asked Bernotat to suspend further transports in their absence. Immediately following their return to duty the fatalities re-commenced. (de Mildt, *In the Name of the People*, p. 352, note 114).

1895 de Mildt, *In the Name of the People*, pp. 129–131.

1896 Bryant, *Confronting the "Good Death"*, pp. 135–144.

roled.[1897] It has been alleged that by 1960 she had resumed her career as a practicing physician, albeit in an unlicenced capacity.[1898]

Ernst Wentzler (1891–1973)[1899] born in Hannoversch Münden, was a paediatrician and director of a private clinic in Berlin.[1900] He joined the NSDAP in 1936, as well as the SA and other Nazi associations, and was physician to the children of a number of prominent Nazis, including Göring, Darré, and perhaps most significantly, Werner Blankenburg (q.v.), Viktor Brack (q.v.), and Hans Hefelmann (q.v.), although it appears he was recommended to the KdF by Leonardo Conti (q.v.).[1901]

Wentzler claimed that he had been visited by Brack in 1939 and asked to become one of the three expert referees for children's "euthanasia". Wentzler, who was an enthusiastic supporter of Binding and Hoche's eugenic proposals, agreed to participate. Thereafter he regularly received forms for assessment until almost the end of the war. "I had the feeling that my activity was something positive, and that I had made a small contribution to human progress," he later commented.[1902] In fact, Wentzler went even further, opening a children's killing ward in his own clinic, and recommending other physicians for "euthanasia" duties, including Wilhelm Bayer and Fritz Kühnke (q.v.), as well as becoming involved in the planning of adult "euthanasia"[1903]

In August 1945 Wentzler returned to his paediatric practice. In 1949, together with Werner Catel (q.v.), he was one of those accused of participation in the killing of children at the Rothenburgsort children's hospital in Hamburg, but the proceedings were abandoned on the grounds that the defendants had acted "without awareness of injustice".[1904] Wentzler was questioned by the courts on numerous occasions regarding his role in the children's "euthanasia" programme but was never formally prosecuted.[1905]

Gerhard Wenzel (1905–?)[1906] [post-war pseudonym, "Martin Rhodus"] was born in Rossbach. Initially a student of economics he switched to medicine, obtaining his degree from the University of Leipzig in 1934. By that time he was already a member of the NSDAP (1932) and SA (1933). In 1935 he was appointed assistant physician at the Uchtspringe mental institution. Like many others recruited by T4 he was conscripted on the outbreak of war, but in May 1941 he was interviewed at the KdF by Hans Hefelmann (q.v.), Richard von Hegener (q.v.), and the director of Uchtspringe, Ernst Beese (q.v.). When asked for his views on "euthanasia" for children considered incurable, Wenzel stated that, based upon his experience, he considered it a "morally defensible solution", provided it was conducted legally. That was good enough for his audience. One month later his service with the *Luftwaffe* was suspended, and he returned to

---

1897 Burleigh, *Death and Deliverance*, p. 267.

1898 de Mildt, *In the Name of the People*, p. 132.

1899 Klee, *Das Personenlexikon zum Dritten Reich*, p. 669.

1900 Aly, Chroust, Pross, *Cleansing the Fatherland*, p. 190.

1901 Friedlander, *The Origins of Nazi Genocide*, p. 44.

1902 Burleigh, *Death and Deliverance*, p. 103.

1903 Friedlander, *The Origins of Nazi Genocide*, p. 53, p. 66.

1904 de Mildt, *In the Name of the People*, p. 352, note 107.

1905 http://www.ushmm.org/museum/exhibit/online/deadlymedicine/profiles/#07-wentzler (Accessed 27 December 2008).

1906 Klee, *Das Personenlexikon zum Dritten Reich*, p. 669.

Uchtspringe. There he claimed to have used every means at his disposal to minimize the number of children murdered, despite Beese's insistence on a more hard line approach.

In mid-1942 Wenzel briefly returned to the *Wehrmacht*, supposedly at Beese's instigation. However, within a few weeks Wenzel was back at Uchtspringe, where he remained until September 1943, at which time he requested and obtained a discharge from Uchtspringe and served out the rest of the war at a military hospital in Göttingen. During the time he had been at Uchtspringe, Wenzel had completed 1,000–1,200 children's medical reports, registering about 80 of those he had examined as "completely incurable". These children, together with 40–50 others who were transported to Uchtspringe after already having been condemned elsewhere, were murdered on Wenzel's killing ward.

In December 1945, Wenzel was arrested on suspicion of his participation in the "euthanasia" programme, but in October 1946 escaped from prison and went into hiding under the name of "Martin Rhodus". He was re-arrested in November 1951, but after a few months detention, was released to await his trial, which began in 1953. Wenzel's defence was simply that at the time the alleged offences had been committed, be believed they were both ethically permissible and perfectly legal. The court believed him.[1907] He was acquitted of all charges, the court quoting with approval the sentiments of a father who had written in response to Ewald Meltzer's 1920's questionnaire on children's euthanasia: "Certainly there are many who deplore the barbarity and heartlessness [of euthanasia], but that is of course the outpourings of a false humanity; the outsider cannot in such cases pass judgement."[1908]

Hilde Wernicke (1899–1947)[1909] was born in Schleswig. She graduated from medical school in Marburg and worked for a time as a general practitioner before taking up employment at Meseritz-Obrawalde in October 1927, where she was soon promoted to the position of Assistant Medical Director. She joined the Nazi party in 1933 and the National Socialist Women's Organisation in 1937. She testified: "In the spring of 1943 Meseritz-Obrawalde was designated to become one of the places for the killing of incurable mental patients. At that time I was to become one of the partakers in this programme. It was a special duty."[1910]

Wernicke participated in the killing of patients at Obrawalde from the time "euthanasia" commenced there until January 1945. She claimed to have become "disgusted with this sort of situation and attempted to leave Obrawald" before any killings actually began there. This was the point at which transports full of sick patients were arriving at the hospital and then leaving for "the East". However:

> When I asked to leave, I was told that during wartime changes of places were not allowed. I suffered a lot under these circumstances. So did my colleagues, especially as rumours increased more and more that these [transported] patients were being killed. At times, relatives informed us about the death of a patient who had been transported "East". There was no way to voice objections. Even old and trusted col-

1907 de Mildt, *In the Name of the People*, pp. 138–140.

1908 Bryant, *Confronting the "Good Death"*, p. 206.

1909 Klee, *Das Personenlexikon zum Dritten Reich*, p. 671.

1910 Susan Benedict and Tessa Chelouche, *Meseritz-Obrawalde: a 'wild euthanasia' hospital of Nazi Germany* (History of Psychiatry, Vol. 19 (1), 2008), p. 71.

leagues were done away with when they spoke up against the action. At times it was rumoured that a colleague had ended up in a concentration camp because he had refused to let go of his patients.

When killing at Obrawalde itself began, Wernicke claimed to have verbally objected—but took no further action. Moreover, she repeatedly protested at her trial that she had never personally killed anybody. Walter Grabowski (q.v.), the director of Obrawalde, had instructed her, and she in turn had instructed the nurses who committed the actual murders. She seemed to think that this chain of command, in which she was only a link, somehow exonerated her from any guilt. It was not an argument likely to impress the court. "My patients were very attached to me", she stated, "and would never be threatened by me."

Wernicke fled from Obrawalde as the Russians arrived to shelter at her father's house. In April 1945 she began to work again as a physician until her arrest in August of that year. Together with a nurse, Helene Wieczorek (q.v.), she was tried in Berlin, where the court found that both defendants had acted from "base motives" and condemned them to death. Found guilty of the murder of more than 600 persons, Wernicke was hanged on 14 January 1947.[1911]

As the only "euthanasia" doctor to be condemned and executed by a (West) German court, it has been suggested that Wernicke had the misfortune to be the subject of the earliest such proceedings to be heard. Certainly it is safe to assume that, based upon subsequent experience, it is unlikely she would have received such harsh punishment if her case had been subject to adjudication a few years later, and it is arguable that she might even have escaped retribution entirely, as so many of her fellow murderers did.[1912]

Hermann Karl Wilhelm Wesse (1912–1989)[1913] was born in Düsseldorf, and studied medicine at the Universities of Cologne and Düsseldorf, finally graduating in 1939. He then worked as an unpaid trainee at the mental hospitals of Bedburg-Hau and Andernach, where he met his future wife, Hildegard (q.v.), who he married in late 1941. He joined the NSDAP in April 1933, and for a short time was also a member of the SA.

At the beginning of 1942 Wesse was employed for two months at the Brandenburg-Görden *Kinderfachabteilung* under Hans Heinze (q.v.), before being sent to a juvenile clinic in Bonn to learn the approved method of exterminating the young. He also spent a very short period as a member of the *Wehrmacht* until his military service was suspended early in 1942. In October of that year he was appointed assistant physician at the Waldniel children's killing centre, where he remained until the facility was closed in May 1943. Thereafter, together with his wife, Wesse spent three months receiving further instruction in the technique of murder from Werner Catel (q.v.) at the children's clinic at the University of Leipzig, where Wesse also obtained his medical doctorate.

The Wesse's were then jointly posted to Uchtspringe, where Hermann was placed in charge of the children's ward. Wesse was recalled to the *Wehrmacht* in December 1943, leaving the children's ward at Uchtspringe in the care of his wife, but his military duties

1911 Ibid., pp. 73–75.

1912 de Mildt, *In the Name of the People*, pp. 96–97.

1913 Klee, *Das Personenlexikon zum Dritten Reich*, p. 671.

were again suspended in April 1944 when he was ordered to Kalmenhof-Idstein to succeed Mathilde Weber (q.v.). Wesse was a dedicated practitioner of "euthanasia", as is evidenced by a letter he wrote to Richard von Hegener (q.v.) on his arrival at Kalmenhof:

> I would like to inform you that, as agreed I have taken over the "*Reich* Committee" ward at Idstein. As there are no "*Reich* Committee" children present at this time, I would be grateful if you could arrange for a speedy transfer of "*Reich* Committee" children to this institution.

Wesse was soon not only murdering children, but adults too as the "euthanasia" programme at Kalmenhof was expanded and he began to kill on his own initiative.[1914]

Wesse remained at Kalmenhof until the end of the war, at which time he was arrested by the Americans, but quickly released, to be re-arrested in September 1946.[1915] In February 1947, Wesse was tried in Frankfurt alongside Weber for their joint activities at Kalmenhof. He was found guilty of murdering "at least twenty-five patients", and sentenced to death, despite his claim that von Hegener had threatened him with "dire consequences" if he did not co-operate.[1916] Shortly afterwards Wesse was back in court, this time in Düsseldorf in connection with his time at Waldniel. He was found guilty of participating in the murder of "at least thirty" children. The court somehow contrived to find sufficient grounds not to impose a second death sentence, instead committing Wesse to life imprisonment. However, he was another to be saved from the executioner by the abolition of the death penalty in West Germany. He was released from prison in September 1966 because of ill-health.[1917] No account appears to have been taken at either of his trials regarding Wesse's victims during the short period he was responsible for the children's ward at Uchtspringe.

Hildegard Maria Elisabeth Wesse (1911–1997)[1918] was born in the town of Strotzbüsch. After obtaining her medical degree, like her future husband she gained experience by working as an unpaid trainee at a number of mental institutions. In July 1941 she was sent to Waldniel, where she was put in charge of the men's ward. As described above, the following October her husband joined her at Waldniel, and after the three months spent together at Leipzig university the couple were ordered to report to Uchtspringe in August 1943, where Hildegard was placed in charge of the women's section.

When, in December 1943, Herman Wesse (q.v.) rejoined the *Wehrmacht*, the director of the institution, Ernst Beese (q.v.), nominated Hildegard to succeed her husband in the children's killing ward. Initially she refused, not because of any objection to "euthanasia", but because she claimed, "after all she was a woman, and Dr Beese did not even know whether she was emotionally able to face up to what was required of her." Beese was insistent, finally stating that if she did not take on the assignment, he would do so himself. It was in order to prevent this, Wesse claimed, that she agreed to accept the

---

1914 de Mildt, *In the Name of the People*, p. 353, note 130.
1915 Ibid., p. 352, note 124.
1916 Bryant, *Confronting the "Good Death"*, p. 141.
1917 de Mildt, *In the Name of the People*, pp. 132–135.
1918 Klee, *Das Personenlexikon zum Dritten Reich*, p. 671.

position in January 1944. She subsequently prepared 400–500 reports, resulting in the murder of about sixty children. At the end of 1944, Beese informed Wesse that adult "euthanasia" was now to commence at Uchtspringe. Again claiming that her conscience would not allow her to leave the killing to the ruthless Beese, with purported reluctance she agreed to take responsibility for the murder of some thirty women patients; she personally administered lethal injections to these victims.

Hildegard Wesse was arrested in September 1946, but it was not until 1953 that her trial began in Göttingen. The delay had served her well, for there was now a totally different political climate to that prevailing at the time of her husband's trials. Her defence was essentially the familiar one; she had believed her deeds to be legal at the time and were ethically permissible. So far as her child victims were concerned, the court swallowed her story whole. She had acted in "good faith" and was therefore to be acquitted "on subjective grounds." However, the murder of the women was not to be so easily dismissed. "If she had (exerted her conscience), then, with her exceptional intelligence, she would have seen that something was not in order," the court ruled. "Entrusting a young assistant doctor with deciding alone about the euthanasia of mental patients is simply indefensible." Any such procedure "must have exceeded the limits of what is legally permissible."[1919] Even so, the court determined that if Wesse was guilty, it was of manslaughter rather than murder, and sentenced her to two years imprisonment. The verdict was never confirmed, since the case against Wesse was dismissed on 27 December 1954.[1920]

Albert Gottlob Widmann (1912–1986)[1921] studied chemistry at the Stuttgart Technical Institute, and after receiving his certificate in chemical engineering in 1936, worked as a research assistant there before obtaining his doctorate in 1938. As a student he had been a member of the *Nationalsozialistischen Kraftfahrkorps* (NSKK – National Socialist Motor Corps), and in 1937 had joined the Nazi Party. He became a member of the SS in 1939. At the Technical Institute Widmann had met Walter Heess (q.v.), and in September 1938 Heess recruited Widmann for the Department of Chemistry in the KTI [(*Kriminaltechnisches Institut der Sicherheitspolizei*—Technical Institute for the Detection of Crime, part of the RKPA (*Reichskriminalpolizeiamt*—Criminal Police)]. A year later the RKPA, headed by Arthur Nebe, became Department V of the RSHA (*Reichssicherheitshauptamt—Reich* Security Main Office), with Widmann in charge of Section VD 2 (chemistry and biology) of the KTI.

When it was still in the planning stage, Nebe informed Widmann that a decision regarding the "euthanasia" programme had been reached, and that the KTI had an important role to play. Widmann claimed that he had been assured by Viktor Brack (q.v.) that the proposed operation was entirely legal, and that he (Widmann) would be absolved from all responsibility. At a subsequent meeting called by Brack, at which Nebe, Hans Hefelmann (q.v.) and Richard von Hegener (q.v.) were probably present, Widmann's advice concerning the most suitable killing agent was sought. After consideration was given to, *inter alia*, morphine, scopolamine, and hydrocyanic acid, Widmann

1919 Bryant, *Confronting the "Good Death"*, p. 208.
1920 de Mildt, *In the Name of the People*, pp. 135–138.
1921 Klee, *Das Personenlexikon zum Dritten Reich*, p. 675.

advised the use of carbon monoxide gas, invisible and odourless, with the somewhat impractical suggestion that this be pumped into hospital wards at night.[1922] The first gassing experiment subsequently took place at Brandenburg in Widmann's presence, as already described. Widmann was then ordered by Nebe to arrange the regular supply of carbon monoxide to the killing centres, since this would not raise any suspicion coming from the KTI, but would do so if emanating from the KdF. Widmann was also involved in supplying the lethal medications used in the children's and "wild euthanasia" actions.[1923]

"Euthanasia" was just the start of Widmann's career as a professional murderer. Shortly before the invasion of the Soviet Union he had discussed the construction of gassing vans with Heess, as detailed above. In the late summer of 1941, accompanied by Widmann Nebe, by then commandant of *Einsatzgruppe B*, was instructed by Himmler to kill the patients at the Novinki asylum, near Minsk, but to use a "more humane" method than shooting. On 18 September 1941, 200 patients from Novinki were brought to a small bath house and murdered using vehicle exhaust gas. A further unsuccessful experiment was conducted with high explosives; 25 patients were locked into two bunkers and the explosives detonated. The latter experiment was not considered a success. As Widmann later testified: "Under these circumstances, killing with gas is preferable because automobiles are available everywhere and because any random room can be used."[1924]

Another experimental killing involving more than 500 mental patients also occurred during the visit of Nebe and Widmann to Byelorussia, this time in Mogilev. A room in the local mental asylum was hermetically sealed and two pipes were driven into the wall. A car was parked outside and one of the pipes connected to the car's exhaust. The car's engine was turned on and the exhaust fumes directed into the sealed room. When after eight minutes the people in the room were still alive, a second car was connected to the other pipe in the wall and both vehicles were operated simultaneously. A few minutes later all of those in the room were dead. Widmann, who monitored the operation, described the events:

> During the afternoon Nebe had the window bricked in, leaving two openings for the gas hose... When we arrived, one of the hoses that I had brought was connected. It was fixed onto the exhaust of a touring car... Pieces of piping stuck out of holes made in the wall, onto which the hose could easily be fitted... After five minutes Nebe came out and said that nothing appeared to have happened. After eight minutes he had been unable to detect any result and asked what should be done next. Nebe and I came to the conclusion that the car was not powerful enough. So Nebe had the second hose fitted onto a transport vehicle which belonged to the regular police. It then took only another few minutes before the people were unconscious. Both vehicles were left running for about another ten minutes.[1925]

The KTI had a section at Sachsenhausen, where in 1944 Widmann was involved in an experiment in which poisoned bullets were used to execute prisoners. At the Nuremberg Medical Trial a report written by one of the defendants, Dr Joachim Mru-

---

1922 de Mildt, *In the Name of the People*, p. 56. In his legitimate capacity as a police official, Widmann had once to reconstruct how two people had died from the inhalation of carbon monoxide fumes. (Burleigh, *Death and Deliverance*, p. 120).

1923 Friedlander, *The Origins of Nazi Genocide*, p. 209.

1924 Wildt, *An Uncompromising Generation*, p. 186.

1925 Kogon, Langbein, Rückerl, *Nazi Mass Murder*, pp. 52–53.

gowsky, was produced which detailed how, in Widmann's presence, each of five condemned men had been shot in the thigh with a bullet filled with aconitine nitrate. Three of the prisoners died in agony. The two survivors were presumably disposed of in more conventional fashion.[1926]

After the war Widmann was briefly held by the Americans before being released to lead an undisturbed existence as a chemist in a paint factory until his arrest in 1959. In October 1962, the Düsseldorf court sentenced Widmann to three years and six months imprisonment for his participation in the Sachsenhausen killings. Five years later he was on trial again, this time in Stuttgart, where he was sentenced to six years and six months in jail for his involvement in "euthanasia" and the experiments at Novinki and Mogilev. By crediting time already served whilst on remand, and on receipt of a donation of 4,000 *Deutschmark* to a charity for the handicapped, the sentence was suspended.

Helene Wieczorek (1905–1947) entered Meseritz-Obrawalde as a student nurse in 1925, and remained there for the rest of her working life. Although never a member of the NSDAP, she did join the National Socialist Women's League. Together with Hilde Wernicke (q.v.) she was recruited for the second phase of "euthanasia" in May or June 1943 by Walter Grabowski (q.v.), who obtained Wieczorek's consent to co-operate with the programme, and had her sign a statement to that effect.[1927]

In 1946 Wieczorek, accused of killing "several hundred" patients at Meseritz-Obrawalde, was the first nurse to stand trial before a West German court. Wieczorek testified:

> Director Grabowski told us we had to help the senior nurses—it was too much for them. We would also have to give the injections. At first I refused and he said that there was no point in my doing so because, being a civil servant of many years standing, I had to perform my duty, especially in times of war. He added that there would be a law stating that the incurable mentally ill persons were to be released from their suffering... I only did my duty and I did everything on the orders of my superiors. Director Grabowski always warned us of the *Gestapo*. He said he would inform the *Gestapo* if we didn't do what he ordered.[1928]

Notwithstanding her defence, the court determined that either together with head nurse Amanda Ratajczak (q.v.) or alone, Wieczorek had administered lethal injections to at least 100 patients. Both she and Wernicke were sentenced to death and executed.[1929] By the time of the duo's appearance in court Ratajczak had already been tried, found guilty, and shot by the Russians.

Gerhard Wischer (1903–1950)[1930] was a psychiatrist born in Berlin. He joined the Nazi party in 1937, and the following year was appointed director of the Waldheim mental hospital. He name appears as a T4 *Gutachter* with effect from 2 August 1941, although he was certainly involved in "euthanasia" planning from an earlier date, since he had

1926 George J Annas and Michael A Grodin, *The Nazi Doctors and the Nuremberg Code: Human Rights in Human Experimentation* (New York: Oxford University Press, 1992), p. 83.

1927 de Mildt, *In the Name of the People*, pp. 179–180.

1928 Susan Benedict and Jochen Kuhla, *Nurses' Participation in the Euthanasia Programs of Nazi Germany* (Western Journal of Nursing Research, Vol. 21(2), 1999), pp. 254–255.

1929 de Mildt, *In the Name of the People*, p. 180.

1930 Klee, *Das Personenlexikon zum Dritten Reich*, p. 682.

been chairman of the *Erbbiologische Bestandsaufnahme* ("Authority for Genetic Registration") since 1938.[1931] He was one of the medical participants in *Sonderbehandlung* 14f13, and was extensively involved in the second phase of "euthanasia".[1932]

Following a meeting in Berlin on 17 August 1943 arranged by Paul Nitsche (q.v.) of a group of carefully chosen psychiatrists considered sympathetic to a resumption of "euthanasia", Wischer, now back at Waldheim, wrote to Nitsche: "...I expect a monthly average of 20–30 patients; there have been no difficulties until now, either with personnel or relatives...New patients are admitted nearly every day, so that we must step lively to keep up." A few weeks later Wischer wrote again: "I...have plenty to do, as almost all new patients from the area between Leipzig, Chemnitz, and Meissen come to me. Of course, I could never accommodate these patients if I were not taking the necessary measures to make room, which are going very smoothly. However, I very much lack the necessary medications." Wischer's problem was apparently soon resolved, for on 29 December he informed Nitsche that "there are many departures that are taken care of smoothly".[1933]

Wischer was arrested in October 1945, tried and sentenced to death in Waldheim in 1950, and executed on 4 November of that year.

Ewald Wortmann [or Worthmann], aka "Hannes Wortmann" (1911–1985) [T4 pseudonym "Dr Friede"][1934] was born in Marne, Schleswig-Holstein. He studied medicine at the universities of Hamburg, Würzburg, Munich, and Berlin. In 1934 he passed his preliminary medical examination in Würzburg, and having completed his degree in 1937, finally received his medical licence in January 1939.[1935] Although he never joined the NSDAP, he became a member of the SA in 1933.

During his time at Würzburg Wortmann had attended Werner Heyde's (q.v.) lectures, as had Aquilin Ullrich (q.v.), and like the latter and Klaus Endruweit (q.v.), Wortmann had been a member of the racially inspired 1938 Bessarabian expedition referred to earlier. Between September 1939 and February 1940 Wortmann worked as a doctor at the Hamburg-Langenhorn prison and for a short time was conscripted into the *Wehrmacht.* In May 1940 he was recruited by T4. When asked by Paul Nitsche (q.v.) about his attitude towards "euthanasia", Wortmann replied that he was not opposed to it in principle.

Initially acting as assistant to Dr Theodor Steinmeyer (q.v.) in selecting patients for "euthanasia" at institutions in southern Germany, in late September or early October 1940, Wortmann was transferred to Sonnenstein, where Horst Schumann (q.v.) briefed him on his duties. The sight of the gassing of patients so shocked him, Wortmann claimed, he immediately asked Schumann for a discharge. Schumann referred him to Heyde, who after attempting to persuade Wortmann to remain with T4, finally agreed to release him from the programme.[1936]

1931 Klee, *"Euthanasie" im NS-Staat,* p. 228.
1932 Friedlander, *The Origins of Nazi Genocide,* p. 145.
1933 Aly, Chroust, Pross, *Cleansing the Fatherland,* p. 82.
1934 Klee, *Das Personenlexikon zum Dritten Reich,* p. 687.
1935 Friedlander, *The Origins of Nazi Genocide,* p. 222.
1936 de Mildt, *In the Name of the People,* p. 350, note 83.

After leaving T4, Wortmann worked at a military hospital in Lodz until rejoining the *Wehrmacht* in the field at the beginning of *Barbarossa*. He became a Russian prisoner of war, returning to Germany in 1950 to reside in Friedrichskoog and work as a general practitioner. Although no criminal proceedings were ever commenced against him, he did act as a witness in the trials of Heinrich Bunke (q.v.), Endruweit, and Ullrich. Although doubts remain about the extent of his "euthanasia" activities (he admitted to writing "comfort letters" to the relatives of murdered victims), Wortmann was the only T4 doctor to at least acknowledge "a certain moral guilt, because I have done nothing to prevent these things."

Anna Wrona (1907–?). Although only a tiny cog in the T4 killing machine, Wrona may be regarded as typical of those ordinary people who committed extraordinarily murderous acts—and received little if any punishment for their crimes.

Born in Bochum and raised as a Catholic, Wrona entered the mental hospital at Johannisthal, near Süchteln, as a student children's nurse in 1927. Having completed her training, she remained at Johannisthal until late 1941. She joined the Nazi Party in 1938 or 1939, but claimed to have ceased her membership after a few months.[1937] In November 1941 she was selected for transfer to the Waldniel institution, where Richard von Hegener (q.v.) informed her of the "euthanasia" project, and asked if she was prepared to participate—which she agreed to do without hesitation, unlike the four colleagues who had preceded her to Waldniel. They had refused to participate in the programme and been returned to Johannisthal.[1938]

In December 1941 Wrona was appointed head nurse at Waldniel. From October 1942 she co-operated with Dr Hermann Wesse (q.v.) in the killing of 25 children. When the Waldniel killing ward was closed in the summer of 1943, Wrona was sent to the mental hospital at Dösen, near Leipzig, where she probably received additional training from Werner Catel (q.v.). She was soon transferred again, this time to the Gross-Schweidnitz institution Unhappy there, in June 1944 she moved yet again, this time to Kalmenhof, where working once more with Wesse, she participated in the murder of an unknown number of children. She remained at Kalmenhof until the end of the war.

Wrona's first trial was held in Frankfurt, where the court concentrated on her activities at Kalmenhof. Whilst employed there she had received a monthly bonus of 30 *Reichsmark* from T4, plus an addition 5 *Reichsmark* "per death" (*pro Sterbefall*). Soon the corpse-related bonus had to be shared 50:50 with nurse Maria Müller (q.v.), another Kalmenhof killer.[1939] Notwithstanding this evidence, the court was unable to conclusively prove that Wrona has actually administered lethal injections, and so found her to be an accomplice to murder rather than a perpetrator of it, and sentenced her to 8 years' imprisonment. The Appeal Court disagreed with the verdict and referred the case back for re-trial; since no additional evidence of her involvement in the Kalmenhof

[1937] Wrona was also a member of the German Labour Front, the National Socialist Welfare Organisation, and the National Socialist Women's League. (de Mildt, *In the Name of the People*, p. 368, note 34).

[1938] de Mildt, *In the Name of the People*, p. 368, note 35.

[1939] Müller had escaped from captivity in 1945 and was never traced. In her absence, and thus her inability to refute the accusation, the defence was able to present her as the true culprit; Wrona, it was claimed, had merely been an uncomfortable bystander. (de Mildt, *In the Name of the People*, p. 352, note 119).

killings could be provided, in February 1949 Wrona was acquitted "for lack of evidence."

Even before that decision had been reached, Wrona had been indicted again, this time in Düsseldorf in October 1948 in connection with her activities at Waldniel. There was no denying the killings she had perpetrated; she herself admitted to murdering the 25 children mentioned earlier with overdoses of Luminal. Her punishment for these crimes was a jail sentence of 4 years. However, the Supreme Court of the British Zone was unable to determine whether Wrona had been truly aware of the nature of her crimes. If she had not, she was to be acquitted. The all-important decision was left to the Düsseldorf court which, not surprisingly, decided in February 1953 in Wrona's favour. She had been indoctrinated to regard her superiors as people "who enjoy unconditional authority and whose official directives are to be obeyed," so the court ruled. She had therefore been led to believe that killing defenceless children was "in accordance with regulations and morally irreproachable."[1940] And so Wrona was once more acquitted "for lack of evidence".[1941]

Adolf Würth (1905–?) was born in Baden-Württemberg. Initially a student of medicine, he switched to anthropology, obtaining his doctorate in Berlin in 1936. Robert Ritter (q.v.) recruited him for his Eugenic and Population Biological Research Station in Tübingen, where Würth joined the team investigating the racial characteristics of Gypsies. Würth never joined the NSDAP, although there was little doubt where his sympathies lay. He was the first to refer to "the final solution of the Gypsy question" in September 1937, and in 1938 noted that "the Gypsy question that we face today is above all a racial question."[1942] In 1940 he was conscripted by the *Wehrmacht.*[1943]

After the war, Würth worked in the Baden-Württemberg Bureau of Statistics until 1970. Although proceedings against Würth and Sophie Ehrhardt (q.v.) commenced in the mid-1980s, they were soon abandoned on the grounds that "the defendants' participation in Ritter's research was a legitimate scientific activity"

Minna Zachow (1893–?) was born in Klein Daberkow, a village in Mecklenburg-Strelitz. She worked first as a housekeeper and later as a saleswoman, before entering the nursing profession in 1924. After service at a sanatorium and mental hospital in Berlin, she was enlisted by T4 in June 1940, and sent initially to Grafeneck, then to Hadamar, and finally to Bernburg, where she remained until April 1942, when she returned to Hadamar. Although never a member of the Nazi party or any related organizations, she had participated in all aspects of "euthanasia".

Zachow was one of four nurses appearing before the court in Frankfurt in January 1948, charged with complicity in the "euthanasia" programme. Although admitting to a guilty conscience for what she had done (and, she claimed, had suffered from at the time), Zachow carried on killing. The court was unable to determine a figure for the

---

1940 de Mildt, *In the Name of the People*, p. 319.

1941 Ibid., pp. 185–187.

1942 http://www.radoc.net/radoc.php?doc=art_e_holocaust_chronology&lang=ry&articles=true (Accessed 29 December 2008).

1943 Friedlander, *The Origins of Nazi Genocide*, p. 251.

murders in which she may have been involved during the first phase of "euthanasia", but conservatively estimated their number at 25 in the second phase. For this, the court deemed a jail sentence of three years and six months appropriate.

Christel Zielke (1913–?) had been a member of the NSDAP since 1932, and had considerable experience of "euthanasia", having worked at Grafeneck, Brandenburg, and the children's killing ward at Niedermarsberg before arriving at Hadamar in time to participate in the second phase of the programme there.[1944] A defendant in the second Hadamar trial, she was found guilty of complicity in at least 25 cases of murder, for which she received a prison sentence of three years and nine months.[1945]

Operatives

Heinrich Barbl (1900–?) was born in Sarleinsbach, Austria. A metalworker who joined the NSDAP and SS after the *Anschluss*, he was posted to Hartheim at the time of its conversion from a children's home to a killing centre. His workplace adjoined the crematorium; there he prepared urns containing ashes of the victims for forwarding to their relatives. No attempt was made to correctly identify the remains, even though Barbl stamped a nameplate for each urn. He performed similar duties for a time at Grafeneck.

In spring 1942 Barbl was posted to Belzec, where he was nicknamed "the idiot", and was in a state of perpetual drunkenness. Calling himself "the plumber", he was responsible for installing the piping in the gas chambers.[1946] Since Belzec was already operational, this must have been during the second phase of killings with the enlarged gas chambers, which were constructed in June-July 1942.[1947] He was also frequently the victim of whippings administered by Christian Wirth (q.v.), who needed little excuse to vent his rage on those he considered inferior.[1948] Whether Barbl was in fact as stupid as he appeared to be is a matter of conjecture; he may well have been using his apparent dim-wittedness to avoid unpleasant duties. Gottlieb Hering (q.v.), who succeeded Wirth as commandant of Belzec in August 1942, refused to allow Barbl to execute prisoners because "he is so daft that he would shoot us, not the Jews."[1949] At the time of the construction of the gas chambers in Sobibor, it was Barbl who carried out the pipe work.[1950] He was present at the testing of the Sobibor gas chamber when 30–40 Jewish women were murdered, concerning which he stated: "Red Cross nurses accompanied the selected women, who were transported by bus. They assisted with undressing."[1951]

Barbl was only in Sobibor between April-June 1942 (which coincides with the time of construction of the first gas chambers there), and was one of many ex-T4 personnel

---

1944 de Mildt, *In the Name of the People*, p. 196.

1945 Ibid., p. 371, note 96.

1946 Jules Schelvis, *Sobibor—A History of a Nazi Death Camp* (Oxford: Berg, 2007), p. 247.

1947 Arad, *Belzec, Sobibor, Treblinka*, p. 73.

1948 Robin O'Neil, *Belzec: Stepping Stone to Genocide* (New York: JewishGen Inc, 2008), p. 47.

1949 Ibid., p. 104.

1950 Ibid., p. 125.

1951 Schelvis, *Sobibor*, p. 101.

subsequently sent to Trieste on the termination of *Aktion Reinhard*. Despite being repeatedly interrogated by the Austrian police, no proceedings were ever brought against him.[1952]

Erich Bauer
Source: Wikimedia Commons / Public Domain

Erich Hermann Bauer (1900–1980)[1953] was born in Berlin, where after an elementary education he trained to become a driver. He served in the German army during the Great War, and was released from French captivity in 1920. From 1923 he was employed as a driver until becoming a Berlin tram-conductor in 1933. He joined the SA in the same year, and the Nazi party shortly afterwards.

Bauer continued with his employment as a tram-conductor until he was recruited in 1940 by T4,[1954] where he worked as a lorry driver and part-time chauffeur, driving doctors around the *Reich*.[1955] So far as is known he had no other direct involvement in "euthanasia" operations. Notwithstanding this apparent lack of any prior participation in mass murder, in early 1942 he was transferred to *Aktion Reinhard* headquarters in Lublin, made an SS *Oberscharführer* and sent to Sobibor, where he remained until the liquidation of the camp in December 1943. Thereafter, together with other *Reinhard* personnel he moved to Italy[1956] as part of *Aktion R*, the continuation of Globocnik's Polish activities under another rather obvious code name.[1957]

At his trial in 1950, Bauer claimed that, although being fully aware of the nature of Sobibor's murderous activities, he himself had never participated in them, acting solely as a lorry driver, collecting and delivering supplies for the camp. There was, however, overwhelming evidence to the contrary. He was identified by Jewish witnesses who had escaped from Sobibor as the camp's "*Gasmeister*", responsible not only for operating the gassing equipment, but countless other acts of cruelty and murder. In fact, he was considered one of the most notorious killers in the camp.

In 1945 Bauer had once more become a prisoner-of-war, this time of the Americans, who released him in 1946 to return to Berlin and take up new employment as a labourer. In 1949 he was recognized by two former Sobibor prisoners, who reported the sighting to the police.[1958] Bauer was arrested and arraigned before a Berlin court in spring 1950. On 8 May of that year he was condemned to death for crimes committed in Sobibor, but since capital punishment had been abolished in the Federal Republic, the sentence was commuted to one of life imprisonment. Over the coming years he ap-

1952 de Mildt, *In the Name of the People*, p. 394, note 39.
1953 Klee, *Das Personenlexikon zum Dritten Reich*, p. 30.
1954 de Mildt, *In the Name of the People*, p. 211.
1955 Burleigh, *Death and Deliverance*, p. 124.
1956 de Mildt, *In the Name of the People*, p. 250.
1957 Joshua D Zimmerman, *Jews in Italy Under Fascist and Nazi Rule, 1922–1945* (Cambridge: Cambridge University Press, 2005), p. 249.
1958 Schelvis, *Sobibor*, p. 247.

peared on occasion as a witness at the trials of former colleagues. On 22 December 1971 Bauer was pardoned and released from custody.[1959][1960]

Heinz Kurt Bolender aka "Heinz Brenner" and "Wilhelm Kurt Vahl" (1912–1966)[1961] was born in Duisburg. He joined the SS *Totenkopfstandarte* ("Death's Head") unit and served at Brandenburg, Sonnenstein, Hadamar, and Hartheim as a *Brenner.* Whilst at Hartheim he began to steal sets of false teeth containing gold fillings and bridges from the corpses of the victims, a practice he was to continue as head of Camp lll in Sobibor, the extermination area containing the camp's gas chambers. He arrived at Sobibor in April 1942, together with Franz Stangl (q.v.), Karl Frenzel (q.v.) and Hubert Gomerski (q.v.).[1962] By then, like many others, he had spent some time in the Soviet Union in winter 1941/42 with the *Organisation Todt* operation. A survivor of the camp, Moshe Bahir, recorded his impression of Bolender:

> It is hard to forget *Oberscharführer* Kurt Bolender, with his athletic body and long hair, who used to go walking half-naked, clad only in training breeches, carrying a long whip with which he brutally lashed the camp prisoners he came upon on his way.[1963]

Bolender's fellow criminal, Erich Bauer (q.v.), testified: "Bolender was in charge of Camp lll."[1964] Bolender himself eventually admitted:

> I spent all my time at Sobibor in Lager lll, supervising the Jewish commandos. It is correct that Jews were gassed in Lager lll. I sorted the *Arbeitshäftlinge* [working prisoners] into groups; one group had to clear out the gas chambers after the gassing had taken place, another had to take the bodies to the graves.

In December 1942, having been found guilty of causing a witness to commit perjury in divorce proceedings, Bolender was sentenced by an SS court to nine months in a SS *Straftlager* (punishment camp). He returned to Sobibor in time to participate in the dismantling of the camp, and served for a time at the Dorohucza and DAW-Lublin labour camps before, like so many of his colleagues, departing for Italy and *Aktion R.* At the war's end Bolender adopted new identities; first "Heinz Brenner", the name by which he had been known in Hartheim, and then "Wilhelm Kurt Vahl". However, unlike many ex-T4 and *Aktion Reinhard* personnel, he did not attempt to contact his family, who after a time, declared him deceased. In the meantime Bolender had found employment as a nightclub doorman, one of several positions he occupied until his arrest in 1961.[1965] In Bolender's home, police found a whip bearing the inlaid monogram "KB"

1959 de Mildt, *In the Name of the People*, pp. 249–250.

1960 A number of sources incorrectly state that Bauer died in Berlin-Tegel prison. The judgement by the Hagen court against Karl Frenzel of 4 October 1985 stated: "'*...Durch Beschluss des Senats von Berlin vom 30. November 1971 wurde die weitere Vollstreckung der Strafe mit Wirkung vom 22. Dezember 1971 im Gnadenwege ausgesetzt.' Bauer ist inzwischen verstorben am 4. Februar 1980.*" ("...By a resolution of the Senate of Berlin of 30 November 1971 the further enforcement of the punishment [against Bauer] was suspended. As an act of clemency a pardon was granted on 22 December 1971. Bauer died on 4 February 1980.") I am grateful to Dick de Mildt for providing me with this information. Although only a detail, the fact that Bauer died a free man is important.

1961 Schelvis, *Sobibor*, p. 248.

1962 Ibid., pp. 84–85.

1963 Miriam Novitch, *Sobibor—Martyrdom and Revolt* (New York: Holocaust Library, 1980), p. 152.

1964 Arad, *Belzec, Sobibor, Treblinka*, p. 193.

1965 de Mildt, *In the Name of the People*, p. 399, note 95.

on the handle. It had been made by Szlomo Szmajzner, a survivor of Sobibor who had been forced to work by and for the SS as a goldsmith.

Initially Bolender denied any knowledge of mass murder, claiming to have been "fighting against the partisans in and around Lublin." It was not a sustainable defence, which Bolender apparently recognized, for on 10 October 1966 he committed suicide, hanging himself in his cell at Hagen.[1966]

Werner Dubois
Source: Wikimedia Commons / Public Domain

Werner Karl Dubois (1913–1971)[1967] was born in Wuppertal, and began his working life as an apprentice to a paintbrush manufacturer. However, before his training had been completed he became unemployed, and instead found a position as an agricultural labourer. He joined the SA in 1933, and in 1936 trained as an assistant driving instructor with the National Socialist Motorist Corps. His application to join the elite SS-*Leibstandarte* Adolf Hitler unit was rejected; instead, in January 1937 he became a truck driver with the SS-*Totenkopfstandarte* Brandenburg. Shortly thereafter he became a member of both the NSDAP and the SS. He was posted to Sachsenhausen concentration camp in March 1938, where he remained until the outbreak of war.[1968]

In autumn 1939, Dubois was assigned to T4, where he worked as a driver of a *Gekrat* vehicle delivering victims from various institutions to Grafeneck, Brandenburg, Hadamar, and Bernburg. He also worked as a *Brenner* and handled the disposal of corpses. Despite having had problems with Horst Schumann (q.v.) at Grafeneck and Bernburg, Dubois remained with the "euthanasia" programme at Hadamar until January 1942, when he joined the *Organisation Todt* mission to Russia. On his return, he reported to T4 in Berlin, where he remained until March or April 1942, when he was ordered to Belzec. There he stayed until May or June 1943, at which time he was transferred to Sobibor.[1969]

Dubois performed a variety of duties at both extermination camps, ranging from supervising the Jewish commando at the gas chambers in Belzec, to being in charge of the wood-gathering detail (*Waldkommando*) at Sobibor. He shed some interesting light on the irrational behaviour of Christian Wirth (q.v.), who in a typical display of uncontrollable rage, once drew a pistol on Dubois for failing to start a lorry's engine, accusing Dubois of sabotage. According to Dubois, when he drew his own pistol in response, Wirth calmed down, walked away, and never raised the incident again.[1970] Dubois seems to have been a relatively honest witness, judged that is by the standard of his colleagues, and there is no reason to doubt the truth of his story. Wirth, it would seem, was a typical bully—when faced with the right kind of determined opposition, he simply

---

1966 Schelvis, *Sobibor*, p. 248.
1967 Klee, *Das Personenlexikon zum Dritten Reich*, p. 120.
1968 de Mildt, *In the Name of the People*, p. 214.
1969 Ibid., p. 219.
1970 Ibid., p. 291.

caved in. Of course, everything depended upon who was doing the opposing. It is unlikely that Wirth would have been so accommodating to any Jewish resistance.

Dubois was present at the time of the uprising in Sobibor, when he was severely wounded. On his recovery he joined *Aktion R* in Italy. Despite finishing the war as an 80 percent invalid, Dubois made no claim for a disability benefit because, he claimed in court: "(I) did not want to reveal the events in Sobibor which had led to (my) injury, or in any other way make false statements in a benefit procedure."[1971]

Dubois was arrested in May 1945 by US authorities, but released in December 1947. He worked as a locksmith, until his appearance in court at Munich in 1963 to account for the crimes he had committed at Belzec. Despite admitting to the murder of six individuals (he had been accused of complicity in the murder of 360,000), he was acquitted on 30 January 1964.[1972] He was immediately re-arrested, this time in connection with his activities at Sobibor. In September 1965 the trial of Dubois and eleven others opened in Hagen. Dubois did not attempt to hide his guilt, admitting:

> Of course I am aware that killings were carried out in the extermination camps. What I did amounts to aiding and abetting. If I am convicted, it is only right. Murder is always murder. In establishing who is guilty, it is my opinion that not only the functionaries should be looked at. Whatever we did, we are all equally guilty. The camp functioned like a chain of separate functions. The whole thing would have collapsed if only one link had broken away. The real "work" during the extermination process was carried out by the *Arbeitsjuden* [work Jews]. They were forced to do it, always in fear of death. All members of the German staff supervised the process and carried the responsibility for the extermination of the Jews. What should be taken into account is that we did not act on our own initiative, but in the context of the *Reich*'s Final Solution to the Jewish problem. We did not have the courage to resist orders; I was faced with the following choice: either be a supervisor in a Jewish camp, or a prisoner in a concentration camp.[1973]

Found guilty of complicity in the murder of at least 15,000 people, Dubois was sentenced to three years' imprisonment.[1974]

Herbert Floss (1912–1943) was born in Reinholdsheim. After leaving school he trained as a textile dyer, but finding suitable employment difficult to obtain, worked at a variety of different jobs. He joined the NSDAP in 1930, the SA in 1931, and the SS in 1935, where he was a member of *Totenkopfsturmbann* Elbe. After serving in Buchenwald he was transferred to Bernburg, where he was employed as a *Brenner.*[1975]

Floss served at all three *Aktion Reinhard* camps, but little is known of his activities at Belzec. In Sobibor he was in charge of the day-to-day routine for a short time, until succeeded by Gustav Wagner (q.v.). Thereafter he assumed the position previously held by Alfred Ittner (q.v.) as collector of the money and valuables of Jews passing before him on their way to the gas chamber. It would appear that Floss was considered something of an expert in the cremation of corpses, which had begun at Sobibor in September or October 1942[1976], and in Belzec in November of that year.[1977] Floss may have had expe-

1971 Ibid., p. 290.
1972 Ibid., pp. 276–277.
1973 Schelvis, *Sobibor*, p. 246.
1974 de Mildt, *In the Name of the People*, p. 292.
1975 http://www.encyclopedie.bseditions.fr/article.php?pArticleId=138 (Accessed 24 September 2009).
1976 Schelvis, *Sobibor*, p. 38.
1977 Arad, *Belzec, Sobibor, Treblinka*, p. 172.

rience of the exhumation and burning of corpses at both of these camps, for Heinrich Matthes, in charge of the extermination area at Treblinka, where the cremation of corpses began in March 1943, stated: "...Floss, who, as I assume was previously in another extermination camp, arrived. He was in charge of the arrangements for cremating the corpses."[1978]

Following the uprising at Sobibor, Floss was assigned to escort a party of 30 Ukrainian guards by train from that camp to Trawniki on 22 October 1943. Near Chelm some kind of dispute presumably arose, in the course of which one of the Ukrainians grabbed Floss' sub-machine gun and fatally wounded him.[1979]

Kurt Hubert Franz (1914–1998)[1980] was born in Düsseldorf, and on leaving school became an apprentice cook. He never finished his training, for influenced by his militaristic and nationalistic step-father, in 1932 joined a violently anti-Semitic youth movement, the *Kyffhäuser-Jugend*. Soon after he volunteered for labour service and spent the next two years working in a variety of state-sponsored camps. A year working as an unpaid butcher's apprentice followed, before in October 1935 he enlisted in the army. Even before he had completed his army service he applied to join the SS. As a member of SS-*Totenkopfstandarte* Thuringia he served at Buchenwald, where he worked as a cook, trained recruits, and acted as a guard.[1981] In autumn 1939, together with two of his Buchenwald colleagues he was transferred to T4 and initiated into the "euthanasia" programme by Werner Blankenburg (q.v.). Franz was initially sent to Grafeneck, subsequently working at Hartheim, Sonnenstein, and Brandenburg. Franz claimed that he worked exclusively as a cook in these killing centres, which given his subsequent record seems inherently unlikely. Either way, in late 1941 or early 1942 he was transferred to the kitchen of the KdF, where he certainly was not directly involved in murder.[1982] Strangely, he never became a member of the Nazi party.

In April 1942, Franz was sent to Belzec, where he was put in charge of the Ukrainian guards. When Franz allegedly refused to take command of the camp's extermination area, Christian Wirth (q.v.) slapped his face for insubordination, but took no further action.[1983] Franz himself described how Wirth duped the arriving Jews with a speech in which he assured them that after their "disinfection", their valuables would be returned to them prior to their being deported onward—at which news the Jews burst into applause.[1984] Since there were so few surviving Jewish witnesses from Belzec, there is no reliable evidence concerning the nature of Franz's behaviour there, but Treblinka provides an all too accurate picture. Having presumably distinguished himself with his brutality in Belzec, Franz was transferred to Treblinka in late August or early September 1942. Initially placed in charge of the Ukrainian guards as at Belzec, Franz was rapidly promoted by Wirth to the position of deputy commandant.

1978 Ibid., p. 174.

1979 Schelvis, *Sobibor*, p. 181.

1980 Klee, *Das Personenlexikon zum Dritten Reich*, p. 162

1981 de Mildt, *In the Name of the People*, p. 213.

1982 Ibid., p. 219.

1983 Ibid., p. 275.

1984 Arad, *Belzec, Sobibor, Treblinka*, p. 70.

Nicknamed "*Lalka*" ("doll" in Polish) because of his immaculate appearance, even judged by the standards of an extermination camp Franz was considered terrifyingly sadistic. To all intents and purposes it was Franz who ruled day-to-day life in the camp, reviewing roll calls, deciding on punishments, and killing at will. He would remove bearded men from transports and ask them if they believed in God. If, as was to be expected, they replied in the affirmative, he would tell each man to hold up a bottle as a target, saying: "If your God indeed exists then I will hit the bottle, and if he does not exist then I will hit you." Joe Siedlicki, a survivor of Treblinka, described Franz's whippings: "He'd give them fifty strokes. They'd be dead at the end. He'd be half dead himself, but he'd beat and beat."[1985] Another inmate, Oscar Strawczinski, wrote of Franz:

> To practice boxing, he would use the head of [a] Jew...He would grab his victim's lapel and strike with the other hand. The victim would have to hold his head straight so that Franz could aim well. And indeed he did it expertly. The sight of the Jew's head after a "training session" of this sort is not difficult to imagine.[1986]

Survivor testimonies are replete with descriptions of Franz's cruelty, which undoubtedly had its foundation in his perverted libido.

Franz was not in Treblinka on the day of the uprising, 2 August 1943 ("In fact he was with his tart in Ostrow", according to Franz Suchomel [q.v.]).[1987] Following the uprising, the then commandant, Franz Stangl (q.v.), was relieved and departed for Italy with Odilo Globocnik, Wirth, and others. Franz became the last commandant of Treblinka, and was responsible for the liquidation of the camp, a task completed in November 1943.[1988] He then joined his comrades in northern Italy, hunting partisans and killing Jews.[1989] After the war he worked for a time as a labourer, then from 1949 as a cook until his arrest in 1959. He had made no attempt to disguise his identity. On searching his apartment at the time of his arrest, the police found a photograph album containing many pictures of his time at Treblinka. The album was captioned "*Schöne Zeiten*" ("*Good Times*").[1990]

Indicted at the first Treblinka trial[1991] (Düsseldorf, October 1964–August 1965), Franz mounted a ludicrous defence, claiming that he had known of the uprising, had arranged for two-thirds of the German guards to be absent at the time, and had deliberately left his submachine-gun in his quarters (hardly surprising if Suchomel's evidence quoted above was accurate).[1992] The court had no difficulty in dismissing Franz's statement, instead determining: "(Franz) ill-treated, punched, beat, and killed when it

---

1985 Sereny, *Into That Darkness*, p. 202.
1986 Arad, *Belzec, Sobibor, Treblinka*, p. 190.
1987 Sereny, *Into That Darkness*, p. 238.
1988 Gutman, *Encyclopedia of the Holocaust*, p. 530.
1989 Alexander Donat (ed), *The Death Camp Treblinka: A Documentary* (New York: Holocaust Library, 1979), p. 277.
1990 Klee, Dressen, Riess, *The Good Old Days*, p. 226.
1991 There were actually three trials of personnel who served at Treblinka—the first, of Josef Hirtreiter, was held at Frankfurt in 1951. The second, major, trial of 10 defendants was held in Düsseldorf in 1964/65. The third, of Franz Stangl, also took place in Düsseldorf in 1970. Hirtreiter's trial is usually disregarded for the purpose of numbering, so references herein to the first and second Treblinka trials are to the trials held in Düsseldorf.
1992 Sereny, *Into That Darkness*, p. 241.

gave him pleasure and when he felt like it. It did not bother him in the least when his dog Barry leapt at helpless Jews at his bidding...and wounded and tore them to pieces in his presence...A large part of the streams of blood and tears that flowed in Treblinka can be attributed to him alone."[1993] He had displayed "almost satanic brutality".

Franz was found guilty of the murder of "at least 300,000 persons", and was given 36 life sentences plus eight years' imprisonment.[1994] He was released from prison in 1993 because of poor health and died in Wuppertal in 1998.

Karl August Wilhelm Frenzel (1911–1996)[1995] was born in Zehdenick, a small town near Berlin. After leaving school in 1926 he first became an apprentice carpenter, but in the difficult economic climate of the time he was forced to find work as an agricultural labourer. Although he claimed to have been a member of the socialist carpenters' union, after a spell in jail following his participation in a demonstration of the unemployed, in August 1930 he joined both the NSDAP and the SA. He was an enthusiastic convert, and through his new Nazi associations found a job with a meat wholesaler, before working in a munitions factory and then as a caretaker.[1996]

In August 1939 Frenzel was conscripted into a *Baubattaillon*, an army building detachment, and on the outbreak of war was sent to the Polish border to dismantle enemy fortifications. At Christmas 1939 Frenzel, who by that time had fathered five children, was discharged from the army as a "*kinderreicher Familienvater*", as already described. Shortly thereafter, thwarted of the pleasures of an army career, at the suggestion of fellow SA members he reported to the KdF in Berlin, who, he was told, were looking for reliable party comrades for a "special assignment". In January 1940, the usual oath of secrecy was administered to Frenzel and 15 other recruits at the KdF by Werner Blankenburg (q.v.) and Viktor Brack (q.v.),[1997] following which Frenzel was dispatched first to Grafeneck to serve as a guard, then briefly to Bernburg, before arriving in Hadamar at the end of 1940. There he helped construct the gas chamber, and was subsequently employed as a *Brenner*. In late 1941 he returned to Bernburg to help dismantle the gassing installation.

Together with other "euthanasia" *Aktion* personnel, Frenzel was recalled to Berlin in mid-April 1942 and instructed to report to Odilo Globocnik in Lublin. Having done so, and having been given the rank of SS-*Oberscharführer*, he was posted to Sobibor, where he arrived on 28 April 1942. Christian Wirth (q.v.) remembered Frenzel from Hadamar, and entrusted him with supervising the completion of the camp's building works.[1998] Thereafter, Frenzel's duties were varied, ranging from participation in the entire extermination process to the maintenance of order and discipline among the prisoners. He was zealous in carrying out these responsibilities, to say the least. Together with Gustav Wagner (q.v.) and Hubert Gomerski (q.v.), he was among the most feared members of the German staff. When Wagner was absent, it was Frenzel who conducted the

---

1993 Klee, Dressen, Riess, *The Good Old Days*, p. 291.
1994 de Mildt, *In the Name of the People*, p. 259.
1995 Schelvis, *Sobibor*, p. 250.
1996 de Mildt, *In the Name of the People*, pp. 209–210.
1997 Ibid., pp. 223–224.
1998 Schelvis, *Sobibor*, p. 36.

"selections" from among arriving prisoners. His was the classic example of the "nobody", who through chance suddenly finds himself a "somebody". The Sobibor survivor, Kurt Ticho, described him as a man who "made decisions to show his superiority. He enforced them through brutality. Mistreating defenceless prisoners made him feel powerful and important."[1999] At Frenzel's post-war trial, Ticho, who was placed in charge of the care of sick inmates despite his complete lack of medical knowledge, testified that on at least three separate occasions Frenzel had ordered the shooting of groups of 12–15 sick prisoners.[2000]

Limitless in his greed and ruthless in the wielding of his powers, survivor's testimonies contain many examples of Frenzel's avarice and cruelty. Like Kurt Franz (q.v.) in Treblinka, he enjoyed administering fearful whippings, which would often result in the recipient's death at the *Lazarett.* When on one occasion he caught an 11 year-old boy stealing a tin of sardines, Frenzel made the boy hold up the tin, and shouted out: "Jews are not allowed to eat foreign sardines." Frenzel then shot the boy.[2001] A survivor stated:

> I knew SS officer Frenzel, who led our building commando. To call him barbaric would be an understatement. He was an outright sadist. Frenzel always held his leather whip at the ready and would strike the prisoners over the head, across the face or any other part of the body, for no reason at all. Many prisoners were permanently crippled or even died as a result.[2002]

Another inmate testified: "(Frenzel) was a sadist and a killer without conscience. His involvement at Sobibor went much beyond the mass exterminations; he committed numerous other crimes as well."

After the uprising in Sobibor on 14 October 1943, when Frenzel was in the camp but managed to evade his planned execution, he was among those dispatched to Italy as part of *Aktion R*. As the war drew to a close he was captured by the Americans, but was quickly released. In 1946 he found employment with a film studio in Göttingen, where he remained until his arrest in 1962. The trial of Frenzel and eleven others who had served at Sobibor commenced at Hagen in September 1965. An unsuccessful attempt to suggest that the murder of tens of thousands of innocent individuals held a spurious legality, as well as proposing some kind of moral equivalence between the actions of the Allies and those of the Nazis, indicated the feebleness of Frenzel's defence:

> As I already pointed out, under the prevailing war conditions, which are now difficult to comprehend, I unfortunately believed that what was going on at Sobibor was lawful. To my regret, I was then convinced of its necessity. I was shocked that just during the war, when I wanted to serve my homeland, I had to be in such a terrible extermination camp. But then I thought very often about the enemy bomber pilots, who surely were not asked whether they wanted to carry out their murderous flights against German people in their homes in such a manner.[2003]

1999 Kurt Ticho, *My Legacy: Holocaust, History and the Unfinished Task of Pope John Paul II* (Wlodawa: Muzeum Pojezierza Leczynsko-Wlodawskiego, 2008), p. 89.

2000 Ibid., p. 96.

2001 de Mildt, *In the Name of the People*, pp. 281–282.

2002 Schelvis, *Sobibor*, pp. 87–88.

2003 Arad, *Belzec, Sobibor, Treblinka,* p. 192. This *tu quoque* type of reasoning, aimed specifically at the Americans and the British, is popular among certain critics of Allied wartime policy, as if the bombing of enemy civilians as part of an overall strategy to win the war was in any way comparable to the mur-

On 20 December 1966, Frenzel was found guilty of complicity in the murder of "at least 150,000" individuals, plus nine specific cases of murder, and was sentenced to life imprisonment. Ten years later the Hagen court ordered Frenzel's case to be reopened, and suspended his sentence. A new trial was ordered, commenced on 5 November 1982, and took almost three years to reach a conclusion. In April 1983, in the midst of the trial, an extraordinary confrontation took place at Frenzel's request between himself and Thomas Toivi Blatt, who as a fifteen-year-old had been a prisoner in Sobibor, and was one of the few survivors of that camp. Winston Churchill once famously remarked: "The Hun is always either at your throat or at your feet," and nothing exemplified that more than Frenzel's performance. "I wanted to apologise to you for all that happened then," he said. "If you would accept my apologies in the name of the victims, it would in some small measure be a comfort to me..." Not surprisingly Blatt ignored the request.[2004]

On 4 October 1985, the original verdict of guilt in connection with mass extermination was upheld; however, most of the charges relating to the individual murders were dismissed on the grounds of conflicting evidence. By that time the memory of survivors was faltering. Frenzel was not detained for long. On 10 January 1990 the court determined that "the defendant is now 78 years of age and... in such a desolate state that, in the court's judgement, enforcing the prison sentence would no longer serve the intentions of the law." And so Frenzel lived out the remainder of his life in the comfort of a retirement home, where he died on 2 September 1996.[2005]

Erich Fuchs

Source: Wikimedia Commons / Public Domain

Erich Fritz Erhard Fuchs (1902–1980)[2006] was a lorry driver and motor mechanic born in Berlin. A member of the German Socialist party since 1928, at the time of the *Machtergreifung* he was working for the Jewish publisher Ullstein. On being branded a "*Judendiener*" ("Jew servant") he quickly joined the Nazi party and the SA, and worked for a number of different organisations, including the National Socialist Welfare Organisation and the *Reich* Air Ministry, until in 1940 he was ordered to report to Brandenburg under labour conscription regulations.[2007] There he was responsible to Irmfried Eberl (q.v.) for the supply and maintenance of vehicles. When Eberl was transferred to Bernburg, Fuchs accompanied him.[2008] There is no evidence that Fuchs was any more than "an interested spectator", as he expressed it, at the killing centres where he worked, although in the light of his subsequent activities, there is reason to question whether he may have been more than just an innocent bystander.

der in cold blood of non-combatants on racial, political, or economic grounds. When the war was won, the bombing immediately stopped; where the Nazis occupied, the killing immediately began.

2004 Thomas Toivi Blatt, *From the Ashes of Sobibor: A Story of Survival* (Evanston: Northwestern University Press, 1997), pp. 235–242.

2005 Schelvis, *Sobibor*, pp. 250–254.

2006 Ibid., p. 254.

2007 de Mildt, *In the Name of the People*, pp. 211–212.

2008 Ibid., p. 221.

In February 1942, Christian Wirth (q.v.) arranged for Fuchs to be transferred to Belzec. It was Fuchs who installed the initial gassing equipment at the first extermination camp. He also installed the false shower heads in the gas chambers, admitting: "The showers were not connected to a water supply, because they were only there as camouflage. You see, the Jews who were to be gassed were told, contrary to the truth, that they were to be bathed and disinfected."[2009] Six weeks later, Fuchs was ordered to Sobibor, then under construction, where he was responsible for the installation of the gassing engine:

> Some time in the spring of 1942 I drove a truck to Lemberg on Wirth's orders and picked up a gassing engine, which I took to Sobibor...We installed the engine on a concrete base and connected the exhaust to the pipeline...A trial gassing was carried out. If my memory serves me right, I think 30 to 40 women were gassed...After about ten minutes the women were dead.[2010]

One month later Fuchs was transferred yet again, this time to Treblinka, where he met up with his old boss, Eberl. As Fuchs testified:

> Subsequently I went to Treblinka. In this extermination camp I installed a generator which supplied electric light for the barracks. The work in Treblinka took me about three to four busy months. During my stay there transports of Jews who were gassed were coming in daily.[2011]

Given his experience with the technique of extermination, it is probable that Fuchs was involved in installing more than lighting.

Fuchs returned briefly to Belzec in late 1942, then, having contributed to all three of the principle *Aktion Reinhard* extermination camps, he again spent a short time at Bernburg, where Eberl was once more in charge after his disastrous spell as commandant of Treblinka. Between December 1942 and February 1943 Fuchs worked at the Wiesloch mental hospital before finally securing his release from T4 to find employment as a lorry driver for a German oil company in Riga.[2012] In February 1945 he joined the *Waffen-SS*, where he served in a tank transport unit. He was made a prisoner of war for a short time, firstly by the Russians, and then by the Americans, before the British briefly utilised his services as a driver/mechanic at the liberated Bergen-Belsen concentration camp, by that time housing displaced persons. Thereafter he found employment as a driver/mechanic in Koblenz until his arrest in 1963.

Fuchs was first indicted in the so-called "Belzec trial", held in Munich in August 1963, accused of participation in the murder of 90,000 persons at that camp. On 30 January 1964, the court decided to drop proceedings against 7 of the 8 accused; one of those discharged was Fuchs.[2013] Even before a final decision had been reached by the Munich Court, the Dortmund prosecutor had filed charges against five of the defendants in the Belzec trial, including Fuchs, in connection with their activities at Sobibor. The so-called "Sobibor trial" commenced in Hagen in September 1965.[2014]

2009 Schelvis, *Sobibor*, p. 99.
2010 Ibid., pp. 100–101.
2011 Arad, *Belzec, Sobibor, Treblinka*, p. 43.
2012 de Mildt, *In the Name of the People*, p. 287.
2013 Ibid., pp. 276–277.
2014 Ibid., p. 279.

Not surprisingly, Fuchs was an unreliable witness, constantly changing his evidence, and attempting to persuade others (including two of his former wives) to amend their testimony. So far as his conduct in the service of mass murder was concerned, he expressed a predictable *mea culpa*, which nevertheless perhaps contained more than a grain of truth:

> This was the supreme cultural disgrace. They were criminals. One should have found another way of disposing of the Jews. On many nights I was unable to sleep. Even today I still see the pictures before me: naked people, naked bodies. People who had committed no crime. It was sly murder. I have often contemplated reporting myself [to the investigating authorities]. I lacked the courage, however. Today I might say that I should have tried to get away from it there [that is from T4 and Aktion Reinhard] earlier. The voice of my conscience tells me today that I should have done many things differently. But at the time I did not know how. After all, at the time, you had to do everything they told you. And finally one got so blunted that one did not feel anything anymore.

Although he had only been in Sobibor for one month, the court determined that Fuchs was guilty of complicity in the murder of "at least 79,000 Jews" there, and sentenced him to a jail term of 4 years. He was never called to account for his actions at Treblinka, nor (like many other *Aktion Reinhard* personnel) for his participation in events at the "euthanasia" killing centres.[2015]

Hubert Gomerski

Source: Hessisches Hauptstaatsarchiv / Signature Abt. 461, Nr. 36346/120

Hubert Gomerski (1911–1999)[2016] was born in Schweinheim, near Aschaffenburg. On leaving school he became an apprentice lathe operator, working as such from 1927 until the outbreak of war, when he was conscripted by the military. He had joined the Nazi party in 1931 and the SS in 1934. In November 1939 he was enrolled into the *Waffen*-SS, serving in Poland. He was transferred to the police reserve in Berlin in January 1941, and from there was instructed to report to T4, with no knowledge of that organization's function—or so he claimed. He was rapidly briefed, sworn to secrecy, and dispatched to Hartheim, where after initially performing administrative tasks, Gomerski became a *Brenner.* He detested the work, and asked to be transferred back to his unit. Instead he was ordered back to Berlin where, according to Gomerski, Gerhard Bohne (q.v.) threatened him with incarceration in a concentration camp, or even death if he refused to carry out orders. Shortly afterwards Gomerski was sent to Hadamar, where he carried out a variety of duties, again including, on occasion, those of a *Brenner.*[2017]

In late April 1942, Gomerski arrived in Sobibor, where he was initially put in charge of a group of Ukrainian guards; subsequently he worked in Camp lll, (where the killing occurred), as well as at the ramp. Here, dressed in a white coat so that those arriving might think he was a doctor, he greeted the sick and incapacitated, assuring them that they would be well looked after. Instead they were transferred to Camp III and shot,

---

2015 Ibid., pp. 287–290.

2016 Schelvis, *Sobibor*, p. 255.

2017 de Mildt, *In the Name of the People*, p. 195.

with Gomerski an active participant in their murder.[2018] He became among the most dreaded SS-men in the entire camp, notorious for his sadistic brutality. At the time of the uprising in Sobibor (14 October 1943), Gomerski had the good fortune to be on leave; had he been present, he would certainly have been among the prisoners' prime targets.[2019]

Gomerski was something of a rarity, possibly being the only individual arraigned in four separate trials for "euthanasia" and *Aktion Reinhard* activities. His first trial opened in Frankfurt in February 1947, and was concerned with his Hadamar history. Deciding that as he had only "occasionally" acted as a *Brenner*, and therefore "from the nature of his activities [could] have gained no indications regarding the true dimension of the killing action", the court acquitted him.[2020] Three years later he was in the dock again, this time in connection with his Sobibor career. After his arrest, Gomerski had testified: "I can only declare that a place by the name of Sobibor is unknown to me."[2021] His subsequent evidence proved the mendacity of that statement.

This time Gomerski was not to be so fortunate. Found guilty of complicity in the murder of an unknown number of people, he was sentenced to life imprisonment. In 1972 the Frankfurt court ordered a retrial, which began a year later and lasted until 1977. At the trial's conclusion Gomerski received a sentence of 15 years' imprisonment for his participation in the murder of at least 150,000 individuals. Unbelievably, the sentence was annulled by the Federal High Court on technical grounds, and a new trial ordered, which began in October 1981. To nobody's great surprise, one month later the proceedings were suspended because of the defendant's "ill health", and were finally abandoned in 1983.[2022]

Despite his claimed poor medical condition, Gomerski managed to survive until 28 December 1999.[2023] In 1978, he had been interviewed by a Dutch psychiatrist and former Auschwitz prisoner, Dr Elie Aron Cohen. Gomerski confessed that he could not deny he had been at Sobibor and had killed prisoners there. But what else could he have done? He did not sleep well, and had disturbing dreams—not of the Jews of Sobibor, but of the injustice the courts had visited upon him.[2024]

2018 Ibid., p. 253.
2019 Arad, *Belzec, Sobibor, Treblinka*, p. 315.
2020 de Mildt, *In the Name of the People*, p. 195.
2021 Schelvis, *Sobibor*, p. 2.
2022 de Mildt, *In the Name of the People*, pp. 252–253.
2023 Schelvis, *Sobibor*, p. 255.
2024 de Mildt, *In the Name of the People*, p. 392, note 28.

Walter Grabowski (1896–1945)[2025] was born in Rosenberg, West Prussia (now Susz, Poland). A sales representative in Stettin, he was an early member of the NSDAP and the SA, joining both in 1926. In 1936 he became a Nazi district leader, and in November 1939, despite his complete lack of medical knowledge or experience, was appointed director of the mental hospital at Schneidemühl with the sole purpose of instituting "euthanasia" there. In 1940/41 he performed the same function at the asylum in Kosten (Koscian). His subsequent record at Meseritz-Obrawalde has already been described.

It is believed that Grabowski committed suicide in 1945.

Siegfried Graetschus
Source: Wikimedia Commons / Public Domain

Siegfried Graetschus (1916–1943) was born in Tilsit. Initially earning his livelihood in agricultural work, he joined the SS in 1935 and the NSDAP in 1936, before being enrolled into T4 in 1939/40. He was ordered to Bernburg, where he allegedly performed clerical duties. Graetschus was among the earliest operatives to be recruited for Belzec, which would perhaps indicate his previously undertaking more than administrative tasks at the "euthanasia" centre.[2026] After then spending a short spell in Treblinka, he arrived at Sobibor in August 1942, where he succeeded Erich Lachmann as head of the Ukrainian guards.[2027]

Graetschus was one of twelve SS-men killed in the Sobibor uprising on 14 October 1943. Exactly who killed him is open to question—several survivors claimed credit for the deed, but it seems that Arkadij Wajspapir and Yehuda Lerner were the likeliest assassins.[2028]

Wilhelm Grossmann (1891–1951)[2029] was born in Frankfurt am Main. Originally a bookkeeper, like many others he fell on hard times in the early 1930s. He joined the NSDAP in 1930 and the SA in 1932. After becoming unemployed in 1931, it was not until January 1933 that he found a position as financial administrator of the Kalmenhof-Idstein mental home. By 1935 he had been promoted to head of administration, and when in 1941 the institution's director was drafted into the army, Grossman informally took on his responsibilities. Grossman was an enthusiastic, if totally unqualified, proponent of children's "euthanasia". He did not limit himself to administrative duties, identifying with the programme to the extent that he threatened children "that he would make angels out of them."

In January 1947, a Frankfurt court adjudged Grossmann to be a "perpetrator", which under the prevailing legislation meant a death sentence. But in the following year the appeal court dismissed the conviction, ruling instead that Grossmann had merely been an "accomplice". In essence the appeal court determined that he had only been obeying

2025 Klee, *Das Personenlexikon zum Dritten Reich,* p. 195.
2026 O'Neil, *Belzec,* p. 118.
2027 Schelvis, *Sobibor,* p. 256.
2028 Ibid., note 41. Lerner refers to killing "Greischutz" with an axe (Novitch, *Sobibor—Martyrdom and Revolt,* pp. 112–113).
2029 Klee, *Das Personenlexikon zum Dritten Reich,* p. 204.

orders. A retrial occurred in 1949, in the course of which Grossman, despite evidence that he had beaten mentally handicapped children with an ox hide whip,[2030] was portrayed as the most caring and compassionate of men, someone who "would never have been prepared on his own initiative to lend his support to the murders." For his role in an unknown number of killings at Kalmenhof, the court deemed a sentence of four years and six months imprisonment appropriate. However, due to claimed ill health, Grossmann never served a single day of the sentence.[2031]

Willy Grossmann (1901–?) was born in Lichtenberg. A member of the NSDAP, he trained to become a nurse in the late 1920s, and was employed in that capacity at the Hubertusburg mental hospital until in either late 1939 or early 1940 he joined the Dresden police and was assigned to T4. After serving first at Sonnenstein (allegedly as the doorman), then at Hadamar, in winter 1941, together with many other T4 personnel, he joined the *Organisation Todt* task force to the Soviet Union. He returned to Sonnenstein for a short period before being dispatched to Trawniki, where he received the training necessary for service at a death camp. Thereafter he served at Treblinka until the liquidation of the camp, when he returned to Berlin. Like other members of *Aktion Reinhard,* in December 1943 he was ordered to Italy, supposedly to participate in anti-partisan activities. Post-war proceedings against him were abandoned because of his ill-health.[2032]

Lorenz Hackenholt
Source: Wikimedia Commons / Public Domain

Laurenzius (Lorenz) Hackenholt (1914–?)[2033] was born in Gelsenkirchen. On leaving school at the age of 14 he became an apprentice bricklayer and worked in construction, until in 1933 he joined the NSDAP and the SS. After two year's military service he became a member of the SS Second *Totenkopf* Brandenburg Division stationed at Oranienburg, and in 1938 served as a driver/mechanic and guard at the nearby Sachsenhausen concentration camp, from where in November 1939, together with a group of other SS-men, he was summoned to the KdF. There the party was interviewed by Viktor Brack (q.v.) and Werner Blankenburg (q.v.), sworn to secrecy, enrolled in the "euthanasia" programme, and dispatched to Grafeneck.

Hackenholt served in all six killing centres from early 1940 until August 1941. He drove the buses that transported the victims as well as working as a *Brenner*, removing the corpses from the gas chamber and cremating them. For a while he was August Becker's (q.v.) chauffeur, and performed the same duty for Brack.[2034] In autumn 1941, Hackenholt was transferred to Odilo Globocnik's staff in Lublin, from where he was sent to Christian Wirth (q.v.) at Belzec. There he showed an obvious mechanical and technical aptitude. After initially acting as a driver

2030 Burleigh, *Death and Deliverance*, p. 267.
2031 de Mildt, *In the Name of the People*, pp. 184–185.
2032 Friedlander, *The Origins of Nazi Genocide*, p. 243.
2033 Klee, *Das Personenlexikon zum Dritten Reich*, p. 215.
2034 de Mildt, *In the Name of the People*, p. 376, note 46.

and procurer of vehicles and equipment, Wirth put Hackenholt in charge of the gas chambers and the motor that supplied them with carbon monoxide fumes. Together with Siegfried Graetschus (q.v.), Hackenholt had converted a Post Office delivery van into a mobile gas chamber by connecting the engine exhaust pipe to the sealed rear compartment. The van was then used to kill the mentally and physically handicapped of the immediate region. Hackenholt was very proud of this killing device, which certainly impressed Wirth enough for him to promote Hackenholt to a position where he could kill on a much more ambitious scale.

The extent of his involvement in mass murder was described by the SS guard Karl Alfred Schluch: "After the Jews entered the gas chambers, the doors were closed by Hackenholt himself or by the Ukrainians subordinate to him. Hackenholt switched on the engine which supplied the gas."[2035] Subsequently Hackenholt went on to design and help construct larger gas chambers at Belzec, Sobibor, and Treblinka. In an example of SS humour, the new gas chamber building at Belzec bore a sign—*Stiftung Hackenholt* (Hackenholt Foundation).

When the decision was taken in late November 1942 to obliterate all traces of *Aktion Reinhard*, Hackenholt was involved in the exhumation and cremation of corpses at both Belzec and Treblinka. He then moved on to the *Alte Flugplatz* camp in Lublin, where he gassed prisoners unfit for further labour, before his final posting to Italy with many of the other T4/*Aktion Reinhard* personnel. In 1944 he was awarded the Iron Cross, Second Class. Schluch described Hackenholt as "an inconsiderate, hard and brutal man, without any sense of honour. I would go so far as to say characterless and indifferent. He drank a lot and was often locked up for it ... He was characterless enough to carry out all orders without question." According to his companions in crime, Hackenholt was killed in spring 1945, but there is a considerable amount of evidence to suggest otherwise.

In 1953 Hackenholt's wife requested that he be officially declared dead so that she could apply for a war widow's pension. She claimed that she had heard nothing from her husband since November or December 1944, at which time he was serving in the SS on the eastern front. Frau Hackenholt later admitted to perjuring herself regarding this statement. She subsequently asserted that she had heard from her husband after the war via an intermediary, but had not seen him. However, as a consequence of his wife's application, Hackenholt was declared dead, with the date of demise being given as 31 December 1945. But according to Erich Bauer's (q.v.) testimony, he had met Hackenholt in 1946 when the latter was living under an assumed name and working as a driver. Others, including Hackenholt's brother, also claimed to have seen him after the war. By the time of the preparatory investigations in connection with the Belzec Trial, there was sufficient evidence for the Munich magistrate to rule Hackenholt "undead"—he was still alive somewhere. An exhaustive investigation began to locate him, but without success; after nearly four years it was abandoned in 1963. Hackenholt was never found.[2036]

2035 Arad, *Belzec, Sobibor, Treblinka,* p. 70.

2036 http://www.mazal.org/archive/documents/Tregenza/Tregenza01.htm (Accessed 25 October 2008).

Gottlieb Hering (1887–1945)[2037] whose career closely paralleled that of Christian Wirth (q.v.), was born in Warmbronn, near Stuttgart. After leaving school at the age of 14, he trained for a career in agriculture; however, after his two-year period of conscripted service in the Kaiser's army was concluded, Hering re-enlisted for a further three year stint in 1909. On his discharge from the army, he joined the Württemberg police force where he remained until his call-up in 1915, following which he served with distinction on the Western Front during the First World War.

Hering rejoined the police force in Stuttgart where, unlike Wirth, he was no closet Nazi, being instead of a decidedly left-wing disposition. But like Wirth he rose to the rank of *Kriminalkommissar* in the *Kripo*. Following the *Machtergreifung* Hering underwent a Damascene conversion, largely to avoid problems with the very individuals he had been prosecuting with such enthusiasm, and became the most fervent of Nazis. He joined the NSDAP in May 1933, continuing with his police duties until in 1939 he too was recruited by T4.[2038] He served as administrative chief at Bernburg and Sonnenstein, before in August 1942 he succeeded Wirth as commandant of the Belzec extermination camp when the latter was made Inspector of all *Aktion Reinhard* camps.[2039] Rudolf Reder described Hering at Belzec:

> [Hering] seldom appeared in the camp; he showed up only in connection with some event. He was a tall thug, strongly built, over forty years of age, with a vulgar look to his face...He was a thorough swine...Once the killing machine broke down. Notified of this, he rode over on horseback, ordered the machine repaired, and didn't let the people out of the airless chambers—let them smother and agonize a few hours more. He squatted down in a rage, shouting and shaking all over. Though he seldom showed up, he was the terror of the SS-men...[2040]

With the liquidation of Belzec in 1943, Hering was appointed commandant of the Poniatowa labour camp, where on 4 November 1943 he supervised the shooting of 14,000 Jews as part of *Aktion Erntefest.* When Wirth was killed in Italy in 1944, Hering succeeded him once more, this time as chief executive of *Aktion R.*

Hering died in mysterious circumstances while awaiting admission to the hospital at Stetten im Remstal on 9 October 1945.[2041] He is rumoured to have been under investigation as a suspected war criminal, suggesting that his death may have been suicide.

2037 Klee, *Das Personenlexikon zum Dritten Reich*, p. 247.

2038 O'Neil, *Belzec*, pp. 30–34.

2039 de Mildt, *In the Name of the People*, p. 379, note 79.

2040 Rudolf Reder, *Belzec* (Oswiecim: Fundacja Judaica Panstwowe Muzeum Oswiecim-Brzezinka, 1999), p. 137.

2041 Klee, *Das Personenlexikon zum Dritten Reich*, p. 247.

Fritz Hirche
Source: Chris Webb - Holocaust Historical Society

Fritz Hirche (1893–1945)[2042] was born in Penzig (now Piensk, Poland). A captain in the *Schutzpolizei* (uniformed police), he joined the NSDAP in 1933, and was a supervisor at Brandenburg and Bernburg, performing the same kind of administrative functions in these killing centres as Christian Wirth (q.v.) did at Grafeneck and elsewhere.[2043] Hirche also served for a time at Hartheim in early 1943,[2044] prior to which he had been in charge of the liquidation office of T4 at the Columbus House in Berlin, responsible for clearing up the paper aftermath of the "euthanasia" programme. A division of this office was responsible for *Aktion Cholm*, referred to earlier.[2045] He served on occasion as a temporary commandant at Belzec, although there is very little information concerning his activities there, other than that he spent most of his time immersed in administrative matters.[2046]

In November 1943 he became head of the *Kripo* in Stralsund, where he committed suicide on 1 May 1945.

Josef ("Sepp") Hirtreiter (1909–1978) was born in Bruchsal. Despite training to become a locksmith, he failed to pass his final examination. Thereafter he worked at a variety of semi-skilled jobs, and joined both the NSDAP and the SA in August 1932.[2047] On the outbreak of war Hirtreiter was working as a mechanic with a transport company. Although of an age to be conscripted, he was declared "indispensable" and thus initially avoided military service. Instead, in October 1940 he was ordered to report to Hadamar, where he claimed to have been employed in the kitchen and on administrative duties.[2048] In summer 1942 he was drafted by the *Wehrmacht*, but after four weeks was back at Hadamar, allegedly to carry on with his paperwork.[2049]

In late summer 1942 Hirtreiter was summoned to Berlin by Christian Wirth (q.v.), who sent him onward to Lublin, thence to Treblinka, for "special duty". Quite why Wirth required the services of a catering assistant/clerk to act as a guard in a death camp is an unanswered question, as is any explanation of Hirtreiter's subsequent behaviour viewed in the light of his alleged previous experience. It is true that he was first placed in charge of the camp kitchen, but he was soon supervising the undressing of deportees *en route* to their death. He then moved to Camp II, where he was noted for terrorising victims as he drove them into the gas chambers with the aid of his specially designed whip. He led the aged, infirm, and small children to the *Lazarett* for execution, and generally behaved with horrific brutality.[2050] At the Treblinka trial held at Düssel-

2042 Ibid., p. 258
2043 Friedlander, *The Origins of Nazi Genocide*, p. 99.
2044 Ibid., p. 206.
2045 Ibid., p. 281.
2046 O'Neil, *Belzec*, p. 30.
2047 de Mildt, *In the Name of the People*, p. 210.
2048 An extraordinarily large number of defendants claimed to have been employed in the Hadamar kitchens at the time that the mass gassings took place there, which suggests either a remarkable quantity of catering staff—or a good deal of questionable testimony.
2049 de Mildt, *In the Name of the People*, p. 221.
2050 Ibid., pp. 254–255.

dorf in 1964/65, a witness in the *Kommando* responsible for sorting clothing described Hirtreiter in action:

> In the women's barracks Hirtreiter lashed out wildly at the women on the left and the right with his whip in order to make them undress even faster. When all the naked women had left the barracks, three babies remained behind...Hirtreiter then grabbed a baby by its feet and repeatedly smashed its head against the barracks wall until it was dead...[2051]

In October 1943 Hirtreiter left Treblinka for Italy and *Aktion R*. In 1945 he was misidentified by the Polish Treblinka enquiry as "Hirtreider", a specialist in killing children without wasting ammunition.[2052] Arrested in July 1946 in Frankfurt on suspicion of involvement in the killing of mental patients at Hadamar, during his interrogation Hirtreiter made mention of his service at a camp called "Malkinia", where Jews were gassed. Nobody made the connection at that time between Malkinia, the junction for trains destined for Treblinka, a few kilometres away, "Hirtreider", and Hirtreiter. In February 1947 the Hadamar charges against him were abandoned for lack of evidence, and Hirtreiter was transferred to an internment camp. The following year he was judged to be a main culprit by a denazification court, found guilty of participation in the gassing of at least 4,000–5,000 Jews at the "concentration camp Malkinia in the vicinity of Warsaw", and sentenced to ten years' hard labour. The newspaper report of this verdict alerted the Frankfurt prosecutor to re-examine the "concentration camp Malkinia" and Hirtreiter's role there. It soon became evident that the camp in question was in fact Treblinka, and that Hirtreiter had been one of a number of particularly vicious guards serving in the camp. Importantly, in his earlier evidence he had named several ex-Hadamar personnel who had also participated in *Aktion Reinhard*. This was to form the basis of subsequent investigations concerning these individuals.[2053]

In March 1951, Hirtreiter was in court again, accused of complicity in the murder of an unknown number of Jews at Treblinka. There was little question of his guilt; he was sentenced to life imprisonment. However, due to ill-health he was released from prison in 1977.[2054]

Hermann Holzschuh (1907–?) was born in Stuttgart. In 1926 he joined the police in Württemberg, and by 1937 had reached the rank of senior assistant detective in the Stuttgart *Kripo*. He joined the NSDAP and SA in 1933, changing from the latter to the SS in 1939. On the recommendation of Jacob Wöger (q.v.), in February 1940 Holzschuh was recruited by T4 to serve as Wöger's deputy and eventual successor at Grafeneck. One year later Holzschuh transferred to Bernburg as chief registrar to Fritz Hirche (q.v.), remaining there until April 1941.[2055] Shortly afterwards he left T4 for *Sipo* and *SD* service in Kiev.[2056]

2051 Ibid., p. 392, note 30.

2052 Reitlinger, *The Final Solution*, p. 136.

2053 de Mildt, *In the Name of the People*, p. 249.

2054 Ibid., p. 255

2055 http://www.landesarchiv-bw.de/stal/grafeneck/grafeneck_holz.htm (Accessed 2 January 2009).

2056 Friedlander, *The Origins of Nazi Genocide*, p. 206.

Otto Horn (1903–?) was born in the Leipzig district. After leaving school he first worked in a factory, then as an agricultural labourer, before in 1926 he entered the nursing profession. He passed his nursing examination at Sonnenstein, then worked in the institutions at Leipzig-Dösen and Arnsdorf, joining the NSDAP in 1937. In 1939 he was drafted by the *Wehrmacht* as a medical orderly, and saw service in the Polish and French campaigns. In September 1940 his unit was transferred from Paris to the *Warthegau*. He was briefly a participant in the campaign against the Soviet Union, but was quickly discharged and ordered to return to Arnsdorf. There he was recruited by T4 and sent once more to Sonnenstein, where he may have arrived after the issue of the "stop" order. After only a short stay at Sonnenstein he was transferred to T4 headquarters in Berlin to perform various administrative tasks.[2057]

In early 1942, Horn was heading eastwards again, this time as a member of the curious *Organisation Todt* mission. In September 1942 he was sent to the Trawniki camp for training, and from there was dispatched to Treblinka the following month. His job at Treblinka was to supervise the Jewish work force responsible for burying (and later cremating) victim's corpses. Horn was described by survivors as a relatively decent man who never mistreated any of the prisoners under his command, and might even have been characterized as being quite friendly toward them. He repeatedly sought a transfer away from Treblinka but without success. He asserted at his post-war trial that after the uprising at the camp in August 1943, he had first gone on leave and then feigned illness, reporting back to T4 in Berlin only around Christmas 1943. He was ordered to Trieste and *Aktion R*, but was soon back at Arnsdorf for a short stay before being called up for military service again. The end of the war found him in the Czech Republic, where he became a Soviet prisoner of war.

Horn was one of those appearing at the Treblinka trial conducted at Düsseldorf in 1964/65. The court determined that he considered "the killing of the Jews and Gypsies as an injustice against humanity, ethics, religion, and criminal law." Moreover, he had not been slow to express his opinion to his comrades, who treated him with derision and shunned his company. Nonetheless, he had been a participant in the murder of hundreds of thousands of Jews, and technically was as guilty as any of the other defendants. Despite this obvious truth the court was prepared to accept Horn's defence that he had participated in the *Aktion* against his will and out of genuine fear for his life. Horn was therefore acquitted on all charges.[2058]

Jakob Alfred Ittner (1907–1976) was born in Kulmbach.[2059] A clerk, he was an early member of the Nazi party, joining in 1926, although he allowed his membership to lapse the following year when he had difficulty in paying his subscriptions. Unemployed, he joined the SA in 1931, and following the Nazis attainment of power, renewed his party membership in the hope it would help him to gain work—which it did. In 1934 his nephew Willy Schneider (q.v.) found him a position with the Foreign Bureau of the NSDAP.[2060] When Schneider was made head of T4's Finance Department, he in-

2057 Ibid., p. 243.
2058 de Mildt, *In the Name of the People*, pp. 268–269.
2059 Schelvis, *Sobibor*, p. 257.
2060 de Mildt, *In the Name of the People*, p. 209.

stalled his uncle as a bookkeeper, where Ittner remained from November 1939 to spring 1942.[2061]

At that point Ittner was ordered to report to Sobibor, where he arrived on 28 April 1942. Before he left Berlin it was made clear to him that he was going to work "in a camp in which Jews were [to be] systematically killed." Ittner was not intended to participate in the killing. His duties were to be bookkeeping and financial administration. However, as he admitted:

> The camp was a large and self-contained organization which had as its mission to kill as many Jews as quickly as possible. It all operated very smoothly, because in each location there were German camp staff preventing any trouble from occurring. Each and every one, in his own position, collaborated to ensure the smooth running of the organization. In a broad sense, all of the camp staff took part in the killings. The mass murder of the Jews was not carried out by one single individual, but by a multitude of SS people. Each one was a small cog in the wheel driving an extermination machine that could work only as long as all of them did. That is why, in my opinion, all the camp guards at Sobibor, regardless of their actual job, carried out the killing of the Jews. I would like to emphasize particularly that on arrival of a transport all other work was abandoned, and the camp staff all took part in the actual extermination process.[2062]

It was Ittner's principal responsibility to collect valuables from the victims as they passed before his counter en route to their death. He thus directly witnessed many of the horrors that occurred as thousands were herded into the gas chambers, often with great brutality.

After a few weeks, Ittner came into conflict with the camp commandant, Franz Stangl (q.v.), when Stangl asked him to buy additional rations for the camp staff out of the funds stolen from the murdered Jews. In a perverted display of honesty (any kind of which was unusual in those serving at Sobibor), Ittner claimed to have refused to use state property for this purpose. Stealing the possessions of the Jews was permissible, killing them acceptable, but Ittner drew the line at not accounting for every penny. As a penance Stangl transferred him to Camp III to supervise the prisoners digging mass graves, extracting gold teeth from corpses, and burying bodies, a gruesome task, and one not at all to Ittner's liking. As he testified: "I saw for myself how the crippled and ailing Jews were shot by the edge of the graves in Lager 3. I used to turn and look away when these executions took place, which is why I have no idea who did the actual shooting. It was worse than barbaric there."[2063] He applied to Stangl for a transfer away from Sobibor, which was refused. Ittner, however, persisted. On his next leave he contacted Friedrich Lorent (q.v.), pointing out that burying Jews was not in accordance with the promises made to him when he was assigned to the camp. Amazingly, those in authority agreed, and shortly afterwards Ittner was back at his desk at T4's Berlin headquarters.

Ittner was one of those arraigned at Hagen in 1965. He had only been present in Sobibor for about three months, and was not identified by any of the survivors, nor was he incriminated by any of his co-defendants. For these reasons, the court considered a

---

2061 Ibid., p. 223.

2062 Schelvis, *Sobibor*, pp. 244–245.

2063 Ibid., p. 257.

prison sentence of four years adequate punishment for involvement in the murder of an "unknown number of Jews, although at least 68,000".[2064]

Robert Emil Franz Xaver Jührs (1911–?)[2065] was born in Frankfurt, and on leaving school became an apprentice cooper. An accident resulting in the loss of the use of one eye forced him to abandon this trade, and thereafter he was employed in a variety of menial jobs.[2066] In 1929 he became a member of an SA drum band (he left the SA in 1935), and in 1930 he joined the NSDAP.[2067] 1940 found Jührs working at the Hessen Labour Exchange as an administrative assistant. In June 1941 he was assigned to Hadamar, where he stayed for about a year, apparently performing similar clerical tasks. There is no direct evidence that he participated in more sinister activities there, although as with others, his subsequent record suggests otherwise.[2068]

In June 1942 Jührs was posted to Belzec, where he was made an SS-*Unterscharführer* and served as guard, both at the ramp and at the *Lazarett*, where he executed sick and disabled deportees. He admitted to carrying out such killings on at least one occasion:

> ...There were Jews that by no means could cover the way to the undressing barrack. Hering [q.v.] gave me an order to shoot these Jews...As I remember, there were seven Jews, men and women who were taken inside the pit...I shot these Jews with a machine gun, as they stood on the edge of the pit...[2069]

Jührs was one of those who supervised the exhumation and burning of hundreds of thousands of bodies at Belzec. By March 1943 much of this task had been completed, and Jührs was transferred to the Dorohucza labour camp, where he remained until November of that year, when the camp was liquidated and all prisoners shot at the Trawniki camp as part of *Aktion Erntefest.* Immediately after this he was in Sobibor, helping to supervise the demolition of the camp and witnessing the execution of the remaining prisoners, although he naturally claimed not to have participated in their shooting.[2070] After a short stay in Berlin, at around Christmas 1943 Jührs joined his comrades in Italy as part of Globocnik's *Aktion R.*

There were at least six locations where investigation into possible criminal activity on Jührs' behalf might have proved fruitful; Hadamar, Belzec, Dorohucza, Trawniki, Sobibor, and La Risiera San Sabba.[2071] In the event, Jührs was considered for inclusion among the accused in the 1947 Hadamar trial held in Frankfurt, but proceedings against him were dropped.[2072] He was a defendant in the 1963 Belzec trial in Munich ,

2064 de Mildt, *In the Name of the People*, pp. 285–287.
2065 Schelvis, *Sobibor*, p. 257.
2066 Friedlander, *The Origins of Nazi Genocide*, p. 238.
2067 de Mildt, *In the Name of the People*, p. 209.
2068 Ibid., p. 221.
2069 Arad, *Belzec, Sobibor, Treblinka*, pp. 71–72.
2070 Schelvis, *Sobibor*, pp. 190–191.
2071 La Risiera was an abandoned rice mill in the San Sabba suburb of Trieste. It was intended to serve as a transit camp for the deportation of Italian Jews to German concentration camps, primarily Auschwitz. However, according to Italian sources, it also served as a killing centre; more than 3,000 Jews and Italian and Yugoslav partisans were murdered there. Erwin Lambert constructed a crematorium at La Risiera in early 1944. (O'Neil, *Belzec*, pp. 197–199).
2072 de Mildt, *In the Name of the People*, p. 221.

accused of complicity in 360,000 cases of murder. Together with 6 of the other 8 defendants, proceedings against Jührs were abandoned in 1964.[2073] The following year he was in the dock again, this time in Hagen as a defendant in the Sobibor trial. Jührs was acquitted on all counts.[2074] No proceedings were ever commenced against him (nor, one suspects, even considered) in respect of his time in Dorohucza, Trawniki, or San Sabba.

It seems almost inconceivable that a self-confessed murderer, who served in six centres where killing occurred virtually every day, could completely evade any form of retribution, other than the short periods he may have spent in custody awaiting trial.

Herbert Kalisch (?–?) was an electrician working for the General Electric Company who was recruited by T4 to install the cabling to the gas chamber and crematorium at both Bernburg and Sonnenstein.[2075] As described above, he was a witness to the gassing of a group of Jews at Brandenburg. He later testified: "The transport that was gassed in the gas chamber of the former prison in Brandenburg on the Havel in about June 1940 contained only Jews, who I would estimate were men and women between the ages of eighteen and fifty-five.[2076]

Kalisch's testimony is important, as it provides evidence that Jews were being gassed on German soil as early as 1940. Whether they were gassed because they were mental patients or simply for being Jews is a moot point.

Alfons Klein (1909–1946) joined the Nazi party and the SA in 1930, and had been employed at Hadamar in an administrative capacity since 1934. In 1939 Fritz Bernotat (q.v.) appointed him director of the institution. Klein was therefore present throughout the entire period Hadamar served as a killing centre. During the first phase of "euthanasia" Friedrich Berner (q.v.) was the dominant personality at Hadamar; Klein's role was to assist in converting the institution into a killing centre and deal with financial matters. However, when Berner left, Bernotat promoted Klein to effectively become overall controller at Hadamar during the second phase of "euthanasia".[2077] Following a meeting in July or August 1944 with Bernotat and Jakob Sprenger, *Gauleiter* of Hesse-Nassau, Klein was informed that forced labourers suffering from incurable tuberculosis were to be transferred to Hadamar. Later he was instructed that these patients were to be killed under the same law that had applied to German insane patients.

Klein was never called to account for his participation in the earlier Hadamar killings, since he had already paid the ultimate price for his subsequent murderous activities. He was a defendant at the first Hadamar trial, held under American jurisdiction at Wiesbaden in October 1945, a judicial process concerned solely with the murder of Soviet and Polish forced labourers during the second phase of "euthanasia". Some idea of Klein's attitude towards these killings can be gleaned from the testimony of a nurse who objected to the murder of helpless patients, and informed Klein that she wanted to stop. He replied: "You lazy cow, if you do, you'll be next."[2078] Klein of course denied such

2073 Ibid., pp. 276–277.
2074 Ibid., p. 294.
2075 Friedlander, *The Origins of Nazi Genocide*, p. 238.
2076 Ibid., p. 278.
2077 Ibid., p. 207, Bryant, *Confronting the "Good Death"*, p. 130.
2078 Burleigh, *Death and Deliverance*, p. 241.

allegations, claiming instead that staff were entirely free to leave Hadamar whenever they so desired. He later withdrew this statement, attesting instead that he was as unable to quit as any of the other personnel.

The court were not impressed. Klein was found guilty of aiding, abetting, and participating in the murder of an unknown number of persons, "but aggregating in excess of 400", and was sentenced to death. He was executed on 14 March 1946.[2079]

Johann Klier (1901–1955)[2080] was born in Stadtsteinach, Bavaria. He became a master baker in 1931, and in 1933 joined the NSDAP and SA. Following a long period of unemployment, he worked at a brass foundry from 1934–1940.[2081] An active member of the party, in October 1940 he was ordered to Hadamar by the regional leadership, and worked there on various construction and engineering tasks. Between the time of his arrival and his departure in June 1942, he claimed never to have been involved in any killings at Hadamar.[2082]

Like others before him, Klier was summoned to T4 headquarters in Berlin in June 1942, sworn to secrecy, and dispatched to Sobibor, where he arrived in August of that year. To begin with he was put in charge of the camp's bakery, but later headed the *Schuhkommando*, the working party responsible for collecting, sorting, and stock-piling the shoes of victims. Klier was another who was on leave when the uprising in Sobibor took place. On the camp's liquidation he was sent to Globocnik's unit in Italy.

After spending nearly four years in an internment camp at Darmstadt, Klier was released in February 1949. By that time his name had cropped up in other investigations into the Sobibor staff. In August 1950 he appeared in the dock in Frankfurt alongside Hubert Gomerski (q.v.), accused of supervising the undressing of prisoners prior to their gassing, and of whipping the prisoners under his command. Klier denied both charges. He had only been responsible for collecting shoes, he claimed, and if he had beaten prisoners it was because he had been "compelled" to do so. Moreover, he had gone out of his way not to inflict any physical harm to his charges in the course of such beatings. He was helped in his defence by the testimony of survivors, who confirmed that he was one of the few SS-men who retained a vestige of humanity. One survivor described Klier as a "benign, portly man," and felt "[he] did not approve of the activities at Sobibor."[2083] Klier had provided Jews with additional rations and, as evidence in later trials disclosed, had even warned some of the impending liquidation of the camp and advised them to escape.[2084] The court found Klier's defence convincing, and acquitted him on both charges.[2085]

2079 http://www.ess.uwe.ac.uk/WCC/hadamar.htm#Alfons%20Klein (Accessed 12 January 2009).

2080 Schelvis, *Sobibor*, p. 258.

2081 de Mildt, *In the Name of the People*, p. 211.

2082 Ibid., p. 221.

2083 Ticho, *My Legacy*, p. 92.

2084 de Mildt, *In the Name of the People*, p. 391, note 21.

2085 Ibid., pp. 251–252.

Erwin Lambert
Source: Wikimedia Commons / Public Domain

Erwin Hermann Lambert (1909–1976)[2086] the "flying master-builder" ("*der fliegende Baumeister*") of T4 and *Aktion Reinhard*, was born in Schildow, near Berlin. After serving apprenticeships as a locksmith and bricklayer, in 1935 he graduated as a master mason.[2087] He joined the NSDAP in March 1933 and had worked for a number of Berlin construction companies, before in late 1939 he was recommended to T4 by the German Labour Front. He was offered the job of independent building foreman for T4, which he turned down on the grounds that he was happy with his then current employment and did not want to be away from his sick mother. His initial refusal notwithstanding, he was enrolled into T4 in January 1940 and immediately put to work on the renovation of Tiergartenstrasse 4, the organisation's head office and provider of its acronym. Subsequently he was responsible for the construction of the gas chambers and crematoria at Hartheim, Sonnenstein, Bernburg, and Hadamar, continuing to ply his trade in these institutions until early 1942, when he was ordered to Treblinka.[2088] There he assisted with the construction of the camp's barracks and fencing, as he testified: "The Treblinka camp was still in the process of construction. I was attached to a building team there."[2089]

In August of that year he returned to Treblinka to supervise the building of the camp's new gas chamber building: "At Treblinka I built the foundations for the large gas chambers."[2090] In the following September or October he was in Sobibor to perform the same task, declaring:

> At that time I was assigned by Wirth [q.v.] to enlarge the gassing structure according to the model of Treblinka...The camp was already in operation, and there was a gassing installation. Probably the old installation was not big enough, and reconstruction was necessary.[2091]

Lambert went on to erect two *Aktion Reinhard* labour camps at Dorohucza and Poniatowa, before returning to Bernburg. In spring 1943 he was in Treblinka for a third time to carry our repairs and new building works. He continued to work for T4 on various building projects, including the recreation centre at Attersee in Austria, until like many other T4 personnel, he was sent to Italy and San Sabba in late 1943, where his expertise in the construction of gas chambers and crematoria was put to good use. After the war he lived quietly in Stuttgart until his arrest in 1962. He was a defendant in the first Treblinka trial in Düsseldorf, where in 1965 he was sentenced to 4 years' imprisonment for his involvement in murder on a grand scale at that camp.[2092]

Three days after the Treblinka trial finished, Lambert was in the dock again, this time at Hagen in connection with his activities at Sobibor. He was adjudged to have

2086 Klee, *Das Personenlexikon zum Dritten Reich*, p. 354.
2087 de Mildt, *In the Name of the People*, p. 213.
2088 Ibid., p. 222.
2089 Arad, *Belzec, Sobibor, Treblinka*, p. 40.
2090 Kogon, Langbein, Rückerl, *Nazi Mass Murder*, p. 132.
2091 Arad, *Belzec, Sobibor, Treblinka*, p. 123.
2092 de Mildt, *In the Name of the People*, p. 267.

been complicit in the murder of at least 79,000 people there, and for that received a jail sentence of 3 years.[2093]

Willi Mentz (1904–1978) was born in Schönhagen in the province of Posen, a region which at that time was within Germany, became Polish in 1919, reverted to Germany as part of the *Warthegau* in 1939, before finally returning to Poland in 1945. Mentz first worked as a labourer, then a miner, before becoming a master-milker.[2094] He joined the Nazi party in 1932, and applied to join the police in early 1940, without success. Instead, his application was passed to the Agricultural Chamber in Münster where, so Mentz claimed, he was offered the position of master-milker at Grafeneck, which he accepted. On arrival there he was informed that "mentally ill patients were killed here in gas chambers and then burned". Thereafter, so he asserted, he stayed at Grafeneck for about eighteen months, during which time he "tended the pigs and the cows." He was then transferred to Hadamar, where he worked as a gardener and handyman until early summer 1942, when, like August Miete (q.v.), he was assigned to Treblinka.[2095] Quite how his supposed previous activities qualified him for service in an extermination camp is yet another unanswered question.

At Treblinka his duties ranged from supervising the *Leichenkommando,* the corpse burying squad, to overseeing the "agricultural unit". However, his principal area of activity was at the so-called *Lazarett* or "hospital", where he was responsible for shooting aged and infirm arrivals, as well as any of the other inmates considered redundant.[2096] Miete's description of the horrors of the *Lazarett* also serve to illustrate Mentz's function there:

> There were always sick and crippled people in the transports...There were also those who had been shot and wounded en route by the SS, policemen, or Latvians who guarded the transports. These ill, crippled, and wounded passengers were brought to the *Lazarett* by a special group of workers. Inside the *Lazarett* they placed or lay these people at the edge of the pit. When all the sick and wounded had been brought, it was my job to shoot them. I fired at the nape of the neck with a 9 mm pistol. Those shot would fall...into the pit...The number of people shot in this way from each transport varied. Sometimes two or three, and sometimes twenty or even more. They included men and women, young and old, and also children...[2097]

Richard Glazar, a survivor of Treblinka, recalled Mentz in action:

> Somehow always unkempt and dishevelled, Willi Mentz, with a black moustache under his nose, is subordinate to Miete in civilian as well as in military life, although he too is a sergeant. In real life he is a dairy farmer, and here he is marksman second class. He is responsible for the routine shootings that take place in the "Infirmary" as the transports arrive. He shoots and shoots, and keeps shooting, sometimes moving on to the next target even when the previous shot had not found its mark and a sentient victim simply slipped into the fires. Messy work.[2098]

After Treblinka, Mentz 's life largely replicated the pattern of many other T4/*Aktion Reinhard* members. He spent a short time in Sobibor before being dispatched to Italy

2093 Ibid., p. 292.

2094 Friedlander, *The Origins of Nazi Genocide*, p. 238.

2095 de Mildt, *In the Name of the People*, p. 223.

2096 Ibid., p. 262.

2097 Arad, *Belzec, Sobibor, Treblinka*, p. 122.

2098 Richard Glazar, *Trap With A Green Fence: Survival in Treblinka* (Evanston: Northwestern University Press, 1999), p. 47.

for *Aktion R*, and at the end of the war was incarcerated for some time in a prisoner-of-war camp before being released in summer 1945 to resume his old trade. He retired in 1952 suffering from tuberculosis and lived on his disability pension until his arrest in June 1960.[2099] Found guilty at the first Treblinka trial of complicity in the murder of at least 300,000 individuals, as well as an additional 25 people. Mentz was sentenced to life imprisonment, plus an additional six years.[2100]

August Miete

Source: Wikimedia Commons / Public Domain

August Wilhelm Miete (1908–?) was born in Westerkappeln and worked at his parent's farm and mill until 1940, at which time he applied to the Agricultural Chamber in Münster to become a settler in the East. It was only then that he joined the Nazi party, probably in order to further his application.[2101] Miete's subsequent career mirrored that of Willi Mentz (q.v.).There were no vacancies for settlers; instead Miete was offered employment at Grafeneck, where he arrived in May 1940 to work on that institution's farm. He remained at Grafeneck until October 1941, when he was sent to Hadamar, initially to work as a manual labourer. From early 1942 he became a *Brenner* at Hadamar, until in June 1942 he was ordered to Berlin, from where he was sent first to Lublin and thence to Treblinka as a newly appointed SS-*Unterscharführer*.[2102] All of the T4 staff serving at the extermination camps were given the nominal rank of *Scharführer* (Sergeant) of varying degrees and wore grey *Waffen*-SS uniforms. Josef Oberhauser (q.v.) described them as "civilians in uniform".[2103]

Like Mentz, at Treblinka Miete was assigned to killing duty in the *Lazarett*, among other tasks. He was notable for the enthusiasm and brutality with which he handled himself. As the court adjudged at his post-war trial:

> Without showing any outward emotion, the defendant gestured the victim to him with his finger, explained to him...that he looked too ill, or too well, that he thought too much or that he was too lazy, and took him along to the *Lazarett*, where the unfortunate received a shot in the back of the neck...

The court estimated that Miete had been personally responsible for killing hundreds, if not thousands in the *Lazarett*. Richard Glazar described him thus:

> Silently, like a ghost, Sergeant August Willi Miete shows up everywhere where someone is giving out, where someone has been marked and branded, where someone can no longer pretend that he is healthy and working at full strength.

Miete had many nicknames in Treblinka, but the most appropriate was the Yiddish, "*Melech Ha-Moves*"—"The Angel of Death".[2104] Another Treblinka survivor, Samuel Willenberg, described Miete in the act of killing a newly arrived young girl from Warsaw:

2099 de Mildt, *In the Name of the People*, p. 261.

2100 Ibid., p. 264.

2101 Ibid., p. 212.

2102 Ibid., p. 223.

2103 O'Neil, *Belzec*, p. 93.

Creeping like a cat, a smile of satisfaction on his dull, fair, moustached face, he stealthily approached the new prey. Reaching her, he pushed her almost gently, almost imperceptibly, as if not wanting to dirty his murderer's hands. He pushed her as a child might push a large ball, prodded her with a stick so she might start rolling by herself. Thus the Angel of Death led her to the rear gate between the two huts abutting the platform. He propelled her toward the sorting yard and the innocent fence at its edge, with its pine branches. Behind it was the *Lazarett*...She vanished behind the fence. A few minutes later we heard a gunshot. Silence, utter silence everywhere. Then Miete strode through the gate in the *Lazarett* fence with its green branches, slipped his pistol into its black holster, and slapped invisible dust from his palms.[2105]

As with many other T4 personnel, Miete was sent to northern Italy in late 1943 to continue killing there. He was arrested in 1960, and in 1964/1965 was one of the defendants in the first Treblinka trial. Found guilty of participating in the murder of 300,000 persons, plus the proven killing of eight additional victims, Miete was sentenced to nine life sentences. He died in prison.[2106]

Gustav Münzberger

Source: Ghetto Fighters House Museum, Israel, Photo Archive

Gustav Münzberger (1903–1977) was born in Weisskirchlitz in the Sudeten, then a region of Czechoslovakia, today part of the Czech Republic. His father was a master carpenter, and having completed his apprenticeship, Münzberger worked in his father's shop. Between 1923–1925 he served in the Czechoslovakian army, before taking over his father's business in 1931. He was again mobilized by the Czech army for a short period at the time of the Sudeten crisis in autumn 1938. Following the absorption of the Sudeten into the *Reich*, he joined the SS in late 1938; two years later he became a member of the Nazi party.[2107]

During the summer of 1940, Münzberger was ordered to report to Sonnenstein. According to Dieter Allers' (q.v.) statement, duty at a killing centre was a matter of completely random selection so far as SS-men were concerned:

None of them, except those they later called the burners, could have got in without their own doing... Münzberger: for heaven's sake, he was a carpenter; why on earth should anyone recruit just him for this work—unless, as he no doubt did, he put in a request for what sounded like a cushy job, just like all of them did. Except the burners—that was perhaps different; they were strictly the troops. They were ordered there, by numbers. Some sergeant picked them out "you and you and you". And you can take my word for it that the sergeant didn't know what he was picking them out for...[2108]

Münzberger claimed that he spent the next two years at Sonnenstein working as a labourer and cook, which given his later record seems improbable.[2109] In 1942 he was ordered to Poland, and arrived at Treblinka in late September of that year. His duties there included standing at the entrance to the building which housed the gas chambers

2104 Glazar, *Trap With A Green Fence*, p. 47.

2105 Samuel Willenberg, *Revolt in Treblinka* (Warsaw: Zydowski Instytut Historyczny, 1992), pp. 61–63. Miete is referred to as "Mitte" throughout this memoir, another example of the problem for survivors of correctly identifying perpetrators by name.

2106 de Mildt, *In the Name of the People*, pp. 260–261.

2107 Friedlander, *The Origins of Nazi Genocide*, pp. 238–239.

2108 Sereny, *Into That Darkness*, p. 84.

2109 de Mildt, *In the Name of the People*, p. 222.

and driving the victims into the chambers with the aid of his pistol and whip.[2110] Towards the end of 1943 he joined his comrades at San Sabba, remaining in Italy until the end of the war.

Münzberger was arrested in July 1963, and was one of those arraigned in the first Treblinka trial. In August 1965 he was found guilty of involvement in the killing of at least 300,000 people, and was sentenced to 12 years' imprisonment. He was released from custody in July 1971. And what motivated him to participate in the murder of those countless men, women and children? In his own words, "gratitude towards the *Führer* for bringing home his Sudeten German *Heimat* into the *Reich*."[2111]

Johann Niemann (1913–1943) was born in Völlen. He became a member of the Nazi party in 1931, and after serving an apprenticeship as a house painter, joined the SS in 1934. After service at the Esterwegen and Sachsenhausen concentration camps, he became a *Brenner* at Bernburg. One of the participants in the *Organisation Todt* mission to the Soviet Union, he was among the earliest arrivals at Belzec, where Christian Wirth (q.v.) appointed him head of Camp II, the extermination section.[2112] Niemann was only at Belzec for a short time before being transferred to Sobibor, where he occasionally performed the duties of deputy commandant before permanently occupying this position in early 1943.

At the time of the revolt in Sobibor Franz Reichleitner (q.v.) was on furlough, leaving Niemann as acting commandant of the camp. Lured into the tailor's barracks on the pretext of trying on a new leather coat, Niemann was killed with an axe by Alexander Sjoebajew.[2113] He was the first of the SS to be executed that day.[2114]

Vinzenz Nohel
Source: Wikimedia Commons / Public Domain

Vinzenz Nohel (1902–1947) was born in Moravia, at that time a part of the Habsburg Empire, today a region in the Czech Republic. He trained as a mechanic but suffered long periods of unemployment. It is unclear whether Nohel was ever a member of the NSDAP, but his brother certainly was, and in addition was a *Brigadeführer* in the SA. In late 1939 or early 1940, the aforementioned brother introduced Nohel to Adolf Kaufmann (q.v.), who offered Nohel a job as a *Brenner* at Hartheim. Nohel accepted the position and the improved salary that it included with alacrity, and remained at Hartheim for as long as it remained a killing centre.[2115]

When interrogated after the war, Nohel described the "*Gaskammer*" *or* "*Gasraum*" at Hartheim, the chamber disguised as a bathroom where people were killed.[2116] He estimated that 20,000 mental patients and 8,000 prisoners from the Mauthausen concen-

2110 Ibid., pp. 264–265.
2111 Ibid., p. 324.
2112 Arad, *Belzec, Sobibor, Treblinka*, p. 28.
2113 Schelvis, *Sobibor*, pp. 161–162.
2114 Ibid., p. 259.
2115 Friedlander, *The Origins of Nazi Genocide*, pp. 233–234.
2116 Kogon, Langbein, Rückerl, *Nazi Mass Murder*, p. 50.

tration camp complex had been gassed at Hartheim.[2117] By the time the killing stopped, Nohel had become an expert in the destruction of human remains. One of the defendants at the Mauthausen trial held under United Stated military jurisdiction at the former Dachau concentration camp in March 1946, Nohel was sentenced to death; he was executed on 27 May 1947.[2118]

Walter Nowak (1921–?) seems the cause of endless confusion. According to Jules Schelvis, an authoritative source, Nowak was employed at Sonnenstein before being transferred to Sobibor, where he was stationed in Camp III, site of the gas chambers. During the post-war interrogation of Nowak's wife she admitted that her husband had been a member of "an SS-Sonderkommando at Sobibor". A search of her house revealed a treasure trove of valuable items "from a Polish camp where Jews from many lands had been burned". Paul Rost (q.v.) confirmed that Nowak had served in Camp lll at Sobibor. Rost stated that he had met Nowak in an American internment camp. On his release in 1947, Nowak "was pursued by the authorities but without success".[2119]

Other sources simply obfuscate the issue. One suggests that in May 1945 a Sobibor guard named Nowak (that is, Walter) was recognized in East Germany by a former camp inmate, Meir Ziss. Nowak was then arrested by Soviet authorities.[2120] If this is the Nowak referred to in a second source, nothing further is known of his fate.[2121] A third source apparently confuses Walter Nowak with another, different Nowak (Anton Julius, 1907–1943), who also served in Sobibor, but was killed in the camp uprising on 14 October 1943.[2122] Yet another source indicates that Walter and Anton Nowak were one and the same.[2123]

It is apparent from all of this that it appears that there may have been two Nowaks at Sobibor—Walter, as described above, fate unknown, and Anton Julius, killed in the uprising. If that was the case, it is known that Anton Julius Nowak was in charge of the haircutting barrack, close to the gas chambers, where from winter 1942 women were forced to undress and have their hair shorn.[2124] Walter Nowak's duties at Sobibor remain unclear.

2117 http://www.scrapbookpages.com/dachauscrapbook/dachautrials/Mauthausen05.html (Accessed 13 January 2008)

2118 http://www.scrapbookpages.com/DachauScrapbook/DachauTrials/Mauthausen01.html (Accessed 13 January 2009).

2119 Schelvis, *Sobibor*, pp. 259–260.

2120 http://www.betemunah.org/blatt/TheMurderers.htm (Accessed 18 November 2008).

2121 de Mildt, *In the Name of the People*, p. 394, note 39. This Nowak was in charge of the undressing barracks at Sobibor, which would appear to be a description of Anton Julius Nowak.

2122 http://www.holocaustresearchproject.org/ar/ARleadership.html (Accessed 18 November 2008).

2123 O'Neil, *Belzec*, p. 328.

2124 Schelvis, *Sobibor*, pp. 259.

Josef Oberhauser
Source: Ghetto Fighters House Museum, Israel, Photo Archive

Josef Oberhauser (1915–1979)[2125] was born in Munich. After leaving primary school he worked on his uncle's farm until in 1934 he volunteered for the army. On his discharge eighteen months later, he became a member of both the NSDAP and the SS-*Totenkopfstandarte* Brandenburg and was stationed at Oranienburg, where he probably served at the Sachsenhausen concentration camp. With the outbreak of war he fought in the Polish campaign as a member of the SS-*Leibstandarte* Adolf Hitler.[2126] When hostilities were concluded he was ordered to Berlin, and from there was sent successively to Grafeneck, Brandenburg, and Bernburg, where in each case he worked as a *Brenner*. He apparently enjoyed his work, boasting as he did during a drinking session at Grafeneck of having burned 20,000 corpses.

In November 1941, Oberhauser was posted to Odilo Globocnik's staff in Lublin. Shortly afterwards he arrived at Belzec, where he became Christian Wirth's (q.v.) right-hand man. Referred to as Wirth's "shadow" by his colleagues, they also alleged that he had been involved in every aspect of the camp's operation. For example, Karl Schluch testified:

> If Oberhauser maintained that he did not participate in the extermination of the Jews in Belzec, or that he did not see the whole operation from beginning to end—from the unloading to the removal of the bodies—, then I say, "try another one!" Oberhauser not only knew the entire running of the extermination operation well but he also took part in it. In my opinion, there is no doubt that Oberhauser was an authoritative person in the killing of the Jews in Belzec camp. The Belzec camp operated for only one reason, and for what Oberhauser did, he was well promoted.[2127]

Oberhauser remained in Belzec until August 1942, when Wirth, now based in Lublin, was promoted to Inspector of all *Aktion Reinhard* camps and took Oberhauser with him as his *aide-de-camp*. In September 1943 Oberhauser loyally followed his master once again, this time to Trieste and *Aktion R.*

The end of the war found Oberhauser in a British prisoner-of-war camp in Austria. He escaped from there and fled to the Soviet occupied zone, where he was arrested and tried in Magdeburg in 1948. He was sentenced to 15 years' imprisonment for his involvement in "euthanasia" and his membership of the SS, but was released under a general amnesty in 1956 to return to Munich and find employment as a waiter in a beer hall. There he was famously filmed by Claude Lanzmann for the film *Shoah*.[2128] Unsurprisingly, Oberhauser was as uncommunicative with Lanzmann as he had been with Gitta Sereny.[2129]

2125 Klee, *Das Personenlexikon zum Dritten Reich*, p. 440.

2126 de Mildt, *In the Name of the People*, p. 213.

2127 O'Neil, *Belzec*, p. 208.

2128 Claude Lanzmann, *Shoah—The Complete text of the Acclaimed Holocaust Film* (New York: 1995), pp. 53–54.

2129 Sereny, *Into That Darkness*, p. 113, footnote.

In August 1963, eight men, among them Oberhauser, were indicted by the Munich prosecutor for crimes committed in Belzec. On 30 January 1964 the case against seven of the men was dropped on the grounds that they had been "acting out of fear for life."[2130] The case against Oberhauser was merely adjourned, however, and a new trial with him as the sole defendant commenced on 18 January 1965, at the conclusion of which he was sentenced to four years and six months' imprisonment for his participation in the murder of hundreds of thousands of Jews at Belzec.

Oberhauser was released after having served half of his sentence to return to the Munich beer hall. In March 1976, a Trieste court sentenced him *in absentia* to life imprisonment for crimes committed in Italy. Oberhauser thus remains not just the only German to be convicted of mass murder at Belzec, but also the only ex-T4 employee to be the subject of court proceedings arising out of three separate but connected *Aktionen*—"euthanasia", *Aktion Reinhard,* and *Aktion R.*[2131]

Franz Reichleitner (1906–1944)[2132] was born in Ried im Traunkreis, Austria. A policeman by profession, he joined the Nazi party in 1935, and became a member of the Linz *Gestapo* in 1937. In 1940 he was assigned to Hartheim as Christian Wirth's (q.v.) subordinate. Franz Stangl (q.v.) and Reichleitner had known each other from CID school. After Reichleitner allegedly succeeded in recommending that Stangl be sent to join him at Hartheim later that year, the two of them shared a room.[2133] Following Wirth's departure from Hartheim, Reichleitner succeeded him there as chief of administration, with Stangl as his deputy.[2134]

When in late August or early September 1942 Stangl was transferred from Sobibor to Treblinka to succeed Irmfried Eberl (q.v.), Reichleitner took over from Stangl as commandant of Sobibor. On one occasion Stangl's wife indirectly suggested to Reichleitner that her husband might have been involved in "the awful things which are being done at Sobibor". Reichleitner hastened to reassure her that it was Wirth who was responsible for the killing of Jews. Stangl's activities in the camp were purely administrative, Reichleitner lied.[2135] Frau Stangl did not apparently quiz Reichleitner about his own duties in the camp.

Moshe Bahir, a survivor of Sobibor, described Reichleitner as

> ...a man in his late forties [sic] with an Austrian accent [who] was always dressed with great elegance and wore gloves. He did not have direct contact with the Jews and the transports. He knew that he could rely on his subordinates, who were very frightened of him. He ran the camp with German precision.[2136]

After the liquidation of Sobibor in November 1943, Reichleitner was transferred to the Trieste region together with other *Aktion Reinhard* personnel. On 3 January 1944 he was killed during a skirmish with the resistance in Fiume, Italy.

2130 O'Neil, *Belzec*, p. 207.
2131 de Mildt, *In the Name of the People*, pp. 276–279.
2132 Klee, *Das Personenlexikon zum Dritten Reich*, p. 485.
2133 Sereny, *Into That Darkness*, p. 53.
2134 de Mildt, *In the Name of the People*, p. 225.
2135 Sereny, *Into That Darkness*, p. 138.
2136 Arad, *Belzec, Sobibor, Treblinka*, p. 188.

Karl Paul Rost (1904–1984) was born in Deutschenbora, near Meissen. On leaving school he trained to be a butcher, before joining the Dresden police force in 1925. He became a member of the NSDAP in 1937 and the SS in December 1940. In May of that year he had been ordered to report to Sonnenstein, where he headed the police squad and transportation command. He also served for a time in Hartheim, before in early 1942 he was dispatched first to Lublin, and then to Sobibor. For a short period he was acting commandant of the overall camp before being put in charge of Camp II, responsible for sorting Jewish property.[2137]

In May 1943 Rost was transferred to Treblinka. Following the liquidation of that camp in autumn 1943, like many others he was sent to northern Italy. After the war he spent a short time in a U.S. prisoner-of-war camp before being released and returning to his family in Dresden. A brief period of Soviet imprisonment ensued before he was released again, to work unmolested in Dresden until his death.

Franz Rum
Source: Ghetto Fighters House Museum, Israel, Photo Archive

Franz Rum (1890–1970) was born in Berlin, and after training to become a waiter worked in this capacity in England, France and Berlin. He served in the military during the Great War before resuming his former profession.[2138] He joined the NSDAP in March 1933, and following the outbreak of war in 1939 and a consequent downturn in business, discussed employment prospects with his customer, Richard von Hegener (q.v.), who offered him a job in the photographic department at T4 headquarters. Shortly afterwards Rum reported to Werner Blankenburg (q.v.), and having been sworn to secrecy, was informed of the "euthanasia" programme. Rum's duties included photocopying patient's records, as a consequence of which he developed an allergic reaction to the fumes from the chemicals being used. Rum therefore applied for a different job, "if possible one in the open air."[2139] So he was posted by T4 to serve as a guard at Treblinka, where he arrived in December 1942.

Rum performed a number of tasks at the extermination camp, including supervising the Jewish commando responsible for the disposal of corpses. He appears to have been a relatively anonymous figure, neither particularly bad nor particularly good, judged by Treblinka standards, although he admitted to regularly using his whip on prisoners. Transferred to Italy with many other *Aktion Reinhard* personnel in November 1943, he was a defendant at the Treblinka trial held in Düsseldorf in 1964/65. Judged to be an accessory to murder in "at least" 100,000 cases, Rum was sentenced to three years' imprisonment, but died before the sentence became final.[2140]

Hans-Heinz Schütt (1908–?) was born in Dummersdorf, Schleswig-Holstein.[2141] His school teacher father had been killed in the Great War, and although Schütt attended

2137 Schelvis, *Sobibor*, p. 261.
2138 Friedlander, *The Origins of Nazi Genocide*, pp. 239–240.
2139 de Mildt, *In the Name of the People*, p. 223.
2140 Ibid., p. 268.
2141 Schelvis, *Sobibor*, p. 261.

the *Gymnasium*, his mother's difficult financial circumstances forced him to leave school early and take up a commercial apprenticeship. After working for a number of companies, in 1931 or 1932 he was employed by the *Deutsche-Nationalen-Handlungsgehilfenverband* (*DHV*), a right-wing German nationalist commercial clerks' union with Nazi associations that was absorbed into the German Labour Front (*DAF*) in 1934. Schütt had joined the SS a year earlier; now he became responsible for the book-keeping at many of the Labour Front's branches. After a dispute with Robert Ley, head of the *DAF*, in 1936 Schütt worked for a time at the Technical University in Berlin, and then at the *Reichs* Office for Grain. The following year, together with the all others in his SS unit, he was registered as a member of the NSDAP. In 1938 he began working for the SS in an administrative capacity, until he was summoned to T4 in the late autumn of 1939.[2142]

Schütt was dispatched to Grafeneck, allegedly without any knowledge of the "euthanasia" project, and was quickly enlightened by Horst Schumann (q.v.), who appointed him deputy chief of the institution's administrative arm. As such, he was responsible for the welfare of approximately 60 T4 functionaries. After Grafeneck closed, Schütt was transferred to Hadamar in early 1941, where he performed similar duties. Following the issue of the "stop order" in August 1941, he worked in the Berlin head office of T4, again as an administrator,[2143] until in April 1942 he was posted to Sobibor, where he remained until mid-August 1942.

In Sobibor Schütt initially continued to act as a book-keeper, paying salaries and dealing with other clerical matters, before succeeding Herbert Floss (q.v.) [who had himself succeeded Alfred Ittner (q.v.)] at the task of collecting valuables from the victims as they passed a cubicle situated at the entrance to the "tube", the enclosed path that led to the gas chambers.[2144] Schütt carried no whip, and apparently took no part in the extermination process, nor was it alleged that he mistreated, still less killed any prisoner. In essence Schütt claimed he was a mere spectator to genocide, and moreover, an unwilling one. At his trial he stated:

> In answer to the question why I was on the ramp when the transports arrived, I declare I was there out of curiosity. I wanted to convince myself of the inhumanity of the *Endlösung* [the "Final Solution"], and to relay my impressions back to Berlin so that I might be released. Under no circumstances did I ever get actively involved at Sobibor.[2145]

Others painted a somewhat less glowing picture of Schütt in Sobibor, alleging his involvement in orgies with Jewish women, but the court found insufficient evidence of this to prove the claim.[2146] What is not in doubt is that he was discharged from Sobibor after a stay of only four months, to return to the *Waffen*-SS. After the war Schütt was briefly arrested in Tübingen in connection with his time at Grafeneck, but it was not until September 1965 that Schütt stood in the dock at Hagen, accused with eleven others of alleged crimes committed at Sobibor. After due process he was acquitted.

2142 de Mildt, *In the Name of the People*, p. 214.
2143 Ibid., pp. 219–220.
2144 Schelvis, *Sobibor*, p. 71.
2145 Ibid., p. 261.
2146 de Mildt, *In the Name of the People*, p. 400, note 132.

As with other *Aktion Reinhard* personnel, Schütt's involvement in "euthanasia" did not form part of the charges brought against him. That they might have is evident from a lengthy letter he wrote to his stepbrother in March 1940, when Schütt was already working at Grafeneck, and which included the following passage: "At the moment I myself am in a special unit [*Sonderkommando*] of which, in greater Germany, perhaps 100 people are aware. You can probably imagine that I am very happy and proud...One day, I may be able to speak to you about all this."[2147]

Otto Stadie (1897–1977) was born in Berlin. After completing his schooling he first worked as a messenger. Thereafter he was employed at a Dr Bernstein's Jewish owned Berlin clinic for skin and venereal diseases. After training as a medical orderly he served in the Great War before moving to Breslau, where he was unemployed for several years. He eventually found employment as a nurse at the Wuhlheide hospital in Berlin in 1927, remaining there until the outbreak of the Second World War, when he was once again conscripted as a medical orderly to serve in France and Poland. He had joined the NSDAP and SA in 1933.[2148]

Stadie was recruited by T4 in 1940 and was sent to Bernburg as a bus driver, accompanying the transports of victims from other mental institutions to the killing centre. After joining the *OT* mission to Russia in winter 1941/42, he was assigned to Treblinka, where he served as Franz Stangl's (q.v.) orderly until July 1943, subsequently relocating to Trieste with his T4 colleagues.

Formerly company commander of the Ukrainian guard unit in Treblinka, Stadie became the camp's chief administrator, responsible for all paperwork as well as personally participating in the unloading of arriving trains, selecting "work Jews", and making mendacious "assurance speeches" to the victims. Although considered relatively decent for a Treblinka SS-man, Stadie was not averse to using his whip and pistol, and had personally ordered the shooting of at least one prisoner. The Treblinka survivor, Richard Glazar, described Stadie thus:

> At a special roll call, the little barrel-shaped staff sergeant Stadie demonstrated what he could do when called upon to fill in for the vacationing Küttner and Franz. Full of rage, he snorted. His cheeks swelled, making his small eyes look all the more evil: "Anyone found with as little as a penny will be severely punished!"[2149]

Stadie was arrested in July 1963, and was a defendant at the first Treblinka trial held in 1964/65 in Düsseldorf, where after due process he received a sentence of seven years' imprisonment for his complicity in the murder of an estimated 300,000 individuals.[2150] He was released in 1969 due to ill health.

2147 Ibid., p. 295–296.

2148 Friedlander, *The Origins of Nazi Genocide*, p. 242.

2149 Glazar, *Trap With A Green Fence*, p. 113.

2150 de Mildt, *In the Name of the People*, pp. 265–266.

Franz Stangl (1908–1971)[2151] was born in Altmünster, Austria. He left school at the age of fifteen, and after serving an apprenticeship, at eighteen and a half years of age became, so he claimed, the then youngest master-weaver in Austria. But prospects in the depressed economic conditions of the time were unfavourable, and in 1931 Stangl decided to join the Austrian police. After two years training he commenced performing the duties of a regular policeman with the Linz constabulary, as well as assisting in the political department of the *Kripo* by investigating the activities of banned political movements (which then included the Communist, National Socialist, and Social Democratic parties). In the turmoil of 1930's Austria, Stangl's abilities were quickly recognized. He was decorated twice and soon reached the rank of *Kriminalbeamter* (CID officer).

In autumn 1935 Stangl was transferred to the political division of the CID in Wels, a small town near Linz. Following the *Anschluss* in March 1938, the Wels political division was absorbed into the Linz *Gestapo*, to where Stangl was re-transferred. By May 1938 Stangl was an official member of both the NSDAP and the SS. There are grounds for believing that he had been a member of the illegal pre-*Anschluss* Austrian Nazi party and SS, but this has never been definitively proven. When recommending Stangl for promotion in 1943, Odilo Globocnik noted that "While still in the Austrian police, he [Stangl] served as an undercover SS-man."[2152]

Stangl claimed to have had a very bad relationship with his *Gestapo* superior, Georg Prohaska, as a result of which, so Stangl alleged, he began to seek ways in which he could get away from Linz.[2153] After lending some assistance to the *Judenreferat* (Jewish department) of the Linz police and their policy of enforced Jewish emigration, an opportunity to escape from his supposed difficulties soon arrived.[2154] According to Stangl, his recruitment to T4 came about as the result of a meeting with his friend, Franz Reichleitner (q.v.), who was already working for that concern. In the course of their discussion Stangl's unhappy personal circumstances came up. Without going into details, Reichleitner told Stangl that he was employed by an organization involved in a project which was "*Geheime Reichssache*" ("Secret *Reich* Business"), and offered to use his influence to get Stangl transferred to what Reichleitner described as "a pleasant job". It is at least conceivable that ambition and desire for material benefit also played their part[2155] and that Stangl personally sought the position rather than relying solely upon Reichleitner's endorsement. It seems unlikely that his relationship with Prohaska could have deteriorated to the extent he claimed, since it is inconceivable Stangl could have been transferred to T4 without Prohaska's support, and probably his recommendation.[2156] Perpetrator testimony is often ambiguous when it is not duplicitous.

2151 Klee, *Das Personenlexikon zum Dritten Reich*, p. 596.

2152 Arad, *Belzec, Sobibor, Treblinka*, p. 168.

2153 At his (Stangl's) trial, Prohaska admitted to having taken an intense dislike to Stangl. However, despite considering Stangl "unprincipled " (a not unreasonable assessment), Prohaska denied having persecuted Stangl in any way. (de Mildt, *In the Name of the People*, p. 375, note 40).

2154 de Mildt, *In the Name of the People*, pp. 216–218.

2155 Friedlander, *The Origins of Nazi Genocide*, p. 207.

2156 Ibid., p. 205.

As described above, in early 1940 Stangl was summoned to Berlin for a meeting with Paul Werner (q.v.), and after a briefing from Viktor Brack (q.v.), in November 1940 was appointed police lieutenant at Hartheim, where he became responsible for the security of the institution and the preparation of certificates containing falsified causes of death. As already mentioned, after Christian Wirth (q.v.) left Hartheim, Reichleitner succeeded him as head of administration, with Stangl as his deputy. Following the departure of Gottlieb Hering (q.v.) from Bernburg, Stangl also briefly became head of administration there, but returned to Hartheim after a few months.[2157]

Again, there is only Stangl's testimony extant regarding the circumstances surrounding his induction into *Aktion Reinhard.* With the official suspension of "euthanasia", he claimed to have been offered a choice in February 1942—either join other T4 personnel in Poland, or be transferred back to Linz, and Prohaska. It was, he suggested, not much of a choice.[2158] So in April 1942 Stangl reported to Lublin, where after an interview with Odilo Globocnik, he was sent to Belzec to see how Wirth had organized mass murder there. One month later he was appointed the first commandant of the new extermination camp at Sobibor. So efficient was he considered at this task that in September 1942 he was transferred to Treblinka to succeed the disastrous Irmfried Eberl (q.v.). Globocnik was sufficiently impressed by Stangl to commend him to Himmler, reporting that he was "the best camp commander and had the most prominent part in the entire *Aktion*."[2159] Richard Glazar described Stangl as Treblinka commandant:

> On top of the sandy rampart a distant figure promenaded and then stopped from time to time...From there he viewed his estate... He didn't carry a heavy bullwhip like all the other SS, but only a light riding crop, and he always wore light-coloured deerskin gloves and a field cap on his head. A few fingers of his right hand hooked into the front of his fitted green uniform jacket...He keeps his distance from everyone and his perspective over everything from above. He rarely comes up to the operation from his headquarters in the lower part of the camp.... When he does appear at roll call, he does so only to look in on events from the sidelines...And then, tapping his riding crop lightly against his boots, he leaves before the end of roll call, without saying a word...He gives the impression of being the lord of the manor...He is in a position where he needs neither to fire a shot himself nor wield a bullwhip.[2160]

After the Treblinka uprising, together with most of the *Aktion Reinhard* personnel, Stangl was transferred to Trieste, ostensibly to organize anti-partisan measures, although it is certain that the killing activities of *Aktion Reinhard* also continued under the somewhat unimaginative title of *Aktion R.*[2161]

At the war's end, Stangl returned to Austria, where he was eventually interned by the Americans as a member of the SS. In late summer 1947, Austrian investigations into activities at Hartheim uncovered Stangl's presence in a prisoner-of-war camp at Glasenbach. He was transferred to a civilian prison in Linz, whence he escaped in May 1948. He made his way to Rome, from where, with the assistance of Bishop Alois Hudal, he obtained a Red Cross *laissez-passer* and a job as an engineer in Damascus. In 1951, he and his family moved from Syria to Brazil, where Stangl worked under his own name in the Volkswagen factory at Sao Bernardo do Campo, near Sao Paulo.

2157 de Mildt, *In the Name of the People*, p. 225.
2158 Ibid., p. 298.
2159 Ibid., p. 299.
2160 Glazar, *Trap With A Green Fence*, p. 46.
2161 Zimmerman, *Jews in Italy Jews in Italy Under Fascist and Nazi Rule*, p. 249.

After a tip off by a former *Gestapo* official, Simon Wiesenthal discovered Stangl's whereabouts and informed the authorities. At the request of the Austrian government, Stangl was arrested by Brazilian police in February 1967. Within four months he had been extradited to Germany, where his trial commenced in Düsseldorf in May 1970. Stangl claimed that although he had been commandant of Treblinka, "I have had nothing to do with the killing of the Jews in the camp",[2162] an assertion the court found no difficulty in dismissing. Found guilty of complicity in the murder of at least 400,000 people at Treblinka,[2163] in December 1970 he was sentenced to life imprisonment. His part in the murder of more than 100,000 people at Sobibor, or of the unknown thousands in whose deaths he was implicated through his involvement in the "euthanasia" programme, did not form part of the judgement. However, the court had little doubt concerning Stangl's character, concluding that "...coupled with ambition, this attitude of expediency and utilitarianism runs like a continuous thread through his biography...The defendant's thinking and acting resulted from his striving to advancement at any price."[2164]

There is no evidence that Stangl was a vicious killer like Wirth, or a psychopathic murderer such as Kurt Franz (q.v.). Indeed it is questionable whether Stangl ever personally killed anyone. He didn't need to. There was no shortage of henchmen to do that for him. Nor was he a fanatical anti-Semite, at least judged by *Gestapo* standards. Rather, he was an efficient organizer and administrator, a man prepared to further his career by following orders and eliminating anybody deemed by the state he served to be unworthy of life. It was said that when he appeared, everyone worked faster, his own men included. Yet when asked whether he hadn't felt that his victims were human, he was only capable of replying: "Cargo. They were cargo."

Stangl died in prison of a heart attack on 28 June 1971.[2165]

Franz Suchomel

Source: Wikimedia Commons / Public Domain

Franz Suchomel (1907–1979)[2166] was born in Krumau, as were the brothers Franz (q.v.) and Josef Wolf. At the time of Suchomel's birth Krumau was a part of the Habsburg Empire, but after 1918 it became a town within the Sudeten region of Czechoslovakia. Like his father, Suchomel was a tailor, taking over his father's business in 1936 and opening a second shop two years later. He served in the Czech army in the late 1920s and again in autumn 1938,[2167] joining the neo-Nazi Sudeten German Party in the same year, although he never became a member of the NSDAP. However, he was a member of the National Socialist Motorist Corps. He was conscripted into the *Wehrmacht* in March 1940, but suspended from duty in November

2162 de Mildt, *In the Name of the People*, p. 298.

2163 At Stangl's trial, the Berlin historian, Dr Wolfgang Scheffler, estimated the total figure for the number of Treblinka dead at 900,000. (Donat, *The Death Camp Treblinka*, p. 14). Stangl, of course, did not arrive at Treblinka until the camp had been operational for five weeks, during which time approximately 312,500 Jews had already been murdered. (Arad, *Belzec, Sobibor, Treblinka*, p. 87).

2164 de Mildt, *In the Name of the People*, p. 300.

2165 Gutman, *Encyclopedia of the Holocaust*, p. 1408. (Sereny, *Into That Darkness*, passim).

2166 Klee, *Das Personenlexikon zum Dritten Reich*, p. 615.

2167 Friedlander, *The Origins of Nazi Genocide*, p. 240.

of that year to return to the manufacture of uniforms in Krumau, where he remained until ordered to report to the KdF in Berlin in early March 1941.[2168]

The extent to which Suchomel was himself involved in his recruitment to T4, if at all, is unclear,[2169] but it seems a remarkable coincidence that so many individuals from a small town in Czechoslovakia (Suchomel, Franz and Josef Wolf, and Franz Wagner [not to be confused with Gustav Franz wagner (q.v).]) found their way into that organization.[2170] Moreover, all of them were associated with photography, for Suchomel too was assigned to the T4 photography section, before being ordered to Hadamar in March 1942 to perform photographic duties there until July 1942, at which time he returned to Berlin.[2171] In view of his background, how Suchomel came to be employed as a photographer at all is mystifying. Even more puzzling is why he was considered suitable material for guard duties at an extermination camp, for after having been given the rank of SS-*Scharführer* like other *Aktion Reinhard* camp functionaries, in August 1942 Suchomel was posted to Treblinka, where his first duty was at the "station", dealing with the incoming transports of deportees, and making free use of his whip and pistol while doing so; later he supervised the women's undressing barrack, and led victims to the "tube", encouraging them to hurry to the showers before the water became cold. Sometimes he even handed out towels to accentuate the lie that the gas chambers were bathhouses. Subsequently he was in charge of the *Goldjuden* ("gold Jews", responsible for handling the valuables of the victims) and the tailor shop. Considered, by Treblinka standards, to have been relatively decent,[2172] in late October 1943 he was ordered to Sobibor. Shortly afterwards he joined his T4 colleagues in Trieste. Following the German surrender and a brief period of confinement as a prisoner-of-war, he was released in August 1945. In 1949 he resumed his profession of master tailor in Altötting, where he also again became the enthusiastic amateur musician he had been in pre-war days.[2173]

Suchomel was arrested in July 1963, one of ten defendants to stand trial in the first Treblinka process. In September 1965 he was convicted of participation in the murder of at least 300,000 people, for which he received a sentence of six years' imprisonment.[2174] Released in 1969, several years later he was secretly filmed whilst being interviewed by Claude Lanzmann for inclusion in Lanzmann's film *Shoah*.[2175] Suchomel was also interviewed by Gitta Sereny in 1971, who clearly had little time for him; she considered that Treblinka had "manifestly been the high point of his life."[2176]

2168 de Mildt, *In the Name of the People*, p. 216.

2169 Suchomel claimed not to know why he had been ordered to T4, nor the reason he had been sent to Hadamar. (Sereny, *Into That Darkness*, p. 56).

2170 Dieter Allers, who was in a better position to know than most, had no doubt about how Suchomel became a T4 operative. "...He had a pal from his home town who was already in there as a photographer and they fixed it between them." (Ibid., p. 80).

2171 de Mildt, *In the Name of the People*, p. 225.

2172 Ibid., p. 266.

2173 Klee, *Das Personenlexikon zum Dritten Reich*, p. 615.

2174 de Mildt, *In the Name of the People*, p. 266.

2175 Lanzmann, *Shoah*, p. 43 ff.

2176 Gitta Sereny, *The German Trauma: Experiences and Reflections 1938–2000* (London: Allen Lane, The Penguin Press, 2000), p. 310.

Friedrich (Fritz) Tauscher (1903–1965) was a lieutenant in the uniformed police assigned to Sonnenstein in 1941–1942, who also served at Brandenburg and Hartheim. He was a member of both the NSDAP and the SS. In late 1942 or early 1943 he was transferred first to Trawniki, then to Belzec.[2177] Other sources suggest that he arrived at Belzec as early as October 1942. However, what is certain is that he was made responsible for the final liquidation of the camp by Gottlieb Hering (q.v.), and became Belzec's *de facto* final commandant.[2178] Tauscher was in charge of the labour camp at Budzyn for a short time in late summer 1943 before becoming the last commandant of the Dorohucza labour camp.[2179] In 1944 he joined his comrades in Italy as a member of *Aktion R.* Tauscher committed suicide in prison in 1965.[2180]

Heinrich Unverhau (1911–?), born in Vienenburg, Lower Saxony, became a plumber's apprentice on leaving primary school, but after suffering an accident which resulted in the loss of his right eye, abandoned that career in 1925. For the next four years he studied at the Königslutter music school, becoming an accomplished musician, and joining the town bands of first Königslutter and then Neuruppin. In October 1932 he became a member of the *Stahlhelm* (a paramilitary nationalist organisation), where he was employed as a musician. When the *Stahlhelm* was incorporated into the SA in autumn 1933, Unverhau was enrolled into the latter organisation; he became a member of the NSDAP in May 1937.

In March 1934, the leader of the Neuruppin mental hospital band offered Unverhau a position as a student nurse at the institution, which he was pleased to accept. He continued to work at Neuruppin until December 1939, when together with other Neuruppin nurses he was ordered to report to T4 in Berlin. There he was instructed about the part he was to play in the "euthanasia" project, sworn to secrecy, and sent to Grafeneck. When Grafeneck closed in December 1940 he was transferred to Hadamar to perform similar duties—escorting victims to the gas chambers, injecting them with sedatives, ventilating the gas chambers, and dealing with the disposal of the bodies and the victim's property.

He too became part of the mysterious *Organisation Todt* mission to Russia. When that ended he returned to Hadamar in March 1942; a few months later he was ordered to report to Lublin, where Christian Wirth (q.v.) assigned him and the colleagues accompanying him to the Belzec extermination camp.[2181] There he found that "exactly the same enterprise had been set up as at Hadamar." Whilst freely admitting as early as 1948 in his testimony to the Tübingen police that "Belzec was an extermination camp for Jews in the district of Tomaszow in Poland", Unverhau claimed to have had nothing to do with the gassings, being principally concerned with the supervision of the undressing barracks and the sorting of clothing.[2182]

2177 Friedlander, *The Origins of Nazi Genocide*, p. 99. p. 204, pp. 206–207.
2178 O'Neil, *Belzec*, p. 183.
2179 Schelvis, *Sobibor*, p. 263.
2180 Ibid., p. 131, note 14. The date of Tauscher's suicide is mis-stated as 1963.
2181 de Mildt, *In the Name of the People*, p. 207.
2182 Ibid., p. 280.

After spending some months in various hospitals suffering from a variety of illnesses, including typhus, Unverhau was sent to Sobibor in June 1943, where he supervised workers at the undressing area in Camp II, the sorting barracks, and the *Waldkommando* (the wood-gathering detail). He remained at Sobibor until he was briefly transferred to Treblinka in September 1943 to assist with the liquidation of that camp. He returned to Sobibor in November for a short time[2183] before at Christmas 1943 he became another of the T4 personnel dispatched to Trieste; he returned to the T4 head office in Berlin in March 1944. One month later he was assigned to an army unit, but was soon readmitted to hospital. He was briefly a prisoner-of-war until his release in September 1945 to resume his career as a musician. By 1952 he was again in gainful employment as a nurse at the Königslutter hospital.

In March 1948, Unverhau was arrested in connection with his activities at Grafeneck, but eventually acquitted on the grounds that he and his co-defendants had reason to believe that threats of execution or incarceration in a concentration camp would be the consequence of non-compliance with the killing programme—although there was no evidence that such threats were ever carried out.[2184] Fifteen years later, Unverhau and seven others were indicted for complicity in the crimes committed at Belzec—in Unverhau's case, the alleged murder of some 360,000 victims. But in January 1964, the Munich court decided not to continue with proceedings against him and six of the other defendants, and Unverhau was again released.[2185]

A few months later Unverhau was on trial for a third time, on this occasion for his activities at Sobibor, and for a third time he was acquitted.[2186] Thus despite involvement at a variety of killing institutions over a period of years (which in fairness he had never attempted to deny), Unverhau received no punishment whatsoever. It should be added that he was considered by former Jewish prisoners to be among the less brutal of the guards at Sobibor, behaving with relative decency, a factor which undoubtedly weighed heavily in his favour with the court.[2187]

Gustav Franz Wagner aka "Günther Mendel" (1911–1980)[2188] was born in Vienna and joined the then illegal Austrian Nazi party in 1931. To avoid arrest for his Nazi-associated activities, he fled to Germany in 1934, where he joined first the SA and then the SS. In 1940 he was ordered to Hartheim, where he worked as a *Brenner*. This proved excellent training for his next posting to Sobibor, where he arrived in March 1942, when the camp was still under construction. At Sobibor he was appointed "*der Spiess*", a sort of second-in-command, in charge of day-to-day camp activities, including those conducted in Camp III, where the gas chambers were situated.[2189]

2183 Schelvis, *Sobibor*, p. 263. Robert Jührs testified that Unverhau was an active participant in the execution of the last Jewish workers at Sobibor at the time of the camp's liquidation. (Schelvis, *Sobibor*, p. 191).

2184 de Mildt, *In the Name of the People*, p. 208.

2185 Ibid., pp. 276–277.

2186 Ibid., p. 279.

2187 Ibid., p. 293.

2188 Schelvis, *Sobibor*, p. 264

2189 Ibid., p. 61

Together with Karl Frenzel (q.v.) and Hubert Gomerski (q.v.), Wagner was considered one of the most cruel and dangerous of the SS-men in the camp. Many thought him the worst of all, "not only the most brutal, but also the shrewdest, and most intelligent of all the SS-men at Sobibor."[2190] Nicknamed *Welfel* (Wolf) by the inmates, there was no limit to his depravity. He was continuously checking the prisoners as they worked, administering beatings and shooting people out of hand. Years later, Regina Zielinski (née Feldman) described to her son a whipping she suffered at Wagner's hands after she was discovered leaning her head against a boiler in the camp laundry in an attempt to ease the pain of an earache:

> [Wagner] administered a severe whipping with his metre-long leather-covered steel whip. Each time the whip snapped around to the front of my mother's body, it left welts and blisters. When he stopped, he told my mother to go back to the laundry and return to work. She was suffering excruciating pain, but because she knew what happened to people who appeared or reported sick, she knew that she must continue working as if nothing were wrong.[2191]

As a consequence of Wagner's vicious beating, Zielinski was forced to have a kidney surgically removed after the war.[2192]

Wagner was another fortunate enough to have been on leave at the time of the uprising in Sobibor. There is no doubt that he would otherwise have been the major quarry. On the other hand, it has been suggested that the uprising might not have succeeded at all had he been in the camp at that time. Wagner was very prescient[2193]

Following the liquidation of Sobibor, like other *Aktion Reinhard* personnel, Wagner was ordered to Italy. In somewhat similar fashion to Franz Stangl (q.v.), after the war he fled via Syria and Lebanon to Brazil, where he lived under the name of Günther Mendel. In May 1978 he was traced by Simon Wiesenthal and his identity was confirmed by Sobibor survivor Stanislav Szmajner. Extradition requests from the governments of Austria, the Federal Republic, Poland, and Israel were rejected by Brazil on appeal. Interviewed by the BBC in 1979, Wagner showed no remorse for the innumerable crimes committed by him in Sobibor, stating: "I had no feelings there ... It was just another job for me. After hours, we have never talked about our work, but we drank and played cards."

On 15 October 1980, Wagner allegedly committed suicide. Szmajner implied that a degree of assistance may have been rendered to help Wagner achieve this goal.[2194]

2190 Ticho, *My Legacy*, p. 110.

2191 Andrew Zielinski, *Conversations With Regina* (Wlodawa: Muzeum Pojezierza Leczynsko-Wlodawskiego, 2008), pp. 89–90

2192 Ibid., p. 173.

2193 The uprising was organised by a newly arrived group of Jewish Soviet soldiers led by Aleksander Aronowich Pechersky. Wagner was overheard telling Frenzel, "*Leg die Russen um*" ("Do the Russians in"). It is possible that he intuitively suspected something was afoot. (Ticho, *My Legacy*, p. 110).

2194 Schelvis, *Sobibor*, p. 264.

Christian Wirth

Source: Wikimedia Commons / CC-BY-SA 3.0 https://creativecommons.org/licenses/by-sa/3.0/deed.en

Christian Wirth (1885–1944)[2195] was born in Oberbalzheim in Württemberg, and trained as a carpenter before undertaking a four year spell in the Imperial army.[2196] He joined the police in 1910, and following the outbreak of hostilities in 1914, served on the Western Front with distinction, being decorated with the rarely awarded gold Military Cross.[2197] After the cessation of hostilities he first worked as a builder before re-entering the police force, where he was noted for his extreme efficiency. He first joined the Nazi party sometime before 1923, but resigned after the Munich beer hall *Putsch*. He rejoined the party in 1931, as well as the SA in 1933 and the SS in 1939. By that time he was a highly regarded *Kriminalkommissar* in the Stuttgart criminal police. A colleague recalled:

> He was a criminologist from head to toe, energetic and methodical, unflinching in his fight against the criminal underworld and was the most hated man in criminal circles...Ruthless to himself, he sacrificed innumerable nights and Sundays to lead the fight against the criminal underworld. Through this he had the trust of the Stuttgart *Kripo*, not only in Württemberg but also further afield...he was always in demand, which earned him a considerable reputation.[2198]

As "the kind of exceedingly efficient officer who is seldom found", Wirth was the ideal man to oversee the first phase of the "euthanasia" programme at the nominated killing centres, and in 1939 he was head-hunted by T4, who clearly knew the kind of man they needed, and were getting.[2199] In October of that year he arrived at Grafeneck; shortly afterwards he was transferred to Brandenburg, where he was appointed administrative director.[2200] Having been present at the first experimental gassing at Brandenburg in late December 1939 or early 1940, Wirth organized the administration not only there, but also at Grafeneck and Hadamar; however, it was at Hartheim that he expanded and refined the killing process in readiness for greater tasks to come.[2201] He made a notorious speech at Hartheim to the T4 recruits:

> Comrades, I've called you here together today in order to inform you about the present position in the castle and what is going to happen from now on. I have been assigned the task of running the castle from now on by the *Reich* Chancellery.[2202] As the boss I am in charge of everything. We must build a crematorium here, in order to burn mental patients from Austria. Five doctors have been chosen who will examine the patients to establish what can or cannot be saved. What can't be saved goes into the crematorium and will be burned. Mental patients are a burden upon Germany and we only want healthy people... Certain

2195 Klee, *Das Personenlexikon zum Dritten Reich*, p. 680.

2196 O'Neil, *Belzec*, p. 30.

2197 Sereny, *Into That Darkness*, p. 81.

2198 O'Neil, *Belzec*, p. 33.

2199 Ibid., p. 33.

2200 Arad, *Belzec, Sobibor, Treblinka*, p. 182.

2201 Friedlander, *The Origins of Nazi Genocide*, p. 203.

2202 Either Wirth made a slip of the tongue in referring to the *Reich* rather than the *Führer* Chancellery, or he was attempting to inflate his importance. A third alternative, of course, is that the witness misheard or misunderstood this phrase.

men will be chosen to work in the crematorium. Above all else, the motto is silence or the death penalty. Whosoever fails to observe this silence will end up in a concentration camp or be shot.[2203]

Having served as a kind of trouble shooter at the killing centres, turning up whenever there was a problem to be resolved, in 1940 Wirth was appointed inspector of all T4 establishments. Franz Stangl (q.v.) described him as

...a gross and florid man...He stayed at Hartheim for several days...and came back often. Whenever he was there, he addressed us daily at lunch. And here it was again, this awful verbal crudity: when he spoke about the necessity for this euthanasia operation, he wasn't speaking in humane or scientific terms...He laughed. He spoke of "doing away with useless mouths" and said that "sentimental slobber" about such people made him "puke".[2204]

Rabidly anti-Semitic and completely without compassion or morality, Wirth was a natural candidate for *Aktion Reinhard*. He was probably involved in "euthanasia" matters in western Poland between September and December 1941, although the precise nature of his activities during this period is obscure; there are claims that he may have been present at some stage at the Chelmno extermination camp, which commenced operations on 8 December 1941. What is certain is that, as already described, following some experimentation with killing methods in the Lublin region, Wirth was responsible for organizing the extermination camp at Belzec, where he officially arrived on 22 December 1941 to be appointed the camp's first commandant. He then supervised the establishment of Sobibor, Treblinka and other marginally less lethal camps, before being promoted to the position of Inspector of all of the *Aktion Reinhard* killing centres on 1 August 1942.[2205] It was almost certainly Wirth who the Treblinka survivor Abraham Krzepicki was describing in this passage:

He was a captain, 50 years old, stout and of medium height. He had puffed-up red cheeks, and a black moustache; he was the very image of the active soldier. He was always full of anger; it is hard to tell whether it was only towards Jews. He used to carry a rubber truncheon in his hand, and he never failed to vent his anger when he passed some Jews. "Idiots!" was his term of insult, and he used to utter it in a squeaking voice...He let out the squeak at the same moment that he started hitting his victim with all the strength and rage of a well-fed man of action...After each blow, he would almost bend down to the ground, like a man cutting grain.[2206]

Krzepicki goes on to relate how after himself suffering a beating from this sadistic officer, "the killer with the red cheeks and the black moustache" began senselessly hitting out with his truncheon at women and children.[2207]

Another Treblinka survivor, Jankiel Wiernik, relating his encounter with a *Hauptmann* (Captain) whilst removing corpses from the gas chambers, was also probably describing Wirth:

We worked under [his] supervision, a medium-sized bespectacled man whose name I do not know. He whipped us and shouted at us. He beat me too, without a stop. When I gave him a questioning look, he

2203 Burleigh, *Death and Deliverance*, p. 124.
2204 Sereny, *Into That Darkness*, p. 54.
2205 de Mildt, *In the Name of the People*, p. 241.
2206 Donat, *The Death Camp Treblinka*, p. 109.
2207 Ibid. p. 112–113.

stopped beating me for a moment and said, "If you weren't the carpenter around here, you would be killed."[2208]

With the virtual completion of *Aktion Reinhard*, together with the other ex-T4 operatives Wirth was sent to Trieste in September 1943, and was shot near Kozina in Istria in May 1944, allegedly by partisans. It is possible that he was in fact killed by his own men; given Wirth's character and the manner in which he treated his subordinates, this seems quite conceivable. Stangl certainly thought so.[2209]

Responsible for the industrialization of mass murder, in a world inhabited by brutal and merciless killers Wirth could lay claim to being the most repugnant of them all. He was a crude, coarse bully, prepared to murder any perceived "enemies of the state" without compunction—indeed with insatiable ferocity. His own subordinates called him "Christian the Terrible".[2210] Having instituted a smoothly functioning series of killing machines at the "euthanasia" centres, it was Wirth who created the system of terror and murder that reigned in the *Aktion Reinhard* camps, thereby empowering his staff with the ability to literally choose between life or death for the inmates. Stangl stated: "It was not possible to save even a child in Treblinka. Wirth gave very specific instructions in this respect";[2211] and "Wirth constantly stressed that those who do no work had to be taken away. Each leader of a working group and each camp commander could send to the Lazarett every prisoner who did not work or behave satisfactorily".[2212] In the words of Franz Suchomel (q.v.): "From my activity in the camps of Treblinka and Sobibor, I remember that Wirth in brutality, meanness, and ruthlessness could not be surpassed...The brutality of Wirth was so great that I personally see it as a perversity".[2213]

Jacob Wöger (1897-?) first worked in local government before serving in the First World War. He joined the Württemberg police force in 1922, and in 1933 became a member of both the NSDAP and the SS. By 1936 he had risen in rank to *Kriminalsekretär* in the Stuttgart *Kripo*; in 1938 he transferred to the *Sipo*, and a year later

---

2208 Ibid., p. 159. The real names of very few of the death camp personnel were known to the prisoners. Some guards were given nick-names (for example, Fritz Küttner was called "Kiewe", Otto Stadie "Fesele" ["barrel"]); others were known only by their first names, actual or assumed, such as Josef Hirtreiter. In his groundbreaking essay, "The Hell of Treblinka", initially published in November 1944 and largely based upon the testimony of survivors, Vasily Grossman was able to identify the monstrous Hirtreiter only as "Sepp". [Vasily Grossman, *The Road*, (London: MacLehose Press, 2010) p. 149]. In some cases surnames were misheard by prisoners; Abraham Krzepicki refers to Max Bielas as "Bieler" (Donat, *The Death Camp Treblinka*, p. 131); Moshe Bahir names Karl Frenzel as "Frantzl" (Novitch, *Sobibor—Martyrdom and Revolt*, p. 153). Richard Glazar wrote: "Almost everyone is given a nickname...We only know their real names from what we have heard. We don't know how they are spelled" (Glazar, *Trap With A Green Fence*, p. 46). Such difficulty in matters of identification could later have potentially serious consequences for defendants, as well as causing a major problem regarding intended or actual prosecutions. See, for example, the case of Ivan Demjanjuk.

2209 Sereny, *Into That Darkness*, p. 262.

2210 Arad, *Belzec, Sobibor, Treblinka*, p. 183.

2211 Donat, *The Death Camp Treblinka*, p. 273.

2212 Arad, *Belzec, Sobibor, Treblinka*, p. 184. The *Lazarett* was a killing site within the camp disguised as a "hospital", where the sick, the incapacitated, the infirm or elderly arrivals, or anybody else an SS-man decided to kill, were shot. The bodies of those dead on arrival were also disposed of here.

2213 Ibid., p. 183.

was recruited by Gerhard Böhne (q.v.) for T4.[2214] Wöger was sent to Grafeneck to act as supervisor. Both Christian Wirth (q.v.) and Gottlieb Hering (q.v.) were members of the Stuttgart police force, as was Wöger's deputy and eventual successor at Grafeneck, Hermann Holzschuh (q.v.), which suggests a high probability of collusion.[2215]

When his tour of duty at Grafeneck was completed, Wöger returned to the Stuttgart police. He was injured in a traffic accident in 1940, which may explain why he was not sent eastwards, like his aforementioned Stuttgart police colleagues, but so far as is known, there was no further involvement on his behalf with T4 in any of its manifestations.[2216]

Franz Wolf
Source: Wikimedia Commons / Public Domain

Franz Wolf (1907–?) was originally a Czech citizen living in the Sudeten town of Krumau who first found employment as a forester. Subsequently, together with his two brothers, he worked in his father's photography shop, until in 1938 his father died and the brothers inherited the business. Wolf had served in the Czech army and was briefly recalled to the colours at the time of the Sudeten crisis. He had joined the pro-Nazi "Sudeten German Party" in 1936; it is unclear whether he joined the NSDAP after Germany annexed the *Sudetenland*, but whether a party member or not, Wolf was conscripted into the *Wehrmacht* in August 1939 and served in Poland and France.[2217]

Another resident of Krumau, Franz Wagner (not to be confused with Gustav Franz Wagner [q.v.]), had been apprenticed to Wolf's father, and worked as head of the T4 photography section in Berlin. It is unclear whether Wolf asked Wagner to recommend him for the position, or Wagner simply acted of his own accord, but in autumn 1940 Wolf's military service was suspended by order of the KdF, and in January 1941 he was ordered to report to Hadamar. There he was assigned the job of photographing naked victims prior to their gassing. In autumn 1941 he worked in the Berlin headquarters of T4, enlarging the photographs he had taken for further study, before moving on to the Heidelberg psychiatric clinic to perform a similar task. He remained in Heidelberg until March 1943,[2218] at which time, together with his brother Josef, who had also been recommended to T4 by Franz Wagner, he was transferred to Sobibor.

At Sobibor, where amongst other supervisory activities he was in charge of a group of 60 women who sorted the luggage of the victims, Wolf behaved in a similar fashion to most of his SS comrades. If considered less brutal than some of them,[2219] nevertheless he was feared by the inmates as a man more than ready and willing to flog the prison-

2214 Friedlander, *Nazi Genocide*, p. 206.

2215 O'Neil, *Belzec*, p. 209.

2216 Friedlander, *The Origins of Nazi Genocide*, p. 208.

2217 de Mildt, *In the Name of the People*, pp. 215–216.

2218 Ibid., p. 224.

2219 The degree of brutality was, of course, relative. As one witness commented: "Anyone who did not continuously shoot or whip [the prisoners] belonged to the 'Good'" (de Mildt, *In the Name of the People*, p. 399, note 106).

ers he commanded, particularly the women. Apart from his physical excesses, he appeared to take particular pleasure in degrading and humiliating his workers, suggesting a probable sexual motive for much of his behaviour.

Josef Wolf was killed during the course of the Sobibor uprising in October 1943, but after serving in Italy together with many other *Aktion Reinhard* personnel, Franz Wolf survived to stand trial in Hagen in 1965, accused of complicity in the murder of at least 39,000 individuals. He claimed to be no anti-Semite and to have been horrified by what he found at Sobibor, but had acted as he did out of fear of the punishment that would surely follow if he had disobeyed orders. Moreover, he had repeatedly sought a transfer away from the death camp. The court was unimpressed by Wolf's evidence, but, not unusually, found extenuating circumstances, ruling that his actions did not compellingly show that he acted out of the racist motives of his superiors:

> He could have acted this way because he believed such behaviour belonged to a proper implementation of orders. If so, it cannot be excluded that he did not act out of his own will to exterminate, but out of the will obediently and readily to be of service to his superiors.

The court thus deemed a sentence of 8 years' imprisonment to be appropriate punishment for Wolf's crimes.[2220]

Ernst Zierke
Source: Ghetto Fighters House Museum, Israel, Photo Archive

Ernst Zierke (1905–1972) was born in Krampe, near Köslin, Pomerania (now Koszalin, Poland). After being expelled from primary school, he found employment as a forest worker before becoming apprenticed to a blacksmith. He worked for a number of agricultural estates prior to becoming unemployed in the late 1920s. He joined both the NSDAP and the SA in August 1930, and was another to eventually enrol as a student nurse at the Neuruppin mental home in 1933.

In late 1939 Zierke was conscripted by T4 and sent to Grafeneck. There he accompanied the transport of patients from other institutions to the killing centre and also worked in the photography department. When Grafeneck was closed, he was transferred to Hadamar, where for most of 1941 he performed similar duties. At the end of 1941 he was in Eichberg for a short time before, together with many other T4 members, he was sent to Russia as part of the suspicious *Organisation Todt* mission. On his return he worked for a short time in Eichberg and Hadamar again, before joining the staff at Belzec in June 1942. Following the closure of that extermination camp, he was transferred to the Dorohucza labour camp in March 1943.[2221] Shortly after the killing of the inmates of Dorohucza as part of *Aktion Erntefest* in November 1943, Zierke was sent to Sobibor to assist with the liquidation of that camp. At the end of December 1943 he joined fellow T4 operatives in northern Italy as part of *Aktion R*.

Arrested by the Americans in May 1945, Zierke was held in detention until 1946. Thereafter he first stood trial in 1948 in Frankfurt. The full extent of his involvement in

2220 de Mildt, *In the Name of the People*, pp. 283–285.

2221 Schelvis, *Sobibor*, pp. 265–266.

T4 was not known at that time, and he was merely accused of photographing Hadamar victims and assisting in their unloading and undressing. Because of his allegedly limited participation, it was impossible to prove that he was aware that he was assisting in a killing process, and he was therefore acquitted "as his activities either lay before the medical examination or consisted of taking pictures."[2222] Thus the court decided, Zierke might have believed "that only the doctor made the final decision on death and that therefore he [the doctor] carried the sole responsibility."[2223] It was a generous interpretation (generous to Zierke, that is), but it paled into insignificance when compared with what was to come.

Fifteen years were to pass before Zierke, now employed in a saw-mill,[2224] was again indicted for his crimes, this time by the Munich prosecutor's office. He was accused of complicity in 360,000 cases of murder at the Belzec death camp. On 30 January 1964 the case against Zierke and six others accused with him was dropped when the court decided that they had not engaged "in particular activities in the sense of the National Socialist regime of terror."[2225] Even before the Munich court had reached its decision, the Dortmund prosecutor filed charges with the Hagen court against 12 former members of the Sobibor staff, including Zierke. Their trial began on 6 September 1965 and lasted 15 months.[2226] At its end, despite having admitted to his participation in the execution of Jewish prisoners after the dismantling of Sobibor had been completed, the court found that Zierke had been compelled to act as he did out of fear for his own life and acquitted him.[2227] So, for his service in three "euthanasia" centres, two *Vernichtungslager*, an almost equally deadly labour camp, and other probable crimes uninvestigated to date, which taken together resulted in the death of tens of thousands of innocent individuals, Zierke received precisely no judicial punishment whatsoever.

Of the 182 individuals listed above of whom particulars are known and who were of mature age at some time during the Nazi era, a purely arbitrary analysis of their dates of birth reveals the following approximate percentages:

Pre 1880 – 6%
1881–1890 – 15%
1891–1900 – 20%
1901–1910 – 44%
Post 1910 – 15%

In 1939, Hitler was fifty years of age. Using the classifications listed, the most junior of the theoreticians were born 1887–1891, hence making them contemporaries of his. The majority of the technocrats were born 1901–1910. Physicians, nurses, and scien-

2222 de Mildt, *In the Name of the People*, pp. 200–201.
2223 Ibid., p. 320.
2224 Ibid., p. 294.
2225 Ibid., pp. 276–277.
2226 Ibid., p. 279.
2227 Ibid., p. 294.

tists were equally divided between those born 1881–1900 and 1901–1910, with a minority born pre-1881, and further minority born post-1910. The great majority of the operatives, the hands-on killers of T4 and *Aktion Reinhard*, were born post-1901. As might be expected, practical mass murder was predominately a young man's occupation.

SOURCES AND BIBLIOGRAPHY

Adams, Mark B (ed), *The Wellborn Science: Eugenics in Germany, France, Brazil, and Russia* (New York: Oxford University Press, 1990).

Adelson, Alan, and Lapides, Robert, *Lodz Ghetto: Inside a Community Under Siege* (New York: Viking, 1989).

Alexander, Leo, *Medical Science Under Dictatorship* (The New England Journal of Medicine 241, 1949) 39–47.

Allen, William Sheridan (ed), *The Infancy of Nazism: The Memoirs of Ex-Gauleiter Albert Krebs 1923–1933* (New York: New Viewpoints, 1976).

Aly, Götz, Chroust Peter, Pross Christian, *Cleansing the Fatherland: Nazi Medicine and Racial Hygiene* (Baltimore: The Johns Hopkins University Press, Baltimore, 1994).

___, and Heim, Susanne, *Architects of Annihilation: Auschwitz and the Logic of Destruction* (London: Phoenix, 2003).

Annas, George J, and Grodin, Michael A, *The Nazi Doctors and the Nuremberg Code: Human Rights in Human Experimentation* (New York: Oxford University Press, 1992).

Arad, Yitzhak, *Belzec, Sobibor, Treblinka: The Operation Reinhard Death Camps* (Bloomington: Indiana University Press, 1987).

___, Gutman, Israel, and Margaliot, Abraham (eds.), *Documents On The Holocaust* (Lincoln: University of Nebraska Press, 1999).

Arendt, Hannah, *Eichmann in Jerusalem: A Report on the Banality of Evil* (New York: Penguin Books, 1977).

Aroneanu, Eugène, *Inside the Concentration Camps: Eyewitness Accounts of Life in Hitler's Death Camps* (Westport: Praeger, 1996).

Aschheim, Steven E, *The Nietzsche Legacy in Germany, 1890–1990* (Berkeley: University of California Press, 1992).

Bacon, Josephine, *The Illustrated Atlas of Jewish Civilization: 4,000 Years of Jewish History* (Royston: Eagle Editions, 1998).

Bartov, Omar (ed), *The Holocaust: Origins, Implementation, Aftermath* (London: Routledge, 2000).

Benedict, Susan, and Kuhla, Jochen, *Nurses' Participation in the Euthanasia Programs of Nazi Germany* (Western Journal of Nursing Research, Vol. 21(2), 1999) 246–263.

___, and Chelouche, Tessa, *Meseritz-Obrawalde: a 'wild euthanasia' hospital of Nazi Germany* (History of Psychiatry, Vol. 19 (1), 2008) 68–76.

___, Caplan, Arthur, Page, Traute Lafrenz, *Duty and 'Euthanasia': the Nurses of Meseritz-Obrawalde* (Nursing Ethics, Vol. 14(6), 2007) 781–794.

Berenbaum, Michael, *Witness to the Holocaust* (New York: HarperCollins, 1997).

___, and Peck, Abraham J (eds.), *The Holocaust and History: The Known, the Unknown, the Disputed, and the Reexamined* (Bloomington: Indiana University Press, 2002).

Bergen, Doris L, *War & Genocide: A Concise History of the Holocaust* (Lanham: Rowman & Littlefield, 2003).

Biesold, Horst, *Crying Hands: Eugenics and Deaf People in Nazi Germany* (Washington: Gallaudet University Press, 1999).

Blatt, Thomas Toivi, *From the Ashes of Sobibor: A Story of Survival* (Evanston: Northwestern University Press, 1997).

Bloxham, Donald, *Genocide on Trial: War Crimes Trials and the Formation of Holocaust History and Memory* (Oxford: Oxford University Press, 2003).

___,*The Final Solution: A Genocide* (Oxford: Oxford University Press, 2009).

Bolaffi, Guido, Bracalenti, Raffaele, Braham, Peter, Gindro, Sandro (eds.), *Dictionary of Race, Ethnicity & Culture* (London: Sage Publications, 2003).

Boobbyer, Philip, *Moral Judgements and Moral Realism in History* (Totalitarian Movements and Political Religions, Vol. 3, No. 2, 2002) 83–112.

Brantlinger, Patrick, *Dark Vanishings: Discourse on the Extinction of Primitive Races, 1800–1930* (Ithaca: Cornell University Press, 2003).

Bridgham, Fred (ed), *The First World War as a Clash of Cultures*, (Rochester: Camden House, 2006).

Brissaud, André, *The Nazi Secret Service* (London: Corgi Books, 1975).

Broberg, Gunnar, and Roll-Hansen, Nils, *Eugenics and the Welfare State: Sterilization Policy in Denmark, Sweden, Norway, and Finland* (East Lansing: Michigan State University Press, 2005).

Browning, Christopher R, *Ordinary Men: Reserve Police Battalion 101 and the Final Solution in Poland* (New York: HarperCollins, 1993).

___, *Fateful Months: Essays on the Emergence of the Final Solution* (New York: Holmes & Meier, 1991).

___, *The Path to Genocide: Essays on Launching the Final Solution* (Cambridge: Cambridge University Press, 1997).

___, *Nazi Policy, Jewish Workers, German Killers* (Cambridge: Cambridge University Press, 2000).

___, *The Origins of the Final Solution: The Evolution of Nazi Jewish Policy, September 1939–March 1942* (London: William Heinemann, 2004).

Bryant, Michael S, *Confronting the "Good Death": Nazi Euthanasia on Trial 1945–1953* (Boulder: University Press of Colorado, 2005).

Bullock, Alan, *Hitler and Stalin: Parallel Lives* (London: Harper Collins, 1991).

Burleigh, Michael, *The Third Reich: A New History* (London: Pan Books, 2001).

___, *Death and Deliverance: 'Euthanasia' in Germany c. 1900–1945* (London: Pan Books, 2002).

___, *Ethics and extermination: Reflections on Nazi genocide*, (Cambridge: Cambridge University Press, 1997).

___, *Eugenic Utopias and the Genetic Present* (Totalitarian Movements and Political Religions: Vol. 1, No. 1, 2000) 56–67.

___, *Political Religion and Social Evil.* (Totalitarian Movements and Political Religions: Vol. 3, No. 2, 2002) 1–60.

___ (ed), *Confronting the Past: New Debates on Modern German History* (London: Collins & Brown Limited, 1996).

___, and Wipperman, Wolfgang, *The Racial State: Germany 1933-1945* (Cambridge: Cambridge University Press, 1991).

Caplan, Arthur L (ed), *When Medicine Went Mad: Bioethics and the Holocaust* (Totowa: Humana Press, 1992).

Cesarani, David (ed), *The Final Solution: Origins and Implementation* (London: Routledge, 1996).

___, (ed), *The Holocaust: Critical Concepts in Historical Studies*, Volume 2 (London: Routledge, 2004).

___, *Eichmann: His Life and Crimes* (London: William Heinemann, 2004).

Chamberlain, Houston Stewart, *Foundations of the Nineteenth Century, Volume I* (Port Chester: Adegi Graphics, 1977).

Conot, Robert E, *Justice at Nuremberg* (New York: Harper & Row,1983).

Crew, David F (ed), *Nazism and German Society, 1933–1945* (London: Routledge, 1994).

Dawidowicz, Lucy S, *The War Against the Jews 1933–1945* (New York: Bantam Books, 1979).

De Brouwer, Anne-Marie, *Supranational Criminal Prosecution of Sexual Violence: the ICC and the practice of the ICTY and the ICTR* (Mortsel: Intersentia 2005).

de Mildt, Dick, *In the Name of the People: Perpetrators of Genocide in the Reflection of Their Post-War Prosecution in West Germany. The 'Euthanasia' and 'Aktion Reinhard' Trial Cases* (The Hague: Martinus Nijhoff Publishers, 1996).

Dobroszycki, Lucjan, *The Chronicle of the Lodz Ghetto* (New Haven: Yale University Press, 1984).

Donat, Alexander (ed), *The Death Camp Treblinka: A Documentary* (New York: Holocaust Library, 1979).

Dowbiggin, Ian, *A Concise History of Euthanasia: Life, Death, God, and Medicine* (Lanham: Rowman & Littlefield, 2007).

Dower, John W, *War Without Mercy: Race and Power in the Pacific War* (New York: Pantheon Books, 1986).

Eckart, Wolfgang Uwe (ed), *Man, Medicine, and the State: The Human Body as an Object of Government Sponsored Medical Research in the 20th Century* (Stuttgart: Franz Steiner Verlag, 2006).

Epstein, Eric Joseph, and Rosen, Philip, *Dictionary Of The Holocaust* (Westport: Greenwood Press, 1997).

Ericsson, Kjersti, and Simonsen, Eva, *On the Border: The contested children of the Second World War* (Childhood, Vol. 15, No.3, 2008) 397–414.

Evans, Richard J, *The Coming of the Third Reich* (London: Allen Lane, 2003).

Evola, Julius, *Eros and the Mysteries of Love: The Metaphysics of Sex* (Rochester: Inner Traditions International Ltd, 1991).

Fraser, Angus, *The Gypsies* (Malden: Blackwell Publishing Ltd., 2003).

Friedlander, Henry, *The Origins of Nazi Genocide: From Euthanasia to the Final Solution* (Chapel Hill: University of North Carolina Press, 1995).

Friedländer, Saul, *The Years of Persecution: Nazi Germany and the Jews 1933–39* (London: Phoenix, 2007).

___, *The Years of Extermination: Nazi Germany and the Jews, 1939–1945* (New York: Harper Collins, 2007).

Friedman, Ina R., *The Other Victims: First-Person Stories of Non-Jews Persecuted by the Nazis* (New York: Houghton Mifflin Harcourt, 1995).

Gallagher, Hugh, *What the Nazi "Euthanasia Program" Can Tell Us About Disability Oppression* (Journal of Disability Policy Studies, Vol.12, No.2, 2001) 96–99.

Gallagher, Hugh Gregory, *By Trust Betrayed: Patients, Physicians, and the License to Kill in the Third Reich* (St. Petersburg: Vandamere Press, 1995).

Gann, L H, and Duignan, Peter, *The Rulers of German Africa, 1884–1914* (Stanford: Stanford University Press, 1977).

Gasman, Daniel, *The Scientific Origins of National Socialism* (New Brunswick: Transaction Publishers, 2004).

Gassert, Philipp, and Steinweis, Alan E (eds.), *Coping With the Nazi Past: West German Debates on Nazism and Generational Conflict 1955–1975* (New York: Berghahn Books, 2007).

Gellately, Robert, *Backing Hitler: Consent and Coercion in Nazi Germany* (Oxford: Oxford University Press, 2002).

___, and Stoltzfus, Nathan (eds.), *Social Outsiders in Nazi Germany* (Princeton: Princeton University Press, 2001).

Giddens, Anthony, and Pierson, Christopher, *Conversations with Anthony Giddens: Making Sense of Modernity*, (Stanford: Stanford University Press, 1998).

Gilbert, G M, *Nuremberg Diary* (New York: Da Capo Press, 1995).

Gilbert, Martin, *The Holocaust: The Jewish Tragedy* (London: Collins, 1986).

Glass, James M, "*Life Unworthy of Life*"—*Racial Phobia and Mass Murder in Hitler's Germany*, (New York: Basic Books, 1997).

Glazar, Richard, *Trap With A Green Fence: Survival in Treblinka* (Evanston: Northwestern University Press, 1999).

Goddard, Henry Herbert, *Feeble-mindedness: Its Causes and Consequences* (New York: Arno Press, 1973).

Goldhagen, Daniel Jonah, *Hitler's Willing Executioners—Ordinary Germans and the Holocaust* (London: Little, Brown and Company, 1996).

Griech-Polelle, Beth, *Image of a Churchman-Resister: Bishop von Galen, the Euthanasia Project and the Sermons of Summer 1941* (Journal of Contemporary History, Vol 36, No.1, 2001) 41–57.

Grodin, Michael A, Annas, George J, Glantz, Leonard H, *Medicine and Human Rights: A Proposal for International Action* (The Hastings Center Report, Vol. 23, Issue 4, 1993).

Grossman, Vasily, *Everything Flows*, (London: Harvill Secker, 2010).

___, *The Road*, (London: MacLehose Press, 2010).

Gutman, Israel (ed), *Encyclopedia of the Holocaust* (New York: Macmillan Publishing Company, 1990).

Gutman, Yisrael, *The Jews of Warsaw 1939–1943* (Bloomington: Indiana University Press, 1989).

___, and Berenbaum, Michael (eds.), *Anatomy of the Auschwitz Death Camp* (Bloomington: Indiana University Press, 1998).

Haar, Ingo, and Fahlbusch, Michael, *German Scholars and Ethnic Cleansing, 1919–1945*, (New York: Berghahn Books, 2005).

Haeckel, Ernst Heinrich Philipp August, *Freedom in Science and Teaching* (Charleston: BiblioBazaar, 2009).

___, *The Riddle of the Universe: At the Close of the Nineteenth Century* (Charleston: BiblioBazaar, 2009).

Hawkins, Mike, *Social Darwinism in European and American Thought 1860–1945: Nature as model and nature as threat* (Cambridge: Cambridge University Press, 1998).

Heim, Susanne, Sachse, Carola, Walker, Mark (eds.), *The Kaiser Wilhelm Society under National Socialism* (Cambridge: Cambridge University Press, 2009).

Hennessy, Peter, *Never Again: Britain 1945–51* (London: Jonathan Cape, 1992).

Herbert, Ulrich (ed), *National Socialist Extermination Policies—Contemporary German Perspectives and Controversies* (New York: Berghahn Books, 2000).

Hilberg, Raul, *The Destruction of the European Jews* (New Haven: Yale University Press, 2003).

___ *Perpetrators Victims Bystanders: The Jewish Catastrophe 1933–1945* (New York: Harper Collins, 1993).

Horsinga-Renno, Mireille, *Cher oncle Georg: La bouleversante enquête d'une femme sur un médecin de la mort impuni* (Strasbourg : La Nuée Bleue, 2006).

Hoskins, Sylvia Anne, *Nurses and National Socialism—a Moral Dilemma: One Historical Example of a Route to Euthanasia* (Nursing Ethics, 12 (1), 2005) 79–91.

Hull, Isabel V., *Absolute Destruction—Military Culture and the Practices of War in Imperial Germany* (Ithaca: Cornell University Press, 2006).

Iredale, Rachel, *Eugenics and its Relevance to Contemporary Health Care*, (Nursing Ethics, 7 (3), 2000), 205–214.

Kater, Michael H, *Doctors Under Hitler* (Chapel Hill: University of North Carolina Press, 2000).

Kershaw, Ian, *The Nazi Dictatorship: Problems and Perspectives of Interpretation* (London: Arnold, 1996).

___, *Hitler 1889–36: Hubris* (London: Allen Lane, The Penguin Press, 1998).

___, *Hitler 1936–45: Nemesis* (London: Allen Lane, The Penguin Press, 2000).

Kevles, Daniel J, *In the Name of Eugenics: Genetics and the Uses of Human Heredity* (Berkeley: University of California Press, 1986).

King, Desmond, *In the Name of Liberalism: Illiberal Social Policy in the United States and Britain* (Oxford: Oxford University Press, 2002).

Klee, Ernst "*Euthanasie*" *im NS-Staat: Die* "*Vernichtung lebensunwerten Lebens*" (Frankfurt am Main: Fischer Taschenbuch Verlag, 1991).

___, *Was sie taten – Was sie wurden: Ärzte, Juristen und andere Beteiligte am Kranken- oder Judenmord* (Frankfurt am Main: Fischer Taschenbuch Verlag, 1986).

___, *Das Personenlexikon zum Dritten Reich – Wer war was vor und nach 1945* (Frankfurt am Main: S.Fischer, 2005).

___, Dressen Willi, Riess Volker (eds.), *The Good Old Days—The Holocaust as Seen by Its Perpetrators and Bystanders* (New York: Konecky & Konecky, 1991).

Klemperer, Victor, *The Language of the Third Reich: LTI—Lingua Tertii Imperii—A Philologist's Notebook* (London, Continuum, 2006).

___, *I Shall Bear Witness: The Diaries of Victor Klemperer 1933–41* (London: Weidenfield & Nicolson, 1998).

___, *To the Bitter End: The Diaries of Victor Klemperer 1942–45* (London: Weidenfield & Nicolson, 1999).

Kocsis, Richard N (ed), *Serial Murder and the Psychology of Violent Crimes: An International Perspective* (Totowa: Humana Press, 2007).

Kogon, Eugen, *The Theory and Practice of Hell* (New York: Berkley Books, 1998).

___, Langbein, Hermann, Rückerl, Adalbert (eds.), *Nazi Mass Murder: A Documentary History of the Use of Poison Gas* (New Haven: Yale University Press, 1993).

Koonz, Claudia, *The Nazi Conscience* (Cambridge Massachusetts: Harvard University Press, 2003).

Kranz, Tomasz, *Extermination of Jews at the Majdanek Concentration Camp* (Lublin: Panstwowe Muzeum na Majdanku, 2007).

Krieger, Richard Alan, *Civilization's Quotations: Life's Ideal* (New York: Algora Publishing, 2002).

Kuechenhoff, Bernhard, *The psychiatrist August Forel and his attitude to eugenics* (History of Psychiatry, No. 19(2), 2008), 215–223.

Kühl, Stefan, *The Nazi Connection: Eugenics, American Racism, and German National Socialism* (New York: Oxford University Press, 2002).

Kynaston, David, *Austerity Britain 1945–51* (London: Bloomsbury, 2008).

Lagerway, Mary D, *Nursing Ethics at Hadamar* (Qualitative Health Research, Vol.9, No.6, 1999), 759–772.

Lamb, David, *Down the Slippery Slope: Arguing in applied ethics* (London: Routledge, 2003).

Lanzmann, Claude, *Shoah—The Complete text of the Acclaimed Holocaust Film* (New York: Da Capo Press, 1995).

Lavi, Shai, "*The Jews are Coming": Vengeance and Revenge in post-Nazi Europe* (Law, Culture and the Humanities, No.1, 2005) 282–301

Law Reports of Trials of War Criminals, Volume ii, *The Belsen Trial,* (London: The United Nations War Crimes Commission, 1947).

Levi, Primo, *The Drowned and the Saved* (London: Abacus Books, 2002).

Levin, Miriam R (ed), *Cultures of Control* (Amsterdam: Harwood Academic Publishers, 2000).

Levy, Richard S (ed), *Antisemitism: A Historical Encyclopedia of Prejudice and Persecution* (Santa Barbara: ABC-CLIO Ltd, 2005).

Lewy, Guenter, *Himmler and the 'Racially Pure Gypsies'* (Journal of Contemporary History, Vol 34(2), 1999) 201–214.

Lifton, Robert Jay, *The Nazi Doctors: Medical Killing and the Psychology of Genocide* (London: Papermac, 1990).

Longerich, Peter, *The Unwritten Order: Hitler's Role in the Final Solution* (Stroud: Tempus Publishing Limited, 2003).

Lusane, Clarence, *Hitler's Black Victims: The Historical Experiences of Afro-Germans, European Blacks, Africans, and African Americans in the Nazi Era* (London: Routledge, 2003).

Lynn, Richard, *Eugenics: A Reassessment* (Westport: Praeger Publishers, 2001).

Lynn, Richard, and Kanazawa, Satoshi, *How to explain high Jewish achievement: The role of intelligence and values* (Science Direct, Personality and Individual Differences, Vol. 44, No.4, 2007) 801–808.

Madley, Benjamin, *From Africa to Auschwitz: How German South West Africa Incubated Ideas and Methods Adopted and Developed by the Nazis in Eastern Europe* (European History Quarterly, Vol. 35 No.3, 2005) 429–464.

Madinger, John, *Money Laundering: A Guide for Criminal Investigators* (Boca Raton: CRC Press, 2006).

Marcuse, Harold, *Legacies of Dachau. The Uses and Abuses of a Concentration Camp, 1933–2001* (Cambridge: Cambridge University Press, 2001).

Masson, David, *Macmillan's Magazine, Vol XII* (London: Macmillan and Co, 1865).

Mazower, Mark, *Hitler's Empire—Nazi Rule in Occupied Europe* (London: Allen Lane, 2008).

McFarland-Icke, Bronwyn Rebekah, *Nurses in Nazi Germany* (Chichester: Princeton University Press, 1999).

Michalczyk, John J (ed), *Medicine, Ethics, and the Third Reich: Historical and Contemporary Issues* (Lanham: Rowman & Littlefield, 1994).

Mitscherlich, Alexander, and Mielke, Fred, *Doctors of Infamy: The Story of the Nazi Medical Crimes* (New York: Henry Schuman, 1949).

Molan, Michael, *Sourcebook on Criminal Law*, (London: Cavendish Publishing Limited, 2001).

Mosse, George Lachmann, *Nazi Culture: Intellectual, Cultural and Social Life in the Third Reich* (Madison: University of Wisconsin Press, 2003).

Mostert, Mark P, *Useless Eaters: Disability as Genocidal Marker in Nazi Germany* (Journal of Special Education. Vol. 36, No. 3, 2002) 155–168.

Müller-Hill, Benno, *Murderous Science: Elimination by Scientific Selection of Jews, Gypsies, and Others in Germany 1933–1945* (Woodbury: Cold Spring Harbor Laboratory Press, 1998).

Nicosia, Francis R, and Huener, Jonathan (eds.), *Medicine and Medical Ethics in Nazi Germany: Origins, Practices, Legacies* (New York: Berghahn Books, 2002).

Novitch, Miriam, *Sobibor—Martyrdom and Revolt* (New York: Holocaust Library, 1980).

Olusoga, David, and Erichsen, Casper, *The Kaiser's Holocaust—Germany's Forgotten Genocide and the Colonial Roots of Nazism* (London: Faber and Faber, 2010).

O'Neil, Robin, *Belzec: Stepping Stone to Genocide* (New York: JewishGen Inc, 2008).

Padfield, Peter, *Himmler: Reichsführer-SS* (London: Papermac, 1990).

Panayi, Panikos (ed), *Weimar and Nazi Germany: continuities and discontinuities* (White Plains: Longman, 2000).

Pearson, Karl (ed), *The Life, Letters and Labours of Francis Galton* (Cambridge: Cambridge University Press, 1914, 1924, 1930).

Perry, Michael W (ed), *G K Chesterton—Eugenics and Other Evils: An Argument Against the Scientifically Organised State* (Seattle: Inkling Books, 2001).

Peters, William, *A Class Divided, Then and Now* (New Haven: Yale University Press, 1987).

Pilgrim, David, *The Eugenic Legacy in Psychology and Psychiatry* (International Journal of Social Psychiatry, Vol.54, No.3, 2008) 272–284.

Pine, Lisa, *Nazi Family Policy 1933–1945* (Oxford: Berg, 1999).

Piper, Franciszek, and Swiebocka, Teresa, *Auschwitz Nazi Death Camp* (Oswiecim: The Auschwitz-Birkenau State Museum in Oswiecim, 1996).

Proctor, Robert N, *Racial Hygiene: Medicine Under the Nazis* (Cambridge Massachusetts: Harvard University Press, 1989).

___ *The Nazi War on Cancer* (Princeton: Princeton University Press, 2000).

Quint, Peter E, *The Imperfect Union: Constitutional Structures of German Unification* (Princeton: Princeton University Press, 1997).

Reder, Rudolf, *Belzec* (Oswiecim: Fundacja Judaica Panstwowe Muzeum Oswiecim-Brzezinka, 1999).

Reggiani, Andrés Horacio, *God's Eugenicist: Alexis Carrel and the Sociobiology of Decline*, (New York: Berghahn Books, 2007).

Reichler, (Rabbi) Max, Blau, (Rabbi) Joel, de Sola Pool (Rev. Dr. D.), *Jewish Eugenics and Other Essays: Three Papers Read Before The New York Board of Jewish Ministers 1915* (New York: Bloch Publishing Company, 1916).

Reitlinger, Gerald, *The Final Solution: The Attempt to Exterminate the Jews of Europe 1939–1945* (Northvale: Jason Aronson Inc, 1987).

___, *The SS, Alibi of a Nation: 1922–1945* (New York: Viking Press, 1957).

Remy, Steven P, *The Heidelberg Myth: the Nazification and Denazification of a German University* (Harvard: Harvard University Press, 2002).

Rieger, Berndt, *Creator of Nazi Death Camps—The Life of Odilo Globocnik* (London: Vallentine Mitchell, 2007).

Roberts, Stephen H, *The House That Hitler Built* (New York: Harper & Brothers, 1938).

Röder, Thomas, Kubillus, Volker, Burwell, Anthony, *Psychiatrists—the Men Behind Hitler: The Architects of Horror* (Los Angeles: Freedom Publishing 1995).

Ronen, Gabriel M, Meaney, Brandon, Dan, Bernard, Zimprich, Fritz, Stögmann, Walter, Neugebauer, Wolfgang, *From Eugenic Euthanasia to Habilitation of "Disabled" Children: Andreas Rett's Contribution* (Journal of Child Neurology, Vol.24, No.1, 2009) 115–127.

Rubenstein, Joshua, Altman, Ilya, *The Unknown Black Book: The Holocaust in the German-Occupied Soviet Territories* (Bloomington: Indiana University Press, 2008), pp. 277–278.

Rürup, Reinhard (ed), *Topography of Terror: Gestapo, SS and Reichssicherheitshauptamt on the "Prinz-Albrecht-Terrain." A Documentation* (Berlin: Verlag Willmuth Arenhövel, 2000).

Samaan, A.E., *From a "Race of Masters" to a "Master Race": 1948 to 1848* (www.CreateSpace.com, 2012)

Sands, Philippe, *East West Street: On the Origins of Genocide and Crimes Against Humanity* (London: Weidenfeld & Nicolson, 2016)

Schafft, Gretchen E., *From Racism to Genocide: Anthropology in the Third Reich* (Urbana: University of Illinois Press, 2004).

Schelvis, Jules, *Sobibor—A History of a Nazi Death Camp* (Oxford: Berg, 2007).

Schmidt, Ulf, Karl Brandt: *The Nazi Doctor—Medicine and Power in the Third Reich* (London: Hambledon Continuum, 2007).

Schmuhl, Hans-Walter, *The Kaiser Wilhelm Institute for Anthropology, Human Heredity, and Eugenics, 1927–1945: Crossing Boundaries* (New York: Springer-Verlag, 2008).

Sereny, Gitta, *Into That Darkness—From Mercy Killing to Mass Murder* (London: Pimlico, 1995).

___, *The German Trauma: Experiences and Reflections 1938–2000* (London: Allen Lane, The Penguin Press, 2000).

Sewell, Dennis, *The Political Gene—How Darwin's Ideas Changed Politics* (London: Picador, 2009).

Shelton, Dinah L., *Encyclopedia of Genocide and Crimes Against Humanity* Vol.1 (New York: Macmillan Reference, 2004).

Shandley, Robert R (ed), *Unwilling Germans—The Goldhagen Debate* (Minneapolis: University of Minnesota Press, 2001).

Sofsky, Wolfgang, *The Order of Terror: The Concentration Camp*, (Princeton: Princeton University Press, 1997).

Soloway, Richard A, *Demography and Degeneration: Eugenics and the Declining Birthrate in Twentieth-Century Britain* (Chapel Hill: University of North Carolina Press, 1995).

Speer, Albert, *Inside the Third Reich* (London: Sphere Books Limited, London, 1971).

Spiro, Jonathan Peter, *Defending the Master Race: Conservation, Eugenics, and the Legacy of Madison Grant* (Burlington: University of Vermont Press, 2008).

Stackelberg, Roderick, *The Routledge Companion to Nazi Germany* (London: Routledge, 2007).

___, and Winkle, Sally Anne, *The Nazi Germany Sourcebook: An Anthology of Texts* (London: Routledge, 2002).

Staub, Ervin, *The Roots of Evil: The Origins of Genocide and Other Group Violence* (Cambridge: Cambridge University Press, 1989).

Stein, Harry, *Buchenwald concentration camp 1937–1945: a guide to the permanent historical exhibition* (Göttingen: Gedenkstätte Buchenwald, Wallstein Verlag, 2004).

Stephenson, Jill, *Women in Nazi Germany* (Harlow: Pearson Education Limited, 2001).

Stone, Dan, *Race in British Eugenics* (European History Quarterly, Vol.31, no.3, 2001) 397–424.

___, *Breeding Superman: Nietzsche, Race and Eugenics in Edwardian and Interwar Britain* (Liverpool: Liverpool University Press, 2002).

___, (ed), *The Historiography of Genocide* (Basingstoke: Palgrave Macmillan, 2008).

Strobl, Gerwin, *The Bard of Eugenics: Shakespeare and Racial Activism in the Third Reich* (Journal of Contemporary History, Vol. 34, no.3, 1999) 323–336.

Ströhle, Andreas, M. D., Wrase, Jana, Ph. D., Malach, Henry, M.D., Gestrich, Christof, Ph.D., and Heinz, Andreas, M.D., *Images in Psychiatry: Karl Bonhoeffer [1868–1948]* (American Journal of Psychiatry,165: May 2008) 575–576.

Strous, Rael D, M.D., *Dr. Irmfried Eberl (1910–1948): Mass Murdering M.D.* (The Israel Medical Association Journal, Vol 11, April 2009) 216–218.

Sumner, William Graham, *The Challenge of Facts and Other Essays*, (New Haven: Yale University Press, 1914).

Szöllösi-Janze, Margit (ed), *Science in the Third Reich* (Oxford: Berg, 2001).

Taylor, Fred (ed), *The Goebbels Diaries 1939–1941* (London: Sphere Books Limited, 1983).

Tenenbaum, Joseph, *Race and Reich: The Story of an Epoch* (Westport: Greenwood Press, 1976).

Thomas, Florian P, Beres, Alana, and Sheveli, Michael I, *"A Cold Wind Coming": Heinrich Gross and Child Euthanasia in Vienna* (Journal of Child Neurology, Vol.21, 2006) 342–348.

Thomson, Matthew, *The Problem of Mental Deficiency: Eugenics, Democracy, and Social Policy in Britain, c.1870–1959* (Oxford: Oxford University Press, 1998).

Ticho, Kurt, *My Legacy: Holocaust, History and the Unfinished Task of Pope John Paul II* (Wlodawa: Muzeum Pojezierza Leczynsko-Wlodawskiego, 2008).

Todorov, Tzvetan, *Facing the Extreme: Moral Life in the Concentration Camps* (London: Phoenix, 1996).

Totten, Samuel, Parsons, William S, Charny, Israel W (eds.), *Century of Genocide: Eyewitness accounts and Critical Views* (New York: Routledge, 2004).

Traverso, Enzo, *The Origins of Nazi Violence* (New York: The New Press, 2003).

Trent, James W, *"Who shall say who is a useful person?" Abraham Myerson's opposition to the eugenic movement* (History of Psychiatry, No. 12, 2001) 33–57.

Trunk, Isaiah, *Judenrat: The Jewish Councils in Eastern Europe under Nazi Occupation* (Lincoln: University of Nebraska Press, 1996).

Turda, Marius, *'A New Religion'? Eugenics and Racial Scientism in Pre-First World War Hungary* (Totalitarian Movements and Political Religions: Vol.7, No. 3, September 2006) 303–325.

___, and Weindling, Paul J (eds.), *"Blood and Homeland". Eugenics and Racial Nationalism in Central and Southeast Europe, 1900–1940* (Budapest: Central European University Press, 2007).

Tusa, Ann, and Tusa, John, *The Nuremberg Trial* (London: Macmillan Publishers Limited, 1984).

van der Vat, Dick, *The Good Nazi: The Life and Lies of Albert Speer* (London: Weidenfeld & Nicolson, 1997).

van Pelt, Robert Jan, *The Case for Auschwitz—Evidence from the Irving Trial* (Bloomington: Indiana University Press, 2002).

von Lang, Jochen, and Sibyll, Claus, *Eichmann Interrogated: Transcripts from the Archives of the Israeli Police*, (New York: Da Capo Press, 1999).

Vos, Aart, "*Wegens bijzondere omstandigheden...*" - *s'Hertogenbosch in bezettingstijd 1940–1945* (Alphen aan de Maas: Uitgeverij Veerhuis, 2008).

Wachsmann, Nikolaus, *KL: A History of the Nazi Concentration Camps* (Little, Brown, London, 2015)

Walker, Mark (ed), *Science and Ideology—A Comparative History* (London: Routledge, 2003).

Waller, James, *Becoming Evil. How Ordinary People Commit Genocide and Mass Killing* (Oxford: Oxford University Press, 2002).

War Cabinet, United Nations War Crimes Commission, *Memorandum of Lord Chancellor to the Cabinet*, (W.P. (44) 294, 1944)

Weikart, Richard, *From Darwin to Hitler. Evolutionary Ethics, Eugenics, and Racism in Germany* (New York: Palgrave MacMillan, 2004).

Weindling, Paul, *Health, Race, and German Politics between National Unification and Nazism, 1870–1945* (Cambridge: Cambridge University Press, 1993).

Weindling, Paul Julian, *Nazi Medicine and the Nuremberg Trials—From Medical War Crimes to Informed Consent* (Basingstoke: Palgrave Macmillan, 2006).

Weiss, Sheila Faith, *Race Hygiene and National Efficiency: The Eugenics of Wilhelm Schallmayer* (Berkeley: University of California Press, 1987).

Welch, David, *Hitler: Profile of a Dictator* (London: Routledge, 2001).

Wells, H G, *Mankind in the Making*, (Charleston: BiblioBazaar, 2006).

___, *Anticipations*, (Charleston: BiblioBazaar, 2007).

Welzer, Harald, *Mass murder and moral code: some thoughts on an easily misunderstood subject* (History of the Human Sciences, Vol.17 Nos 2/3, 2004) 15–32.

Wertham, Fredric, *A Sign for Cain: An Exploration of Human Violence*, (New York: Paperback Library, 1969.

Weyers, Wolfgang, Ackerman, A. Bernard (ed), *Death of medicine in Nazi Germany: Dermatology and Dermatopathology Under the Swastika* (Philadelphia: Lippincott-Raven Publishers, 1998)

Wildt, Michael, *An Uncompromising Generation—The Nazi Leadership of the Reich Security Main Office* (Madison: University of Wisconsin Press, 2009).

Willenberg, Samuel, *Revolt in Treblinka* (Warsaw: Zydowski Instytut Historyczny, 1992).

Wilson, Kathleen, *The Island Race: Englishness, empire and gender in the eighteenth century* (London: Routledge, 2003).

Wistrich, Robert S, *Who's Who in Nazi Germany* (London: Routledge, 1995).

Wyman, David S (ed), *The World Reacts to the Holocaust* (Baltimore: The John Hopkins University Press, 1996).

Zielinski, Andrew, *Conversations With Regina* (Wlodawa: Muzeum Pojezierza Leczynsko-Wlodawskiego, 2008).

Zimbardo, Philip, *The Lucifer Effect: How Good People Turn Evil* (London: Rider Books, 2007).

Zimmerer, Jürgen, *The Birth of the Ostland out of the Spirit of Colonialism: a Postcolonial Perspective on the Nazi policy of Conquest and Extermination, Patterns of Prejudice*, Vol.39, No 2, 2005 197–219.

Zimmerman, Joshua D, *Jews in Italy Under Fascist and Nazi Rule, 1922–1945* (Cambridge: Cambridge University Press, 2005).

GLOSSARY

Ahnenerbe - Literally, "Ancestral Heritage". Institution for the study of the cultural history and archeology of the Aryan race.

Aktion Erntefest - Operation Harvest Festival. Mass shooting of Jews at Majdanek and other camps in November 1943.

Aktion Reinhard - code name for the extermination of the Jews of Poland and other countries and the theft of their belongings.

Aktion R - code name for the activities of former *Aktion Reinhard* personnel in Northern Italy following the closure of the Polish extermination camps they had been operating.

Anschluss - the enforced annexation of Austria by Germany in March 1938.

Einsatzgruppe (*Einsatzgruppen der Sicherheitspolizei und des SD*) - Death squads, responsible for the murder of millions of Jews and others in Eastern Europe. Divided into smaller units, *Einsatzkommandos* or *Sonderkommandos.*

Freikorps - Extremist right-wing paramilitary groups active at the time of the Weimar Republic.

14f13 - Extension of the "euthanasia" programme to concentration camp prisoners.

Gekrat (*Gemeinnützige Krankentransport GmbH*) - Organisation responsible for transporting "euthanasia" victims to the killing centres.

Generalgouvernement - that portion of German occupied Poland not incorporated into the *Reich.*

Gestapo - (*Geheime Staatspolizei*) - Secret State Police.

Gutachter - Supposed medical experts who determined the fate of individuals based upon a completed *Meldebogen.*

Häftlinge - Prisoner

Judenrat - Jewish Council.

KdF (*Kanzlei des Führers der NSDAP*) - Chancellery of the Führer of the Nazi party. Agency principally dedicated to handling Adolf Hitler's personal affairs. T4 operated under the umbrella of the KdF.

Kripo (*Kriminalpolizei*) - Criminal Police.

KTI (*Kriminaltechnisches Institut der Sicherheitspolizei*) - Technical Institute for the Detection of Crime.

KZ (*Konzentrationslager*) - Concentration Camp.

Lebensborn - Literally "Fount of Life". Organisation established with the goal of producing racially approved children.

Luminal - phenobarbitone, a barbiturate drug.

Machtergreifung - the 1933 Nazi seizure of power.

Meldebogen - Questionnaires to be completed in respect of all hospital patients potentially targeted for "euthanasia".

Muselmann (pl. Muselmänner—literally 'Muslim' from their perceived appearance. Slang term for concentration camp inmates dying from starvation and exhaustion and resigned to their inevitable selection and death.

NSDAP (*Nationalsozialistische Deutsche Arbeiterpartei*) - The National Socialist German Workers Party, usually abbreviated to "Nazi".

Organisation Todt - Entity primarily responsible for civil and military engineering projects,

RAG (*Reichsarbeitsgemeinschaft Heil- und Pflegeanstalten*") - Organisation for registering inmates of mental homes.

RSHA (*Reichssicherheitshauptamt*) - *Reich* Security Main Office. Organisation created in 1939 by Heinrich Himmler for overall responsibility in connection with all policing and security matters.

SA (*Sturmabteilung*) - Storm Troop. The paramilitary wing of the Nazi party.

SD (*Sicherheitsdienst des Reichsführers-SS*) - Security Service of the SS.

SiPo (*Sicherheitspolizei*) - Security Police. A combination of the *Gestapo* and the *Kripo.*

Sonderbehandlung - Special treatment. A Nazi euphemism for murder.

SS (*Schutzstaffel*) - Protection Detachment. Initially a small guard unit, later to grow under Heinrich Himmler to one of the largest Nazi organisations, responsible for countless continent-wide atrocities.

Stiftung (*Gemeinnützige Stiftung für Anstaltspflege*) - the organizational arm of T4, responsible for choosing sites, installing extermination facilities therein, selecting personnel, and dealing with economic affairs.

T4 - Tiergartenstrasse 4. Abbreviation of the Berlin address of the organisations responsible for administering the "euthanasia" programme and certain extermination centres.

Volk, Völkisch - literally "folk, folksy", but in Nazi terminology meaning members of a race or tribe, specifically people whose bloodline was German.

Warthegau - Former Polish territory annexed to Germany in 1939.

WVHA (*Wirtschafts-Verwaltungshauptamt*) - SS Main Economic and Administrative Department, responsible for the business and economic affairs of the SS, including the staffing and organization of concentration camps.

ZVST (*Zentralverrechnungsstelle Heil- und Pflegeanstalten*) - Central Clearing Office for major financial matters, including the financing of T4.

Veronal - phenobarbitone, a barbiturate drug.

INDEX

KNOWN T4 PERSONNEL WHO SERVED IN *AKTION REINHARD* CAMPS

Name - SS Rank - T4 Service – *Aktion Reinhard* Camp

Arndt, Kurt - Scharführer - Hadamar/Treblinka

Bär, Rudolf - Scharführer - Bernburg/Brandenburg/Treblinka
Bärbl, Heinrich - Rottenführer- Hartheim/Grafeneck/Belzec/Sobibor
Bauch, Ernst - Rottenführer- Bernburg/Sonnenstein/Sobibor
Bauer, Hermann Erich - Oberscharführer - T4/Sobibor
Becher, Werner - Unterscharführer - Sonnenstein/Sobibor
Beckmann, Rudolf - Oberscharführer - Hartheim/ Sobibor
Beulich, Max - Scharfüher - Sonnenstein/Sobibor
Biela, Max - Scharführer - Bernburg/Brandenburg/Treblinka
Blauroch, Emil - Scharführer - Sonnenstein/Sobibor
Bolender, Kurt - Oberscharführer- Brandenburg/Hadamar/Hartheim/Sonnenstein/Sobibor
Bootz, Helmut - Scharführer - Bernburg/Grafeneck/Treblinka/Sobibor
Börner, Gerhardt -Scharführer - Sonnenstein/Sobibor
Borowski, Werner - Untersturmführer - Bernburg/Belzec/Treblinka
Bredow, Paul -Scharführer - Grafeneck/Hartheim/Sobibor/Treblinka
Bree, Max - Scharführer - Grafeneck/Hadamar/Sobibor/Treblinka

Dachsel, Arthur - Oberwachtmeister - Sonnenstein/Belzec/Sobibor
Dietz, Erich - Scharfuhrer - Sonnenstein/Sobibor
Dubois, Werner +* - Oberscharführer - Brandenburg/Bernburg/Hadamar/Grafeneck/Belzec/Sobibor

Eberl, Irmfried (Dr). Obersturmführer - Bernburg/Brandenburg/Treblinka
Eisold, Johannes - Scharführer - Sonnenstein/Treblinka

Felfe, Hermann - Scharführer - Grafeneck/Sonnenstein/Treblinka
Fichtner, Erwin -Scharführer - Bernburg/Belzec
Floss, Herbert - Scharführer - Bernburg/Belzec/Sobibor/Treblinka
Forker, Albert - Scharführer - Sonnenstein/Sobibor/Treblinka
Franz, Kurt Hubert ** - Untersturmführer - Brandenburg/Grafeneck/Sonnenstein/Belzec/Treblinka
Frenzel, Karl + - Oberscharführer - Bernburg/Grafeneck/Hadamar/Sobibor

Fuchs, Erich +* - Unterscharführer - Bernburg/Brandenburg/Belzec/Sobibor/Treblinka

Getzinger, Anton - Oberscharführer - Hartheim/Sobibor
Girtzig, Hans * - Scharführer - Grafeneck/Hartheim/Belzec/Sobibor
Gley, Heinrich * - Oberscharführer - Grafeneck/Sonnenstein/Belzec
Gomerski, Hubert - Oberscharführer – Hartheim/Hadamar/Sobibor
Goetzinger, Anton - Oberscharführer - Hartheim/Sobibor
Graetschus, Siegfried - Oberscharführer - Bernburg/Treblinka/Sobibor
Gringers, Max (Karl) - Scharführer – Bernburg/Hadamar/Hartheim/Belzec
Grömer, Josef (Ferdl) - Sturmmann - Hartheim/ Sobibor
Grossmann, Willi - Scharführer - Hadamar/Sonnenstein/Treblinka
Groth, Paul - Unterscharführer - Hartheim/Belzec/Sobibor

Hackel, Emil - Scharfuhrer - Sonnenstein/Sobibor
Hackenholt, Lorenz – Hauptscharführer - Grafeneck/Sonnenstein/Belzec/Sobibor/ Treblinka
Hengst, August - Scharführer - Bernburg/Brandenburg/Treblinka
Hering, Gottleib -Hauptsturmführer - Bernburg/Brandenburg/Hartheim/Sonnenstein/ Belzec
Hirche, Fritz -Scharführer – Bernburg/Hartheim/Belzec
Hirtreiter, Josef - Scharführer - Hadamar/Treblinka
Hödl, Franz - Unterscharführer - Hartheim/Sobibor
Horn, Otto ** - Scharführer - Sonnenstein/Treblinka

Ittner, Alfred + -Oberscharführer - T4/Sobibor

Jührs, Robert + -Unterscharführer - Hadamar/ Belzec/Sobibor

Kainer, Erwin - Scharführer - Hadamar/Treblinka
Kamm, Rudolf -Scharführer - Sonnenstein/Belzec/Sobibor
Kilminsky, Otto -Scharführer – Hadamar/Belzec
Klier, Johann (Josef) - Unterscharführer - Hadamar/Sobibor
Klos, Walter -Scharführer - Sonnenstein/Belzec
Konrad, Josef (Fritz) -Scharführer - Grafeneck/Sonnenstein/Sobibor
Kraschewski, Fritz -Scharführer - Grafeneck/Hadamar/Belzec

Lambert,Erwin Hermann ** - Unterscharführer - Bernburg/Hadamar/Hartheim/ Sonnenstein/Sobibor/Treblinka
Ludwig, Karl Emil -Scharführer - T4/Sobibor/Treblinka.

Matthes, Heinrich ** -Scharführer - T4/Treblinka/Sobibor
Mätzig, Willi -Scharführer - Bernburg/Brandenburg/Treblinka/Sobibor
Mentz, Willi ** - Scharführer - Grafeneck/Hadamar/Treblinka/Sobibor
Michel, Hermann - Oberscharführer - Hartheim/Sobibor/Treblinka
Miete, August ** - Scharführer - Grafeneck/Hadamar/Treblinka
Münzberger, Gustav ** - Scharführer - Sonnenstein/Treblinka

Nowak, Walter (Anton) - Scharführer - Sonnenstein/Sobibor.

Oberhauser, Josef * - Oberscharführer - Bernburg/Brandenburg/Grafeneck/Belzec/
Sobibor/Treblinka

Pötzinger, Karl - Scharführer - Bernburg/Brandenburg/Treblinka/Sobibor

Rehwald, Wenzel (Fritz) - Unterscharführer - Bernburg/Hadamar/Hartheim/
Sonnenstein/Sobibor
Richter, Kurt (Karl) - Scharführer - Hartheim/Sonnenstein/Sobibor/Treblinka
Reichtleitner, Franz - Hauptsturmführer - Hartheim/Sobibor
Rost, Paul -Scharführer - Hartheim/Sonnenstein/Sobibor/Treblinka
Rum, Franz Albert ** - Scharführer - T4/Treblinka

Scharfe, Herbert - Scharführer - Sonnenstein/Treblinka
Schemmel, Franz - Scharführer - Hartheim/Sonnenstein/Belzec/Treblinka
Schiffner, Karl -Scharführer - Sonnenstein/Belzec/Sobibor/Treblinka
Schluch, Karl * - Scharführer - Grafeneck/Hadamar/Belzec
Schmidt, Fritz - Scharführer - Bernburg/Sonnenstein/Treblinka.
Schulz, Emanuel - Scharfuhrer - Grafeneck/Hadamar/Sonnenstein/Trebinka
Schütt, Heinz-Hans +** - Scharführer - Grafeneck/Hadamar/Sobibor.
Schwarz, Gottfried - Untersturmführer - Bernburg/Grafeneck/ Belzec/Sobibor
Seidel, Kurt - Oberwachmann – Sonnenstein/Treblinka
Sporleder, Erich - Scharführer - Brandenburg/Bernburg/Sobibor
Stadie, Otto - Scharführer - Bernburg/Treblinka
Stangl, Franz Paul - Hauptsturmführer - Bernburg/Hartheim/Sobibor/Treblinka
Stengelin, Erwin - Unterscharführer – Hadamar/Treblinka/Sobibor
Steubl, Karl -Scharführer - Hartheim/Sobibor
Suchomel, Franz ** - Scharführer - Hadamar/Treblinka/Sobibor.

Tauscher, Fritz - Oberscharführer - Brandenburg/Hartheim/Sonnenstein/Belzec

Unverhau, Heinrich +* - Oberscharführer - Grafeneck/Hadamar/Belzec/Sobibor

Vallaster, Erich Josef - Scharführer - Hartheim/Belzec/Sobibor.
Vey, Kurt - Scharfuhrer - Sonnenstein/Belzec/Sobibor

Wagner, Gustav Franz - Oberscharführer – Hartheim/Sobibor.
Wendland, Willi - Scharführer - Sonnenstein/ Sobibor
Wirth, Christian - Sturmbannführer - Brandenburg/Bernburg/Grafeneck/ Sonnenstein/ /Hadamar//Hartheim/Belzec/Sobibor/Treblinka
Wolf, Franz + - Unterscharführer - Hadamar/Sobibor
Wolf, Josef - Unterscharführer - T4/Sobibor

Zanker, Hans -Scharführer - Sonnenstein/Belzec
Zaspel, Fritz, - Unterscharführer – Sonnenstein/Sobibor
Zierke, Ernst +* - Unterscharführer - Grafeneck/Hadamar/Sonnenstein/Belzec/Sobibor

*Prosecuted for crimes committed in Belzec
+Prosecuted for crimes committed in Sobibor
**Prosecuted for crimes committed in Treblinka

Staff between all three Aktion Reinhard camps was interchangeable, but for geographic reasons occurred more frequently between Sobibor and Belzec.

Excluded from this schedule are extermination camp personnel who had not been initially recruited by T4 and served at either a "euthanasia" centre or at T4 head office. Only two of the SS-Garrison at Belzec, and seven at Sobibor, were actually bona fide SS men. The remainder were civilians (policemen, nurses and tradesmen) in SS uniform.

WITH THANKS TO ROBIN O'NEIL

***ibidem*-Verlag**
Melchiorstr. 15
D-70439 Stuttgart
info@ibidem-verlag.de

www.ibidem-verlag.de
www.ibidem.eu
www.edition-noema.de
www.autorenbetreuung.de